ANNUAL REVIEW OF NEUROSCIENCE

ANNUAL REVIEW OF NEUROSCIENCE

VOLUME 9, 1986

W. MAXWELL COWAN, *Editor*
Salk Institute for Biological Studies

ERIC M. SHOOTER, *Associate Editor*
Stanford University School of Medicine

CHARLES F. STEVENS, *Associate Editor*
Yale University School of Medicine

RICHARD F. THOMPSON, *Associate Editor*
Stanford University

ANNUAL REVIEWS INC.   4139 EL CAMINO WAY   PALO ALTO, CALIFORNIA 94306 USA

ANNUAL REVIEWS INC.
Palo Alto, California, USA

International Standard Serial Number: 0147-006X
International Standard Book Number: 0-8243-2409-9

Annual Review and publication titles are registered trademarks of Annual Reviews Inc.

Annual Reviews Inc. and the Editors of its publications assume no responsibility for the statements expressed by the contributors to this *Review*.

TYPESET BY AUP TYPESETTERS (GLASGOW) LTD., SCOTLAND
PRINTED AND BOUND IN THE UNITED STATES OF AMERICA

Annual Review of Neuroscience
Volume 9, 1986

CONTENTS

(continued)   v

vi   CONTENTS (*continued*)

SOME RELATED ARTICLES IN OTHER *ANNUAL REVIEWS*

From the *Annual Review of Biochemistry*, Volume 55 (1986)

Molecular Properties of Voltage-Sensitive Sodium Channels, W. A. Catterall
Neuropeptides: Multiple Molecular Forms, Metabolic Pathways, and Receptors,
D. R. Lynch and S. H. Snyder

From the *Annual Review of Cell Biology*, Volume 1 (1985)

Intermediate Filaments, P. M. Steinert and D. A. D. Parry
Cell Migration in the Vertebrate Embryo, J. P. Thiery, J. L. Duband, and G. C.
Tucker
Acetylcholine Receptor Structure, Function, and Evolution, R. M. Stroud and
J. Finer-Moore

From the *Annual Review of Medicine*, Volume 37 (1986)

The Autonomic Nervous System in Congestive Heart Failure, G. S. Francis and
J. N. Cohn

From the *Annual Review of Pharmacology and Toxicology*, Volume 26
(1986)

Pharmacology of Thyrotropin-Releasing Hormone, A. Hirota, M. A. Carino,
and H. Lai
Potential Animal Models for Senile Dementia of Alzheimer's Type, with Emphasis
on AF64A-Induced Cholinotoxicity, I. Hanin and A. Fisher
Electrophysiological Correlates of the Neurotoxicology of Sensorimotor Systems,
H. E. Lowndes and T. L. Baker
Molecular Pharmacology of Botulinum Toxin and Tetanus Toxin, L. L. Simpson

From the *Annual Review of Physiology*, Volume 48 (1986)

Cell Adhesion Molecules in Neural Histogenesis, G. M. Edelman
Genes Encoding Mammalian Neuroendocrine Peptides: Strategies Towards Their
Identification and Analysis, K. E. Mayo, R. M. Evans, and G. M. Rosenfeld
Neural Grafting in the Aged Rat Brain, F. H. Gage and A. B. Björklund
Neuronal Receptors, S. H. Snyder
Growth Hormone Releasing Factors, M. C. Gelato and G. R. Merriam
Substance P and Neurotensin: Their Roles in the Regulation of Anterior Pituitary
Function, N. Aronin, R. Coslovsky, and S. E. Leeman

From the *Annual Review of Psychology,* Volume 37 (1986)

ERRATUM

ANNUAL REVIEW OF NEUROSCIENCE, Volume 8 (1985)

In *Phototransduction in Vertebrate Rods*, by E. A. Schwartz

On page 355, the sentence at the top of the page should read, "Cytoplasm normally has a large capacity for buffering changes in both Ca concentration and pH."

Ann. Rev. Neurosci. 1986. 9 : 1–26

ARTIFICIAL INTELLIGENCE AND THE BRAIN:
Computational Studies of the Visual System

Shimon Ullman

Massachusetts Institute of Technology, Artificial Intelligence Laboratory, Cambridge, Massachusetts, and Department of Applied Mathematics, The Weizmann Institute of Science, Rehovot, Israel

INTRODUCTION

Scope of the Review

Neuroscientists and researchers in the field of artificial intelligence share a common general goal: both try to understand the processes that underlie thinking and intelligent behavior.

The methods are, of course, very different. Neuroscientists study empirically the physiology, anatomy, and chemistry of the nervous system and behavior. Researchers in artificial intelligence use computational methods to study information processing aspects of tasks that are characteristic of the human brain, such as visual perception, motor control, and natural language processing.

Given the enormous complexity of many biological nervous systems, and in particular the human brain, it seems that a combination of empirical and computational studies would be required to understand their functions. This paper describes some aspects of artificial intelligence research that may prove relevant to the understanding of certain brain functions in one area—the domain of visual perception. I review advances in the computational studies of vision and discuss their possible relevance to the processing of visual information in the primate brain.

The review does not cover other areas of artificial intelligence research that also bear relevance to the study of the brain. These areas include for example, the study of motor control and parallel processing by networks of

1

neuron-like elements and the study of learning. Finally, the review includes computational studies of vision, such as Land's theory of lightness computation, that were developed outside the framework of artificial intelligence research.

The Goal of Computational Vision

The computational study of vision should not be regarded as an attempt to model biological mechanisms directly. The goal of such studies is to understand the visual system at what Marr & Poggio (1977) have termed "the computational level." Marr & Poggio's point was that an information processing system as complex as the primate visual system must be studied and understood on more than a single level. In addition to the study of the biological mechanisms that build up the system, such as the neurons and their interconnections, it is important to study the information processing aspects of the system: What computations are employed, and why?

Studies in computational vision proceed by developing and evaluating workable solutions to problems solved routinely by the visual system in everyday perception. For example, one of the problems the visual system is faced with is stereo vision: how to combine the separate images of the two eyes into a single three-dimensional representation of the environment. Computational studies of stereo vision attempt to develop schemes that can actually solve this problem, and at the same time are consistent with the available psychophysical, physiological, and anatomical data concerning stereo vision. Although the schemes developed in this manner are not necessarily identical to the processes employed by the visual system, it is hoped that the insight gained in studying the nature of the problems and the range of possible solutions will prove useful to the understanding of biological systems that are engaged in the solution of similar problems. The gaining of such insights may in fact prove crucial to the understanding of the system's functions. Empirical studies of the visual cortex over the last 20 years or so have demonstrated the difficulties in inferring the functional role of units in the visual system from physiological studies alone. A computational theory of visual information processing may provide useful tools for making such inferences.

In the next sections I review the current state of computational vision in five areas that seem to have potential relevance to the study of the primate visual system. The areas are (a) edge detection and zero crossings, (b) channels and multiscale analysis, (c) visual motion, (d) lightness perception, and (e) selective visual attention. An important area in computational vision that is not covered in this review is stereo vision, since it has been reviewed recently in this series (Poggio & Poggio 1984).

EDGE DETECTION AND ZERO CROSSINGS

DOG Filtering of the Image

The first processing stages of the incoming image are performed already at the retinal level. An extensive research effort starting with Kuffler (1952) and Barlow (1953) has provided a rather detailed picture of the operations performed by the retina at the output level (the ganglion cells). It is known that the retina contains a number of different classes of ganglion cells. The exact definition of the classes depends on the species and the classification method. Based on physiological responses, for example, ganglion cells in the cat retina have been divided into X, Y, and W cell types (Enroth-Cugell & Robson 1966, Cleland & Levick 1974; for a review see Lennie 1980). A broadly similar classification was suggested for the primate retina (e.g. De Monasterio 1978). I therefore use the therms X-like and Y-like units to refer also to primate retinal ganglion types.

The spatial arrangement of retinal ganglion receptive fields was described originally as composed of concentric center-surround antagonistic subfields, and having two complementary types: the on- and off-center (Kuffler 1952). More quantitative studies of receptive field shapes have shown that the spatial organization of both X- and Y-type receptive fields can be approximated by the difference of antagonistic two-dimensional Gaussian functions (Rodieck 1965, Rodieck & Stone 1965, Enroth-Cugell & Robson 1966). This approximation is often referred to as the DOG (for "difference of Gaussians") model for the spatial organization of retinal receptive fields. Using this approximation, the conclusion is that on its way to the visual cortex the incoming image passes through a set of DOG-shaped filters. Obvious questions that arise then are what is the functional role of this filtering operation, and how is the filtered image used at the next levels of processing in the primary visual cortex?

The Primal Sketch Notion

Although much is known about the physiology and anatomy of the visual cortex, the answers to the above questions are not entirely clear. Starting with the studies of Hubel & Wiesel (1962, 1968), different types of cortical units have been discovered and their properties studied in detail. It seems fair to conclude, however, that from a functional standpoint most of the questions remain unanswered. It is still unclear what operations are performed on the retinal output in the subsequent processing stages, and why.

Computational studies of vision approach this problem from a different direction. The questions they ask are, what operations are theoretically useful to perform on the incoming image, and how do these operations correspond to the observed properties of the visual system?

Two main answers have been suggested to this question: one on a more general level, the second more specific. The suggested general answer has been outlined in David Marr's influential paper "Early Processing of Visual Information" (Marr 1976). Marr has argued that the initial input to the visual system, which is a large grey-level intensity array, is too large and unstructured to permit efficient manipulation at subsequent processing stages. He suggested that the goal of the first stage of visual information processing is to produce a primitive but rich description of the grey-level changes in the image in such a manner that all subsequent computations could be implemented as manipulations of the primitive descriptors. The early description of the image was termed in this theory the *primal sketch*.

What is the primal sketch comprised of? A series of subsequent studies (Marr & Poggio 1979, Marr, Poggio & Ullman 1979, Marr & Hildreth 1980, Marr & Ullman 1981) have advanced a more specific suggestion: The initial analysis of the image includes the localization of significant intensity changes at a number of scales, based in part on the detection of zero crossings in the retinal output. The remainder of this section describes the notion of zero crossings, their use for edge detection, and their possible relation to processing in the primary visual cortex. The multiple scale issue is examined in the subsequent section.

Edges Are Useful Primitives

For the current discussion, edges can be defined as locations in the image where the light intensity changes significantly from one level to another. A representation of the edges in the image proved a useful first step in many image processing applications. Due to the practical importance of this computation, substantial research effort has been devoted in the engineering field of image processing to the development of fast and reliable edge detection methods (see Davis 1975 for a review).

The use of edge information to represent the incoming image is not the only approach that has been suggested. Alternatives include, for example, description of the image in terms of Fourier-like or Gabor-like components. In actual applications, however, edge-based methods are by far the most popular.

Perhaps the main reason that edge-based representations of images proved useful is that intensity edges often have physical significance. They usually correspond to discontinuities in physical properties, such as depth, surface orientation, or reflectance properties. The analysis of intensity edges is therefore a valuable step in using the image to infer physical properties of the surrounding environment. The usefulness of edge information is also supported by the ability of human observers to use edge and contour information. It is often possible to recognize objects from a simplified line drawing that makes explicit only boundaries and contours in the image. Such a simplified representation often preserves the essential information in the original image, despite the fact that the two are markedly different in terms of the underlying intensity distribution.

From DOGs to Edges: The Use of Zero Crossings

Physiological studies tell us that the initial processing of the incoming image includes a filtering stage with DOG-shaped receptive fields. Computational arguments suggest the importance of edge analysis. Are there any connections between these two approaches? It was suggested by Marr & Hildreth (1980) that the two may be in fact closely related.

The key argument in Marr & Hildreth's analysis is that zero crossings in the DOG-filtered image correspond closely to intensity edges in the original image. Roughly speaking, it is possible to perform an edge detection operation by first passing the image through DOG-shaped receptive fields, and then locating zero crossings in the filtered image.

To see the relations among DOG-filtering, edge detection, and zero crossings, consider first the problem of detecting edges in a one-dimensional signal. An edge is, by definition, a significant change in the signal. The change in the signal can be measured by its derivative. It is plausible, therefore, to locate edges by detecting local maxima (or minima) in the signal's derivative. Alternatively, it is possible to locate zero crossings (sign changes) in the second derivative, since zero crossings in the second derivative correspond to peaks in the first.

One procedure for edge detection is therefore to perform a second-derivative operation, and then locate the zero crossings in the second derivative. A problem that arises in the procedure outlined above is that differentiation tends to amplify the high-frequency noise in the original signals. To counteract this effect, the high frequency components of the signal must be attenuated. This can be achieved by smoothing the signal prior to the differentiation operation. Marr & Hildreth (1980) have argued that this can be done optimally by using a Gaussian function. The procedure for edge detection is therefore amended as follows:

1. smooth the original signal by passing it through a Gaussian-shaped filter,
2. perform a second derivative operation,
3. locate the zero crossing.

The next observation is that steps 1 and 2 can be collapsed into a single operation. It can be shown that Gaussian smoothing followed by a second

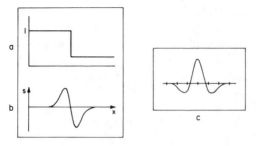

Figure 1   The use of a DOG-shaped receptive field for edge detection. Figure 1b is the result of passing the (one-dimensional) signal 1a through 1c. A zero crossing in 1b corresponds to the edge in 1a.

derivative operation is entirely equivalent to a single step, in which the signal passes through the second derivative of a Gaussian.

The final observation is that the second derivative of a Gaussian shown in Figure 1c is almost identical in shape to the difference of Gaussians (Marr & Hildreth 1980, Figure 11).

The conclusion from the above discussion is, therefore, that DOG-like receptive fields can be used for edge detection. This procedure is shown in Figure 1 for a one-dimensional signal. Figure 1b shows the result of passing 1a through the "receptive field," 1c. It can be seen that the zero crossing in 1b indicates the position of the edge in the original signal, 1a. The discussion so far has been for a one-dimensional signal. A simple possible extension to two dimensions is to use receptive fields shaped as the difference of two 2-D Gaussians. Under certain restrictions (examined in Marr & Hildreth 1980), edges in the 2-D image would be detected in an entirely analogous manner: the image is passed through DOG-shaped filters, and the zero crossings in the output are then used to localize edges in the original image. A 2-D example is shown in Figure 2.

Possible Implications for the Visual Cortex

The analysis so far leads to the general suggestion that following the retinal operation the next step is to locate and represent a map of the zero crossings in the output. Since at the level of the lateral geniculate nucleus (LGN) the units are similar in their responses to X-like and Y-like retinal ganglion cells, the hypothesis is that zero-crossing detection is performed at the primary visual cortex.

Let us consider briefly how zero crossings may be detected by the mechanisms of the visual cortex. The neural image transmitted from the retina to the visual cortex via the LGN is carried by units of two complementary types, the on-center and off-center units. In a biological

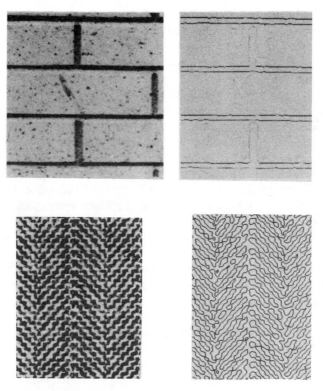

Figure 2 Zero crossing representation of a 2-D image. On the *left*: original images (brick wall, cloth). On the *right*: the zero crossings of the filtered images.

system the existence of such complementary types is not surprising. Mathematically, DOG filtering of the image can produce both positive and negative values. The transmittance of significant negative values presents a problem, since the maintained activity of ganglion cells and LGN units is rather low and cannot decrease significantly. A possible solution is to split the signal into two parts: the positive part is carried by the activity of on-center units, and the negative by the activity of off-center units.

Consider next the result of passing an intensity edge through a DOG filter, as shown in Figure 1b. This output has a zero crossing at the location of the edge, flanked by two peaks, one positive, the other negative. This pattern implies that in terms of neural activity a zero crossing will be flanked by two peaks of activity of on-center units on one side and off-center units on the opposite side. The activity peaks could be used for the reliable detection of zero crossings. Marr & Hildreth (1980) have suggested a simple scheme that relies on the combination of the two peaks. This scheme is illustrated in Figure 3. When two adjacent units, one on-center, the other off-center, are active simultaneously, they indicate the existence of a zero crossing running between them.

Recent empirical evidence has suggested, however, that this particular implementation of a zero crossing detector is probably incorrect. One line of evidence comes from Schiller's (1982) APB experiments. In these experiments, APB infusion to the retina of the rhesus monkey was used to block selectively the responses of retinal on-center units. The "AND-scheme" described above depends on the combined activity of on- and off-center units and is therefore expected to malfunction following the APB treatment. The experimental results did not follow this expectation. Many of the units (primarily units that responded to dark edges prior to the APB application) were entirely unaffected by the elimination of the on-center responses by the APB. The responses of other units (primarily units responding to light edges) were markedly changed, or eliminated completely.

A modified mechanism for zero-crossing detection was therefore proposed by T. Poggio (1983). In the modified scheme, a zero crossing is defined by the activity of on-center units and the absence of activity in neighboring on-center units. In this scheme a zero-crossing detector is constructed from on-center subunits only. A similar detector can be constructed exclusively from off-center subunits. This mechanism uses inhibition, or a veto

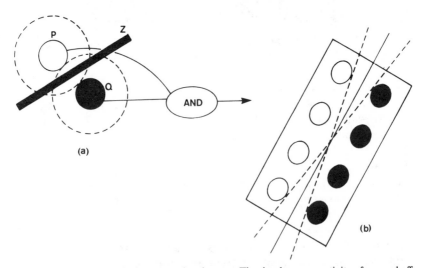

Figure 3   (*a*) A hypothetical zero crossing detector. The simultaneous activity of on- and off-center units (indicated by the "and" gate) indicates the existence of a zero crossing *z* between them. (*b*) By using two rows of units, the detector responds to a restricted range of orientations (indicated by the *dashed lines*). From Marr & Hildreth (1980).

mechanism, of the type suggested by Barlow & Levick (1965). The zero crossing is detected in this case by the avoidance of inhibition, logically equivalent to an AND-NOT operation, rather than the AND operation in the Marr-Hildreth scheme. Torre & Poggio (1978) have shown that a veto operation of the type required by the AND-NOT mechanism can be obtained by interactions between pairs of synaptic inputs on a patch of the dendritic membrane of a neuron. One of the synapses in such a pair uses shunting inhibition to effectively block the neighboring excitatory input. A zero crossing detector of this type could rely exclusively on on- or off-center subunits, or combine on- and off-center subunits independently on different dendritic branches.

Two final comments regarding the differences between the AND and AND-NOT models are (a) that only in the AND-NOT implementation does inhibition play a major role and (b) that the AND-NOT model seems also more consistent with recent psychophysical results by Watt & Morgan (1983).

In conclusion, it is suggested that intensity changes are detected and represented by the visual cortex. These intensity changes include zero crossings in the LGN output, which may be detected by a mechanism of the AND-NOT type. Finally, it should be emphasized that even if the above suggestions are roughly correct, the primary visual cortex is probably involved in other functions as well.

ANALYSIS AT MULTIPLE SCALES

Spatial Frequency Channels in Human Vision

In 1968, Campbell & Robson found psychophysical evidence for the existence within the human visual system of separate channels, tuned to different spatial frequencies. The multiple channel notion received further support in a large number of psychophysical studies. This body of evidence is not discussed here (for a review, see e.g. Braddick et al 1978). The neurophysiological correlate of these findings has not been entirely clarified, but some physiological and anatomical evidence has been raised in support of the multiple channel view. It has been shown, for example, that cortical cells vary in their receptive field size (Hubel & Wiesel 1968) and in their spatial frequency tuning (De Valois et al 1982), and a 2DG study in the cat has suggested a columnar organization of spatial frequency specificity in striate cortex (Tootell et al 1981).

Although the psychophysical evidence supporting the multiple channel view seems fairly strong, many questions remain unanswered. In particular, the fundamental question of the functional significance of the channels is still unclear.

The Functional Role of Multiple Channels in Image Analysis

Proposed explanations for the role of multiple channels can be divided into two classes: decomposition theories and multiscale computations.

The decomposition theories suggest that the role of the multiple channels is to perform a decomposition of the image into more elementary functions. The image is viewed as a two-dimensional function that is decomposed into elementary functions such as Fourier-like or Gabor-like components. Mathematically, this means that the image $I(x, y)$ is expressed as a sum $\sum C_n g_n(x, y)$, where $g_n(x, y)$ are a fixed family of basis functions, and C_n are coefficients that depend on the image $I(x, y)$. The activity of units in the primary visual cortex can be considered according to this view as representing the coefficients C_n.

An alternative view that emerges from the computational study of vision is that multiscale computations provide a powerful tool in the analysis of images. For example, when searching for edges in the image, it proved difficult to detect reliably all the significant edges in the image using an operator (such as the DOG discussed in the previous section) of a fixed size. Improved results are obtained by using a family of operators, having similar shapes but different sizes. Multiscale computations proved useful in a variety of other visual processes such as stereopsis (Marr & Poggio 1979), the computation of surface orientation (Terzopoulos 1983), simple pattern recognition tasks (Rosenfeld 1980), and others (for a review see Rosenfeld 1983).

There are two main seasons for using multiscale computations in vision. The first may be called the physical reason. Significant features in the image can occur at a variety of scales. Some shadow edges, for instance, may spread over a considerable spatial extent, whereas changes at object boundaries are usually sharply localized. Performing similar computations at different resolution scales has proved useful in dealing with features of different scales in the image.

The second reason for using multiscale image analysis is computational efficiency. The efficiency of multiscale computations was discovered and analyzed extensively in recent years in connection with the numerical solution of differential equations (for a review see Hackbusch & Trottenberg 1982). The basic idea is rather simple. The computations involved in solving such problems often use an iterative procedure applied to a grid of points. In image analysis applications, for instance, the 2-D image is often covered by a regularly spaced grid of points. The computation then proceeds in parallel across the image. At each stage, each point in the grid updates its current value as a function of its own value and the value of neighboring points in the grid. A major consideration in such parallel computations concerns the scale of the grid. If the image is covered with a fine grid (that is, the distance between neighboring points is small), the computation is in general more accurate but also slower, because many steps are required to propagate information from one location to another. Coarser grids are generally faster but less accurate. Multigrid methods have been developed to combine optimally the advantages of different scales. Since the visual system performs some of its computations by networks of interconnected elements that resemble in certain respects the grids used by mathematicians, it is conceivable that multiscale processes are employed to combine high resolution with high processing rates.

Multiscale computation theories may not be entirely incompatible with the decomposition idea, but they do lead to different general expectations. For example, decomposition theories favor linear processing at the early stages, whereas multiscale edge analysis, for instance, requires more nonlinear interactions. Decomposition theories expect the channels to be rather independent, at least at the early processing stages, while multiscale computations allow strong interactions among channels.

In conclusion, it should be noted that the theories of multiscale computation that come out of computational vision are not yet detailed enough to allow specific neurophysiological predictions. They do provide, however, a potential framework for understanding the apparently fundamental role of multiple scale channels in vision.

It can be hoped that research in this area, which is currently a central topic in the computational study of vision, will lead to a better theoretical understanding of multiscale processing in vision, and consequently to new insights regarding channels in human vision. The areas where such theories seem most likely to emerge first are edge detection, motion computation, and stereoscopic vision. The first two are discussed in this review; for multiscale theories of stereo vision see Marr & Poggio (1979), Poggio & Poggio (1984).

THE ANALYSIS OF VISUAL MOTION

When objects move and change shape, or when the observer moves with respect to the surrounding environment, the retinal image undergoes complex transformations. These dynamic image transformations provide valuable information about objects in the visible environment and their motion in space relative to the observer. It is not surprising, therefore, that motion perception plays a central role in both biological and machine vision. Computational studies of motion perception have been concerned with two main problems: the measurement of visual motion and the subsequent interpretation of visual motion. In this section I discuss primarily the first problem, and comment briefly on the second.

The Problem of Measuring Visual Motion

The first problem faced by a visual system that analyzes visual motion is that the motion of elements and regions in the image is not given directly, but must be computed from more elementary measurement. The initial registration of light by the photoreceptors (or by an electronic camera) can be thought of as producing an array of light intensity values that change with time. The first problem, therefore, is to produce motion measurements from these time-varying intensity values. This problem of measuring visual motion turned out to be considerably more difficult than anticipated initially. Our knowledge of the biological mechanisms that measure visual motion is incomplete. From a computational standpoint, we still lack efficient and reliable methods for measuring visual motion. In the section below I briefly review computational schemes that have been proposed for the measurement of visual motion by the human visual system and point out open problems and potential implications for neurophysiological studies.

Two Main Schemes for Measuring Visual Motion

Psychophysical evidence accumulated in recent years has promoted the view that two different mechanisms are involved in the process of motion detection and measurement. The terms "short-range" and "long-range" processes are commonly used for the two mechanisms (Braddick 1980). The short-range mechanism measures continuous motion, or motion presented discretely with spatial displacements up to about 15 min of arc (in the center of the visual field) and temporal intervals of less than about 60–100 msec. The long-range mechanism can process larger displacements and temporal intervals. Under the appropriate spatial and temporal presentation parameters, it can "fill in" the gaps in a discrete presentation of stimuli even when the stimuli are separated by up to several degrees of visual angle, and by long temporal intervals (400 msec and more). Under ideal conditions, the resulting motion, termed "apparent" or "beta" motion, can be perceptually indistinguishable from continuous motion.

The distinction between the two systems is more fundamental than their difference in range. The short-range process operates more directly on changes in the local light intensity distribution. The long-range process proceeds by first identifying image features such as terminators, corners, blobs, regions, etc, and then matching these features over time (Ullman 1979).

THE SHORT-RANGE PROCESS OF MOTION MEASUREMENTS   Computational work and the modeling of biological motion detection systems have led to two main schemes for short-range motion measurement: correlation techniques, and gradient methods. In the correlation scheme, motion detection is achieved by comparing the outputs of two detectors of light increments at two adjacent positions. The output at position p_1 and time t is compared with that of position p_2 at time $t - \delta t$ (low-pass temporal filtering can be employed instead of the temporal delay). Several variations of this scheme have been proposed as models for biological systems. The first is obtained by multiplying the two values, i.e.

$$D(p_1, t) \times D(p_2, t - \delta t),$$

where D denotes the output of the subunits (Figure 4a). If a point of light moves from p_2 to p_1 in time equal to δt, then it will cause a light increment at p_2, and after an interval of δt a similar increment at p_1; therefore, the above product will be positive. In an array of such detectors, the average output will be essentially equivalent to a cross-correlation of the inputs. This scheme provides a successful model for the overall optomotor behavior of various insects in response to motion in their visual fields.

An alternative method along the same general line is the "And-Not" scheme proposed by Barlow & Levick (1965) for the directionally selective units in the rabbit's retina. Since Barlow & Levick found evidence for

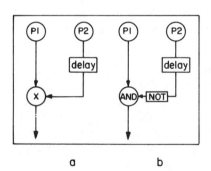
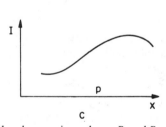

a   b   c

Figure 4   Motion detection schemes. (*a*) and (*b*): The delayed comparison scheme. P_1 and P_2 are detectors that respond transiently to a spot of light. (*a*) For a spot moving to the left at the appropriate speed, the responses of P_1 and P_2 coincide, yielding a positive output of the combined unit. (*b*) The veto scheme. Motion from P_2 to P_1 produces no response, since the delayed response from P_2 cancels the response from P_1. (*c*) The idea behind the gradient method. At point p the intensity profile has a positive slope. If the stimulus now moves to the left, the intensity value I at p increases. If the stimulus moves to the right $I(p)$ decreases. The sign of the temporal change in $I(p)$ thus signals the direction of motion. The speed of motion can be determined from measurements of the slope and the temporal change.

inhibitory interactions within the directionally selective mechanism, they proposed a model in which the motion detector computes the logical AND of $D(p_1, t)$ and NOT $D(p_2, t - \delta t)$ (Figure 4b). In this scheme, a motion from p_2 to p_1 will be "vetoed" by the delayed response from p_2, whereas motion from p_1 to p_2 will produce a positive response. A model of this type has been described for the visual system of the fly (Poggio & Reichardt 1976), and an elegant synaptic mechanism that can implement these computations has been proposed (Torre & Poggio 1978).

A different scheme for motion measurement has been proposed by Marr & Ullman (1981) as a model for motion analysis by cortical simple cells. This scheme is based on the combined measurement of intensity gradients and temporal changes, as shown schematically for a one-dimensional example in Figure 4c. Suppose that at point p the intensity profile (intensity I as a function of position x) has a positive slope. Clearly, if the intensity profile now moves to the left, the intensity value I at p increases, and for a motion to the right, $I(p)$ decreases. The sign of the temporal change in $I(p)$ thus signals the direction of motion. Furthermore, the speed of motion can be determined from measurements of the slope and the temporal change. This scheme is especially useful near an object's edges (or at the zero crossings of the retinal image discussed above), where relatively steep intensity gradients are induced. According to this model, the intensity profile is analyzed by sustained x-like units, and the temporal change by transient, y-like units. Computer experiments have shown that this gradient scheme can successfully recover motion information from image sequences, but evidence is at present insufficient to determine whether a scheme of this type is employed by cortical simple cells.

The integration of motion measurements   The motion of edge segments can be determined by using one of the schemes outlined above. An additional problem arises, however, since the motion of edges and contours cannot be determined on the basis of purely local measurements. The underlying reason is the "aperture problem" illustrated in Figure 5. If the motion is to be detected by a unit that is small compared to the overall contour, then the only information that can be extracted is the motion component perpendicular to the local orientation of the element. Motion along the element would be invisible. To determine the motion completely, a second stage that combines the local measurements, either in local neighborhoods or along the contour, is required (Figure 5).

Computational aspects of the integration problem have been studied by Hildreth and Ullman (Hildreth 1984, Ullman & Hildreth 1983). The scheme proposed by Hildreth determines the velocity field from the initial measurements by minimizing the variation in the velocity field along image

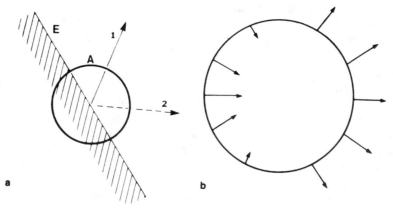

Figure 5 The aperture problem. (*a*) The direction of motion of a one-dimensional profile cannot be recovered uniquely by a unit that is small compared to the moving element. Looking at the moving edge *E* through the aperture *A*, it is impossible to determine whether the actual motion is, for example, in direction 1 or 2. (*b*) A schematic representation of local motion measurements performed at the edge of a moving disk. The disk is translating from left to right. The local measurements give only the motion components perpendicular to the local orientation. A subsequent processing stage that integrates these local measurements can infer the correct motion.

contours. Mathematically, the scheme is defined in the following manner. For each point along an image contour, the initial motion measurements give the motion components perpendicular to the contour. Because of the aperture problem, the tangential components remain invisible. These perpendicular measurements are therefore insufficient to determine uniquely the velocity field along a contour. From all the velocity fields that are compatible with the initial measurements, the scheme determines the one that minimizes the overall variation $\int (\partial V/\partial s)_{ds}^2$, where V is the 2-D velocity field and the integral is taken along the contour. Hildreth has developed a method for computing this minimal variation motion field, and has shown that in many instances the results are similar to motion patterns perceived by human observers under similar conditions (Hildreth 1984).

Possible neurophysiological implications These computational studies have potential implications to neurophysiological studies. They suggest that following the initial motion measurement a subsequent stage is required to combine the local measurements either within a local neighborhood or along contours. Motion-selective units in the primary visual cortex of the cat and monkey do not seem suitable for the integration stage. Such units are selective to the orientation as well as the movement direction of the stimulus. To activate such a unit optimally the stimulus must have the orientation preferred by the unit, and move in the preferred direction. In contrast, a unit involved in the integration phase is expected to exhibit more dissociation between the effects of orientation and the direction of movement. Ideally, an "integration unit" would exhibit specificity for direction, provided that the stimulus contains a sufficient range of orientations. It was suggested on the basis of these considerations (Ullman 1983) that area MT in the rhesus monkey may be involved in the motion-integration process. Recent electrophysiological experiments have supported this suggestion. Movshon et al (1984) have discovered promising candidates for "integration units" in area MT. The units were tested with stimuli composed of two oriented gratings. Unlike units in area 17, these units responded best when the stimulus as a whole moved in the preferred direction, irrespective of the orientation of the component gratings.

THE LONG-RANGE MOTION MEASUREMENT PROCESS   The apparent motion phenomena mentioned above illustrate the visual system's capacity to establish visual motion by matching tokens over considerable spatial distances and temporal intervals. In perceiving continuous motion between successively presented elements, the visual system is faced with the difficult problem of establishing a "correspondence" between the elements of the two presentations. That is, for each element in the first frame, its counterpart in the subsequent frame must be located. A simple example of the correspondence problem between two successively presented frames is illustrated in Figure 6. The *filled circles* in the figure represent the first frame, the *unfilled circles* the second. There are two possible one-to-one matches between the elements of the two frames, leading to two possible patterns of perceived motion: horizontal motion (Figure 6a), and diagonal motion (Figure 6b).

In this example the match is only two-way ambiguous. In the more general case each frame may contain not only two, but many elements arranged in complex figures, and a correspondence has then to be

a   b

Figure 6   The correspondence problem. Two frames are shown in a brief temporal succession. The *filled dots* in (a) represent the first frame, the *unfilled* the second. (In the actual presentation all the dots are identical in appearance.) Two one-to-one matches are possible, leading to two possible patterns of perceived motion: horizontal (a) or along the diagonals (b). In perceiving apparent motion, the visual system is effectively deciding how to match successively presented elements.

established between the figures. The rules governing the correspondence process in human vision are now partially understood, as a result of a number of studies (Kolers 1972, Ullman 1979).

Less is known about this long-range process at the neurophysiological level. One obvious question is just where in the visual system can apparent-motion phenomena be observed in the response of single units. Units (for example, in areas V1 and superior temporal sulcus or MT of the monkey) could be tested for apparent-motion responses by flashing bars at stationary locations and using relatively wide separations (for example, wider than the largest center size of simple cells at the tested eccentricity). Initial results (Newsome et al 1982) suggest that units in MT may indeed play a role in the perception of long-range apparent motion. If long-range motion units can be identified, it may become possible to go a step further and investigate the relationship between the psychophysically established correspondence rules and their neurophysiological correlates.

The Recovery of Three-Dimensional Structure from Motion

A remarkable capacity of the human motion perception system is its ability to recover the 3-D shape of moving objects even when the objects are unfamiliar and when each static view of the scene contains no 3-D information. The first systematic investigation of this remarkable capacity was carried out in the Wallach & O'Connell study of the kinetic depth effect (Wallach & O'Connell 1953). In these experiments, an unfamiliar object was rotated behind a translucent screen; the shadow of its orthographic projection was observed from the other side of the screen. (In orthographic projection, the object is projected by parallel light rays that are perpendicular to the screen.) In most cases, the viewers were able to give a correct description of the hidden object's 3-D structure and motion in space, even when each static view was unrecognizable and contained no 3-D information. The original kinetic depth effect employed primarily wire-frame objects that project as sets of connected lines. Later studies established that 3-D structure can be perceived from displays consisting of unconnected elements in motion and in continuous, as well as apparent, motion.

The computational problems underlying the recovery of 3-D structure from motion have been investigated in a number of studies. The main issues they explored were the conditions under which the structure-from-motion problem has a unique solution, and the development of algorithms for the recovery of structure from motion. The main conclusion has been that when the changing image is induced by rigid objects in motion, the 3-D structure of the objects is determined uniquely by the 2-D transformations of the image (Ullman 1979, Longuet-Higgins & Prazdny 1980).

The recovery of 3-D structure from motion is an example of a visual capacity that is relatively well understood from a computational stand-point, but is virtually unexplored so far at the neurophysiological level. It would be of interest in future studies to attempt to identify locations within the visual system that participate in this process. If such locations can be identified, the insight gained from the computational studies may prove useful in the empirical investigation of these areas.

In summary, in the area of visual motion there have been a number of fruitful interactions between computational studies and experimental research. Reasonable directions for further exploration include: (a) the test-ing of specific models of motion-detecting units, (b) further studies of the integration stage in the short-range process, (c) exploratory studies of the possible sites of the long-range motion correspondence and the structure-from-motion processes.

LIGHTNESS PERCEPTION

The study of lightness is concerned with the perception associated with different shades of grey in the image. A satisfactory theory of lightness perception should be able to predict, for example, which of any given pair of surfaces in an image would appear to a human observer as lighter or darker. The task may appear quite straightforward: it seems reasonable to assume that the surface that reflects more light toward the observer should appear lighter. It turns out, however, that there is no simple relation between the perceived lightness at a point and the light intensity at that point. The light intensity coming from a region A in the image may be considerably higher than the intensity coming from region B, yet A may appear considerably darker than B.

That perceived lightness is not directly related to image intensity is in fact not surprising. The visual system is not a light-measuring device: it is concerned not with light intensities but with the physical properties of the surrounding surfaces. This observation has been the starting point of Land's retinex theory of lightness and color perception (Land 1959, 1983 Land & McCann 1971). According to this view, lightness perception is the end result of a process that attempts to recover surface reflectances from the intensity distribution in the image. The problem faced by the visual system in recovering reflectance values is that the image intensity at a point is determined by a number of factors that are difficult to disentangle. Illumination, surface orientation, and surface reflectance combine in an intricate manner to produce the image intensity at a point. The visual system has access only to the final product, the image intensities, from which reflectance values are to be recovered.

The problem considered by Land was a simplified version of the general problem described above. In his formulation the surfaces were assumed to form a "Mondrian": a collection of flat, coplanar surface patches, with a constant reflectance within each patch. Under these conditions the image intensity $I(x, y)$ can be expressed by the product $E(x, y)R(x, y)$, where $E(x, y)$ is the illumination and $R(x, y)$ the reflectance, i.e. the fraction of the illumination reflected back by the surface. Land's key observation was that $E(x, y)$ and $R(x, y)$ usually behave quite differently. The illumination in the scene $E(x, y)$ tends to vary gradually and smoothly as a function of position, whereas the reflectance $R(x, y)$ changes abruptly at surface boundaries. This distinction certainly holds for the Mondrian world, and it is probably also a reasonable assumption under many natural conditions.

The idea is therefore to decompose the intensity distribution into two components: a smooth one that varies gradually over the entire image, which is due to the ambient illumination; and a component that represents reflectance values, which is constant within patches and changes abruptly at patch boundaries.

Land & McCann (1971) also suggested an algorithm to carry out the above decomposition. A different algorithm that performs essentially the same computation was proposed by Horn (1974). These algorithms will not be described here; details can be found in the above citations and in Land (1983).

The Retinex and Color Perception

The discussion so far has been about lightness perception in achromatic images. Land also proposed the retinex theory as a basis for color perception. According to this view the retinex computation outlined above is carried out in a number of independent channels. The channels can be, for instance, the three primary colors, red, green, and blue, each carrying out a retinex computation on a different portion of the spectrum. (The terms "red," "green," and "blue" are used for convenience, the three different receptor types have in fact rather broad and overlapping absorption spectra.) The three computations carried out in parallel result in three numbers for each point, called "designators." Perceived color at a point is determined by the designator triplet. Land has shown that this procedure predicts perceived colors under large variations in scene illumination, at least in the Mondrian world.

A certain modification to the above scheme is necessary because pure color channels do not seem to exist in the primate visual system. The different colors are not kept separate in independent channels, but are mixed already at the retinal ganglion level. This mixing does not pose a serious problem to the retinex computation, since the computation can be carried out on linear combinations of the primary colors. If the primary colors are denoted by R, G, and B, the retinex computation can be carried out, for example, on quantities such as $R - G$ (the difference between the red and green channels) or $R + G + B$ (the sum of all channels). This means that smooth variations in the quantity $R - G$, for instance, are ignored, whereas sharp transitions are maintained and used to compute one of the three designators.

Neurophysiological Implications of the Retinex Color Theory

A major requirement of the retinex computation is the ability to detect relatively sharp transitions within each one of the color channels. What type of unit can be expected for this task? As discussed above, a center-surround DOG can be used efficiently for detecting sharp intensity transitions. A similar mechanism could be used in the color domain. To detect sharp transitions in a "red" channel, for instance, one can use a DOG-shaped receptive field with both center and surround responding selectively in the red region of the spectrum. A channel that detects sharp transitions in the quantity $R - G$ is again expected to have a DOG center-surround organization, but with both center and surround responding to both red and green. The center should be excited by red and inhibited by green light, and the surround should have the opposite response, excited by green and inhibited by red. Units with such color response characteristics have been called *double-opponent* cells. Because the different color channels are not kept separate, the retinex theory gives rise to the expectation that double opponent cells of this type would provide the basis for color computation. In contrast to this expectation, most color-selective cells at the level of the retina, LGN, and layer $4C\beta$ of the macaque primary visual cortex have a different color organization (Wiesel & Hubel 1966, Hubel & Wiesel 1968). These cells have predominantly one of three types of chromatic organization: color opponent cells (type I), color opponent center-only (type II), and broad-band (type III). Marr (1982) has noted this discrepancy between the computational theory and the experimental findings. He described the most predominant type of chromatic cells (type I) as "extremely difficult to understand" and "impossible to fit" to the general framework discussed above (Marr 1982, p. 262). It has been an interesting development, therefore, when an abundance of double-opponent cells were discovered in the cytochrome oxidase rich blobs in area 17 of the monkey (Livingstone & Hubel 1984). In this blob system, which appears to be involved in the analysis of color, the main cell types are red-green double opponent, yellow-blue double opponent, and broad-band cells. These cell types provide an adequate basis for a retinex-like color computation. By "retinex-like" I mean that the details of Land's retinex computation, many of which are in dispute, are not crucial for the implications discussed above. Lightness and color computation is therefore an example in which computation considerations provided useful guidance for electrophysiological studies and a framework for the analysis and interpretation of empirical findings.

SELECTIVE VISUAL ATTENTION

Functional and Spatial Parallelism

Unlike most digital computers, the processing of visual information by the brain involves extensive parallel computations. There are two types of parallelism employed by the visual system: functional and spatial. *Functional parallelism* means that different computations are applied simultaneously to the same location. The X- and Y-type system, of retinal ganglion cells, for example, form two systems operating in parallel, each one covering the entire visual field. Similarly, current views of the visual cortex (e.g. Zeki 1978a,b) suggest that different visual areas in the extrastriate cortex process different aspects of the input (such as color, motion, and binocular disparity) at the same location simultaneously, thereby achieving functional parallelism. Spatial parallelism means that the same operations are applied simultaneously to different spatial locations. The operations performed by the retina and the primary visual cortex, for example, fall under this category. The high degree of parallelism employed by the visual system is often suggested as a general explanation for its efficiency. Compared to the components used in electronic computers, the processing rate by individual neurons is slow, yet the performance of the system as a whole far surpasses the capacities of any artificial system. It seems, however, that a high degree of spatial parallelism is not always desirable Certain computations of spatial information appear to be inherently sequential in the sense that they can be performed better if the processing can be directed at different stages in the computation to different locations in the visual field. In their book, *Perceptrons*, Minsky & Papert (1969) have established that certain spatial relations such as connectivity or inside/outside relations cannot be computed at all by a class of parallel pattern recognition devices called perceptrons. These computations can be performed, however, if the perceptrons are augmented in a manner that allows them to direct the computation first to one location, then to another. The ability to restrict the computation to selected regions in the visual field plays a similar role in dealing with what Minsky & Papert have called the "figure in the context" problem. They have found that it is often a straightforward task from a computational standpoint to recognize a simple figure irrespective of its location, size, and shape. A triangle of any shape, for instance, can be recognized by using the facts that it contains exactly three straight lines and no free ends. The recognition problem becomes significantly more complicated, however, when there is more than a single figure in the visual field. It then becomes necessary to distinguish between the features, such as straight line segments and free ends, that belong to the figure under consideration, and similar features that may be in close spatial proximity to it but belong to different shapes. Again, the problem can be approached by structuring the processing in space in such a manner that the computations are applied selectively to the objects of interest.

Sequential Processing and Selective Visual Attention

The conclusion is that from a theoretical standpoint there seems to be a strong requirement to incorporate in the processing of visual information a mechanism for directing the processing selectively to different locations at different times (see also Ullman 1984). This general conclusion also appears to fit with recent psychophysical and physiological findings. Psychophysically, phenomena related to the sequential processing of selected locations have been studied under the heading of "selective visual attention." These studies received a new impetus recently when it was shown by the studies of Treisman (Treisman & Gelade 1980), Posner (1980), Julesz (1981), and others that selective attention phenomena play a crucial role even in simple and immediate perceptual tasks. Physiological experiments with alert animals have begun to suggest mechanisms that may be involved in selective visual attention (e.g. Mountcastle 1976, Fuster & Jervey 1981, Richmond & Sato 1982, Wurtz et al 1982). A theory of the physiological basis of selective visual attention has recently been proposed by Crick, who hypothesized that it may be controlled by the activity of the thalamic reticular complex (Crick 1984).

Our understanding of the functional role and the physiological mechanisms of selective visual attention is at present rudimentary. This is a fundamental research area in the study of vision that will call for a close cooperation between computational and empirical studies. If selective attention mechanisms play a role in the early stages of visual information processing, then the activity of units in alert animals that perform visual tasks would be difficult to characterize exclusively in terms of the optimal stimuli to which the units are tuned. The response of the units would depend not only on the stimulus configuration but also on the location in space and the particular computation being performed. Under such conditions a close cooperation between theory and experimental work seems indispensable, both for the design of experiments and the interpretation of the results.

CONCLUDING REMARKS

The Relevance of Computational Vision to Empirical Vision Research

The problem addressed by computational vision and by empirical vision research may seem at a first glance to be quite different. In computational vision research the main question one asks is how it is possible to perform a given visual task. Empirical vision research, on the other hand, is concerned with the question of how the task is actually performed by a given biological system. It nevertheless appears that in many cases the two disciplines are likely to produce converging results. Results obtained in computational studies may prove relevant to the neurosciences for three reasons:

1. The intricacy of the tasks under study: The programming of computer systems to perform even limited visual tasks turned out to be a formidable challenge. The typical situation for many problems in computational vision is not that many possible solutions present themselves, but that it is exceedingly difficult to find even one workable solution. Under such conditions, when the set of possible solutions is constrained, the solution discovered by computational methods is likely to bear relevance to the processes employed by the visual system.

2. The use of common constraints: Computational studies of vision have found that vision problems are usually underconstrained, and their solution requires therefore the use of certain constraints. A typical example is Land's lightness computation. The effects of surface reflectance and ambient illumination seem intractably confounded when only their final product, the image intensity distribution, is given. The effects can nevertheless be often distinguished by taking into account the differences in the typical behavior of reflectance and illumination changes. The use of assumptions and constraints of this type is widespread in computational vision [e.g. rigidity constraint in motion perception (Ullman 1979), continuity in stereo computation (Marr & Poggio 1979), smoothness of the velocity field (Hildreth 1984)]. These constraints are often inherently necessary for performing the task in question and are therefore likely to apply equally well to artificial devices and to biological visual systems.

3. Many of the schemes developed in computational vision attempt to incorporate as much as possible known constraints from empirical research. As an example, the use of zero crossings for representing intensity changes performs an edge detection operation using known receptive field shapes. Although computational studies are not aimed at direct modeling of biological mechanisms, the fact that empirical constraints are taken into

account increases the likelihood that methods developed in computational vision would prove relevant to the neurosciences.

Levels of Interaction Between Computational Vision and the Neurosciences

One can distinguish three levels of specificity in the interactions so far between computational vision and the neurosciences. At the most specific level, a computational theory may suggest a model for a specific mechanism that can be tested empirically. Examples of this type include the zero crossing detector proposed by Marr & Hildreth (1980) and the motion-detecting unit proposed by Marr & Ullman (1981) at a more general level, computational theories may be useful in identifying processing stages that are also performed by biological visual systems, but were not apparent from epirical research. An example of this level is the study of the motion integration problem (Adelson & Movshon 1982, Hildreth 1984, Ullman & Hildreth 1983). This study has identified the requirement for a motion integration stage subsequent to the initial motion measurements at the level of the primary visual cortex. Unlike the previous category, the conclusions were not specific enough to imply a specific mechanism. They did suggest, however, an additional processing stage to look for and they proved useful in the design and interpretation of both psychophysical experiments and physiological studies in area MT of the macaque monkey.

Finally, there is still a more general level that may be called the *module level*. This includes the identification of more-or-less independent processing modules within the visual system and the overall gross architecture of the system. Work at this level is not directly related to physiological studies at the single unit level, but may prove relevant, for example, in understanding the role of different visual areas and their interconnections. Marr's notion of the primal sketch is an example at this level. The primal sketch suggests neither a specific mechanism nor a precise set of processing stages. It does suggest, however, a role for a module within the visual system that is concerned with the symbolic representation of intensity changes in the image.

Another example is the structure from motion process (Ullman 1979, Longuet-Higgins & Prazdny 1980). It was shown that 3-D structure can be recovered on the basis of motion information alone, independent of other sources of 3-D information. This independence raises the possibility that a structure-from-motion computation may be associated with a yet unknown region of the visual system. Physiological and lesion techniques could perhaps be used to explore the existence of such a mechanism.

Finally, in the near future the interactions between computational vision and the neurosciences will probably take place primarily at the level of process and module, and less at the level of specific mechanisms. Computational studies are useful in identifying necessary or likely modules and processing stages, but these are usually not immediately translatable in any obvious manner to biological mechanisms.

ACKNOWLEDGMENT

I thank Dr. E. Hildreth for her comments and C. Weintraub for preparing the manuscript.

Literature Cited

Adelson, E. H., Movshon, J. A. 1982. Phenomenal coherence of moving visual patterns. *Nature* 300:523–25

Barlow, H. B. 1953. Summation and in-hibition in the frog's retina. *J. Physiol.* 119:69–88

Barlow, H. B., Levick, W. R. 1965. The mechanisms of directionally selective units in rabbits' retina. *J. Physiol.* 178:477–504

Braddick, O. J. 1980. Low-level and high-level processes in apparent motion. *Philos. Trans. R. Soc. London B* 290:137–51

Braddick, O., Campbell, F. W., Atkinson, J. 1978. Channels in vision: Basic aspects. *Handb. Sensory Physiol.* 8:3–38

Cleland, B. G., Levick, W. R. 1974. Brisk and sluggish concentrically organized ganglion cells in the cat's retina. *J. Physiol.* 240:421–56

Crick, F. 1984. Function of the thalamic reticular complex: The searchlight hypothesis. *Proc. Natl. Acad. Sci. USA* 81:4586–90

Davis, L. 1975. A survey of edge detection techniques. *CGVIP* 4:248–70

De Monasterio, F. M. 1978. Properties of concentrically organized X and Y ganglion cells of macaque retina. *J. Neurophysiol.* 41:1394–1417

De Valois, R. L., Albrecht, D. G., Thorell, L. G. 1982. Spatial frequency selectivity of cells in macaque visual cortex. *Vis. Res.* 22:545–59

Enroth-Cugell, C., Robson, J. G. 1966. The contrast sensitivity of retinal ganglion cells of the cat. *J. Physiol.* 187:517–52

Fuster, J. M., Jervey, J. P. 1981. Inferotemporal neurons distinguish and retain behaviorally relevant features of visual stimuli. *Science* 212:952–55

Hackbusch, W., Trottenberg, U., eds. 1982. Multigrid methods. *Lect. Notes Math.* 960. New York: Springer-Verlag

Hildreth, E. C. 1984. The computation of the velocity field *Proc. R. Soc. London B* 221:189–220

Horn, B. K. P. 1974. Determining lightness from an image. *CGVIP* 3:277–99

Hubel, D. H., Wiesel, T. N. 1962. Receptive fields, binocular interaction, and functional architecture in the cat's visual cortex. *J. Physiol.* 160:106–54

Hubel, D. H., Wiesel, T. N. 1968. Receptive fields and functional architecture of monkey striate cortex. *J. Physiol.* 195:215–43

Julesz, B. 1981. Textons, the elements of texture perception, and their interactions. *Nature* 290:91–97

Kollers, P. A. 1972. *Aspects of Motion Perception.* New York: Pergamon

Kuffler, S. W. 1952. Neurons in the retina: Organization, inhibition and excitation problems. *Cold Spring Harbor Symp. Quant. Biol.* 17:281–92

Land, E. H. 1959. Experiments in color vision. *Sci. Am.* 200:84–99

Land, E. H., McCann, J. J. 1971. Lightness and retinex theory. *J. Optic Soc. Am.* 61(1):1–11

Land, E. H. 1983. Recent advances in retinex theory and some implications for cortical computations: Color vision and the natural image. *Proc. Natl. Acad. Sci.* 80:5163–69

Lennie, P. 1980. Parallel visual pathways: A review. *Vision Res.* 20:561–94

Livingstone, M. S., Hubel, D. H. 1984. Anatomy and physiology of a color system in the primate visual cortex. *J. Neurosci.* 4(1):309–56

Longuet-Higgins, H. C., Prazdny, K. 1980. The interpretation of a moving retinal image. *Proc. R. Soc. London B* 208:385–97

Marr, D. 1976. Early processing of visual information. *Philos. Trans. R. Soc. London B* 275:483–524

Marr, D. 1982. *Vision.* San Francisco: Freeman

Marr, D., Hildreth, E. 1980. Theory of edge detection. *Proc. R. Soc. London B* 207:187–217

Marr, D., Poggio, T. 1977. From understanding computation to understanding neural circuitry. *Neurosci. Res. Program Bull.* 15(3):470–88

Marr, D., Poggio, T. 1979. A computational theory of human stereo vision. *Proc. R. Soc. London B* 204:301–28

Marr, D., Poggio, T., Ullman, S. 1979. Bandpass channels, zero-crossings, and early visual information processing. *J. Opt. Soc. Am.* 70:868–70

Marr, D., Ullman, S. 1981. Directional selectivity and its use in early visual processing. *Proc. R. Soc. London B* 211:151–80

Minsky, M., Papert, S. 1969. *Perceptrons.* Cambridge, Mass./London: MIT Press

Mountcastle, V. B. 1976. The world around us: Neural command functions for selective attention. The F. O. Schmitt Lecture in Neuroscience 1975. *Neurosci. Res. Program Bull.* 14(Suppl.):1–47

Movshon, J. A., Adelson, E. H., Gizzi, M. S., Newsome, W. T. 1984. The analysis of moving visual patterns. In *Pattern Recognition Mechanisms,* ed. C. Chagas, R. Gattar, C. G. Gross. Rome: Vatican Press

Newsome, W. T., Mikami, A., Wurtz, R. H. 1982. *Abstr. 12th Meet. Soc. for Neurosci.,* Minneapolis, p. 812

Poggio, G., Poggio, T. 1984. The analysis of stereopsis. *Ann. Rev. Neurosci.* 7:379–412

Poggio, T. 1983. Visual algorithms. In *Physical and Biological Processing of Images,* ed. O. J. Braddick, A. C. Sleigh, pp. 128–53. Berlin: Springer-Verlag

Poggio, T., Reichardt, W. 1976. *Q. Rev. Biophys.* 9:377–438

Posner, M. I. 1980. Orienting of attention. *Q. J. Psychol.* 32:3–25

Richmond, B. J., Sato, T. 1982. Visual responses of inferior temporal neurons are modified by attention to different stimuli dimensions. *Soc. Neurosci. Abstr.* 8:812

Rodieck, R. W. 1965. Quantitative analysis of cat retinal ganglion cell response to visual stimuli. *Visual Res.* 5:583–601

Rodieck, R. W., Stone, J. 1965. Analysis of receptive fields of cat retinal ganglion cells. *J. Neurophysiol.* 28:833–49

Rosenfeld, A. 1980. Quadtrees and pyramids for pattern recognition. *Proc. Int. J. Conf. Pat. Recog., Miami,* pp. 802–7

Rosenfeld, A., ed. 1983. *Multiresolution Image Processing and Analysis.* New York: Springer-Verlag

Schiller, P. H. 1982. Central connections of the retinal ON and OFF pathways. *Nature* 297(5867):580–83

Terzopoulos, D. 1983. Multilevel computation processes for visual surface reconstruction. *Comput. Graph. Vis. Image Process.* 24:52–96

Tootell, R. B., Silverman, M. S., De Valois, R. L. 1981. Spatial frequency columns in primary visual cortex. *Science* 214(13):813–15

Torre, V., Poggio, T. 1978. *Proc. R. Soc. London Ser. B* 202:409–16

Treisman, A., Gelade, G. 1980. A feature integration theory of attention. *Cognitive Psychol.* 12:97–136

Ullman, S. 1979. *The Interpretation of Visual Motion.* Cambridge, Mass.: MIT Press

Ullman, S. 1983. The measurement of visual motion. *Trends Neurosci.* 6(5):177–79

Ullman, S. 1984. Visual routines. *Cognition* 18:97–159

Ullman, S., Hildreth, E. C. 1983. The measurement of visual motion. See Poggio 1983, pp. 154–76

Wallach, H., O'Connell, D. N. 1953. *J. Exp. Psychol.* 45:205–17

Watt, R. J., Morgan, M. J. 1983. The recognition and representation of edge blur: Evidence for spatial primitives in human vision. *Vis. Res.* 23(12):1465–77

Wiesel, T. N., Hubel, D. H. 1966. Spatial and chromatic interactions in the lateral geniculate body of the rhesus monkey. *J. Neurophysiol.* 29:1115–56

Wurtz, R. H., Goldberg, M. E., Robinson, D. L. 1982. Brain mechanisms of visual attention. *Sci. Am.* 246(6):124–35

Zeki, S. M. 1978a. Functional specialization in the visual cortex of the rhesus monkey. *Nature* 27:423–28

Zeki, S. M. 1978b. Uniformity and diversity of structure and function in rhesus monkey prestriate visual cortex. *J. Physiol.* 277:273–90

Ann. Rev. Neurosci. 1986. 9: 27–59

NEUROTRANSMITTER RECEPTOR MAPPING BY AUTORADIOGRAPHY AND OTHER METHODS*

Michael J. Kuhar,[*,†] *Errol B. De Souza,*[*,†] *and James R. Unnerstall*[†,‡]

*Laboratory of Neuroscience, Addiction Research Center, National Institute on Drug Abuse, Baltimore, Maryland 21224
† Departments of Neuroscience, Pharmacology and Experimental Therapeutics, Psychiatry and Behavioral Sciences, Johns Hopkins University School of Medicine, Baltimore, Maryland 21205
‡ Experimental Therapy Branch, National Institute of Neurological and Communicative Disorders and Stroke, Bethesda, Maryland 20205

INTRODUCTION

Localizing receptors by microscopic methods has become as feasible and popular as neurotransmitter and enzyme histochemistry has been for many years. A main reason for this has been the development and adaptation of microscopic autoradiographic methods in parallel with the biochemical binding techniques that have so dramatically advanced the study of receptors at the molecular level. Another obvious impetus for advance is that these methods are very useful and provide the investigator with powerful tools for examining the nervous system.

Our goal in this chapter is to summarize the principles, methods, and uses of receptor localization or "receptor mapping." The main emphasis is on light microscopic autoradiography, since this approach dominates the field at this time. However, we discuss other methods, both explicitly and implicitly, since certain principles must underlie any and every experimental approach that is taken. Various aspects of receptor mapping have been reviewed previously (Kuhar 1981, 1982a,b, 1983, 1985).

27

Uses of Receptor Mapping

Before delving into experimental issues, a discussion of the uses of receptor maps is worthwhile. The study of receptors has often been within the discipline of pharmacology, at least historically. Perhaps this is the reason that receptor mapping has often been considered a pharmacological tool.

MECHANISM OF DRUG ACTION   Receptor maps provide powerful insights into the mechanism of drug action because the maps help to identify those tissue elements which have the receptors and therefore those elements which are affected by drug administration. Consider the popular example of opiate drugs. Opiate administration causes a variety of physiological effects, including analgesia, pupillary constriction, respiratory depression, and suppression of visceral reflexes. The only way this wide range of physiological action can be fully understood is to associate the physiological effects with receptors in various neuronal circuits that mediate the effects (Kuhar 1978). In other words, finding out which neuronal circuits have the receptors and where they are in the circuit provides a picture (pun intended) that helps to clarify the mechanism of action of the drugs.

CHEMICAL ORGANIZATION OF BRAIN   Because drug receptors are most often really receptors for endogenous biochemicals such as neurotransmitters, receptor maps also delineate neurotransmitter systems. Many neurotransmitter-containing neuronal pathways have been mapped in the CNS and periphery. Receptor maps help to complete our view of this organization and have been especially informative in identifying receptor subtypes. While the distribution of receptors often overlaps the distribution of nerve terminals containing the related neurotransmitter, a surprising finding has been that receptors are often found where neurotransmitters are not, or are at least low in concentration. Some areas of the brain seem to be enriched with many different kinds of receptors. These areas, which are mainly sensory and limbic, may be uniquely chemosensitive (Kuhar 1981). Also, receptor binding sites have not been found in some areas where a neurotransmitter is in fact found (see below). While this may mean that receptors are simply absent from some neurotransmitter-rich areas (altering receptor levels would be another way to regulate neuronal activity), it may also mean that we simply have not learned how to detect them in these regions. They may be of low affinity for available ligands, be low in number, or be unavailable for binding under the experimental conditions utilized. Although all of these issues can be troublesome, they are at the same time examples of interesting, new issues that are often raised by the utilization of a new technique.

Receptor maps are also useful guides in a variety of related types of experiments. For example, it is more reasonable to study the effects of iontophoretically applied drugs on the activity of single units in areas where receptors are found in high concentrations. Also, the direct injections of drugs and a study of subsequent behaviors are best performed when drug injections are done in well-characterized brain regions.

NEUROPATHOLOGY   Yet another potential use of receptor mapping is in neuropathology. It is known from biochemical studies that receptors are changed in a variety of neuropsychiatric disorders (Olsen et al 1980). It is possible, perhaps likely, that receptor changes may mediate or may be a primary cause of some of these disorders. Since receptor changes can occur in small regions of the brain and since these changes may occur in parallel with other morphological changes, receptor mapping can be a powerful tool in exploring neuropathology. For example, several recent studies have utilized human tissue obtained at autopsy (Whitehouse et al 1983, Uhl & Kuhar 1984). These studies have shown changes in receptors in well-defined areas, and these changes have many important implications.

Mismatch Between Neurotransmitter and Receptor

The light microscopic distribution of receptors does not precisely parallel the distribution of neurotransmitters. Whereas this at first seems surprising, it is clear that there should not be a total match between the histochemical distribution of neurotransmitter and the distribution of receptors. Neurotransmitters are distributed throughout certain neurons, while receptors are distributed throughout the receptive neuron, which is a different cell with a different spatial distribution. Hence there should not be a gross match in distribution. If there were a way to examine only nerve terminal neurotransmitters and synaptic receptors, then the match might be closer. In our laboratory, we observed the lack of matching in our earliest studies (Kuhar & Yamamura 1975, 1976, Simantov et al 1977).

Although the degree of mismtach is often exaggerated, it seems that a mismatch does indeed exist. There appears to be a neurotransmitter where there is no receptor, and a receptor where there is no neurotransmitter. The reasons for this are unknown but could involve receptor regulation or ontogenetic factors. An interesting and relevant notion is that the brain is an endocrine organ where a neurotransmitter is released that acts at distant receptors. Neurotransmitters are found in cerebrospinal fluid. In any case, results of receptor-mapping studies are at least compatible with the view that neurohormones can act at some distance from their site of release in the brain. Also, these "unmatched" receptors would be available for drugs.

The mismatch between receptor and neurotransmitter is often exaggerated for several reasons. The main reason may be that we do not label and therefore do not reveal all the receptors. This is particularly true when we are using agonists as ligands. Receptors can change conformation and thus change the binding affinity for agonists. Therefore, low affinity agonist states of the receptor will not be labeled efficiently because agonist ligands will diffuse away rapidly from low affinity sites. This has been well studied for both muscarinic cholinergic receptors and β-adrenergic receptors (Wamsley et al 1984b, Rainbow et al 1984b). Fortunately, antagonists can be found that label all the sites with a high affinity and thus reveal even low-affinity agonist states of the receptor. The situation with the GABA receptor is interesting and somewhat more complex. Muscimol, a GABA agonist, binds to a GABA receptor with high affinity, and its distribution has been studied by autoradiography (see references in Table 2). However, there are no high-affinity muscimol sites in some places where there are high concentrations of GABAergic nerve terminals, such as the deep cerebellar nuclei. Yet, benzodiazepine receptors, which are believed to be associated with the functional GABA receptor supramolecular complex, are found in the deep cerebellar nuclei, and their density is increased in neuropathological tissue where the number of Purkinje cell afferents to the nuclei is reduced (P.J. Whitehouse and co-workers, in preparation). Another reason for the mismatch is the uneven quenching of tritium emissions in various brain regions (see below). Thus, the situation can be complex, and an evaluation of the extent of the "mismatch problem" can only be made after the properties of the receptors, and the technical problems, are thoroughly understood.

Issues of Resolution and Sensitivity

All of the uses mentioned above ultimately depend on the fundamental advantages of microscopic receptor mapping; these are increased anatomical resolution and greatly increased sensitivity of measurement (at least when compared to biochemical approaches). Because one can quantitatively assess the distribution of receptors at the microscopic level, one can measure receptors in small regions—a not so easy task with biochemical assays. The increase in sensitivity can be orders of magnitude greater than that in biochemical studies. Thus, receptor mapping will be useful in any situation requiring measurement in small regions or in situations where overall quantities (but not necessarily densities) of receptors are low. An example of such a situation is studying the axonal flow of receptors; although it can be done biochemically, light microscopic autoradiography provides a sensitive method of measurement and a degree of anatomical resolution not available by other means (Kuhar & Zarbin 1984, Zarbin et al 1983b). The issue of anatomical resolution is indeed an important one and is discussed further below in the section on ultrastructural approaches.

TWO IMPORTANT PRINCIPLES

The usual sequence for localizing receptors by autoradiography is to label or tag the receptor of interest with a radiolabeled ligand, in an intact animal or tissue section, and then to generate an autoradiogram by an appropriate means, which reveals the receptor bound ligand. These two steps of selective labeling and appropriate visualization are, in our experience, the two main factors that are critically necessary for experimental success.

The first principle of selective or preferential labeling of receptors has been facilitated by the development of receptor binding techniques (Yamamura et al 1985, Snyder & Bennett 1976). Over the past decade, specific binding ligands have been identified for many drug and neurotransmitter receptors. The vast majority of these ligands have been thoroughly characterized pharmacologically so that their relative specificity is understood. The availability of these ligands form a basis for receptor binding studies as well as for receptor mapping. A thorough understanding of the criteria for receptor identification (Burt 1985, Snyder & Bennett 1976) is necessary for anyone working in this area.

The second principle, appropriate visualization, is important because of the many technical pitfalls that have been encountered. Because the receptor is not visualized directly but, rather, indirectly with the ligand (drug, antibody, etc) that binds to it, the autoradiogram or image must be produced in such a way that the diffusion of ligand away from receptor is prevented or minimized. It does no good, or is even misleading, if a receptor is appropriately labeled but the ligand is lost from the tissue during preparation for obtaining the image by autoradiography or other means. This is not a significant problem when irreversibly binding ligands are used. However, the majority of available ligands bind in a reversible manner and autoradiography must be done in a way that takes into account the diffusible nature of the signal. The literature on this topic is not extensive, but it is informative and very helpful (Appleton 1964, Stumpf & Roth 1966, Roth & Stumpf 1969, Young & Kuhar 1979a, Barnard 1979, Barnard et al 1979).

AUTORADIOGRAPHIC METHODS

In Vivo Labeling Autoradiography

The term "in vivo receptor labeling" is applied to the procedure in which receptors are labeled in intact living tissues in vivo after systemic administration of the ligand or drug. If the ligand has a very high affinity for the receptor, tracer quantities of the drug can be injected into the animal, and after a relatively short time interval, the drug is carried to the brain by the blood, diffuses into the brain, and binds to the receptors. The nonreceptor-bound drug is then removed from the brain and other tissues by various excretory processes. The high affinity of the drug for the receptor causes a retention of the drug on or in the vicinity of the receptor molecule.

Detailed strategies for carrying out in vivo labeling experiments have been described (Kuhar 1982a,b, Kuhar 1983). A critical factor is that the binding or the regional accumulation that one observes in the brain must be proven to be associated with receptors rather than with some nonspecific binding site. One must satisfy the criteria for receptor binding in all situations. Some examples of in vivo receptor binding studies are listed in Table 1.

After the conditions for labeling a receptor in vivo have been identified, autoradiograms can be generated to provide information on the distributions of receptor binding sites at the microscopic level. As mentioned above, the autoradiograms can only be produced by methods that minimize diffusion of drug from the binding site. Thaw mounting or dry mounting of cryostat-cut sections to emulsion-coated slides in the dark has been successfully used for this purpose (Stumpf & Roth 1966, Kuhar 1985). Dry mounting has the advantage that diffusion is less of a problem, but it has the disadvantage of being considerably more laborious. Only after the autoradiograms are formed can the slide-mounted tissue sections be treated in such a way that the ligand can be lost from the tissue. Technical details of these procedures have been described in many publications (Appleton 1964, Stumpf & Roth 1966, Roth & Stumpf 1969, Kuhar & Yamamura 1975, Kuhar 1982a,b, 1983, 1985).

In vivo labeling autoradiography has been important in this field. Most recently, PET scanning images of receptors have been produced because of the feasibility of labeling certain neurotransmitter receptors in vivo. Perhaps the main advantage of in vivo labeling is that receptors can be labeled and then imaged by various noninvasive imaging techniques such as PET (Raichle 1978). Nevertheless, in vivo labeling procedures have serious limitations. One is that only ligands with high affinity for receptors can be used successfully. This explains why in vitro labeling autoradiography has been much more productive in terms of examining different kinds of receptors. Also, the binding conditions that are so critically important to control for specificity of labeling cannot be modified or controlled very easily in vivo. In vivo labeling is somewhat expensive as well, in that an entire animal must be loaded with radioactive ligand even when only small portions of a tissue, such as the brain, are actually needed in an experiment to obtain information. In vitro labeling autoradiography was a significant step in being able to use a larger number of ligands with greater pharmacological specificity.

Table 1   Examples of central nervous system receptors visualized after in vivo labeling

Receptor	Ligand	Reference
Muscarinic cholinergic	^{3}H-quinuclidinylbenzilate	Kuhar & Yamamura 1974, 1975, 1976
	^{125}I-(R)-3-quinuclidinyl-4-iodobenzilate	Eckelman et al 1984
	^{123}I-(R)-3-quinuclidinyl-4-iodobenzilate	Eckelman et al 1984
Nicotinic cholingeric	^{3}H-α-bungarotoxin	Silver & Billiar 1976
	^{125}I-α-bungarotoxin	Hunt & Schmidt 1978a,b
		Miller et al 1982
Opiates	^{3}H-etorphine	Schubert et al 1975
		Atweh & Kuhar 1977a–c, 1983
	^{3}H-diprenorphine	Pert et al 1975, 1976
		Atweh & Kuhar 1977a–c
		Murrin et al 1980
		Pearson et al 1980
		Wamsley et al 1982
		Seeger et al 1984
	^{11}C-carfentanyl	Frost et al 1985
D2-dopamine and S2-serotonin	^{3}H-spiperone	Hollt & Schubert 1978
		Kuhar et al 1978
		Klemm et al 1979
		Laduron et al 1978
		Murrin et al 1979
		Murrin & Kuhar 1979
	^{11}C-N-methylspiperone	Wagner et al 1983
		Wong et al 1984
	76-Br-bromospiroperidol	Maziere et al 1984
Benzodiazepine (central-type)	^{3}H-flunitrazepam	Mohler et al 1980, 1984
		Ciliax et al 1984
	^{3}H-Ro15-1788	Mohler et al 1984
	^{3}H-Ro15-4513	Mohler et al 1984
	^{11}C-Ro15-1788	Hantraye et al 1984

In Vitro Labeling Autoradiography

In in vitro labeling procedures, slide-mounted tissue sections are incubated with radioactive ligands so that the receptors are labeled under very controlled conditions. Following the labeling, the slide-mounted tissue sections can be rinsed to remove nonreceptor-bound drug and improve the specific to nonspecific binding ratios. Autoradiograms can then be produced in various ways. Again, an important issue is that these autoradiograms must be produced in ways that minimize diffusion of drug away from receptor and maintain register between tissue and autoradiographic image for the best possible anatomical resolution. The details of many of these procedures have been presented elsewhere (Roth et al 1974, Kuhar 1985, Young & Kuhar 1979a, Rotter et al 1979a, Herkenham & Pert 1982, Barnard 1979). Examples of in vitro receptor binding studies are listed in Table 2.

In vitro labeling procedures have many advantages over in vivo labeling procedures. The in vitro labeling provides greater specificity and efficiency. Inhibitors can be added to prevent metabolism of the labeling ligand, and ligands that will not cross the blood brain barrier in in vivo labeling procedures can be used. Various receptor distributions can be easily compared, because it is possible to use different ligands with consecutive sections of tissue. Also, this approach has a relatively low cost because an entire animal does not have to be labeled with drug. It is also possible to carry out studies with human postmortem tissue.

An important advantage in the in vitro labeling procedures is that one can carry out labeling and preparation of tissue for autoradiography and then remove the sections from the slide and measure the receptor binding directly by liquid scintillation counting. Thus, extensive preliminary studies are possible where the optimal conditions for labeling receptors can be defined. Also, the pharmacological specificity of relatively nonspecific ligands can be enhanced, sometimes by adding excess nonradioactive drug to block sites of labeling that are not of primary interest. Before sectioning the tissues and thaw-mounting them on slides, investigators have used cardiac perfusions containing low concentrations of fixative to help preserve the tissues (Rotter et al 1979a, Young & Kuhar 1979a). In many cases, the addition of sucrose as a cryoprotectant has been important.

After labeling, washing, and drying, the mounted tissue sections can be fixed with gaseous formaldehyde (Herkenham & Pert 1982) or directly apposed to dry emulsion to generate autoradiograms. Herkenham & Pert (1982) have proposed the use of defatting procedures after vapor fixation along with dipping sections into wet emulsion. Even though this procedure results in a significant loss of ligand from the tissue (Kuhar & Unnerstall 1982), it may have special advantages in some cases in which formaldehyde links the ligand to the receptor in significant quantities. Often, dry techniques are used, where the tissue sections are placed in contact with dry rather than wet emulsion.

Young & Kuhar (1979a) used a coverslip technique, described by Roth and co-workers (1974), that has been very successful. Although the coverslips are fragile and difficult to use, this general approach is extremely valuable. Recently, we have utilized Aclar plastic (Allied Chemical Corp.) as a substitute for the glass coverslips, with the improvement that the plastic coverslips are not fragile like the glass. Also, the availability of tritium-

Table 2   Examples of central nervous system receptor visualized after in vitro labeling autoradiography

Receptor	Ligand	Reference
Muscarinic cholingeric	^{3}H-propylbenzilycholine mustard	Rotter et al 1979a,b
		Kuhar et al 1981
	^{3}H-quinuclidinylbenzilate	Wamsley et al 1981, 1984a
		Cortes et al 1984a
		Nonaka & Moroji 1984
		Seybold & Elde 1984
	^{3}H-N-methylscopolamine	Wamsley et al 1980, 1984b
		Palacios 1982
	^{3}H-pirenzepine	Yamamura et al 1983b
		Wamsley et al 1984a
Nicotinic cholinergic	^{125}I-α-bungarotoxin	Hunt & Schmidt 1978a,b
		Arimatsu et al 1978
		Segal et al 1978
		Clarke et al 1984
	^{3}H-acetylcholine	Clarke et al 1984
	^{3}H-nicotine	Clarke et al 1984
		London et al 1985
D2-dopamine	^{3}H-spiperone	Palacios et al 1981b
		Altar et al 1984
		Palacios 1984
	^{3}H(-)sulpiride	Gehlert & Wamsley 1984
	^{125}I-LSD	Nakada et al 1984
		Engel et al 1984
α_1-adrenergic	^{3}H-prazosin	Rainbow & Biegon 1984
		Unnerstall et al 1985
	^{125}I-HEAT	Jones et al 1983
	^{125}I-BE 2254	Palacios 1984
α_2-adrenergic	^{3}H-para-aminoclonidine	Young & Kuhar 1980a
		Probst et al 1984
		Seybold & Elde 1984
		Unnerstall et al 1984
	^{3}H-rauwolscine	Unnerstall et al 1985
	^{3}H-RX781094	Unnerstall et al 1985
β-adrenergic($_1$ and $_2$)	^{3}H-dihydroalprenolol	Palacios & Kuhar 1980
	^{125}I-pindolol	Rainbow et al 1984b
	^{125}I-cyanopindolol	Palacios 1984
S1-serotonin	^{3}H-serotonin	Young & Kuhar 1980c
		Biegon et al 1982
		Marcinkiewicz et al 1984
	^{3}H-8-OH-N,N-dipropyl-2-aminotetralin	Marcinkiewicz et al 1984
	^{3}H-LSD	Young & Kuhar 1980c
		Meibach et al 1980
		Palacios et al 1983

Table 2—*continued*

Receptor	Ligand	Reference
S2-serotonin	^{3}H-LSD	Young & Kuhar 1980
		Meibach et al 1980
		Seybold & Elde 1984
	^{125}I-LSD	Nakada et al 1984
		Engel et al 1984
	^{3}H-ketanserin	Slater & Patel 1983
		Palacios 1984
	^{3}H-spiperone	Palacios et al 1981b
		Altar et al 1984
		Palacios 1984
Opiate	^{3}H-dihydromorphine	Young & Kuhar 1979a
		Goodman & Snyder 1982b
		Moon Edley et al 1982
		Maurer et al 1983
		Quirion et al 1983b
		Moskowitz & Goodman 1984
		Seybold & Elde 1984
		Tempel et al 1984
		Unnerstall et al 1983
	^{3}H-diprenorphine	Young & Kuhar 1979a
		Wamsley et al 1982
	^{3}H-D-ala[2]-methioninamide[5]-enkephalin	Young & Kuhar 1979a
	^{3}H-naloxone	Herkenham & Pert 1980, 1982
		Wise & Herkenham 1982
		Lewis et al 1983, 1984
		Herkenham et al 1984
	^{3}H-D-ala[2]-D-leu[5]-enkephalin	Duka et al 1981
		Goodman & Snyder 1982b
		Herkenham & Pert 1982
		Moon Edley et al 1982
		Quirion et al 1983b
		Moscowitz & Goodman 1984
	^{3}H-etorphine	Duka et al 1981
	^{3}H-ethylketazocine	Goodman & Snyder 1982b
	^{3}H-bremazocine	Goodman & Snyder 1982b
		Maurer et al 1983
		Foote & Maurer 1983
	^{3}H-D-ala[2]-MePhe[4], Gly-ol[5]-enkephalin	Maurer et al 1983
	^{3}H-dynorphin A	Lewis et al 1984
	^{3}H-DAGO	Quirion et al 1983b
	^{3}H-DTLET	Quirion et al 1983b
	^{125}I-D-ala[2]-MePhe[4]-enkephalin	Goodman et al 1980

Table 2—*continued*

Receptor	Ligand	Reference
Opiate (*continued*)	^{125}I-D-ala^2-D-leu^5-enkephalin	Goodman et al 1980
GABA (high-affinity)	^{3}H-muscimol	Palacios et al 1980, 1981c
		Penney et al 1981
		Pan et al 1983
	^{3}H-GABA	Wilkin et al 1981
		Bowery et al 1984
(low-affinity)	^{3}H-bicuculline	Olsen et al 1984
Benzodiazepines	^{3}H-flunitrazepam	Young & Kuhar 1979b, 1980b
(central type)		Young et al 1981
		Lo et al 1983
	^{3}H-clonazepam	Richards & Mohler 1984
	^{3}H-DMCM	Richards & Mohler 1984
	^{3}H-Ro15-1788	Richards & Mohler 1984
(peripheral type)	^{3}H-Ro5-4864	Gehlert et al 1983
		Anholt et al 1984
	^{3}H-PK11195	Benavides et al 1983
Picrotoxin/barbiturate site (associated with GABA-benzodiazepine receptors)	^{35}S-*t*-butyl-bicyclo-phosphothionate	Gee et al 1983
H1-histamine	^{3}H-mepyramine	Palacios et al 1979, 1981d
Glutamate	^{3}H-glutamate	Halpain et al 1983, 1984
		Monaghan et al 1983
		Greenamyre et al 1983, 1984
	^{3}H-AMPA	Rainbow et al 1984c
Kainic acid	^{3}H-kainic acid	Foster et al 1981
		Henke et al 1981
		Berger & Ben-Ari 1983
		Unnerstall & Wamsley 1983
Adenosine	^{3}H-cyclohexyladenosine	Lewis et al 1981
		Goodman & Snyder 1982a
Glycine	^{3}H-strychnine	Zarbin et al 1981
Angiotensin II	^{125}I-[Ile^5] angiotensin II	Gehlert et al 1984a,b
		Mendelsohn et al 1984
Bombesin	^{125}I-Tyr^4-bombesin	Wolf et al 1983a
Bradykinin	^{3}H-bradykinin	Manning & Snyder 1983
Cholecystokinin	^{3}H-CCK$_8$	Van Dijk et al 1981, 1984
	^{125}I-CCK$_{33}$	Zarbin et al 1983a
Corticotropin-releasing factor	Nle^{21}, ^{125}I-Tyr^{32}-ovine CRF	De Souza et al 1984
		De Souza & Kuhar 1985
Insulin	^{125}I-insulin	Young et al 1980d

Table 2—*continued*

Receptor	Ligand	Reference
Neurotensin	^{3}H-neurotensin	Young & Kuhar 1981
		Palacios & Kuhar 1981
		Uhl & Kuhar 1984
Oxytocin	^{3}H-oxytocin	Brinton et al 1984b
Thyrotropin-releasing hormone	^{3}H-TRH	Rostene et al 1984
	^{3}H-methyl-TRH	Palacios 1983
	^{3}H-(^{3}-Me-His^2)-TRH	Pilotte et al 1984
Somatostatin	^{125}I-Tyr^{11}-somatostain	Tran et al 1984
		Leroux & Pelletier 1984
		Uhl et al 1985
	^{125}I-(leu^8-D-Trp^{22}, Tyr^{25})somatostatin-28	Tran et al 1984
		Leroux & Pelletier 1984
		Uhl et al 1985
Substance K	^{125}I-substance K	Mantyh et al 1984b
	^{125}I-kassinin	Mantyh et al 1984b
Substance P	^{125}I-substance P	Shults et al 1982
		Rothman et al 1984
	^{3}H-substance P	Quirion et al 1983a
		Mantyh et al 1984a,b
		Maurin et al 1984
	^{125}I-physalaemin	Wolf et al 1983b
Vasopressin	^{3}H-arginine vasopressin	Yamamura et al 1983a
		Baskin et al 1983
		Van Leeuwen & Wolters 1983
		Brinton et al 1984a
Vasoactive intestinal peptide	^{125}I-VIP	Besson et al 1984
		De Souza et al 1985

sensitive sheet film has been another important technical advance (Palacios et al 1981a, Penney et al 1981), particularly because the sheet film images are readily analyzed by computerized techniques (Goochee et al 1980, Kuhar et al 1984, Palacios et al 1981a, Altar et al 1984).

In vitro autoradiography has been an important advance in this field, but it still has significant limitations. Although one can use ligands with a lower affinity than are necessary for in vivo labeling autoradiography, low affinity sites are still not easily identified, because the dissociation rate of ligand is rapid enough that there is a large diffusion of ligand away from receptor during the washing necessary after the incubation step. Another limitation is that the preparation of tissue, which involves dissection and freezing, may destroy some receptors. These preparative steps might also produce

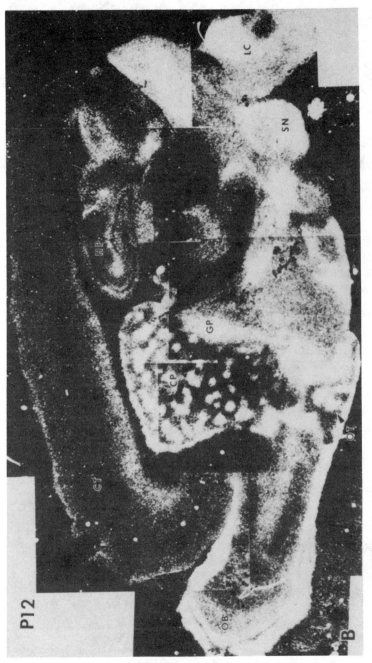

Figure 1A   Example of receptor autoradiogram prepared by in vitro labeling procedures (ligand in parentheses): Opiate receptor distribution ([3H]-dihydromorphine) in sagittal section of rat brain 12 days after birth (darkfield photo from Unnerstall et al 1983).

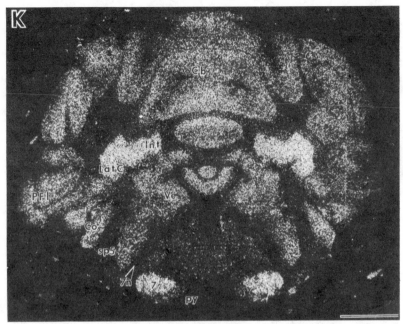

Figure 1B Example of receptor autoradiogram prepared by in vitro labeling procedure: CRF receptor distribution (^{125}I-CRF) in rat cerebellum and medulla (darkfield photo, courtesy of E. De Souza).

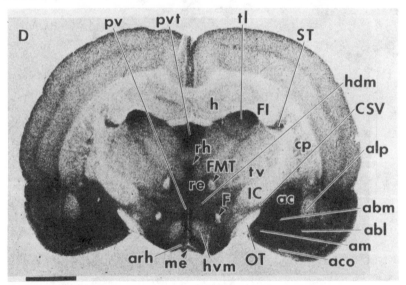

Figure 1C Example of receptor autoradiogram prepared by in vitro labeling procedure: α-2 Adrenergic receptor (^{3}H-para-aminoclonidine) in rat forebrain (brightfield photo from Unnerstall et al 1984).

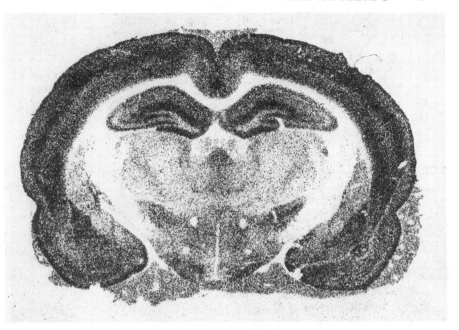

Figure 1D Example of receptor autoradiogram prepared by in vitro labeling procedure: Benzodiazepine receptor distribution (^{3}H-flunitrazepam) in rat forebrain (brightfield photo from Palacios et al 1981a).

morphological defects in the tissue, caused by the freezing or by other steps. Also, the techniques discussed so far are really light microscopic and lack the ultrastructural resolution necessary for a complete picture of the distribution of receptors at the level of membranes. Although some of these problems such as low affinity sites and loss of receptors during preparation do not appear to be significant problems, at least in our studies, the lack of ultrastructural resolution is quite important and is discussed below.

QUANTITATIVE ANALYSES OF AUTORADIOGRAPHS

One of the great advantages of autoradiography in general is that it is quantitative (Perry 1964, Pryzbylski 1970, Dormer 1973, Rogers 1979). However, only recently have quantitative autoradiographic techniques been fully applied to receptor mapping. A problem with the quantitation of autoradiograms is that grain densities in nuclear emulsions or optical densities in film are not linearly related to the content of radioactivity in the specimen. Usually, grain density and optical density increase more slowly than tissue radioactivity. Eventually the image on the film is saturated. For example, this has been found to be true for ^{3}H-Ultrofilm, the photographic film most commonly used in receptor-mapping studies (Ehn & Larsson 1979). A variety of laboratories have utilized this film or other emulsions and studied their properties with radioactive materials (Palacios et al 1981a, Alexander et al 1981, Rainbow et al 1982, Penney et al 1981, Unnerstall & Kuhar 1981, 1982, Pan et al 1983, Geary & Wooten 1983, Geary et al 1985). Because the relationship between the signal in the emulsion or film and the content of radioactivity in the specimen is not a simple one and will vary depending on the emulsion and perhaps on some other experimental variables, the quantitative interpretation of auto-radiograms will require the utilization of autoradiographic standards along with experimental tissues.

There are several kinds of radioactive standards. Both carbon-14 and tritium standards are commercially available. Some individual investi-gators have prepared their own standards using radiolabeled methacrylate or by mixing radioactive compounds with tissues (Reivich et al 1969, Sokoloff et al 1977, Alexander et al 1981, Penney et al 1981, Unnerstall & Kuhar 1981, Unnerstall et al 1982, Pan et al 1983, Geary & Wooten 1983, Geary et al 1985). The basic procedure is to prepare standards containing varying quantities of radioactivity, expose the standards to produce an autoradiogram, and then relate the quantity of radioactivity in the standard to the grain density or optical density in the autoradiogram. If an adequate number of data points are available, an accurate relationship between optical or grain density and radioactivity can be derived. Various mathematical functions have been used that adequately describe the relationship; choice of function depends on the length of exposure of the autoradiograms, the range of data studied, and the sensitivity of the instruments utilized to measure optical density/grain density. For example, investigators using the 2-deoxyglucose techniques have utilized a third-order polynomial function to describe the relationship between radioac-tivity levels and optical density in film (Reivich et al 1969, Sokoloff et al 1977, Goochee et al 1980). Other investigators studying receptor mapping have found a fourth-order polynomial equation to be useful (Geary & Wooten 1983, Pan et al 1983), while others have utilized a power function (Unnerstall & Kuhar 1981, Unnerstall et al 1982, Kuhar et al 1984).

The β-emissions from tritium have a relatively low energy, and this causes some problems when using tritium standards. The problem, which is also discussed in more detail below in the section on tritium quenching, is that there is a significant absorption of the β-emissions by the specimen itself. Thus, tritium standards prepared with plastic may not have the same degree of absorption of β-emissions as those prepared with tissues. Because this could lead to a significant error in calculations, one must calibrate the standards in such a way as to correct for differences in self absorbtion between the material used for the standard and the tissue being studied (Reivich et al 1969, Alexander et al 1981, Geary & Wooten 1983, Pan et al 1983, Geary et al 1985).

In some experiments, particularly where there is a uniform distribution of radioactivity in the sample, the samples themselves (or similarly prepared ones) can be used as internal standards. For example, tissues can be labeled and autoradiograms can be produced. Then the tissues can be scraped from the slides and the radioactivity in the specimens measured directly. One can then directly relate grain density in emulsions with the radioactivity content of tissues and thereby construct useful standard curves.

Thus, quantitation is not always a simple procedure. Appropriate standards will be required and the relationship between grain density in the emulsion and the radioactivity in the specimen must be determined. Often, precise quantitation may not be required and semiquantitative studies, simply comparing optical densities or grain densities, will be adequate.

GRAY/WHITE QUENCHING WITH TRITIUM IN BRAIN

The gray/white quenching problem occurs because gray matter causes a smaller absorption of the β-emissions from tritium than does white matter. Thus, even though gray matter and white matter could have the same content of radioactivity, the grain density in the emulsion will be lower for white matter because of the relatively greater absorption of the low-energy beta emissions from tritium. It has been known for some time that there can be significant absorption of beta emissions by different parts of tissues (Dormer 1973, Perry 1964, Pryzbylski 1970), but the issue of tritium absorption in receptor autoradiography has mainly been addressed only recently, with the increasing use of tritium-sensitive film (Alexander et al 1981, Orzi et al 1983, Herkenham & Sokoloff 1984, Taylor et al 1984, Geary et al 1985).

The gray/white quenching problem can be quite bothersome. The difference in optical density can be as great as a factor of two due to the greater absorption of beta emissions by white matter. This can become an even more difficult problem, since white and gray matter are often mixed in varying quantities in various brain regions. Thus, it is important that investigators have some understanding of this problem and are aware of potential strategies for dealing with it.

While gray/white quenching can be a problem with tritium, substantial evidence indicates that this quenching does not occur with beta emissions from carbon-14 and with some of the emissions from iodine-125. One way to cope with the tritium-quenching problem therefore is to avoid the use of tritium. Unfortunately, carbon-14 labeled compounds have too low a specific activity to be used as receptor-binding ligands. However, iodine-125–labeled ligands are available for some receptors and can be used effectively. Not only does iodine-125 appear not to have the gray/white quenching problem to a significant degree, iodine-125-labeled compounds are available in very high specific activities and therefore require shorter exposure times than comparable tritium-labeled compounds.

If one cannot avoid the use of tritium, there are some other ways either to calculate the degree of regional quenching so as to make corrections in quantitative studies or to treat the tissues so as to reduce or eliminate the regional quenching problem. The latter approach involves defatting tissues such that the gray/white differential quenching problem is eliminated. Herkenham & Sokoloff (1984) have found that defatting tissues in ethanols and xylene eliminates or reduces the problem of gray/white quenching in labeled slide-mounted tissue sections previously fixed with gaseous formaldehyde. One problem with this approach is that all of the radioactive ligand is not covalently linked to the receptor site by the formaldehyde, and various ligands are linked in varying degrees. Thus, the defatting process, which eliminates the quenching problem, may leach significant quantities of ligand from the tissue section (Kuhar & Unnerstall 1982). Therefore, the ability to use the defatting procedure may be dependent on the ligand that one uses and on whether or not it is leached from the tissue uniformly or from subtypes of receptors or preferentially from certain regions. The defatting approach is, nevertheless, important and may be very useful, especially when using tritiated ligands.

As mentioned above, another approach for dealing with the differential gray/white quenching in brain is to calculate correction factors for the brain region of interest. One approach that is believed to result in a uniform labeling of brain regions has been to infuse animals with ^{3}H-3-O-methylglucose. This allows for direct testing of regional quenching or absorption of radioactive emissions in autoradiographic experiments (Alexander et al 1981, Orzi et al 1983). Yet another method is to soak slide-mounted tissue sections with concentrations of radioactive compounds that distribute throughout the tissue. These sections can then be dried and exposed to emulsion to provide an autoradiogram that will reflect the differing degree of quenching in the different regions. Different quantities of radioactivity could be soaked into the sections, with the possibility of constructing standard curves for each brain region of interest (Taylor et al 1984). As expected, these soaking procedures do not reveal a gray/white quenching when carbon-14 or when iodine-125 is used (Taylor et al 1984).

Another approach to determine the relative degree of quenching is to thaw-mount tissue sections on microscope slides that have been previously coated with radioactivity. The emissions must pass through the tissue and will be absorbed as some function of tissue density. Thus, images of these slides with tissue sections will reveal regional quenching. It has been found that gray/white absorption differences determined by this procedure are similar to those determined by other procedures. Again, gray/white differences in absorption were found only with tritium and not with carbon-14 or iodine-125 (Taylor et al 1984). This procedure is relatively simple and is easily applied to any isotope or any tissue.

Because of the serious problems with tritium, iodine-125–labeled ligands are obviously very desirable. However, using iodine-125 in place of tritium could lead to problems not encountered with tritium. Because of the low energy of beta emissions from tritium, infinite sample thickness is just a few microns, and the autoradiographic image density will not depend on tissue section thickness, at least where section thickness is greater than 4 or 5 μ. However, with iodine-125, infinite thickness is experimentally greater, and therefore variations in section thickness, including those due to microtome error, will affect the density of the image. Thus, unformity of section thickness becomes a consideration when using iodine-125. The decay scheme of iodine-125 is fairly complicated, and it appears that several emissions create the autoradiographic image (Rogers 1979). Apparently X-rays are important with this isotope.

In any case, quantitating autoradiograms is not trivial. Not only must one prepare calibrated standards for use along with experimental tissues, but one must be careful of varying absorption of emissions within the tissue sample studied. Fortunately, several approaches can be used for examining whether or not tissue specimens will cause significant absorption of emissions under various experimental conditions. A well-designed experiment will include these tests. As has been pointed out many times in the past, autoradiographic experiments must also include controls for positive and negative chemography (Rogers 1979). Fortunately, semiquantitative methods in which one directly compares grain densities in emulsions may be adequate in many experiments. Many of these issues have been discussed previously (Kuhar & Unnerstall 1985).

OTHER APPROACHES FOR VISUALIZING RECEPTORS

Immunohistochemistry

Because receptor binding techniques have developed primarily with radioactive ligands as tools (in fact the very existence and progress of the field seems tied to the development and availability of high affinity, high specific activity ligands), autoradiography has been the obvious method of choice for microscopic studies. The availability of antibodies for receptors would provide an important alternative in immunohistochemical techniques.

The antibodies could be used in several ways. For qualitative light microscopy, fluorescent markers (e.g. fluorescein or rhodamine dye) can be covalently attached to the antibodies. For light or electron microscopy, an enzymatic reaction product can be deposited locally; the only commonly employed enzyme is peroxidase, which is linked to the receptor by an antibody bridge. For electron microscopy, electron-dense labels can be attached to the antibodies: ferritin, hemocyanase, or colloidal gold are possible labels.

Important successes in localizing receptors by immunohistochemistry have already been reported (Daniels & Vogel 1975, Bourgeois et al 1971, Barnard et al 1979, Lentz et al 1977, Hourani et al 1974, Swanson et al 1983). Immunohistochemical techniques are faster than autoradiography, since emulsion development is not required. Also, these techniques may be useful under experimental conditions where tissue morphology is better preserved. A sometimes important gain with immunohistochemical studies is that the cell bodies containing receptors are clearly delineated (Swanson et al 1983, Richards et al 1984); this is not the case with autoradiography. The identification of the specific cells producing specific receptors is a major problem in the field, and has often been approached indirectly with the use of regional studies, neurotoxins, lesions, and tissues from disease states.

Ultrastructural techniques with antibodies would have better electron microscopic resolution than those with autoradiography (see Barnard 1979 for additional discussion and references). However, although efforts have been made to quantitate immunohistochemical approaches, they do not seem to be as readily quantitative as autoradiography.

Imaging Receptors in Vivo: Scanning Techniques

Positron emission tomography (PET) is a technique providing a computerized image reconstruction where the distribution of positron-emitting compounds are revealed. In this technique, a ligand containing a radioactive isotope that decays by emitting a positron (positive electron) is injected into an animal or human. If the ligand is suitable, then receptors in the brain will be preferentially labeled by this in vivo labeling technique, discussed above. Emitted positrons will almost immediately combine with an electron, and the two are mutually annihilated. The annihilation results in the emission of two gamma rays that go out in very nearly opposite directions, pass through the tissue, and are detected outside of the subject by a circular array of detectors. The signals are transformed by computer, which rapidly reconstructs the spatial distribution of the radioactivity and displays the resulting image on a cathode-ray screen. The great advantage of this approach is that it is a noninvasive method whereby receptors can be quantitatively and spatially detected in humans. This provides a much larger population for receptor measurement than is possible with postmortem tissues. Since receptors may be involved in certain neuropsychiatric disorders, the possibility of assessing receptors in vivo in humans, in various clinical populations, before, during, and after onset of symptoms or drug treatment, provides a very unique and important opportunity to assess the importance of receptors in these disorders. Although this area of research is just beginning, a large effort is being exerted in many laboratories. To date, only a few papers have been published, but they clearly indicate the feasibility of this approach (Maziere et al 1981, Wagner et al 1983, Eckelman et al 1984, Hantraye et al 1984). By this approach a decline in dopamine and serotonin receptors with age has been noted (Wong et al 1984).

A disadvantage with PET scanning is that it is expensive and requires a large team with a complicated technology. A cyclotron is necessary to produce the radioactive isotope, and nuclear chemists are required to synthesize the short half-life isotopes into receptor specific ligands rapidly. The PET scanning equipment is also expensive and requires skilled personnel. Another difficulty, compared to autoradiographic methods, is that PET scans are of lower resolution; a resolution unit is on the average just one cubic centimeter. Nevertheless, despite some limitations, great opportunities for information are available by PET scanning.

Another imaging technique is single photon imaging. It is in some ways perhaps more desirable than PET, but it lacks the resolution that PET has and it has some other technical problems. It has, however, been used successfully to image muscarinic cholinergic receptors in vivo (Eckelman et al 1984), and it may become a widely used tool in the future.

ULTRASTRUCTURAL APPROACHES

Most of the discussion so far has focused on light microscopic autoradiography. Because the light microscope cannot resolve membranes, an electron microscopic method will be necessary for localizing receptors. We hope that these methods will provide a high degree of resolution, as well as be technically simple and reproducible. Since most ligands available for studying receptors are radiolabeled, electron microscopic autoradiography has been the obvious method of choice. It has been used where irreversible ligands exist (Hunt & Schmidt 1978a, Mohler et al 1980, Kuhar et al 1981). However, even electron microscopic autoradiographic methods have limits in resolution (Salpeter et al 1969), and it is clear that additional methods are desirable for localizing receptors to specific membranes, such as the postsynaptic membrane at a synapse. Other approaches, possibly immunocytochemical ones (Lentz & Chester 1977, Goldsmith et al 1979), are feasible and will provide increased resolution (see above). Easy, reproducible, high resolution methods for localizing receptors are greatly needed.

SUMMARY AND FUTURE DIRECTIONS

Great advances have been made in localizing receptors by microscopic and other methods over the last ten years. In our. laboratory the first neurotransmitter receptor autoradiogram in brain was developed in February of 1974. Several hours of cautious data evaluation were required before we could accept that we had localized a pharmacologically relevant binding site (the muscarinic cholinergic receptor). Since then, many factors have substantially and dramatically advanced the field. These include the development of new ligands and the identification of new receptors, the development of new equipment, the dramatic development of in vitro labeling techniques, the availability of the easily used tritium sensitive sheet film, the application of computerized image analysis techniques, and most recently, the use of PET and single photon scanning in receptor mapping. After a decade's experience of always waiting days to months for the development of a suitable autoradiographic image, the production of an image in minutes by PET is indeed remarkable. Shen Nung, a mythical figure from the history of Chinese medicine, is said to have had the power of crystal vision whereby he could make his body transparent and observe the sites of action of various herbal medicines. Receptor mapping by autoradiography and PET scanning makes this mythology a reality.

The future holds great promise. Binding ligands have been developed not only for other receptors but also for enzymes and other sites (see Table 3). Thus, the binding-imaging approach will be applicable to a wide range of problems. Antibodies will become an important tool in the near future and we hope will provide rapid and high resolution localization. The appropriate use of and the pitfalls in quantitative autoradiography are becoming known and appreciated. Computerized image analysis techniques for rapid quantitation are being developed. PET scanning has the power to give receptor studies a clinical focus, where the potential seems great. Finally, ultrastructural studies, though difficult, will probably improve in the future.

Figure 2 Early painting of Shen Nung. Emperor Shen Nung is regarded as a founder of Chinese medicine. One of his many accomplishments was the identification of hundreds of medicinal herbs. Related to this, he supposedly possessed the supernatural power of crystal vision, which rendered his body transparent and enabled him to observe the site of action of each drug. Receptor mapping by autoradiography or PET scanning or other techniques makes such mythology a reality.

Table 3 Examples of central nervous system enzymes and other sites visualized by autoradiography

Binding site	Ligand	Reference
Angiotensin converting enzyme	³H-captopril	Strittmatter et al 1984
Carboxypeptidase E (enkephalin convertase)	³H-GEMSA	Lynch et al 1984
Protein kinase C	³H-phorbol 12, 13-dibutyrate	Nagle & Blumberg 1983 Murphy et al 1983
Monoamine oxidase	³H-MPTP	Parsons & Rainbow 1984 Javitch et al 1984 Wieczorek et al 1984
	³H-pargyline	Parsons & Rainbow 1984
Calcium channels	³H-nitrendipine	Murphy et al 1982 Gould et al 1985 Quirion 1983
	³H-PN 200-110	Cortes et al 1983, 1984b
	³H-PY 108-068	Cortes et al 1984b
	³H-desmethoxy-verapamil	Ferry et al 1984
Dopamine uptake sites	³H-mazindol	Javitch et al 1985
Noradrenaline uptake sites	³H-desmethylimi-pramine	Biegon & Rainbow 1982, 1983a
	³H-mazindol	Javitch et al 1985
Serotonin uptake sites	³H-imipramine	Biegon & Rainbow 1983b Grabowsky et al 1983 Fuxe et al 1983
	³H-nitroimipramine	Rainbow & Biegon 1983
Choline uptake sites	³H-hemicholinium	Rainbow et al 1984a
Aspartic/glutamic acid uptake sites	³H-D-aspartate	Parsons & Rainbow 1983
Catecholamine storage sites	³H-reserpine	Murrin et al 1977

Literature Cited

Alexander, G. M., Schwartzman, R. J., Bell, R. I., Yu, J., Renthal, A. 1981. Quantitative measurement of local cerebral metabolic rate for glucose utilizing tritiated 2-deoxyglucose. *Brain Res.* 223:59–67

Altar, C. A., Walter, R. J., Neve, K. A., Marshall, J. F. 1984. Computer-assisted video analysis of ³H-spiroperidol binding autoradiographs. *J. Neurosci. Methods* 10:173–88

Anholt, R. R. H., Murphy, K. M. M., Mack, G. E., Snyder, S. H. 1984. Peripheral-type benzodiazepine receptors in the central nervous system: Localization to olfactory nerves. *J. Neurosci.* 4:593–603

Appleton, T. C. 1964. Autoradiography of soluble labeled compounds. *J. R. Micro. Soc.* 83:277–81

Arimatsu, Y., Seto, A., Amano, T. 1978. Localization of alpha-bungarotoxin binding sites in mouse brain by light and electron microscopic autoradiography. *Brain Res.* 147:165–67

Atweh, S. F., Kuhar, M. J. 1977a. Autoradiographic localization of opiate receptors in rat brain. I. Spinal cord and lower medulla. *Brain Res.* 124:53–68

Atweh, S. F., Kuhar, M. J. 1977b. Autoradiographic localization of opiate receptors in rat brain. II. The brainstem. *Brain Res.* 129:1–12

Atweh, S. F., Kuhar, M. J. 1977c. Autoradiographic localization of opiate receptors in rat brain. III. The telencephalon. *Brain Res.* 134:393–405

Atweh, S. F., Kuhar, M. J. 1983. Distribution and physiological significance of opioid receptors in the brain. *Br. Med. Bull.* 39:47–52

Barnard, E. A. 1979. Visualization and counting of receptors at the light and electron microscope levels. In *The Receptor, General Principles and Procedures*, ed. R. D. O'Brien, 1:247–310. New York: Plenum

Barnard, E. A., Dolly, J. O., Lang, B., Lo, M., Shorr, R. G. 1979. The application of specifically-acting toxins to the detection of functional components common to peripheral and central synapses. *Adv. Cytopharmacol.* 3:409–35

Baskin, D. G., Petracca, F., Dorsa, D. M. 1983. Autoradiographic localization of specific binding sites for ^{3}H-(Arg^8) vasopressin in the septum of the rat brain with tritium-sensitive film. *Eur. J. Pharmacol.* 90:155–57

Benavides, J., Quarteronet, D., Imbault, F., Malgouris, C., Uzan, A., et al. 1983. Labeling of "peripheral-type" benzodiazepine binding sites in rat brain by using ^{3}H-PK11195, an isoquinoline carboxamide derivative: Kinetic studies and autoradiographic localization. *J. Neurochem.* 41:1744–50

Berger, M., Ben-Ari, Y. 1983. Autoradiographic visualization of ^{3}H-kainic acid receptor subtypes in the rat hippocampus. *Neurosci. Lett.* 39:237–42

Besson, J., Dussaillant, M., Marie, J.-C., Rostene, W., Rosselin, G. 1984. *In vitro* autoradiographic localization of vasoactive intestinal peptide (VIP) binding sites in the rat central nervous system. *Peptides* 5:339–40

Biegon, A., Rainbow, T. C. 1982. Quantitative autoradiography of ^{3}H-desmethylimipramine binding sites in rat brain. *Eur. J. Pharmacol.* 82:245–46

Biegon, A., Rainbow, T. C. 1983a. Localization and characterization of ^{3}H-desmethylimipramine binding sites in rat brain by quantitative autoradiography. *J. Neurosci.* 3:1069–76

Biegon, A., Rainbow, T. C. 1983b. Distribution of imipramine binding sites in the rat brain studied by quantitative autoradiography. *Neurosci. Lett.* 37:209–14

Biegon, A., Rainbow, T. C., McEwen, B. S. 1982. Quantitative autoradiography of serotonin receptors in rat brain. *Brain Res.* 242:197–204

Bourgeois, J. P., Tsuji, S., Boquet, P., Pilot, J., Ryter, A., et al. 1971. Localization of the cholinergic receptor protein by immunofluorescence in eel electroplax. *FEBS Lett.* 16:92–94

Bowery, N. G., Price, G. W., Hudson, A. L., Hill, D. R., Wilkin, G. P., et al. 1984. GABA receptor multiplicity: Visualization of different receptor types in the mammalian CNS. *Neuropharmacology* 23:219–31

Brinton, R. E., Gee, K. W., Wamsley, J. K., Davis, T. P., Yamamura, H. I. 1984a. Regional distribution of putative vasopressin receptors in rat brain and pituitary by quantitative autoradiography. *Proc. Natl. Acad. Sci. USA* 81:7248–52

Brinton, R. E., Wamsley, J. K., Gee, K. W., Wan, Y. P., Yamamura, H. I. 1984b. ^{3}H-Oxytocin binding sites in the rat brain demonstrated by quantitative light microscopic autoradiography. *Eur. J. Pharmacol.* 102:365–67

Burt, D. R. 1985. Criteria for receptor identification. In *Neurotransmitter Receptor Binding*, ed. H. I. Yamamura, S. J. Enna, M. J. Kuhar, 2:41–60. New York: Raven

Ciliax, B. J., Penney, J. B., Young, A. B. 1984. *In vivo* [^{3}H] flunitrazepam binding: Characterization and changes after striatal lesions. *Soc. Neurosci. Abstr.* 10:533

Clarke, P. B. S., Schwartz, R. D., Paul, S. M., Pert, C. B., Pert, A. 1984. Autoradiographic comparison of ^{3}H-nicotine, ^{3}H-acetylcholine and ^{125}I-alpha-bungarotoxin binding to rat brain. *Soc. Neurosci. Abstr.* 10:733

Cortes, R., Supavilai, P., Karobath, M., Palacios, J. M. 1983. The effects of lesions in the rat hippocampus suggest the association of calcium channel blocker binding sites with specific neuronal population. *Neurosci. Lett.* 42:249–54

Cortes, R., Probst, A., Palacios, J. M. 1984a. Quantitative light microscopic autoradiographic localization of cholinergic muscarinic receptors in the human brain: Brainstem. *Neuroscience* 12:1003–26

Cortes, R., Supavilai, P., Karobath, M., Palacios, J. M. 1984b. Calcium antagonist binding sites in the rat brain: Quantitative autoradiographic mapping using the 1,4-dihydropyridines ^{3}H-PN200-110 and ^{3}H-PY108-068. *J. Neural Trans.* 60:169–97

Daniels, M. P., Vogel, Z. 1975. Immunoperoxidase staining of alpha-bungarotoxin binding sites in muscle endplates shows distribution of acetylcholine receptors *Nature* 254:339–41

De Souza, E. B., Kuhar, M. J. 1985. Corticotropin-releasing factor receptors in pituitary and brain: Autoradiographic identification. *Res. Publ. Assoc. Res. Neurol. Mental Dis.* 63: In press

De Souza, E. B., Perrin, M. H., Insel, T., Rivier, J., Vale, W. W., et al. 1984. Corticotropin-releasing factor receptors in rat forebrain: Autoradiographic identification. *Science* 224: 1449–51

De Souza, E. B., Seifert, H., Kuhar, M. J. 1985. Vasoactive intestinal peptide (VIP) receptor localization in rat forebrain by autoradiography. *Neurosci. Lett.* 56: 113–20

Dormer, P. 1973. Quantitative autoradiography at the cellular level. *Molec. Biol. Biochem. Biophys.* 14: 347–93

Duka, Th., Schubert, P., Wuster, M., Stoiber, R., Herz, A. 1981. A selective distribution pattern of different opiate receptors in certain areas of rat brain as revealed by *in vitro* autoradiography. *Neurosci. Lett.* 21: 119–24

Eckelman, W. C., Reba, R. C., Rzeszotarski, W. J., Gibson, R. E., Hill, T. et al. 1984. External imaging of cerebral muscarinic acetylcholine receptors. *Science* 223: 291–93

Ehn, E., Larsson, B. 1979. Properties of an antiscratch-layer-free x-ray film for the autoradiographic registration of tritium. *Sci. Tools* 26: 24–29

Engel, G., Muller-Schweinitzer, E., Palacios, J. M. 1984. 2-[125Iodo]LSD, a new ligand for the characterization and localization of 5-HT$_2$ receptors. *Naunyn-Schmiedebergs Arch. Pharmacol.* 325: 328–36

Ferry, D. R., Joll, A., Gado, C., Glossman, H. 1984. (−)^{3}H-desmethoxy-verapamil labeling of putative calcium channels in brain: Autoradiographic distribution and allosteric coupling to 1,4-dihydro pyridine and diltiazem binding sites. *Naunyn-Schmiedebergs Arch. Pharmacol.* 327: 183–87

Foote, R. W., Maurer, R. 1983. Kappa opiate binding sites in the substantia nigra and bulbus olfactorius of the guinea pig as shown by *in vitro* autoradiography. *Life Sci.* 33(Suppl. I): 243–46

Foster, A. C., Mena, E. E., Monaghan, D. T., Cotman, C. W. 1981. Synaptic localization of kainic acid binding sites. *Nature* 289: 73–75

Frost, J. J., Wagner, H. N., Dannals, R. F., et al. 1985. Imaging opiate receptors in the human brain by positron tomography. *J. Comput. Tomogr.* 9: 231–36

Fuxe, K., Calza, L., Benfenati, F., Zini, I., Agnati, L. F. 1983. Quantitative autoradiographic localization of ^{3}H-imipramine binding sites in the brain of the rat: Relationship to ascending 5-hydroxytryptamine neurone systems. *Proc. Natl. Acad.*

Sci. USA 80: 3836–40

Gee, K. W., Wamsley, J. K., Yamamura, H. I. 1983. Light microscopic autoradiographic identification of picrotoxin/barbituate binding sites in rat brain with ^{35}-t-butyl-bicyclophosphotionate. *Eur. J. Pharmacol.* 89: 323–24

Geary, W. A. III, Toga, A. W., Wooten, G. F. 1985. Quantitative film autoradiography for tritium: Methodological considerations. *Brain Res.* 337: 99–108

Geary, W. A. III, Wooten, G. F. 1983. Quantitative film autoradiography of opiate agonists and antagonist binding in rat brain. *J. Pharmacol. Exp. Ther.* 225: 234–40

Gehlert, D. R., Speth, R. C., Wamsley, J. K. 1984a. Autoradiographic localization of angiotensin II receptors in the rat brain and kidney. *Eur. J. Pharmacol.* 98: 145–46

Gehlert, D. R., Speth, R. C., Healy, D. P., Wamsley, J. K. 1984b. Autoradiographic localization of angiotensin II receptors in the rat brainstem. *Life Sci.* 34: 1565–71

Gehlert, D. R., Wamsley, J. K. 1984. Autoradiographic localization of ^{3}H-sulpiride binding sites in the rat brain. *Eur. J. Pharmacol.* 98: 311–12

Gehlert, D. R., Yamamura, H. I., Wamsley, J. K. 1983. Autoradiographic localization of "peripheral" benzodiazepine binding sites in the rat brain and kidney using ^{3}H-Ro 5-4864. *Eur. J. Pharmacol.* 95: 329–30

Goldsmith, P. C., Cronin, M. J., Weiner, R. I. 1979. Dopamine receptor sites in the anterior pituitary. *J. Histochem. Cytochem.* 27: 1205–7

Goochee, C., Rasband, W., Sokoloff, L. 1980. Computerized densitometry and color coding of ^{14}C-deoxyglucose autoradiographs. *Ann. Neurol.* 7: 359–70

Goodman, R. R., Snyder, S. H. 1982a. Autoradiographic localization of adenosine receptors in rat brain using ^{3}H-cyclohexyladenosine. *J. Neurosci.* 2: 1230–41

Goodman, R. R., Snyder, S. H. 1982b. K opiate receptors localized by autoradiography to deep layers of cerebral cortex: Relation to sedative effects. *Proc. Natl. Acad. Sci. USA* 79: 5703–7

Goodman, R. R., Snyder, S. H., Kuhar, M. J., Young, W. S. III. 1980. Differential localization of delta and mu opiate receptor localization by light microscopic autoradiography. *Proc. Natl. Acad. Sci. USA* 77: 6239–43

Gould, R. J., Murphy, K. M. M., Snyder, S. H. 1985. Autoradiographic localization of calcium channel antagonist receptors in rat brain with ^{3}H-nitrendipine. *Brain Res.* 330: 217–23

Grabowsky, K. L., McCabe, R. T., Wamsley,

J. K. 1983. Localization of ³H-imipramine binding sites in rat brain by light microscopic autoradiography. *Life Sci.* 32: 2355–61

Greenamyre, J. B., Young, A. B., Penney, J. B. 1983. Quantitative autoradiography of L-³H-glutamate binding to rat brain. *Neurosci. Lett.* 37: 155–60

Greenamyre, J. B., Young, A. B., Penney, J. B. 1984. Quantitative autoradiographic distribution of L-³H-glutamate binding sites in rat central nervous system. *J. Neurosci.* 4: 2133–44

Halpain, S., Parsons, B., Rainbow, T. C. 1983. Tritium film autoradiography of sodium-independent glutamate binding sites in rat brain. *Eur. J. Pharmacol.* 86: 313–14

Halpain, S., Wieczorek, C. M., Rainbow, T. C. 1984. Localization of ³H-glutamate receptors in rat brain by quantitative autoradiography. *J. Neurosci.* 4: 2247–58

Hantraye, P., Kaijima, M., Prenant, C., Guibert, B., Sastre, J., et al. 1984. Central-type benzodiazepine binding sites: A positron emission tomography study in the baboon's brain. *Neurosci. Lett.* 48: 115–20

Henke, H., Beaudet, A., Cuenod, M. 1981. Autoradiographic localization of specific kainic acid binding sites in pigeon and rat cerebellum. *Brain Res.* 219: 95–105

Herkenham, M., Moon Edley, S., Stuart, J. 1984. Cell clusters in the nucleus accumbens of the rat, and the mosaic relationship of opiate receptors, acetylcholinesterase and subcortical afferent terminations. *Neuroscience* 11: 561–93

Herkenham, M., Pert, C. B. 1980. *In vitro* autoradiography of opiate receptors in rat brain suggests loci of "opiatergic" pathways. *Proc. Natl. Acad. Sci. USA* 77: 5532–36

Herkenham, M., Pert, C. B. 1982. Light microscopic localization of brain opiate receptors: A general autoradiographic method which preserves tissue quality. *J. Neurosci.* 2: 1129–49

Herkenham, M., Sokolof, L. 1984. Quantitative receptor autoradiography: Tissue defatting eliminates differential self-absorption of tritium radiation in gray and white matter of brain. *Brain Res.* 321: 363–68

Hollt, V., Schubert, P. 1978. Demonstration of neuroleptic receptor sites in mouse brain by autoradiography. *Brain Res.* 151: 149–53

Hourani, B. T., Torain, B. F., Henkart, M. P., Carter, R. L., Marchesi, V. T., et al. 1974. Acetylcholine receptors of cultured muscle cells demonstrated with ferritin-alpha-bungarotoxin conjugates. *J. Cell. Sci.* 16: 473

Hunt, S. P., Schmidt, J. 1978a. The electron-microscopic autoradiographic localization of alpha-bungarotoxin binding sites within the central nervous system of the rat. *Brain Res.* 142: 152–59

Hunt, S., Schmidt, J. 1978b. Some observations on the binding pattern of alpha-bungarotoxin in the central nervous system of the rat. *Brain Res.* 157: 213–32

Javitch, J. A., Strittmatter, S. M., Snyder, S. H. 1985. Differential visualization of dopamine and norepinephrine uptake sites in rat brain using ³H-mazindol autoradiography. *J. Neurosci.* In press

Javitch, J. A., Uhl, G. R., Snyder, S. H. 1984. Parkinsonism-inducing neurotoxin, N-methyl - 4 - phenyl - 1,2,3,6 - tetra-hydropyridine: Characterization and localization of receptor binding sites in rat and human brain. *Proc. Natl. Acad. Sci. USA* 81: 4591–95

Jones, L. S., Gauger, L. L., Davis, J. N. 1983. Brain alpha-1-adrenergic receptors: Suitability of ¹²⁵I-HEAT as a radioligand for *in vitro* autoradiography. *Eur. J. Pharmacol.* 93: 291–92

Klemm, N., Murrin, L. C., Kuhar, M. J. 1979. Neuroleptic and dopamine receptors: Autoradiographic localization of ³H-spiperone in rat brain. *Brain Res.* 169: 1–9

Kuhar, M. J. 1978. Histochemical localization of opiate receptors and opioid peptides. *Fed. Proc.* 37: 153–57

Kuhar, M. J. 1981. Autoradiographic localization of drug and neurotransmitter receptors in the brain. *Trends Neurosci.* 4: 60–64

Kuhar, M. J. 1982a. Localization of drug and neurotransmitter receptors in brain by light microscopic autoradiography. In *The Handbook of Psychopharmacology*, ed. S. D. Iverson, L. L. Iverson, S. H. Snyder, 1: 299–320. New York: Plenum

Kuhar, M. J. 1982b. Localizing drug and neurotransmitter receptors *in vivo* with tritium-labeled tracers. In *Receptor-Binding Radiotracers*, ed. W. C. Eckelman, 1: 37–50. Boca Raton, Fla.: CRC

Kuhar, M. J. 1983. Autoradiographic localization of drug and neurotransmitter receptors. In *Methods in Chemical Neuroanatomy, Handbook of Chemical Neuroanatomy*, ed. A. Bjorklund, T. Hökfelt, 1: 398–415. Amsterdam: Elsevier

Kuhar, M. J. 1985. Receptor localization with the microscope. In *Neurotransmitter Receptor Binding*, ed. H. I. Yamamura, S. J. Enna, M. J. Kuhar, pp. 153–76. New York: Raven. 2nd ed.

Kuhar, M. J., Murrin, L. C., Malouf, A. T., Klemm, N. 1978. Dopamine receptor binding *in vivo*: The feasibility of autoradiographic studies. *Life Sci.* 22: 203–10

54   KUHAR, DE SOUZA & UNNERSTALL

Kuhar, M. J., Taylor, N., Wamsley, J. K., Hulme, E. C., Birdsall, N. J. M. 1981. Muscarinic cholinergic receptor localization in brain by electron microscopic autoradiography. *Brain Res.* 216:1–9

Kuhar, M. J., Unnerstall, J. R. 1982. In vitro receptor autoradiography: Loss of label during ethanol dehydration and preparative procedures. *Brain Res.* 244:178–81

Kuhar, M. J., Unnerstall, J. R. 1985. Quantitative receptor mapping by autoradiography: Some current technical problems. *Trends Neurosci.* 8:49–53

Kuhar, M. J., Whitehouse, P. J., Unnerstall, J. R., Loats, H. 1984. Receptor autoradiography: Analysis using a PC-based imaging system. *Soc. Neurosci.* 10:558

Kuhar, M. J., Yamamura, H. I. 1974. Light microscopic autoradiographic localization of cholinergic muscarinic sites in rat brain. *Proc. Soc. Neurosci.* 4:29

Kuhar, M. J., Yamamura, H. I. 1975. Light autoradiographic localization of cholinergic muscarinic receptors in rat brain by specific binding of a potent antagonist. *Nature* 253:560–61

Kuhar, M. J., Yamamura, H. I. 1976. Localization of cholinergic muscarinic receptors in rat brain by light microscopic autoradiography. *Brain Res.* 110:229–43

Kuhar, M. J., Zarbin, M. A. 1984. Axonal transport of muscarinic cholingergic receptors and its implications. *Trends Pharmacol. Sci.*, pp. 53–54 (Suppl.)

Laduron, P. M., Janssen, P., Leysen, J. E. 1978. Spiperone: A ligand of choice for neuroleptic receptors. 2. Regional distribution and in vivo displacement of neuroleptic drugs. *Biochem. Pharmacol.* 27:317–21

Lentz, T. L., Chester, J. 1977. Localization of acetylcholine receptors in central synapses. *J. Cell. Biol.* 75:258–67

Lentz, T. L., Mazurkiewicz, J. F., Rosenthal, J. 1977. Cytochemical localization of acetylcholine receptors at the neuromuscular junction by means of horseradish peroxidase-labeled alpha-bungarotoxin. *Brain Res.* 132:423–42

Leroux, P., Pelletier, G. 1984. Radioautoradiographic localization of somatostatin-14 and somatostatin-28 binding sites in the rat brain. *Peptides* 5:503–6

Lewis, M. E., Khachaturian, H., Watson, S. I. 1983. Comparative distribution of opiate receptors and three opioid peptide neuronal systems in rhesus monkey central nervous system. *Life Sci.* 33(Suppl. I):239–42

Lewis, M. E., Patel, J., Moon Edley, S., Marangos, P. J. 1981. Autoradiographic visualization of rat brain adenosine using N^6-cyclohenyl (^{3}H)adenosine. *Eur. J.*

Pharmacol. 73:109–10

Lewis, M. E., Young, E. A., Houghten, R. A., Akil, H., Watson, S. J. 1984. Binding of ^{3}H-dynorphin A to apparent K opioid receptors in deep layers of guinea pig cerebral cortex. *Eur. J. Pharmacol.* 98:149–50

Lo, M. M. S., Niehoff, D. L., Kuhar, M. J., Snyder, S. H. 1983. Autoradiographic differentiation of multiple benzodiazepine receptors by detergent solubilization and pharmacological specificity. *Neurosci. Lett.* 39:37–44

London, E. D., Waller, S. B., Wamsley, J. K. 1985. Autoradiographic localization of ^{3}H-nicotine binding sites in the rat brain. *Neurosci. Lett.* 53:179–84

Lynch, D. R., Strittmatter, S. M., Snyder, S. H. 1984. Enkephalin convertase localization by [^{3}H]guanidinoethylmercaptosuccinic acid autoradiography: Selective association with enkephalin-containing neurons. *Proc. Natl. Acad. Sci. USA* 81:6543–47

Manning, D. C., Snyder, S. H. 1983. ^{3}H-Bradykinin receptor localization in spinal cord and sensory ganglia—evidence for a role in primary afferent function. *Soc. Neurosci. Abstr.* 9:590

Mantyh, P. W., Hunt, S. P., Maggio, J. E. 1984a. Substance P receptors: Localization by light microscopic autoradiography in rat brain using [^{3}H] SP as the radioligand. *Brain Res.* 307:147–65

Mantyh, P. W., Maggio, J. E., Hunt, S. P. 1984b. The autoradiographic distribution of kassinin and substance K binding sites is different from the distribution of substance P binding sites in rat brain. *Eur. J. Pharmacol.* 102:361–64

Marcinkiewicz, M., Verge, D., Gozlan, H., Pichat, L., Hamon, M. 1984. Autoradiographic evidence for the heterogeneity of 5-HT_1 sites in the rat brain. *Brain Res.* 291:159–63

Maurer, R., Cortes, R., Probst, A., Palacios, J. M. 1983. Multiple opiate receptor in human brain: An autoradiographic investigation. *Life Sci.* 33(Suppl. I):231–34

Maurin, Y., Buck, S. H., Wamsley, J. K., Burks, T. F., Yamamura, H. I. 1984. Light microscopic autoradiographic localization of ^{3}H-substance P binding sites in rat thoracic spinal cord. *Life Sci.* 34:1713–16

Maziere, M., Comar, D., Godot, J. M., Collard, P., Cepeda, C., et al. 1981. In vivo characterization of myocardium muscarinic receptors by positron emission tomography. *Life Sci.* 29:2391–97

Maziere, B., Loc'h, C., Hantraye, P., Guillon, R., Duquesnoy, N., et al. 1984. ^{76}Br-bromospiroperidol: A new tool for quantitative in vivo imaging of neuroleptic receptors. *Life Sci.* 35:1349–56

Meibach, R. C., Maayani, S., Green, J. P. 1980. Characterization and radioautography of ^{3}H-LSD binding by rat brain slices *in vitro*: The effect of 5-hydroxytryptamine. *Eur. J. Pharmacol.* 67:371–82

Mendelsohn, F. A. O., Quirion, R., Saavedra, J. M., Aguilera, G., Catt, K. J. 1984. Autoradiographic localization of angiotensin II receptors in rat brain. *Proc. Natl. Acad. Sci. USA* 81:1575–79

Miller, M. M., Silver, J., Billiar, R. B. 1982. Effects of ovariectomy on the binding of [^{125}I]-alpha-bungarotoxin (2.2 and 2.3) to the suprachiasmatic nucleus of the hypothalamus: An *in vivo* autoradiographic analysis. *Brain Res.* 247:355–64

Mohler, H., Battersby, M. K., Richards, J. G. 1980. Benzodiazepine receptor protein identified and visualized in brain tissue by a photoaffinity label. *Proc. Natl. Acad. Sci. USA* 77:1666–70

Mohler, H., Siegart, W., Richards, J. G., Hunkeler, W. 1984. Photoaffinity labeling of benzodiazepine receptors with a partial inverse agonist. *Eur. J. Pharmacol.* 102:191–92

Monaghan, D. T., Holets, V. R., Toy, D. W., Cotman, C. W. 1983. Anatomical distributions of four pharmacologically distinct ^{3}H-L-glutamate binding sites. *Nature* 306:176–79

Moon Edley, S., Hall, L., Herkenham, M., Pert, C. B. 1982. Evolution of striatal opiate receptors. *Brain Res.* 249:184–88

Moskowitz, A. S., Goodman, R. R. 1984. Light microscopic autoradiographic localization of mu and sigma opioid binding sites in the mouse central nervous system. *J. Neurosci.* 4:1331–42

Murphy, K. M. M., Gould, R. J., Oster-Granite, M. L., Gearhart, J. D., Snyder, S. H. 1983. Phorbol ester receptors: Autoradiographic identification in the developing rat. *Science* 222:1036–38

Murphy, K. M. M., Gould, R. J., Snyder, S. H. 1982. Autoradiographic visualization of ^{3}H-nitrendipine binding sites in rat brain: Localization to synaptic zones. *Eur. J. Pharmacol.* 81:517–19

Murrin, L. C., Coyle, J. T., Kuhar, M. J. 1980. Striatal opiate receptors: Pre- and postsynaptic localization. *Life Sci.* 27:1175–83

Murrin, L. C., Enna, S. J., Kuhar, M. J. 1977. Autoradiographic localization of ^{3}H-reserpine binding sites in rat brain. *J. Pharmacol. Exp. Ther.* 203:564–74

Murrin, L. C., Gale, K., Kuhar, M. J. 1979. Autoradiographic localization of neuroleptic and dopamine receptors in the caudate-putamen and substantia nigra: Effects of lesions. *Eur. J. Pharmacol.* 60:229–35

Murrin, L. C., Kuhar, M. J. 1979. Dopamine

receptors in the rat frontal cortex: An autoradiographic study. *Brain Res.* 177:279–85

Nagle, D. S., Blumberg, P. M. 1983. Regional localization by light microscopic autoradiography of receptors in mouse brain for phorbol ester tumor promoters. *Cancer Lett.* 18:35–40

Nakada, M. T., Wieczorek, C. M., Rainbow, T. C. 1984. Localization and characterization by quantitative autoradiography of ^{125}LSD binding sites in rat brain. *Neurosci. Lett.* 49:13–18

Nonaka, R., Moroji, T. 1984. Quantitative autoradiography of muscarinic cholinergic receptors in the rat brain. *Brain Res.* 296:295–303

Olsen, R. W., Reisine, T., Yamamura, H. I. 1980. Neurotransmitter receptors: Biochemistry and alterations in neuropsychiatric disorders. *Life Sci.* 27:801–8

Olsen, R. W., Snowhill, E. W., Wamsley, J. K. 1984. Autoradiographic localization of low affinity GABA receptors with ^{3}H-bicuculline methochloride. *Eur. J. Pharmacol.* 99:247–48

Orzi, F., Kenney, C., Jehlo, J., Sokoloff, L. 1983. Measurement of local cerebral glucose utilization with 2-deoxyglucose in the rat. *J. Cereb. Blood Flow Metab.* 3:577–78

Palacios, J. M. 1982. Autoradiographic localization of muscarinic cholinergic receptors in the hippocampus of patients with senile dementia. *Brain Res.* 243:173–75

Palacios, J. M. 1983. Autoradiographic visualization of receptor binding sites for thyro-tropin-releasing hormone in the rodent brain. *Eur. J. Pharmacol.* 92:165–66

Palacios, J. M. 1984. Light microscopic autoradiographic localization of catecholamine receptor binding sites in brain. Problems of ligand specificity and use of new ligands. In *Catecholamines: Neuropharma-cology and Central Nervous System—Theoretical Aspects*, pp. 73–84. New York: Liss

Palacios, J. M., Kuhar, M. J. 1980. Beta-adrenergic-receptors localization by light microscopic autoradiography. *Science* 208:1378–80

Palacios, J. M., Kuhar, M. J. 1981. Neurotensin receptors are located on dopamine-containing neurones in rat midbrain. *Nature* 294:587–89

Palacios, J. M., Niehoff, D. L., Kuhar, M. J. 1981a. Receptor autoradiography with tritium-sensitive film: Potential for computerized densitometry. *Neurosci. Lett.* 25:101–5

Palacios, J. M., Niehoff, D. L., Kuhar, M. J. 1981b. ^{3}H-spiperone binding sites in brain: Autoradiographic localization of multiple receptors. *Brain Res.* 213:277–89

Palacios, J. M., Probst, S., Cortes, R. 1983. The distribution of serotonin receptors in the human brain: High density of ³H-LSD binding sites in the raphe nuclei of the brainstem. *Brain Res.* 274:150–55

Palacios, J. M., Wamsley, J. K., Kuhar, M. J. 1981c. High affinity GABA receptor: Autoradiographic localization. *Brain Res.* 222:285–308

Palacios, J. M., Wamsley, J. K., Kuhar, M. J. 1981d. The distribution of histamine-Hl receptors in the rat brain: An autoradiographic study. *Neuroscience* 6:15–38

Palacios, J. M., Young, W. S. III, Kuhar, M. J. 1979. Autoradiographic localization of Hl-histamine receptors in brain using ³H-mepyramine: Preliminary studies. *Eur. J. Pharmacol.* 58:295–304

Palacios, J. M., Young, W. S. III, Kuhar, M. J. 1980. Autoradiographic localization of gamma-aminobutyric acid (GABA) receptors in rat cerebellum. *Proc. Natl. Acad. Sci. USA* 77:670–75

Pan, H. S., Frey, K. A., Young, A. B., Penney, J. B. 1983. Changes in ³H-muscimol binding in substantia nigra, entopeduncular nucleus, globus pallidus and thalamus after striatal lesions as demonstrated by quantitative receptor autoradiography. *J. Neurosci.* 3:1189–98

Parsons, B., Rainbow, T. C. 1983. Quantitative autoradiography of sodium-dependent ³H-D-aspartate binding sites in rat brain. *Neurosci. Lett.* 36:9–12

Parsons, B., Rainbow, T. C. 1984. High affinity binding sites for ³H-MPTP may correspond to monoamine oxidase. *Eur. J. Pharmacol.* 102:375–77

Pearson, J., Brandeis, L., Simon, E., Hiller, J. 1980. Radioautography of binding of tritiated diprenorphine to opiate receptors in the rat. *Life Sci.* 26:1047–52

Penney, J. B., Pan, H. S., Young, A. B., et al. 1981. Quantitative autoradiography of ³H-muscimol binding in rat brain. *Science* 214:1036–38

Perry, R. P. 1964. Quantitative autoradiography. In *Methods in Cell Physiology*, ed. D. M. Prescott, 1:305–26. New York: Academic

Pert, C. B., Kuhar, M. J., Snyder, S. H. 1975. Autoradiographic localization of the opiate receptor in rat brain. *Life Sci.* 16:1849–54

Pert, C. B., Kuhar, M. J., Snyder, S. H. 1976. The opiate receptor: Autoradiographic localization in rat brain. *Proc. Natl. Acad. Sci. USA* 73:3729–33

Pilotte, N. S., Sharif, N. A., Burt, D. A. 1984. Characterization and autoradiographic localization of TRH receptors in sections of rat brain. *Brain Res.* 293:372–76

Probst, A., Cortes, R., Palacios, J. M. 1984. Distribution of alpha-2 adrenergic receptors in the human brainstem: An autoradiographic study using ³H-p-amino-clonidine. *Eur. J. Pharmacol.* 106:477–88

Pryzbylski, R. J. 1970. Principles of quantitative autoradiography. In *Introduction to Quantitative Cytochemistry*, ed. G. L. Wied, G. F. Bahr, pp. 477–505. New York: Academic

Quirion, R. 1983. Autoradiographic localization of a calcium channel antagonist, ³H-nitrendipine binding site in rat brain. *Neurosci. Lett.* 36:267–71

Quirion, R., Shults, C. W., Moody, T. W., Pert, C. B., Chase, T. N., et al. 1983a. Autoradiographic distribution of substance P receptors in rat central nervous system. *Nature* 303:714–16

Quirion, R., Zajac, J. M., Morgat, J. L., Roques, B. P. 1983b. Autoradiographic distribution of mu and delta opiate receptors in rat brain using highly selective ligands. *Life Sci.* 33(Suppl. I):227–30

Raichle, M. E. 1978. Quantitative in vivo autoradiography with positron emission tomography. *Brain Res. Rev.* 1:47–68

Rainbow, T. C., Biegon, A. 1983. Quantitative autoradiography of ³H-nitroimipramine binding sites in rat brain. *Brain Res.* 262:319–22

Rainbow, T. C., Biegon, A. 1984. Quantitative autoradiography of ³H-prazosin binding sites in rat forebrain. *Eur. J. Pharmacol.* 102:195–96

Rainbow, T. C., Bleisch, W. V., Biegon, A., McEwen, B. S. 1982. Quantitative densitometry of neurotransmitter receptors. *J. Neurosci. Methods* 5:127–38

Rainbow, T. C., Parsons, B., Wieczorek, C. M. 1984a. Quantitative autoradiography of ³H-hemicholinium-3 binding sites in rat brain. *Eur. J. Pharmacol.* 102:195–96

Rainbow, T. C., Parsons, B., Wolfe, B. B. 1984b. Quantitative autoradiography of β₁- and β₂-adrenergic receptors in rat brain. *Proc. Natl. Acad. Sci. USA* 81:1585–89

Rainbow, T. C., Wieczorek, C. M., Halpain, S. 1984c. Quantitative autoradiography of binding sites for ³H-AMPA, a structural analogue of glutamic acid. *Brain Res.* 309:173–77

Reivich, M., Jehle, J. W., Sokoloff, L., Ketz, S. S. 1969. Measurement of regional cerebral blood flow with antipyrine in awake cats. *J. Appl. Physiol.* 27:296–300

Richards, J. G., Mohler, H. 1984. Benzodiazepine receptors. *Neuropharmacol.* 23:233–42

Richards, G., Mohler, H., Haefely, W. 1985. Benzodiazepoine receptors and their ligands. In *Mechanism of Drug Action Series*, ed. G. N. Woodruff. London: MacMillan

Rogers, A. W. 1979. *Techniques of Autoradiography*. Amsterdam: Elsevier

Rostene, W. H., Morgat, J.-L., Dussaillant, M., Rainbow, T. C., Sarrieau, A., et al. 1984. *In vitro* biochemical characterization and autoradiographic distribution of ³H-thyrotropin-releasing hormone binding sites in rat brain sections. *Neuroendocrinology* 39:81–86

Roth, L. J., Diab, I. M., Watanabe, M., Dinerstein, R. J. 1974. A correlative radioautographic, fluorescent, and histo-chemical technique for cyto-pharmacology. *Mol. Pharmacol.* 10:986–98

Roth, L. J., Stumpf, W. E. 1969. *Autoradiography of Diffusible Substances*. New York: Academic

Rothman, R. B., Herkenham, M., Pert, C. B., Liang, T., Cascieri, M. A. 1984. Visualization of rat brain receptors for the neuropeptide, Substance P. *Brain Res.* 309:47–54

Rotter, A., Birdsall, N. J. M., Burgen, A. S. V., Field, P. M., Hulme, E. C., et al. 1979a. Muscarinic receptors in the central nervous system of the rat. I. Technique for autoradiographic localization of the binding of ³H-propylbenzilylcholine mustard and its distribution in the forebrain. *Brain Res. Rev.* 1:141–66

Rotter, A., Birdsall, N. J. M., Field, P. M., Raisman, G. 1979b. Muscarinic receptor in the CNS of the rat. II. Distribution of binding of [³H] propylbenzilylcholine in the midbrain and hindbrain. *Brain Res. Rev.* 1:167–84

Salpeter, M. M., Bachman, L., Salpeter, E. E. 1969. Resolutions in electron-microscope autoradiography. *J. Cell Biol.* 41:1–12

Schubert, P., Hollt, V., Herz, A. 1975. Autoradiographic evaluation of the intracisternal distribution of ³H-etorphine in the mouse brain. *Life Sci.* 16:1855–56

Seeger, T. F., Sforzo, G. A., Pert, C. B., Pert, A. 1984. *In vivo* autoradiography: Visualization of stress-induced changes in opiate receptor occupancy in the rat brain. *Brain Res.* 305:303–11

Segal, M., Dudai, Y., Amsterdam, A. 1978. Distribution of alpha-bungarotoxin cholinergic nicotinic receptor in rat brain. *Brain Res.* 148:105–19

Seybold, V. S., Elde, R. P. 1984. Receptor autoradiography in thoracic spinal cord: Correlation of neurotransmitter binding sites with sympathoadrenal neurons. *J. Neurosci.* 4:2533–42

Shults, C. W., Quirion, R., Jensen, R. T., Moody, T. W., O'Donohue, T. L., et al. 1982. Autoradiographic localization of substance P receptors using ¹²⁵I-Substance P. *Peptides* 3:1073–75

Silver, J., Billiar, R. B. 1976. An autoradio-

graphic analysis of [³H]-alpha-bungarotoxin distribution in the rat brain after intraventricular injection. *J. Cell Biol.* 71:956–63

Simantov, R., Kuhar, M. J., Uhl, G. R., Snyder, S. H. 1977. Opioid peptide enkephalin: Immunohistochemical mapping in rat central nervous system. *Proc. Natl. Acad. Sci. USA* 74:2167–71

Slater, P., Patel, S. 1083. Autoradiographic distribution of serotonin-2 receptors in rat brain. *Eur. J. Pharmacol.* 92:297–98

Snyder, S. H., Bennett, J. P. Jr. 1976. Neurotransmitter receptors in the brain: Biochemical identification. *Ann. Rev. Physiol.* 38:153–75

Sokoloff, L., Reivich, M., Kennedy, C., Des Rosiers, M. H., Patlak, C. S., Pettigrew, K. D., Sakurada, O., Shinohara, M. 1977. The ¹⁴C deoxyglucose method for the measurement of local cerebral glucose utilization. *J. Neurochem.* 28:897–916

Strittmatter, S. H., Lo, M. M. S., Javitch, J. A., Snyder, S. H. 1984. Autoradiographic visualization of angiotensin converting enzyme in rat brain with ³H-captopril: Localization to a striatonigral pathway. *Proc. Natl. Acad. Sci. USA* 81:1599–1603

Stumpf, W. E., Roth, L. G. 1966. High resolution autoradiography with dry-mounted, freeze-dried frozen sections. Comparative study of six methods using two diffusible compounds. ³H-estradiol and ³H-mesobilirubinogen. *J. Histochem. Cytochem.* 14:274–87

Swanson, L. W., Lindstrom, J., Tzartos, S., Schmued, L. C., O'Leary, D. D. M., et al. 1983. Immunohistochemical localization of monoclonal antibodies to the nicotinic acetylcholine receptor in chick midbrain. *Proc. Natl. Acad. Sci. USA* 80:4532–36

Taylor, N. R., Unnerstall, J. R., Mashal, R. D., De Souza, E. B., Kuhar, M. J. 1984. Receptor autoradiography: Coping with regional differences in autoradiographic efficiency with tritium. *Soc. Neurosci.* 10:557

Tempel, A., Gardner, E. L., Zukin, R. S. 1984. Visualization of opiate receptor upregulation by light microscopic autoradiography. *Proc. Natl. Acad. Sci. USA* 81:3893–97

Tran, V. T., Uhl, G. R., Perry, D. C., Manning, D. C., Vale, W. W., et al. 1984. Autoradiographic localization of somatostatin receptors in rat brain. *Eur. J. Pharmacol.* 101:307–9

Uhl, G. R., Kuhar, M. J. 1984. Chronic neuroleptic treatment enhances neurotensin receptor binding in human and rat substantia nigra. *Nature* 309:350–52

Uhl, G. R., Tran, V., Snyder, S. H., Martin, J. B. 1985. Somatostatin receptors: Distribution in rat central nervous system and human frontal cortex. *J. Comp. Neurol.* In press

Unnerstall, J. R., Fernandez, I., Oressanz, L. M. 1985. The alpha adrenergic receptor: Radiohistochemical analysis of functional characteristics and biochemical differences. *Pharmacol. Biochem. Behav.* 22: 859–74

Unnerstall, J. R., Kopajtic, T. A., Kuhar, M. J. 1984. Distribution of alpha-2 agonist binding sites in the rat and human central nervous system: Analysis of some functional, anatomic correlates of the pharmacologic effects of clonidine and related agents. *Brain Res. Rev.* 7: 69–101

Unnerstall, J. R., Kuhar, M. J. 1981. Benzodiazepine receptors are coupled to a subpopulation of GABA receptors: Evidence from a quantitative autoradiographic study. *J. Pharmacol. Exp. Ther.* 218: 797–804

Unnerstall, J. R., Molliver, M. E., Kuhar, M. J., Palacios, J. M. 1983. Ontogeny of opiate binding sites in the hippocampus, olfactory bulb and other regions of the rat forebrain by autoradiographic methods. *Dev. Brain Res.* 7: 157–69

Unnerstall, J. R., Nieholf, D. L., Kuhar, M. J., Palacios, J. M. 1982. Quantitative receptor autoradiography using ³H-ultrofilm: Application to multiple benzodiazepine receptors. *J. Neurosci. Meth.* 6: 59–73

Unnerstall, J. R., Wamsley, J. K. 1983. Autoradiographic localization of high affinity ³H-kainic acid binding sites in the rat forebrain. *Eur. J. Pharmacol.* 86: 361–71

Van Dijk, A., Gillessen, D., Mohler, H., Richards, J. G. 1981. Autoradiographical localization of cholecystokinin-receptor binding in rat brain and pancreas using ³H-CCK as radioligand. *Br. J. Pharmacol.* 74: 858P

Van Dijk, A., Richards, J. G., Trzeciak, A., Gillessen, D., Mohler, H. 1984. Cholecystokinin receptors: Biochemical demonstration and autoradiographic localization in rat brain and pancreas using ³H-cholecystokinin as radioligand. *J. Neurosco.* 4: 1021–33

Van Leeuwen, F., Wolters, P. 1983. Light microscopic autoradiographic localization of ³H-arginine-vasopressin binding sites in the rat brain and kidney. *Neurosci. Lett.* 41: 61–66

Wagner, H. N., Burns, H. D., Dannals, R. F., Wong, D. F., Langstrom, B., et al. 1983. Imaging dopamine receptors in the human brain by positron tomography. *Science* 221: 1264–66

Wamsley, J. K., Gehlert, D. R., Roeske, W. R., Yamamura, H. I. 1984a. Muscarinic antagonist binding site heterogeneity as evidenced by autoradiography after direct labeling with ³H-QNB and ³H-pirenzipine. *Life Sci.* 34: 1395–1402

Wamsley, J. K., Lewis, M. S., Young, W. S. III, Kuhar, M. J. 1981. Autoradiographic localization of muscarinic cholingeric receptors in rat brainstem. *J. Neurosci.* 1: 176–91

Wamsley, J. K., Zarbin, M. A., Birdsall, J. M., Kuhar, M. J. 1980. Muscarinic cholingeric receptors: Autoradiographic localization of high and low affinity agonist binding sites. *Brain Res.* 200: 1–12

Wamsley, J. K., Zarbin, M. A., Kuhar, M. J. 1984b. Distribution of muscarinic cholinergic high and low affinity agonist binding sites: A light microscopic autoradiographic study. *Brain Res. Bull.* 12: 233–43

Wamsley, J. K., Zarbin, M. A., Young, W. S. III, Kuhar, M. J. 1982. Distribution of opiate receptors in the monkey brain: An autoradiographic study. *Neuroscience* 7: 595–613

Whitehouse, P. J., Wamsley, J. K., Zarbin, M. A., Price, D. L., Tourtellote, W. W., et al. 1983. Amyotrophic lateral sclerosis: Alterations in neurotransmitter receptors. *Ann. Neurol.* 14: 8–16

Wieczorek, C. M., Parsons, B., Rainbow, T. C. 1984. Quantitative autoradiography of ³H-MPTP binding sites in rat brain. *Eur. J. Pharmacol.* 98: 453–54

Wilkin, G. P., Hudson, A. L., Hill, D. R., Bowery, N. G. 1981. Autoradiographic localization of GABA_B receptors in rat cerebellum. *Nature* 294: 584–87

Wise, S. P., Herkenham, M. 1982. Opiate receptor distribution in the cerebral cortex of the rhesus monkey. *Science* 218: 387–89

Wolf, S. S., Moody, T. W., O'Donohue, T. L., Zarbin, M. A., Kuhar, M. J. 1983a. Autoradiographic visualization of rat brain binding sites for bombesin-like peptides. *Eur. J. Pharmacol.* 87: 163–64

Wolf, S. S., Moody, T. W., Quirion, R., Schults, C. W., Chase, T. N., et al. 1983b. Autoradiographic visualization of substance P receptors in rat brain. *Eur. J. Pharmacol.* 91: 157–58

Wong, D. F., Wagner, H. N. Jr., Dannals, R. F., Links, J. M., Frost, J. J., et al. 1984. Effects of age on dopamine and serotonin receptors measured by positron tomography in the living human brain. *Science* 226: 1393–96

Yamamura, H. I., Enna, S. J., Kuhar, M. J. 1985. *Neurotransmitter Receptor Binding.* New York: Raven

Yamamura, H. I., Gee, K. W., Brinton, R. E., Davis, T. P., Hadley, M., et al. 1983a. Light microscopic autoradiographic visualization of ³H-arginine vasopressin binding sites in rat brain. *Life Sci.* 32: 1919–24

Yamamura, H. I., Wamsley, J. K., Deshmukh, P., Roeske, W. R. 1983b. Differential light microscopic autoradiographic localization of muscarinic cholinergic receptors in the brainstem and spinal cord of the rat using ^{3}H-pirenzipine. *Eur. J. Pharmacol.* 91:147–49

Young, W. S. III, Kuhar, M. J. 1979a. A new method for receptor autoradiography ^{3}H-opioid receptor labeling in mounted tissue sections. *Brain Res.* 179:255–70

Young, W. S. III, Kuhar, M. J. 1979b. Autoradiographic localization of benzodiazepine receptors in brains of humans and animals. *Nature* 280:393–95

Young, W. S. III, Kuhar, M. J. 1980a. Noradrenergic alpha-1 and alpha-2 receptors: Light microscopic autoradiographic localization. *Proc. Natl. Acad. Sci. USA* 77:1696–1700

Young, W. S. III, Kuhar, M. J. 1980b. Radiohistochemical localization of benzodiazepine receptors in rat brain. *J. Pharmacol. Exp. Ther.* 212:337–46

Young, W. S. III, Kuhar, M. J. 1980c. Serotonin receptor localization in rat brain by light microscopic autoradiography. *Eur. J. Pharmacol.* 62:237–39

Young, W. S. III, Kuhar, M. J. 1981. Neurotensin receptor localization by light microscopic autoradiography in rat brain. *Brain Res.* 206:273–85

Young, W. S. III, Kuhar, M. J., Roth, J., Brownstein, M. J. 1980d. Radiohistochemical localization of insulin receptors in the adult and developing rat brain. *Neuropeptides* 1:15–22

Young, W. S. III, Niehoff, D. L., Kuhar, M. J., Beer, B., Lippa, A. S. 1981. Multiple benzodiazepine receptor localization by light microscopic radiohistochemistry. *J. Pharmacol. Exp. Ther.* 216:425–30

Zarbin, M. A., Innis, R. B., Wamsley, J. K., Snyder, S. H., Kuhar, M. J. 1983a. Autoradiographic localization of cholecystokinin receptors in rodent brain. *J. Neurosci.* 3:877–906

Zarbin, M. A., Palacios, J. M., Wamsley, J. K., Kuhar, M. J. 1983b. Axonal transport of beta-adrenergic receptors. *Mol. Pharmacol.* 24:341–48

Zarbin, M. A., Wamsley, J. K., Kuhar, M. J. 1981. Glycine receptor: Light microscopic autoradiographic localization with ^{3}H-strychnine. *J. Neurosci.* 1:532–47

Ann. Rev. Neurosci. 1986. 9:61–85

NMR SPECTROSCOPY OF BRAIN METABOLISM IN VIVO

James W. Prichard

Department of Neurology, Yale University School of Medicine, New Haven, Connecticut 06510

Robert G. Shulman

Department of Molecular Biophysics and Biochemistry, Yale University School of Medicine, New Haven, Connecticut 06510

INTRODUCTION[1]

Nuclear magnetic resonance (NMR) methods applicable to the study of functioning brain in situ are new in neuroscience, and many neuroscientists may not be familiar with them. Although we provide more explanatory material than reviews in this series usually contain, many readers may wish to consult other works (Gadian 1982, Moore 1984) that provide access to a wider range of basic NMR literature as well as introductions to NMR theory for scientists in other disciplines. For an excellent discussion of in vivo NMR studies of muscle, which are not covered here, the reader is referred to a recent review (Radda et al 1984).

NMR spectroscopy is possible because some atomic nuclei act like tiny bar magnets when placed in a magnetic field. They line up with or against the field and can be excited in a controlled way by irradiation with radio frequency energy. During relaxation from the excitation, they emit radio frequency signals that contain a great deal of information about the molecules they are in. The whole process is practical with samples ranging from crystalline solids to living people, though only compounds in or near

---

[1] Abbreviations: ADP, adenosine diphosphate; ATP, adenosine triphosphate; NAA, *N*-acetyl aspartate; PCr, phosphocreatine; pHi, intracellular pH; Pi, inorganic phosphate; T, Tesla (unit of magnetic field strength equal to 10,000 gauss).

0147–006X/86/0301–0061$02.00

the millimolar range can be detected in vivo. It is thought to be harmless for the great majority of human subjects (Budinger & Cullander 1983, NRPB 1983, Saunders & Smith 1984). Instruments implementing it are comparable in cost to electron microscopes and modern x-ray diagnostic equipment.

The NMR phenomenon was discovered independently in two laboratories in 1946 (Bloch et al 1946, Purcell et al 1946). It was of such clearly far-reaching importance that Felix Bloch and Edward Purcell received the Nobel Prize for Physics in 1952. NMR methods for the study of test tube-size samples have become steadily more productive over the last four decades. By the late 1970s, improvements in magnet technology made similar methods applicable to much larger samples, including the human body. A short history of these developments is available (Andrew 1984).

Current NMR research on living animals and humans has two branches that are related but must not be confused. NMR *imaging* uses spatial resolution of some strong signal, usually that of the hydrogen nuclei in water molecules, to make pictures of anatomical structure (Bydder 1984). The images are more detailed than those obtained by x-ray methods, including computed tomography (CT scanning), and making them involves no ionizing radiation. Imaging methods based on ^{23}Na (Maudsley & Hillal 1984) and ^{31}P (Maudsley et al 1984) are being developed. NMR *spectroscopy*—the subject of this review—uses much weaker signals from phosphorus, carbon 13, nonwater hydrogen, and some other nuclei to obtain quantitative information about specific compounds present in the tissue, with relatively crude spatial resolution. Imaging can be done with weaker and less homogeneous magnetic fields than those needed for spectroscopy. Because it is somewhat less demanding technically and is analogous to CT scanning, its development for in vivo use has been faster. In the most visible example of this, imaging is already in widespread use for routine medical diagnosis; spectroscopy is unlikely to reach that stage for several more years. The most advanced instruments now available can obtain both images and spectra, but each modality has its own optimum requirements, and the means of combining localization of signal with spectral information are still being developed (Aue et al 1984, Bendall & Gordon 1983, Bottomley et al 1985).

There are two principal reasons to do NMR spectroscopy in vivo. The first is purely scientific—one wishes to study properties of tissue which are sensitive to tissue disruption. Metabolic studies free of agonal artifact on organs in their natural hormonal, neural, and hemodynamic environments can be expected to produce new understanding of how complex organisms function. The other reason is largely medical—one wishes to study otherwise inaccessible human tissues, both to investigate pathophysiological processes which may be unique to humans and, eventually, to provide information useful in the management of individual patients. Phosphate energy stories, lactate concentrations, and intracellular pH in the human brain are good examples of NMR-measurable variables which are important for both purposes and cannot be studied easily by any other technique.

PHOSPHORUS (^{31}P) STUDIES

The first NMR measurements possibly including signals from living brain were ^{31}P spectra (Chance et al 1978). These were obtained from an anesthetized mouse inserted into an 18 mm NMR tube and studied in a conventional spectrometer with a radio frequency coil that surrounded the entire head; muscle probably contributed much of the signal under these circumstances. Brain spectra reliably uncontaminated by signals from other tissues first became possible when surface coils were developed (Ackerman et al 1980), and nearly all of the results reviewed below were obtained with them. However, NMR spectra can be obtained from more sharply localized regions of the body by techniques which rely on manipulation of radio frequency fields (Bendall & Gordon 1983) or magnetic field gradients (Aue et al 1984, Bottomley et al 1985).

Animals

The first NMR observations of living brain with a surface coil (Ackerman et al 1980) were made on an anesthetized rat and demonstrated that ^{31}P spectra from that organ resemble the one in Figure 1, which is from rabbit. At least 8 resonances were present; 5 could be assigned with confidence to the particular compounds indicated by the labeling in Figure 1. The resonances labeled "sugar phosphates" and "phosphodiesters" were later shown to be more variable with species and age (see below) than the others; they both probably contain contributions from several compounds, but these have not been firmly identified. An eighth, very broad resonance was attributed to relatively immobile phosphates in bone. It interferes with quantitation of the other resonances, which are superimposed on it. An effective method for eliminating it is pre-saturation (Ackerman et al 1984b).

Ackerman and colleagues (1980) noted that the concentration ratio of PCr to ATP derived from their spectrum was 1.93, which is higher than the highest values obtained by chemical assay of freeze-clamped brain samples. They also estimated that in vivo ADP and Pi concentrations were probably much lower than those measured by destructive analytical techniques. If these things are true, the phosphorylation state of brain tissue reflected by the expression $[ATP]/([ADP][HPO_4^{2-}])$ must be well above the generally accepted value of 3400 M^{-1}. More recently, workers using perfused heart (Matthews et al 1982) and rabbit brain in vivo (Prichard et al 1983) assumed that the creatine kinase reaction was at equilibrium and calculated the concentration of ADP participating in it from quantities measurable in ^{31}P spectra; in both cases the calculated values were 20–30 μM. There is also indirect biochemical evidence that chemical assay gives values higher than are present in vivo (Veech et al 1979). Contamination of chemical measurements by agonal change is a plausible explanation for the

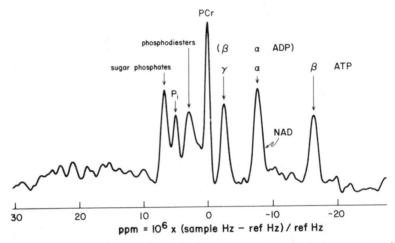

Figure 1 Phosphorus spectrum of rabbit brain. The seven labeled resonances are present in brain spectra from other species, including humans. The "ATP" resonances contain contributions from other nucleotide di- and triphosphates. Labels in parentheses indicate that ADP phosphates resonate at nearly the same frequencies as the ATP alpha and gamma phosphates; their contribution to the spectrum is undetectable due to the low concentration of ADP. Resonances from nicotinamide adenine dinucleotide (NAD) and the alpha phosphate of ATP are too close to be resolved. The "sugar phosphates" and "phosphodiesters" resonances are probably composite signals from phosphomono- and phosphodiester configurations in several compounds; the largest contribution to the former may be from phosphoethanolamine (see text). The formula on the axis explains the calculation of "chemical shift"—a measure of small but characteristic differences in the resonant frequencies of nuclei of the same species in different molecular environments; these small shifts are the basis of the chemical selectivity of NMR spectroscopy. "Ref Hz" is the frequency of some resonance chosen as the reference point, in this case that of PCr; "sample Hz" is the frequency of any other resonance in the spectrum. Chemical shifts in parts per million are independent of magnetic field strength and therefore facilitate comparison of data from different spectrometers. By convention, chemical shift numbers are plotted increasing to the left, which is the "downfield" direction of decreasing magnetic field strength. The variations between 10 and 30 ppm are noise. The spectrum was made from 512 scans obtained in 5 min from a paralyzed, pump-ventilated rabbit under nitrous oxide analgesia, using a 4 cm surface coil in an Oxford TMR 32/200 spectrometer operating at a magnetic field of 1.9 T. Reprinted with permission from *Neurology* 37:781.

discrepancy; because the reactions involved are fast, it could occur even during very rapid freezing of tissue. However, it is conceivable that both the NMR and chemical estimates are right, as they would be if much ADP present in vivo is bound to protein or otherwise sequestered from the pool exposed to creatine kinase.

Once the practical utility of surface coils was established, the way was open for development of animal models in which some aspects of pathological brain function could be studied in vivo. Several groups (Ackerman et al 1984a, Bottomley et al 1982, Decorps et al 1984, Delpy et al 1982, Hilberman et al 1984, Naruse et al 1984, Naruse et al 1983, Prichard et al 1983, Thulborn et al 1982) validated ^{31}P spectroscopy for this purpose by showing that it could detect changes in phosphate energy stores, Pi, and pHi known to occur during a variety of metabolic stresses. The area of a resonance is proportional to the concentration of the compound producing it; changes in area reflect changes in concentration directly. The resonant frequency of Pi is sensitive to pH, and since the intracellular fluid volume of brain is large relative to other compartments containing Pi, it can be used to measure pHi of the brain.

Thulborn et al (1982) observed a simultaneous decline in PCr and ATP, a rise in Pi, and tissue acidification in Mongolian gerbil brain ipsilateral to carotid occlusion. With the surface coil over the contralateral hemisphere or cerebellum, the changes were much less pronounced. The NMR changes correlated well with cerebral edema estimated from the specific gravity of grey matter in brains removed 1 hr after the occlusion and with histological signs of cell damage. This study was the first systematic demonstration that data from in vivo NMR methods and conventional measurements can supplement each other in analysis of a specific pathophysiological problem.

Prichard et al (1983) studied the behavior of cerebral phosphate energy stores and pHi in paralyzed, pump-ventilated rabbits during hypoglycemia, hypoxia, and status epilepticus, simultaneously with measurements of conventional physiological variables. Insulin shock caused PCr and ATP to fall and Pi and pHi to rise within minutes of the disappearance of the electroencephalogram. All of these changes reversed when glucose was given. Blood pressure, electrocardiogram, and arterial pO_2, pCO_2, and pH remained within normal limits throughout the experiment. Similarly, hypoxic hypoxia and bicucculine-induced seizures caused the expected changes in the same range of NMR and physiological variables. This study showed that the physiological measurements were routinely feasible with the animal in the spectrometer and that the NMR measurements were unperturbed by such experimental arrangements.

Decorps et al (1984) showed that radio frequency surface coils permanently fixed to the skulls of rats remained usable for as long as a month, enabling repeated observations from exactly the same region of brain to be made on different days. They also observed the time course of reversible cerebral acidosis and loss of phosphate energy stores following intraperitoneal injection of sublethal doses of potassium cyanide, and found that it correlated well with the behavioral abnormalities of freely moving rats given the same doses.

Hilberman et al (1984) demonstrated the usefulness of line-fitting routines for quantitation of individual resonances in a group of overlapping ones. This is an especially important problem for NMR studies done in vivo, which must be carried out at relatively low (circa 2 T) magnetic field strengths and always yield spectra with broad lines. In experiments on hypoxia in dogs, they found that the phosphodiester resonance accounted for nearly 40% of the total ^{31}P signal. It was relatively more intense than in published spectra from adult rodents and their own unpublished spectra from newborn puppies. They reported preliminary observations on chloroform-methanol-HCl extracts of dog brain which suggest that the signal in the region around 2 ppm in the in vivo spectrum is predominately from phosphodiester-containing phospholipids.

Assignments in the phosphodiester region may be important for future NMR studies of cerebral development and pathology. The few published spectra from brains of adult humans have an intense resonance there, whereas those from neonates do not (see below). Perchloric acid extracts of guinea pig brain identified glycerol 3-phosphorylcholine, glycerol 3-phosphorylethanolamine, and certain of their metabolites as principal sources of the phosphodiester signal (Glonek et al 1982). It now appears that species, age, and method of extraction must all be considered if correct assignments are to be made. Moreover, NMR signals obtained in vivo may originate from mobile portions of large polymers as well as from small, rapidly tumbling molecules. Glycogen is 100% detectable by ^{13}C spectroscopy (Sillerud & Shulman 1983); the presence in the ^{31}P spectrum of signals from relatively mobile phospholipid components of membranes would not be surprising.

A study of bicuculline-induced status epilepticus in rabbits (Petroff et al 1984) confirmed and extended preliminary observations by the same group (Prichard et al 1983). The PCr/Pi ratio fell 50% and pHi fell to 6.7–6.9 from control values near 7.1 during the first hour of status. These remained depressed for up to 3 hr, despite the virtual disappearance of intense seizure discharge after 1 hr. In all animals, ATP remained in the normal range throughout the experiments. Repeat doses of bicuculline demonstrated that the brain retained its capacity to mount massive electrical seizures under these circumstances. Calculations based on the assumption of equilibrium in the creatine kinase reaction indicated that the cerebral acidosis was responsible for most of the PCr decline. A later ^{1}H study documented a persistent rise in brain lactate caused by the seizure discharge (O. A. C. Petroff et al, in preparation).

Measurement of cerebral pHi in vivo is a special capability of NMR spectroscopy. Invasive methods are available for use with experimental animals (see Petroff et al 1985 for references), and positron emission tomographic methods are under development for use in humans (Brooks et al 1984, Rottenberg et al 1984, Syrota et al 1983). In the foreseeable future, however, the accuracy, safety, and relative simplicity of the ^{31}P NMR method ensure that it will be widely employed for both experimental and clinical research. Because existing NMR titration data on Pi were not fully appropriate for work on brain, a study was done to establish suitable constants for conversion of the Pi–PCr chemical shift difference (ΔPi) to pH in the Henderson-Hassalbalch equation (Petroff et al 1985). From new titration data, the relation

$$pH = 6.77 + \log[(\Delta Pi - 3.29)/(5.68 - \Delta Pi)]$$

was obtained. Mg affected the constants appreciably only in concentrations above 2.5 mM, which is well above estimates of free Mg concentration in brain. PCr was a satisfactory internal chemical shift reference down to circa pH 6.5, below which titration of PCr introduced a progressively larger error. Both for this reason and because PCr is usually depleted in metabolic states which cause severe tissue acidosis, some external reference is necessary for pHi measurements in such states.

Cerebral pHi values calculated using the new formula were 7.14 ± 0.04 (SD) and 7.13 ± 0.03, respectively, for paralyzed, mechanically ventilated rabbits and rats under nitrous oxide analgesia. These values are toward the alkaline end of the rather wide range reported by workers using destructive analytical methods.

Magnetization Transfer Experiments

It is possible to "label" nuclei having the same resonant frequency by irradiating them selectively with radio frequency energy, which briefly changes their magnetization state. If such nuclei are transferred enzymatically from the molecules they were in at the time of irradiation to other molecules at a rate comparable to the lifetime of the changed magnetic state, their presence in the second molecular population may be detectable at its (different) resonant frequency. Under favorable conditions, the rate constants of an enzyme-catalyzed reaction and the fluxes through it can be calculated from this measurement in living systems. The principles of such experiments and results obtained in various preparations have been reviewed recently (Alger & Shulman 1984).

Only two studies using magnetization-transfer techniques in living brain have been published, both done on rats (Balaban et al 1983, Shoubridge et al 1982). The results, which are somewhat different, raise interesting issues for future work. Shoubridge et al (1982) found a unidirectional ATP synthetase rate of 0.33 μM/gm wet weight/sec. This is only 10–15% of the rate in yeast and perfused heart. These workers also reported unidirectional fluxes through creatine kinase of 1.64 and 0.68 μM/g wet weight/sec, respectively, for the forward (PCr hydrolysis) and reverse reactions. Since PCr participates in no other known reaction, the fluxes must actually be equal during any period of stable PCr concentration. The results obtained therefore imply that some ATP is in a metabolic compartment which does not contain creatine kinase, or that ATP participates in other reactions with total fluxes amounting to a substantial fraction of the flux through creatine kinase. The latter explanation has recently been shown to be the correct one in perfused heart (Ugurbil et al 1984). Proof of either explanation in brain would advance understanding of cerebral bio-chemistry. The problem was complicated by the report of Balaban et al (1983), who found equal forward and reverse fluxes of about 2 μM/gm wet weight/sec. They used a different magnetization transfer technique, a different anesthetic, and a coil that included the entire head of the rat, so that the signals may have come in part from muscle.

Magnetization transfer experiments are difficult to do, but they provide important, unique kinetic information about the function of some enzymes in their natural cellular environment. Beyond the resolvable discrepancies in presently available data lies the further question of how the measurable fluxes behave in different functional states of the brain; results from other systems show that magnetization transfer techniques open a window on some aspects of enzyme regulation in vivo (Alger & Shulman 1984).

Humans

The first ^{31}P spectra from the living human brain were obtained in a study of newborn infants, most of whom suffered from some degree of perinatal brain damage (Cady et al 1983). The same group later published additional observations on normal as well as abnormal infants (Hope et al 1984). A practical result of this work is that the PCr/Pi ratio appears to be of use both for assessment of tissue damage and for prediction of outcome after metabolic stress. The ratio was 1.35 ± 0.22 (SD) in six normal infants. Lower ratios found in some infants after severe asphyxia returned toward normal as the infants' clinical conditions improved. In others, falling ratios over the several days after birth asphyxia heralded death or neurological impair-ment. An infant who had persistently low ratios for the first 26 days of life developed multiple porencephalic cysts during the same period. Low ratios rose in some infants after mannitol infusion, which presumably reduced cerebral edema. A trend toward an inverse relation between PCr/Pi and pHi noted in the first study by interested readers (Petroff & Prichard 1983) was reinforced in the second; the most likely explanation for it is an increase in the proportion of extra- to intracellular fluid, as would be expected in developing porencephaly and some kinds of cerebral edema. A prominent phosphomonoester resonance was a consistent feature of all but the most abnormal spectra; data from several sources suggest that it is more prominent in infant than adult brains of both humans and rats and may therefore reflect metabolic conditions in the rapidly growing brain. The compounds responsible for it have not been identified with certainty. Extract work suggested ribose-5-phosphate as a possible source (Glonek et al 1982), but Hope et al (1984), referring to unpublished data, expressed the view that phosphoethanolamine is a more likely assignment.

Another group confirmed the presence of an intense phosphomonoester resonance in newborn human infants and marshalled reasons for its assignment principally to the phosphoryl esters of choline and ethanolamine (Younkin et al 1984). They did not find a correlation of PCr/Pi ratio or other features of the ^{31}P spectrum with age or clinical condition, possibly because their series was small.

By the end of 1984, two groups had published ^{31}P spectra from brains of adult humans (Bottomley et al 1984, Radda et al 1984). All seven peaks in Figure 1 were identifiable in these spectra, but the phosphodiester resonance was relatively more intense. As noted above, this difference between adult and infant brains within the same species appears to exist in humans and dogs, but not in rodents, whereas the phosphomonoester resonance may be more prominent in the infant in some species. When these resonances are firmly assigned and correlated with species and age, they may well be a rich source of new information about brain development under normal and pathological conditions.

CARBON (^{13}C) STUDIES

Due to inherent properties of ^{13}C, the chemical shift range of resonances in carbon spectra of organic compounds is very wide; it is some 200 ppm, compared to about 10 and 30 ppm, respectively, in ^{1}H and ^{31}P spectra. The difference is reflected in the chemical shift axes of Figures 1–3. Its practical significance is that ^{13}C spectra from organic samples are likely to contain resonances which are well resolved (separated) from each other, so that precise measurements on the compounds or the chemical groups they represent are possible. However, ^{13}C signals are weaker than ^{31}P ones for three reasons: First, the inherent signal strength of ^{13}C is only about one quarter that of [31]P. Next, whereas ^{31}P is nearly 100% abundant in nature, only 1.1% of carbon nuclei are the magnetic isotope ^{13}C; the rest are ^{12}C, which are not magnetic and give no NMR signal. Finally, magnetic interaction of ^{13}C with nearby ^{1}H nuclei reduces the detectable signal from ^{13}C by splitting its resonances into multiple smaller ones; the interaction can be removed by selective radio frequency irradiation of the ^{1}H nuclei, but this involves the potential risk of tissue heating. The total effect of these factors is an obstacle to ^{13}C spectroscopy of living systems, but not an insurmountable one. A sizable body of literature has accumulated on ^{13}C metabolic studies on cell suspensions, perfused organs, and, more recently, organs in situ; this work has been reviewed (Alger & Shulman 1984).

In our laboratory, ^{13}C spectroscopy has been done on rat and rabbit brain in vivo (K. L. Behar et al, in preparation). Figure 2 illustrates an experiment in which rabbit brain was observed before and after intravenous infusion of 1-^{13}C glucose combined with hypoxia. In *A*, the natural abundance spectrum from the brain contains resonances from ^{13}C atoms in carboxyl, olefinic, and methylenic configurations, mostly in lipid molecules. In *B*, additional resonances from the alpha and beta anomers of 1-^{13}C glucose in brain are evident. The spectrum (*C*) created by subtracting spectrum *A* from spectrum *B* reveals resonances from the methyl carbon of lactate and carbons 2, 3, and 4 of glutamate and glutamine. Data from the whole experiment showed that the ^{13}C lactate signal rose, fell, and rose again in the course of two descents into hypoxia, while the signals from the amino acids rose steadily. These results showed that detection of brain metabolites enriched with ^{13}C is practical in vivo. That being true, the low natural abundance of ^{13}C is a fortunate circumstance in that it allows turnover studies of concentrated metabolites to be done. The experiment illustrated in Figure 2 was done at 1.9 Tesla—a magnetic field strength that is available for research on humans. Metabolic studies similar to the illustrated experiment could be done in humans, but still more recent spectroscopic developments suggest that combined ^{13}C and ^{1}H methods

Figure 2 Carbon 13 spectra of rabbit brain. Naturally abundant [13]C (1.1% of total carbon), principally in carboxyl, olefinic, and methylenic bond configurations, is responsible for the three most prominent resonances in spectrum *A*. After intravenous infusion of 1-^{13}C glucose and induction of hypoxia, resonances from the alpha and beta anomers of the infused glucose appeared (*B*). Subtraction of spectrum *A* from spectrum *B* revealed other new resonances (*C*) due to flow of ^{13}C into the methyl carbon of lactate (LAC) and carbons 2, 3, and 4 of glutamate and glutamine, labeled GLX to indicate that signals from the two compounds are not resolved from each other. The lower-case delta beneath the axis is commonly used to mean "chemical shift." The rabbit was maintained as described in the legend of Figure 1 and studied in the same spectrometer, using a 2.3 cm surface coil and proton decoupling; data collection time for spectra *A* and *B* was 20 min.

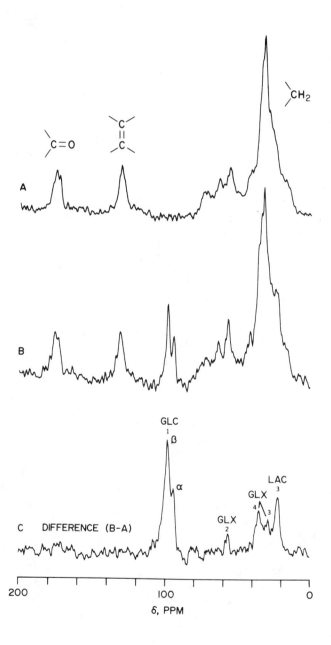

will be more effective, for reasons given below in the section on editing techniques.

HYDROGEN (^{1}H) STUDIES

The NMR signal from ^{1}H is inherently much stronger than that from any other nucleus, and nearly all concentrated metabolites contain ^{1}H nuclei which in principle could be used to identify them in ^{1}H spectra. Until recently, two problems prevented this from being done in living systems. First, the signal from ^{1}H nuclei in water is so strong that the weaker signals from most other compounds cannot be selectively detected in its presence. In studies of non-living systems by ^{1}H spectroscopy, the problem is solved by dehydrating the sample and resuspending it a ^{1}H-free solvent to achieve molecular mobility. Since 1983, techniques for eliminating the water resonance have been successfully applied to living systems in our laboratory, with the results discussed below. Second, the chemical shift range of ^{1}H is narrow, so that resonances from many different compounds overlap each other; this prevents useful measurement of most of them. We and our colleagues have developed procedures for editing spectra at the time of acquisition so that several metabolites can be selectively and quantitatively detected; these procedures are discussed in the next section.

Figure 3A is an unmodified ^{1}H spectrum of living rat brain, entirely dominated by the resonance from water protons. (This signal is the basis of NMR imaging; variations in its intensity and rates of relaxation in different parts of the body provide a high degree of contrast.) Figure 3B shows the result of applying the simplest technique for eliminating the water resonance. Selective irradiation of water protons at their resonant frequency temporarily destroyed their orderly alignment with the magnetic field, which is a necessary condition for generating an NMR signal. Immediately after this "saturating" radiation, before most of the water protons had become realigned with the magnetic field, spectrum B was acquired. The intensity of the water resonance is reduced several hundred times compared to A: consequently, resonances from other metabolites are visible on its upfield (viewer's right) side. That region of the same spectrum was further processed to straighten its baseline and is displayed at greater amplification in C. Virtually all of the resonances that can be seen in Figure 3C are from specific concentrated metabolites or groups of metabolites having ^{1}H nuclei in similar chemical configurations; signal-to-noise ratios in ^{1}H spectra are so much greater than those in ^{31}P and ^{13}C spectra that the baseline noise evident in Figures 1 and 2 is nearly invisible. Resonances in Figure 3C that have been assigned are labeled.

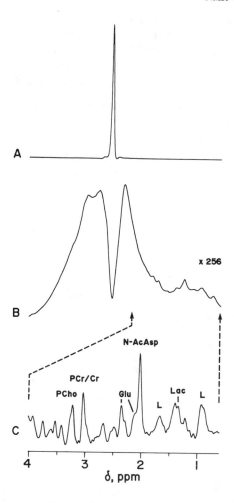

Figure 3   Proton spectra of rat brain. The large resonance from water protons (*A*) was greatly reduced by presaturation (*B*), and further processing revealed several much smaller resonances from other compounds on the upfield (*right*) side of it (*C*). The chemical shift scale pertains only to spectrum *C*. Resonances have so far been assigned to a pool of choline-containing compounds (PCho), PCr and creatine together (PCr/Cr), glutamate (Glu), *N*-acetylaspartate (*N*-AcAsp), lactate (Lac), and unidentified lipids (L). However, nearly all of the peaks, including the unlabeled ones, are metabolite resonances; noise variations are comparable in amplitude to the tic marks on the chemical shift axis and are much smaller in proportion to the metabolite signals than in the ^{31}P and ^{13}C spectra of Figures 1 and 2. These spectra were accumulated in 2.5 min and, like all in vivo spectra in the following Figures, were obtained from paralyzed, pump-ventilated rats maintained under nitrous oxide analgesia and studied with surface coils in a Bruker WH360 spectrometer.

Assignment of resonances to particular compounds or chemical groups depends on procedures that are illustrated in Figure 4. Spectrum A is from a dehydrated perchloric acid extract of freeze-clamped rat brain resuspended in D_2O; sharp resonance lines from numerous water-soluble metabolites are present. These can be identified in such extracts in three principal ways:

1. The extract spectrum is compared to spectra from a known pure compound, PCr for instance, in the same solvent. If resonances from the known compound have the same relative intensities and chemical shifts from some common reference point as resonances found in the extract spectrum, there is a high liklihood that the compound is present in the extract. In a complex spectrum, however, this test may not be conclusive, and additional ones are necessary.
2. A pure compound is added to the extract. Precise augmentation by this procedure of resonances already present in the extract spectrum is strong evidence that the compound was present in the original material.
3. Finally, selective removal of particular sets of resonances from the extract spectrum by exposure of the original material to a specific enzyme provides evidence which is independent of and complementary to that from procedure 2. The assignments in Figure 4A are based on a large body of previous work of this kind. Many of the less intense resonances in the spectrum have not yet been assigned.

Figure 4B is a spectrum from excised but not extracted rat brain tissue in D_2O; the water proton resonance was suppressed by presaturation. The four most prominent resonances correspond to the ones assigned to PCr/Cr, NAA, and lactate in spectrum A. They are broader because the metabolites are less mobile in the excised tissue than in the extract. The lactate resonance is more intense because the excised tissue was not frozen quickly in situ to minimize agonal change. Lipid resonances are present because lipids were not extracted.

The spectrum of Figure 4C is from the brain of a living, hypoxic rat. The major resonance lines are again identifiable by comparison with spectra A and B, though all resonances are much broader due to inhomogeneities of the magnetic field caused by the animal's body. In every respect except the hypoxia, the spectra in Figures 4C and 3C were obtained under the same experimental conditions; the most prominent difference between them is the greater intensity of the lactate resonance in Figure 4C, attributable to hypoxia. This kind of manipulation of a resonance by a well-understood metabolic stress is the final step in validation of a spectroscopic technique for in vivo use. Studies from our laboratory (Behar et al 1983, 1984) have shown that reversible elevations of the lactate resonance by hypoxia are detectable at both 8.4 T—the magnetic field strength used in the studies of

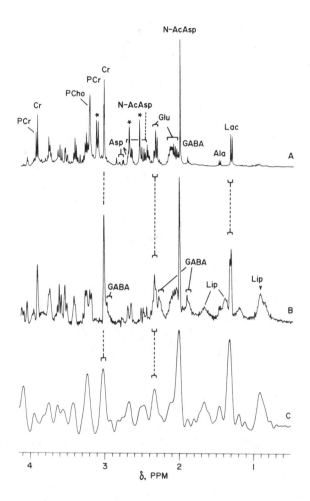

Figure 4 Proton spectra of perchloric acid extract of rat brain resuspended in D$_2$O (*A*), excised but not extracted rat brain (*B*), and rat brain in vivo (*C*). Assigned resonances are from the same compounds labeled in Figure 3, as well as creatine (Cr), aspartate (Asp), gamma-aminobutyric acid (GABA), alanine (Ala), unidentified lipids (Lip), and extraction reagents (*). Because of the narrow line widths characteristic of extract spectra, the complex line structure caused by signals from ^{1}H in different parts of the same molecule can be appreciated. See text for explanation of the differences among the three spectra. Animal preparation for spectrum *C* was as described for Figure 3, and all spectra were made in the Bruker WH360 spectrometer. Reprinted with permission from *Proc. Natl. Acad. Sci. USA* 80:4945.

Figures 3 and 4—and 1.9 T, which is available in instruments physically large enough for human work.

Bottomley et al (1985) showed that ^{1}H spectra could be obtained from the living human brain in an instrument operating at 1.5 T when the water proton resonance was removed by presaturation, as in Figure 3. Their most important result was the demonstration that the anatomical source of the spectrum can be controlled by a pulsed magnetic field gradient and surface coil detection method. Resonances from NAA and pools of compounds containing choline and creatine are clearly evident in their published spectrum. Lactate would surely be detectable in such spectra if it were elevated, even with the reduced spectral resolution of the 1.5 T magnetic field. However, selective detection of amino acids other than NAA and accurate quantitation of anything at the field strengths available for human work will require the editing techniques described in the next section.

Combined ^{1}H and ^{31}P observation in vivo is possible with a surface coil double-tuned for use in both frequency ranges. Figure 5 depicts the course of such a combined study of hypoglycemic encephalopathy in the rat (Behar et al 1985). The ^{31}P spectra (E–H) show virtually complete loss of phosphate energy stores and a corresponding increase in Pi caused by insulin, followed by recovery after glucose infusion. Proton spectra (A–D) documented the fall of glutamate and rise of aspartate, which are known to occur during profound hypoglycemia as glutamate metabolized via the aspartate amino transferase reaction becomes a principal source of carbon for the tricarboxylic acid cycle (Siesjo & Agardh 1983).

EDITING TECHNIQUES

Any measurement method dependent on detection of specific compounds will be confounded by samples containing many compounds which yield signals that are close together. NMR spectroscopy is no exception. The problem is especially great in vivo, since the usual biochemical strategy of simplifying the signal source by physical or chemical fractionation of the sample is not available. Several remarkably successful methods for simplifying in vivo spectra have been developed quite recently.

Homonuclear Decoupling

The magnetic interaction, or coupling, between nearby nuclei can be used to distinguish the signal of a single compound from many resonances in the same spectral region. The principle is subtraction of an unmodified spectrum from one acquired under conditions that decouple some particular interaction between neighbouring nuclei. The only signals in the

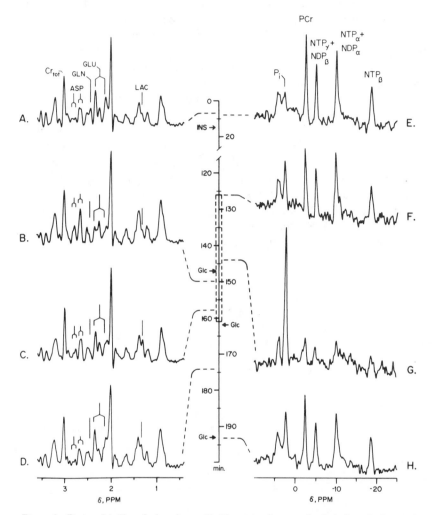

Figure 5 Proton (*A–D*) and phosphorus (*E–H*) spectra from rat brain before, during, and after profound insulin-induced hypoglycemia. The temporal relation of spectra is shown on the time line in the center; electroencephalographic activity was absent during the period indicated by *dotted lines* flanking the time line; glucose was given at times indicated by *arrows* labeled "Glc." Labeling of resonances is as in the preceding Figures, except for glutamine (GLN) and the composite resonance from PCr and creatine (Cr$_{tot}$), and designation of nucleotide di- and triphosphate resonances as "NDP" and "NTP" to indicate some contribution from bases other than adenine. See text for description of the metabolic changes. Animal preparation and spectrometer were the same as for Figure 3. Reprinted with permission from *J. Neurochem.* 44:1045.

resulting difference spectrum will be from the compound in which the decoupled interaction is normally present. The first case we consider is that of homonuclear decoupling where the interacting nuclei are both protons. Figure 6 illustrates the mechanism and result of this procedure, as applied to selective detection of glutamate in the ^{1}H spectrum of a rat brain in vivo. Two spectra were built up from sets of scans taken with and without irradiation of the protons bonded to the beta carbon of glutamate; the effect of the irradiation was to induce a transient change in the magnetization state of the protons bonded to the alpha and gamma carbons by uncoupling them from the beta-carbon protons. The difference spectrum (A) contains resonances from the alpha- and gamma-carbon protons, whereas no glutamate signal could be detected in either the unmodified spectrum (B) or the irradiated one (not shown). So far, resonances from alanine, taurine, gamma-aminobutyric acid and lactate as well as glutamate have been isolated this way (Rothman et al 1984). Quantitative studies of these compounds in vivo are therefore possible.

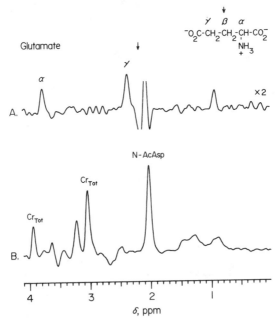

Figure 6   Homonuclear editing of rat brain proton spectrum to yield resonances from protons bonded to the alpha and gamma carbons of glutamate (A). These resonances are not detectable in the raw ^{1}H spectrum (B). Single-frequency irradiation of the beta-carbon protons (*inset*) was given at the frequency indicated by the *arrow* above spectrum A; see text for explanation. Animal preparation and spectrometer were the same as for Figure 3. Reprinted with permission from *Proc. Natl. Acad. Sci. USA* 82:1633.

Heteronuclear Decoupling

Compounds in which magnetic coupling exists between nearby nuclei of different species can be selectively detected by an editing procedure analogous to selective homonuclear decoupling. The heteronuclear procedure enables an entirely different kind of measurement in which the observed nucleus is ^{1}H and the irradiated one is ^{13}C. With proton observe–carbon decouple editing, the sensitivity of the ^{1}H resonance is retained while the ^{13}C label is followed. This makes possible a far more powerful form of ^{13}C enrichment experiment than the one illustrated in Figure 2. The principle is presented schematically in Figure 7. Three populations of lactate molecules in a magnetic field are represented by the three formulas: In one (*center*), the methyl carbon is ^{12}C, and exerts no magnetic influence on the protons bonded to it; in the other two populations (*left* and *right*), the methyl carbons are ^{13}C, half of them aligned spin up and half spin down with respect to the magnetic field. If 50% of total lactate methyl carbons were ^{13}C, the portion of the ^{1}H spectrum where the signals from lactate methyl protons resonate would look like the mock spectrum labeled "coupled." A central resonance from protons bonded to ^{12}C is flanked by two resonances that are shifted away from the central one by the magnetic effect of the up and down spins of ^{13}C on the protons bonded to them. If the sample were irradiated at the resonant frequency of ^{13}C, the side resonances would collapse into the central one as shown in the "^{13}C-decoupled" mock spectrum because the effect of the magnetic carbons on their attached protons would be removed. The areas of the single resonance in the

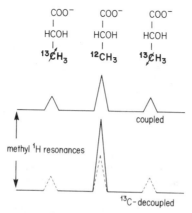

Figure 7 Schematic illustration of ^{1}H observe-^{13}C decouple method of measuring fractional labeling of lactate methyl carbon with ^{13}C. See text for explanation.

decoupled spectrum and the central resonance in the coupled one would be measures, respectively, of total lactate in the sample and the proportion of it with ^{12}C in the methyl position; subtraction would give the fraction of the lactate pool labeled at that position with ^{13}C. In principle, the method can be used in living systems to follow the flow of ^{13}C through any detectable metabolite pool that can be enriched with it. Direct observation of ^{13}C as in Figure 2 can detect only the labeled fraction of a compound. Indirect measurement of ^{13}C labeling by the method of Figure 7 no: only provides full labeling information about the compound, it does so by detecting the much stronger ^1H signal and hence is much more sensitive.

The first use of the proton observe–carbon decouple method on a living system was in a study of yeast metabolism (Sillerud et al 1981). It has recently been shown to be effective for measurement of fractional labeling of lactate and glutamate in living rat brain (Rothman et al 1985). Figure 8, taken from that study, shows the equivalent of a decoupled spectrum (C), the difference spectrum (B) obtained by subtracting the coupled spectrum from the decoupled one, and the time course of ^{13}C labeling of glutamate at the C4 position (A).

Selective Excitation

A method much more effective than presaturation for eliminating the water proton resonance from ^1H spectra was developed by spectroscopists working with test tube samples (Hore 1983), and it has recently been adapted for observations with surface coils on whole animals (Hetherington et al 1985). It works by minimizing excitation of the water

---

Figure 8   Glutamate labeling with ^{13}C in rat brain. *A*. The amplitude (arbitrary units) of the proton resonance of $[4\text{-}^{13}CH_2]$ glutamate is plotted as a function of time after the start of a $[1\text{-}^{13}C]$ glucose intravenous infusion, which began at time 0 and continued throughout the experiment. ^1H observe–^{13}C decouple spectra were acquired with single frequency ^{13}C decoupling centered at the ^{13}C resonance of $[4\text{-}^{13}CH_2]$ glutamate. Each data point represents the sum of two ^{13}C difference spectra accumulated over a 162 sec time interval (*horizontal bar*). *Vertical bars* represent the amplitude uncertainty due to root mean square noise in the spectrum. The total concentrations of glutamate determined from the $[4\text{-}^{12+13}CH_2]$ glutamate resonance amplitude in the $(^{12}C + ^{13}C)$ subspectrum remained constant during the infusion.

B. ^1H observe–^{13}C decouple difference spectrum of $[4\text{-}^{13}CH_2]$ glutamate. The spectrum is the sum of two difference spectra centered at 22 min after the start of the $[1\text{-}^{13}C]$ glucose infusion. With the assumption that the concentration of the creatine + PCr (Cr$_{tot}$) represented by the resonance at 3.03 ppm in spectrum *C* was 10.5 μM/gm wet weight, the concentration of $[4\text{-}^{13}CH^2]$ glutamate was calculated to be about 1.5 mM.

C. The sum of the two $(^{12}C + ^{13}C)$ subspectra that are components of the two ^{13}C difference spectra whose sum is spectrum *B*. The resonance of $[4\text{-}^{12+13}CH_2]$ glutamate is at 2.35 ppm (*dashed line*). Reprinted with permission from *Proc. Natl. Acad. Sci. USA* 82: 1633.

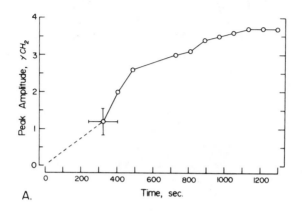

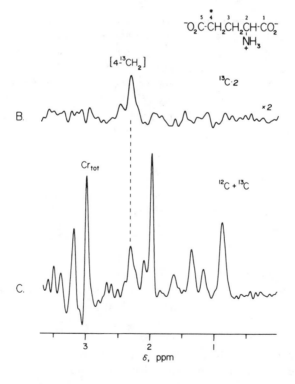

protons rather than reducing the response of the water proton signal to excitation, as presaturation does. A pulse sequence called "1331" (from the relative durations of some of its components) can be delivered in such a way as to provide nearly zero excitation at the frequency of the water protons, while excitation maxima occur on either side of it at frequencies that can be optimized for specific resonances of interest. The technique can be combined with other editing methods.

Hetherington et al (1985) also showed that another powerful and versatile selective excitation technique can be used for surface coil measurements on intact animals. Its developers (Morris & Freeman 1978) named it the DANTE (Delays Alternating with Nutations for Tailored Excitation) sequence after their quite soberly described recognition of the resemblance between the magnetization trajectory it induces and the route of Dante Aligheri's tour of Purgatory in the company of Virgil. The method works by substituting a series of short excitation pulses for one long pulse, so that excitation occurs a little at a time rather than all at once. The resonant frequency of a nucleus is the frequency at which it precesses around the axis of the magnetic field. Continued precession in the intervals between the short pulses causes loss of phase coherence among the excited nuclei, the dephasing being greater the farther the resonant frequency of a nuclear species is from the transmitter frequency. Because of this, only nuclei precessing at a rate very near the transmitter frequency experience a cumulative effect of successive pulses in the series. The result is excitation highly selective for a narrow portion of the spectrum. The technique has a variety of uses and is compatible with other editing methods.

PROSPECTS

Selective NMR spectroscopy of the living brain began in 1980 with the introduction of surface coils. This review was completed in January of 1985. Most of the studies mentioned in it belong to an early development and validation period in the history of an observational method with a great deal of potential for generating major advances in the study of normal and deranged brain metabolism. As a whole, the work reviewed here shows that in vivo NMR measurements agree well with measurements made by other techniques; major unexpected artifacts or distortions have not appeared. The capability of in vivo NMR to get information unobtainable any other way can therefore be exploited with cautious confidence.

Spectrometers operating in the range of 2–5 T are becoming available for in vivo work on human and animal brain. At such field strengths it is possible to measure ATP, PCr, Pi, and pHi in the ^{31}P spectrum, which may also yield important information about some phosphomono- and phosphodiesters. In the ^1H spectrum, NAA and elevated lactate can be measured in humans and these plus glutamate, glutamine, taurine, alanine, and gamma-aminobutyric acid in animals. Prospects are excellent for extending these capabilities through editing techniques. The proton observe–carbon decouple technique for measuring the flow of ^{13}C through pools of lactate, glutamate, and some other metabolites can provide abundant new data about the rates at which these turn over under normal and pathological conditions. From such data glycolytic and respiratory rates can be estimated in some cases. Most of these measurements require only a few minutes of acquisition time, and there is no known barrier to repeating them as often as desired in the same subject.

The greatest problem presently facing users of in vivo NMR methods is localization of the anatomical source of the signals. With a single exception (Bottomley et al 1985), all of the studies reviewed here relied entirely on coil placement to achieve anatomical selectivity. Although much useful information can be obtained that way, vigorous current research is likely to provide improved localization methods in the near future. An important biological fact favoring such progress is the nearly exclusive localization of NAA to brain, where it is more concentrated in gray than white matter. Since this compound is the source of the most intense resonance in the ^1H spectrum, it can be used to evaluate the success of localization procedures.

ACKNOWLEDGMENTS

Studies from our own laboratory were supported by grants from the United States Public Health Service (AM 27121 and GM 30287), the National Science Foundation (PCM 8402670), and the Esther A. and Joseph Klingenstein Fund.

Literature Cited

Ackerman, J. J. H., Berkowitz, B. A., Deuel, R. K. 1984a. Phosphorus-31 NMR of rat brain in vivo with bloodless perfluorocarbon perfused rat. *Biochem. Biophys. Res. Commun.* 119:913–19

Ackerman, J. J. H., Evelhoch, J. L., Berkowitz, B. A., Kichura, G. M., Deuel, R. K., Lown, K. S. 1984b. Selective suppression of the cranial bone resonance from 31P NMR experiments with rat brain in vivo. *J. Magn. Reson.* 56:318–22

Ackerman, J. J. H., Grove, T. H., Wong, G. G., Gadian, D. G., Radda, G. K. 1980. Mapping of metabolites in whole animals by 31P NMR using surface coils. *Nature* 283:167–70

Alger, J. R., Shulman, R. G. 1984. NMR studies of enzymatic rates in vitro and in vivo by magnetization transfer. *Q. Rev. Biophys.* 17:83–124

Andrew, E. R. 1984. A historical review of NMR and its clinical applications. *Brit. Med. Bull.* 40:115–19

Aue, W. P., Muller, S., Cross, T. A., Seelig, J. 1984. Volume-selective excitation. A novel approach to topical NMR. *J. Magn. Reson.* 56:350–54

Balaban, R. S., Kantor, H. L., Ferretti, J. A. 1983. In vivo flux between phosphocreatine and adenosine triphosphate determined by two-dimensional phosphorus NMR. *J. Biol. Chem.* 258:12787–89

Behar, K. L., den Hollander, J. A., Petroff, O. A. C., Hetherington, H., Prichard, J. W., Shulman, R. G. 1985. The effect of hypoglycemic encephalopathy upon amino acids, high energy phosphates, and pHi in the rat brain in vivo: Detection by sequential 1H and 31P NMR spectroscopy. *J. Neurochem.* 44(4):1045–55

Behar, K. L., den Hollander, J. A., Stromski, M. E., Ogino, T., Shulman, R. G., Petroff, O. A. C., Prichard, J. W. 1983. High-resolution 1H nuclear magnetic resonance study of cerebral hypoxia in vivo. *Proc. Natl. Acad. Sci. USA* 80:4945–48

Behar, K. L., Rothman, D. L., Shulman, R. G., Petroff, O. A. C., Prichard, J. W. 1984. Detection of cerebral lactate in vivo during hypoxemia by 1H NMR at relatively low field strengths (1.9 Tesla). *Proc. Natl. Acad. Sci. USA* 81:2517–19

Bendall, M. R., Gordon, R. E. 1983. Depth and refocusing pulses designed for multipulse NMR with surface coils. *J. Magn. Reson.* 53:365–85

Bloch, F., Hansen, W. W., Packard, M. E. 1946. Nuclear induction. *Phys. Rev.* 69:127

Bottomley, P. A., Edelstein, W. A., Foster, T. H., Adams, W. A. 1985. In vivo solvent suppressed localized hydrogen nuclear magnetic resonance (NMR): A new window to metabolism? *Proc. Natl. Acad. Sci. USA* 82:2148–52

Bottomley, P. A., Hart, H. R., Edelstein, W. A., Schenk, J. F., Smith, L. S., Leue, W. M., Mueller, O. M., Reddington, R. W. 1984. Anatomy and metabolism of the normal human brain studied by magnetic resonance at 1.5 Tesla. *Radiology* 150:441–46

Bottomley, P. A., Kogure, K., Namon, R., Alonso, O. F. 1982. Cerebral energy metabolism in rats studied by phosphorus nuclear magnetic resonance using surface coils. *Magnet. Reson. Imag.* 1:81–85

Brooks, D. J., Lammertsma, A. A., Beaney, R. P., Leenders, K. L., Buckingham, P. D., Marshall, J., Jones, T. 1984. Measurement of regional cerebral pH in human subjects using continuous inhalation of 11CO2 and positron emission tomography. *J. Cerebral Blood Flow Metab.* 4:458–65

Budinger, T. F., Cullander, C. 1983. Biophysical phenomena and health hazards of in vivo magnetic resonance. In *Clinical Magnetic Resonance Imaging*, ed. A. R. Margulis, C. B. Higgins, L. Kaufman, L. B. Crooks, pp. 303–20. San Francisco: Radiol. Res. Educ. Found.

Bydder, G. M. 1984. Nuclear magnetic resonance imaging of the brain. *Br. Med. Bull.* 40:170–74

Cady, E. B., Costello, A. M., Dawson, M. J., Delpy, D. T., Hope, P. L., Reynolds, E. O. R., Tofts, P. S., Wilkie, D. R. 1983. Noninvasive investigation of cerebral metabolism in newborn infants by phosphorus nuclear magnetic resonance spectroscopy. *Lancet* 2:1059–62

Chance, B., Nakase, Y., Bond, M., Leigh, J. S., McDonald, G. 1978. Detection of 31P nuclear magnetic resonance signals in brain by in vivo and freeze-trapped assays. *Proc. Natl. Acad. Sci. USA* 75:4925–29

Decorps, M., Lebas, J. L., Leviel, J. L., Confort, S., Remy, C., Benabid, A. L. 1984. Analysis of brain metabolism changes induced by acute potassium cyanide intoxication by 31P NMR in vivo using chronically implanted surface coils. *FEBS Lett.* 168:1–6

Delpy, D. T., Gordon, R. E., Hope, P. L., Parker, D., Reynolds, E. O. R., Shaw, D., Whitehead, M. D. 1982. Noninvasive investigation of cerebral ischemia by phosphorus nuclear magnetic resonance. *Pediatrics* 70:310–13

Gadian, D. G. 1982. *Nuclear Magnetic Resonance and Its Applications to Living Systems.* Oxford: Clarendon

Glonek, T., Kopp, S. J., Kot, E., Pettegrew, J. W., Harrison, W. H., Cohen, M. M. 1982. P-31 nuclear magnetic resonance analysis of brain: The perchloric acid extract spectrum. *J. Neurochem.* 39:1210–19

Hetherington, H. P., Avison, M. J., Shulman, R. G. 1985. 1H homonuclear editing of rat brain using semi-selective pulses. *Proc. Natl. Acad. Sci. USA* 82:3115–18

Hilberman, M., Subramanian, V. H., Haselgrove, J., Cone, J. B., Egan, J. W., Gyulai, L., Chance, B. 1984. In vivo time-resolved brain phosphorus nuclear magnetic resonance. *J. Cerebral Blood Flow Metab.* 4:334–42

Hope, P. L., Cady, E. B., Tofts, P. S., Hamilton, P. A., Costello, A. M., Delpy, D. T., Chu, A., Reynolds, E. O. R. 1984. Cerebral energy metabolism studied with phosphorus NMR spectroscopy in normal and birth-asphyxiated infants. *Lancet* 2:366–70

Hore, P. J. 1983. Solvent suppression in fourier transform nuclear magnetic resonance. *J. Magn. Reson.* 55:283–300

Matthews, P. M., Bland, J. L., Gadian, D. G., Radda, G. K. 1982. A 31P-NMR saturation transfer study of the regulation of creatine kinase in the rat heart. *Biochim. Biophys. Acta* 721:312–20

Maudsley, A. A., Hilal, S. K. 1984. Biological aspects of sodium-23 imaging. *Br. Med. Bull.* 40:165–66

Maudsley, A. A., Hilal, S. K., Simon, H. E., Wittekoek, S. 1984. In vivo MR spectroscopic imaging with P-31. *Radiology* 153:745–50

Moore, W. S. 1984. Basic physics and relaxation mechanisms. *Br. Med. Bull.* 40:120–24

Morris, G. A., Freeman, R. 1978. Selective excitation in Fourier transform nuclear magnetic resonance. *J. Magn. Reson.* 29:433–62

Naruse, S., Horikawa, Y., Tanaka, C., Hirakawa, K., Nishikawa, H., Watari, H. 1984. In vivo measurement of energy metabolism and the concomitant monitoring of encephalogram in experimental cerebral ischemia. *Brain Res.* 206:370–72

Naruse, S., Takada, S., Koizuka, I., Watari, H. 1983. In vivo 31P NMR studies on experimental cerebral infarction. *Jpn. J. Physiol.* 33:19–28

NRPB 1983. Revised guidance on acceptable limits of exposure during nuclear magnetic resonance clinical imaging. (Statement by National Radiological Protection Board). *Br. J. Radiol.* 56:974–77

Petroff, O. A. C., Prichard, J. W. 1983. Cerebral pH by NMR. *Lancet* 2:105–6

Petroff, O. A. C., Prichard, J. W., Behar, K. L., Alger, J. R., Shulman, R. G. 1984. In vivo phosphorus nuclear magnetic resonance spectroscopy in status epilepticus. *Ann. Neurol.* 16:169–77

Petroff, O. A. C., Prichard, J. W., Behar, K. L., Alger, J. R., den Hollander, J. A., Shulman, R. G. 1985. Cerebral intracellular pH by 31P nuclear magnetic resonance spectroscopy. *Neurology* 35:781–88

Prichard, J. W., Alger, J. R., Behar, K. L., Petroff, O. A. C., Shulman, R. G. 1983. Cerebral metabolic studies in vivo by 31P NMR. *Proc. Natl. Acad. Sci. USA* 80:2748–51

Purtell, E. M., Torrey, H. C., Pound, R. V. 1946. Resonance absorption by nuclear magnetic moments in a solid. *Phys. Rev.* 69:37–38

Radda, G. K., Bore, P. J., Rajagopalan, B. 1984. Clinical aspects of 31P NMR spectroscopy. *Br. Med. Bull.* 40:155–59

Rothman, D. L., Behar, K. L., Hetherington, H. P., Shulman, R. G. 1984. Homonuclear 1H double resonance difference spectroscopy of the rat brain in vivo. *Proc. Natl. Acad. Sci. USA* 81:6330–34

Rothman, D. L., Behar, K. L., Hetherington, H. P., den Hollander, J. A., Bendall, M. R., Petroff, O. A. C., Shulman, R. G. 1985. 1H observed 13C decoupled spectroscopic measurements of lactate and glutamate in the rat brain in vivo. *Proc. Natl. Acad. Sci. USA* 82:1633–37

Rottenberg, D. A., Ginos, J. Z., Kearfort, K. J., Junck, L., Bigner, D. D. 1984. In vivo measurement of regional brain tissue pH using positron emission tomography. *Ann. Neurol.* 15:S98–S102 (Suppl.)

Saunders, R. D., Smith, H. 1984. Safety aspects of NMR clinical imaging. *Br. Med. Bull.* 40:148–54

Shoubridge, E. A., Briggs, R. W., Radda, G. K. 1982. 31P NMR saturation transfer measurements of the steady state rates of creatine kinase and ATP synthetase in the rat brain. *FEBS Lett.* 140:288–92

Siesjo, B. K., Agardh, C. D. 1983. Hypoglycemia. In *Handbook of Neurochemistry*, ed. A. Lajtha, pp. 353–79. New York: Plenum. 2nd ed.

Sillerud, L. O., Alger, J. R., Shulman, R. G. 1981. High-resolution proton NMR studies of intracellular metabolites in yeast using 13C decoupling. *J. Magn. Reson.* 45:142–50

Sillerud, L. O., Shulman, R. G. 1983. High-resolution 13C nuclear magnetic resonance studies of glucose metabolism in Escherichia coli. *Biochemistry* 22:1087–94

Syrota, A., Castaing, M., Rougement, D., Berridge, M., Baron, J. C., Bousser, M. G., Pocidalo, J. J. 1983. Tissue acid-base balance and oxygen metabolism in human cerebral infarction studied with positron emission tomography. *Ann. Neurol.* 14:419–28

Thulborn, K. R., duBoulay, G. H., Duchen, L. W., Radda, G. 1982. A 31P nuclear magnetic resonance in vivo study of cerebral ischemia in the gerbil. *J. Cerebral Blood Flow Metab.* 2:299–306

Ugurbil, K., Maidan, R. R., Petein, M., Michurski, S. P., Cohn, J. N., From, A. H. L. 1984. NMR measurements of myocardial CK rates by multiple saturation transfer. *Circulation* 70:(Suppl. II-84)

Veech, R. L., Lawson, J. W. R., Cornell, N. W., Krebs, H. A. 1979. Cytosolic phosphorylation potential. *J. Biol. Chem.* 254:6538–47

Younkin, D. P., Delivoria-Papadopoulos, M., Leonard, J. C., Subramanian, V. H., Eleff, S., Leigh, J. S., Chance, B. 1984. Unique aspects of human newborn cerebral metabolism evaluated with phosphorus nuclear magnetic resonance spectroscopy. *Ann. Neurol.* 6:581–86

Ann. Rev. Neurosci. 1986. 9:87–119

CYCLIC GMP CASCADE
OF VISION

Lubert Stryer

Department of Cell Biology, Sherman Fairchild Center,
Stanford University School of Medicine, Stanford, California 94305

Introduction

Retinal rods cells are intriguing and challenging objects of inquiry. They
display three striking properties. First, rods have attained the ultimate in
sensitivity. A rod cell can be triggered by a single photon. Second, they can
detect incremental stimuli over a very wide span of background light levels.
Rods adapt over more than a 10^5-fold range of background illumination
and preserve amplitude information. Third, rods are very reliable detectors.
The amplitude and timing of their output signals are highly reproducible. A
most attractive feature of rods is that much of their transduction machinery
is packaged in a discrete portion of the cell, the outer segment, which can
readily be detached and purified. The rod outer segment is a marvel of
eukaryotic simplicity. It has only one role, visual transduction, and so the
molecules mediating excitation and adaptation are present in a highly
concentrated form, separate from the energy-generating and protein-
synthesizing machinery of the inner segment. This gift of nature is being
explored by a broad spectrum of scientists, from spectroscopists to
biochemists to electrophysiologists. The pieces of the puzzle are just now
coming together. It is a time of ferment and intellectual excitement. In this
article, I review a key facet of visual excitation in vertebrate photoreceptors,
the cyclic GMP cascade. Several excellent reviews and monographs present
other aspects and perspectives of this rich field (Miller 1981, Fein & Szuts
1982, O'Brien 1982, Lamb 1984, Chabre 1985, Schwartz 1985, Dahlem
Conference 1985).

The structural design of rod outer segments is that of a dual membrane
system. A stack of some 2000 disks is surrounded by a plasma membrane
(Figure 1). The disks are flattened membranous sacs that arise by
invagination of the plasma membrane at its base, near the cilium

87

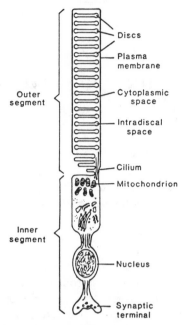

Figure 1   Schematic diagram of a retinal rod cell. Typical dimensions of the outer segment are 60 μm length × 6 μm diameter in frogs and 60 μm length × 1 μm diameter in mammals. The centers of disks are 30 nm apart. The distance between the cytosolic face of one disk and that of the adjacent disk is about 15 nm. (From O'Brien 1982.)

connecting the outer and inner segments of the cell. The disk membranes and the plasma membrane are physically separate and have distinct roles. The disks absorb light and convert photons into amplified changes in the concentration of transmitter molecules. The plasma membrane in turn converts these chemical changes into electrical signals. Excitation is triggered by the photoisomerization of the retinal chromophore of rhodopsin. This conformational transition in a single photoreceptor protein molecule leads to the transient closure of hundreds of sodium channels in the plasma membrane and a block of the entry of more than 10^6 Na^+ ions. The resulting hyperpolarization is conveyed to the synapse at the other end of the rod and communicated to other cells of the retina.

Cyclic GMP, Not Calcium Ion, Is Likely to Be the Transmitter

What is the nature of the transmitter that carries the excitation signal from the disks to the plasma membrane? Two candidates have elicited much interest: calcium ion and, more recently, cyclic GMP (cGMP). It was proposed a decade ago that calcium ions taken up by disks in the dark are released into the cytosol on illumination (Yoshikami & Hagins 1973). The released calcium ions would then diffuse to the plasma membrane and directly close many sodium channels. This model was supported by the finding that the introduction of calcium into rod outer segments (ROS) leads to the closure of sodium channels (Hagins 1979). However, it has recently been shown that calcium closes channels more slowly than does light (Yau & Nakatani 1985). Furthermore, ROS depleted of calcium respond to light, as do ROS treated to contain a high level of buffer for calcium in their cytosol (Matthews et al 1985, Cote et al 1985). Hence, changes in cytosolic calcium concentration are not required for visual excitation. The most direct evidence comes from recent electrophysiological studies of excised patches of plasma membrane of ROS (Fesenko et al 1985, Nakatani & Yau 1985). The striking result is that sodium channels are opened by the addition of cGMP to the cytosolic side of this excised patch of plasma membrane, whereas calcium ion has no effect. These experiments strongly suggest that calcium ion is not the excitatory transmitter in vision. They simultaneously show that cGMP is central to visual excitation.

A transmitter role for cGMP seemed likely even before these incisive patch-clamp experiments because of the convergence of several lines of biochemical and electrophysiological research (reviewed in Miller 1981). Interest in cGMP was stimulated by the finding that the level of cyclic nucleotides in ROS is regulated by light (Bitensky et al 1971). It was then found that ROS contain an unusually high concentration (~ 70 μM) of cGMP, and that light leads to a rapid, amplified decrease in cGMP level by activating a cGMP phosphodiesterase (Goridis et al 1974, Woodruff et al 1977, Bitensky et al 1981). Interest in this cascade was heightened by the observation that a single photoexcited rhodopsin can lead to the hydrolysis of more than 10^5 molecules of cGMP in ROS suspensions (Yee & Liebman 1978). At the same time, electrophysiological studies revealed that cGMP depolarizes the plasma membrane of ROS within milliseconds of being injected intracellularly (Miller & Nicol 1979). Moreover, cGMP was found to increase the latency of the light-induced hyperpolarization. These experiments suggested that cGMP keeps sodium channels open in the dark and that light closes them by lowering the concentration of cGMP (Figure 2). The link between photoexcited rhodopsin and activated phosphodiesterase was then found. The photoactivation of the phosphodiesterase is mediated by transducin, a protein that cycles between an inactive GDP state in the dark and an active GTP state on illumination (Fung et al 1981, Stryer et al 1981b). Deactivation involves the interplay of two inhibitory proteins, rhodopsin kinase and arrestin, with photoexcited rhodopsin (Kühn 1984, Zuckerman et al 1985). These excitatory and inhibitory

(a)

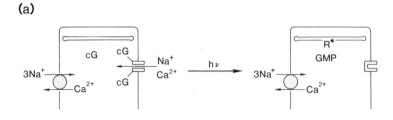

(b)

Figure 2 Cyclic GMP controls sodium channels in the plasma membrane. (a) In the dark, a high level of cGMP in the cytosol opens sodium channels in the plasma membrane. Na^+ and Ca^{2+} enter the outer segment through these channels. Ca^{2+} is extruded in exchange for Na^+ by a Na^+–Ca^{2+} exchanger. On illumination, photoexcited rhodopsin triggers a cascade that results in the hydrolysis of cGMP to GMP. The lowered level of cGMP closes sodium channels. Ca^{2+} continues to be extruded from the outer segment by the exchanger. (b) Flow of information in the cGMP cascade. R* is photoexcited rhodopsin, T-GTP is the activated form of transducin, and PDE* is the activated form of the phosphodiesterase.

proteins of the cGMP cascade account for about 90% of the total protein of the rod outer segment, consistent with the notion that they are at the heart of the process of visual excitation.

My review of this cascade begins with the sodium channels of the plasma membrane. I next consider the changes in cGMP level produced by light. Following this I discuss the excitatory and inhibitory proteins of the cascade and their interactions. I then turn to homologies between vision and hormone action. One of the most intriguing recent findings is that transducin and the G proteins of the adenylate cyclase belong to the same family of signal-coupling proteins. The light-triggered transducin cycle exemplifies a mode of amplification that recurs in nature. It also seems likely that the triggering events in rod and cone vision are similar and that the same fundamental plan is used in vertebrates, molluscs, and arthropods.

Cyclic GMP Directly Opens Sodium Channels

Most of what we know about the light-controlled conductance of ROS comes from electrophysiological studies. Measurements of extracellular voltage gradients (Hagins et al 1970) and of the voltage across the plasma membrane (Tomita 1970) first revealed that light leads to a graded hyperpolarization of rods resulting from a block in the entry of sodium ions into the outer segment. Since then, much has been learned from measurements of the membrane current of outer segments, which provide information not accessible through voltage recordings. In the suction electrode technique (Baylor et al 1979a,b), an outer segment projecting from a slice of retina is gently drawn into the tip of a glass micropipet. A snug fit between the electrode wall and outer segment membrane gives a high resistance seal that makes it possible to measure ion flows across the plasma membrane using a transducer connected to the suction electrode. In the dark, the membrane current is about 25 pA for a toad outer segment 60 μm in length and 6 μm in diameter. Most of this dark current is due to the flow of Na^+ into the outer segment. Thus, the 25 pA current means that about 1.5×10^8 Na^+ flow into an outer segment in the dark in one second. The membrane surface area is 1200 μm^2, and so the current density is 1.3×10^5 Na^+ per μm^2 per second. This sodium influx is suppressed by a bright light flash and is halved by a dim light flash resulting in the formation of 35 photoexcited rhodopsin molecules. Rods give a graded hyperpolarization rather than an all-or-none action potential.

Responses to very dim light flashes show that a single photoexcited rhodopsin molecule leads to a 1 pA decrease in the dark current, which arises from a block of the inflow of 6.2×10^6 Na^+ ions per second. This conductance decrease is localized to a region of plasma membrane within about 3 μm of the excited disk (Lamb et al 1981). Plots of the photocurrent (the blockage of dark current) versus time are rounded for dim light flashes. In the single-photon limit, the time between the light flash and the maximal closure of Na^+ channels is 1 to 2 sec. There is very little time or amplitude jitter in response to single photons. The low ratio of the standard deviation to the amplitude of the single-photon response indicates that the number of transmitter molecules contributing to the single-photon response is greater than 100.

Measurements of fluctuations in membrane current have provided insight into the transduction process and the nature of the sodium channel. Three types of fluctuations have been observed; they can be regarded as *rhodopsin noise, transmitter noise,* and *channel noise.* In complete darkness, spontaneous fluctuations having the same amplitude (1 pA) and time course as single-photon responses occur on average about once per 50 sec (Baylor et al 1980). These events probably arise from the thermal isomerization of the retinal chromophore of rhodopsin. Continuous noise with a smaller amplitude (~ 0.2 pA) and a power spectrum curve extending to about 1 Hz is also observed. This portion of the noise is likely to be due to fluctuations in the concentrations of activated species of the transduction chain following rhodopsin but preceding the sodium channels themselves (Baylor et al 1980). Recent patch-clamp studies (Detwiler et al 1982, Bodoia

& Detwiler 1984, Gray & Attwell 1985) have revealed a higher frequency noise component with a very small amplitude that probably comes from the random opening and closing of sodium channels in the dark. This noise indicates that a channel has a unit current of about 4 fA (1.2×10^4 Na^+ per sec) and a mean open time of 2 ms (Gray & Attwell 1985, Zimmerman & Baylor 1985). Thus, about 24 Na^+ ions traverse the channel while it is open. The effect of light is to reduce the rate of opening of these channels with no effect on their rate of closing. A 4 fA unit current means that a 1 pA single-photon response is due to the closure of some 250 channels.

A major advance in our understanding of the molecular basis of vision has come from recent patch-clamp studies of excised fragments of ROS plasma membrane (Fesenko et al 1985, Nakatani & Yau 1985). An advantage of this preparation is that the membrane patch is oriented so that its cytosolic side faces the external solution. Moreover, the excised piece of plasma membrane is free from the cell, which makes it feasible to assay for direct effects of applied substances. The exciting finding is that cGMP directly opens the sodium channels, whereas calcium ion has virtually no effect. ATP and GTP also have no effect and are not required for the opening of channels by cGMP, thus indicating that the action of cGMP is not mediated by a kinase. Also, 5'-GMP, cyclic AMP, and the 2', 3' isomer of cGMP do not alter the conductance. The opening of channels by cGMP is reversible: the channels open when cGMP is applied and stay open until it is flushed away. Opening is half-maximal at 30 μM cGMP and depends on cGMP concentration in a cooperative manner, with a Hill coefficient of 1.8 (Fesenko et al 1985) (Figure 3). This dependence of channel opening on cGMP concentration is unaffected by calcium over the concentration range of 10 nM to 1 mM. The current-voltage relationship and ionic selectivity of this cGMP-controlled channel are similar to that of the light-regulated channel in intact rods, indicating that the cGMP-controlled channel is likely to be identical to the light-controlled channel.

Noise analyses of membrane currents in excised patches have provided additional insight into the characteristics of this channel. The elementary event, most likely the opening of a single channel, has a magnitude of 3 fA (1.9×10^4 Na^+ ions per sec) at a membrane potential of -30 mV, in good agreement with the value obtained from noise analyses of intact cells. The power spectrum of the fluctuations in membrane current at 100 μM cGMP is well-described by a Lorentzian function with a cut-off frequency at about 300 Hz, which is equivalent to a time constant of 0.5 msec (Fesenko et al 1985). Opening and closing rates for the channel can be inferred from the power spectrum cut-off frequency if one assumes that the channel exists in two states, a closed one and an open one containing bound cGMP. A 0.5

msec time constant would arise from a channel opening rate of 1.8×10^3 s^{-1} and a closing rate of $0.2 \times 10^3 \, s^{-1}$, given that about 90% of the channels are open in 100 μM cGMP. The reciprocal of this closing rate, 5 msec, is equal to the average time that a channel stays open. Most likely, a channel closes when cGMP dissociates, which gives the channel an opportunity to test again the cGMP concentration and decide whether or not to re-open. Why should the sodium channel stay open for 5 msec instead of say 5 μsec or 500 msec? If the channel stayed open for 500 msec, rods would not be able to respond in times of a few milliseconds to bright light pulses. Frogs would have lots of trouble catching flies if the sodium channels in their ROS stayed open for such a long interval! However, a very short open time, such as 5 μsec, would be disadvantageous for a different reason. The ratio of open to closed channels is equal to the ratio of the opening to the closing rate. The inferred opening rate of the channel, $1.8 \times 10^3 \, s^{-1}$, is within an order of magnitude of the diffusion-controlled limit of about $10^4 \, s^{-1}$ estimated from $k = 10^8 \, M^{-1} \, s^{-1} \times [cGMP]$, where $[cGMP] = 100 \, \mu$M. Thus, a much smaller portion of the channels would be open if the channel open time was 5 μsec instead of 1 msec. Moreover, a 5 μsec open time would render that channel more vulnerable to changes in cGMP level not triggered by photoexcited rhodopsin. In fact, the closing rate of the channel is nicely matched to the kinetic characteristics of the light-activated phosphodiesterase system.

The density of sodium channels in the plasma membrane can now be estimated. A dark current of 25 pA would arise from about 8300 channels,

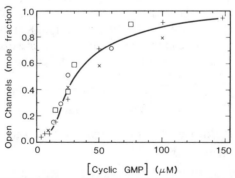

Figure 3 Cyclic GMP directly opens sodium channels in an excised patch of the plasma membrane. The fraction of open channels is shown as a function of the cGMP concentration. Measurements of four different patches are shown ($+$, in 10 nM Ca^{2+}; the others in 0.1 mM Ca^{2+}). The line is calculated for cooperative binding with half-opening at 30 μM cGMP and a Hill coefficient of 1.8. (From Fesenko et al 1985.)

each with a current flow of 3 fA. The plasma membrane of frog ROS has an area of 1200 μm^2, and so the density of open channels in the dark is about 7 per μm^2. However, only a fraction of the sodium channels are open in the dark under normal physiological conditions. In very low calcium media, the dark current increases 10 to 20-fold, probably because of an increase in the cGMP level. Hence the total channel density is likely to be of the order of 100 to 200 per μm^2. The unit conductance of these ROS sodium channels, 75 fS based on 3 fA at -40 mV, is more than 100-fold smaller than that of the electrically excitable sodium channel (~ 10 pS) and the acetylcholine receptor channel (~ 30 pS) (Hille 1984). The ROS has many channels with a small unit conductance rather than a few with a large conductance. This design would enable ROS to convey amplitude information and minimize time and amplitude jitter in response to small numbers of photons.

Biochemical studies of ion fluxes in vesicles derived from ROS membranes have also revealed the presence of channels directly controlled by cGMP (Caretta et al 1979, Capovilla et al 1983, Koch & Kaupp 1985). The vesicles studied probably contained plasma membrane in addition to disk membranes, the major component. Fluxes of Na^+, K^+, Rb^+, and Ca^{2+} were activated by the addition of cGMP at concentrations of about 50 μM, the physiological level. Chloride did not traverse this channel, showing that it is cation specific. The ionic selectivity of the cGMP-opened channel in these vesicles resembles that of the light-regulated channel in the plasma membrane, which has the following relative conductances: Li^+, 1.4; Na^+, 1.0; K^+, 0.8; Rb^+, 0.6; and Cs^+, 0.15 (Hodgkin et al 1985). The sodium channel in the plasma membrane, like the one in these vesicles, is appreciably permeable to Ca^{2+}. About 5% of the dark current is contributed by the inward flow of Ca^{2+} through the sodium channel of the plasma membrane (Yau & Nakatani 1984, Gold 1985). The cation channels in these vesicles, like those in the plasma membrane, are opened by cGMP in the absence of ATP and GTP. Moreover, the channels in the vesicles respond rapidly (<40 msec) to the addition of cGMP (Caretta & Cavaggioni 1983). Thus, the cGMP-activated vesicle channel is tantalizingly similar to the sodium channel in the plasma membrane.

Cyclic GMP is Rapidly Hydrolyzed and Synthesized Following a Light Pulse

Cyclic GMP is formed from GTP by guanylate cyclase and is hydrolyzed to 5'-GMP by the phosphodiesterase. The level of cGMP in ROS depends on the rates of these opposed reactions. The cytosolic level of cGMP could also be regulated by the binding of this nucleotide to specific proteins. It has been known for some time that light leads to a decrease in the cGMP level of intact ROS by activating the phosphodiesterase (Goridis et al 1974,

Kilbride & Ebrey 1979, Woodruff & Bownds 1979). A key question is whether changes in cGMP level elicited by illumination occur rapidly enough for cGMP to be a transmitter. Biochemical and electrophysiological studies of purified suspensions of frog outer segments still attached to the mitochondria-rich ellipsoid of the inner segment (called o.s.-i.s.) have provided a clear-cut answer (Cote et al 1984). This preparation behaves very much like rods in intact retinas in its excitation and adaptation characteristics and has more stable nucleotide levels than do outer segments severed from the adjoining ATP-generating region. The cGMP content of these o.s.-i.s. in 1 mM Ca^{2+} (a physiological extracellular level) in the dark is 1 per 100 rhodopsins, which is equivalent to a cGMP concentration of 60 μM in the cytosol. Light flashes exciting 10^6 rhodopsins per rod lead to a 20% decrease in cGMP level, as does continuous illumination bleaching 10^7 rhodopsins per second. In low calcium media, the cGMP level is 2.6-fold higher than in normal Ringer solution, and the light-induced decrease in cGMP level is also larger. In 1 nM Ca^{2+}, the cGMP level drops 71% during continuous illumination. The kinetics of cG decrease following illumination were then compared with the kinetics of closure of sodium channels in the plasma membrane as measured with a suction electrode. The important result was that the light-induced decrease in cGMP preceded the suppression of dark current: 10% of the cGMP was hydrolyzed within 50 msec after a flash that bleached 400 rhodopsins per rod (Figure 4). At this light intensity, the latency of channel closure was 60 msec and the time required for complete suppression of the membrane current was 480 msec. Thus, the decrease in cGMP level occurs rapidly enough for cGMP to be the transmitter in visual excitation. However, this study poses the question, which I consider below, as to how a relatively small change in the total cGMP level of ROS can lead to a large change in the sodium conductance of the plasma membrane.

The flux of cGMP in intact retinas has been determined from measurements of ^{18}O incorporation into the α-phosphoryl group of guanyl nucleotides following hydrolysis of cGMP in $[^{18}O]H_2O$ (Goldberg et al 1983). This powerful isotopic technique makes it possible to detect cGMP synthesis and hydrolysis in the absence of a net change in cGMP level. Retinas were illuminated with 10 msec light flashes at a frequency of 4 s^{-1} for 20 s, and the ^{18}O content of their nucleotides was measured. These experiments involving the intact system revealed that cGMP but not cyclic AMP turns over rapidly, even in the dark, and that the flux of cGMP is markedly increased by light. The half-time for turnover of cGMP in the absence of illumination is 1.5 s. The cGMP flux increased five-fold over a 1000-fold range in incident light intensity, whereas there was little change in cGMP level. These results led to the proposal that it is the synthesis or hydrolysis of cGMP, rather than the cGMP concentration per se, that is directly coupled to visual excitation. It was suggested that protons generated in the hydrolysis of cGMP serve a transmitter role in visual excitation. This motion was supported by the finding that perfusion of ROS with permeant acids and bases leads to changes in the dark current (Mueller & Pugh 1983). However, it was then found that the introduction into ROS of high concentrations of imidazole, an effective proton buffer, does not block excitation (Yoshikami & Hagins 1985). The involvement of protons is also ruled out by the finding that the cytosolic pH changes by less than 0.002 on illumination, as monitored by the fluorescence of 6-carboxy-fluorescein. The effects of permeant acids and bases on the dark current can now be explained in terms of their effects on the sodium-calcium antiporter.

The ^{18}O labeling experiments clearly demonstrate that guanylate cyclase keeps up with the phosphodiesterase over a broad range of light levels so that there is no sustained, large decrease in cGMP concentration following a light pulse. However, they do not provide information concerning transient changes in cGMP concentration in the vicinity of the disk that absorbed a photon. It should also be noted that the concentration of free cGMP in the cytosol may be substantially less than the total concentration of cGMP. Some of the cGMP may be bound to high-affinity sites on the

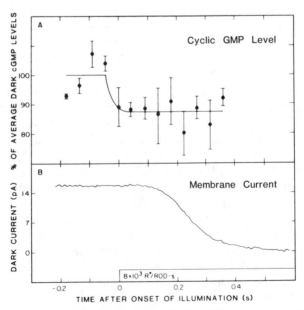

Figure 4 The decrease in cGMP concentration following illumination precedes the closing of sodium channels in the plasma membrane. (From Cote et al 1984.)

phosphodiesterase and other proteins of the ROS (Yamazaki et al 1980, Capovilla et al 1983). Thus, a 10% decrease in total cGMP following a light pulse (as in Figure 4) may correspond to a substantially larger change in the level of free cGMP in the cytosol. The cooperative opening of the sodium channel by cGMP (Figure 3) (Fesenko et al 1985) suggests that this channel is designed to detect changes in cGMP concentration. Furthermore, the fraction of channels open in an excised patch of plasma membrane exposed to 60 μM cGMP is much higher than that observed with intact ROS in the dark, which have a total cGMP level of 60 μM. The dark current of ROS can be markedly increased by the injection of cGMP or by the addition of isobutylmethylxanthine (IBMX), a phosphodiesterase inhibitor that leads to higher cGMP levels (Miller & Nicol 1979, Lipton & Dowling 1981, Capovilla et al 1983, MacLeish et al 1984, Matthews et al 1985, Cobbs & Pugh 1985). A simple interpretation of these data is that the free cytosolic cGMP level in the dark is only about 5 μM, with the other 55 μM cGMP bound to sites that do not release cGMP in times of less than a second.

Cyclic GMP and Calcium Levels in the Cytosol Are Reciprocally Regulated

The cGMP level in rod outer segments is influenced by calcium ion as well as by light (Lipton & Dowling 1981). Conversely, the cytosolic level of calcium ion is influenced by cGMP. The molecular basis of the interplay of calcium and cGMP is now becoming understood. The cGMP level increases in low Ca^{2+} and decreases in high Ca^{2+} because both guanylate cyclase and the phosphodiesterase are affected (Lolley & Racz 1982). The catalytic activity of the phosphodiesterase increases two-fold when the Ca^{2+} level is increased (Kawamura & Bownds 1981), whereas that of guanylate cyclase is halved at high Ca^{2+} levels (Fleischman & Denisevich 1979). The effect of Ca^{2+} on cGMP level provides a simple explanation for the finding that the dark current is very much increased at low Ca^{2+} levels and is suppressed at high Ca^{2+} levels.

According to the calcium hypothesis, the cytosolic Ca^{2+} level is very low in the dark and increases markedly upon illumination. This prediction seemed to be supported by the finding that a single photon leads to the efflux of more than 10^4 Ca^{2+} from ROS as measured with a calcium-sensitive extracellular electrode (Yoshikami et al 1980, Gold & Korenbrot 1980) and by laser microprobe analyses (Schroder & Fain 1984). This efflux was interpreted to indicate that light releases Ca^{2+} from disks and that the resulting increase in cytosolic Ca^{2+} increases the rate of extrusion of Ca^{2+} from the ROS by the sodium-calcium antiporter. However, recent experiments provide important new data concerning calcium fluxes and lead to a very different conclusion (Yau & Nakatani 1984, 1985, Gold 1985). A key finding is that Ca^{2+} rapidly enters ROS through the sodium channel in the dark. The inward flux is about 1.5 pA, which is equivalent to 4.7×10^6 Ca^{2+} per second. An equal efflux of Ca^{2+} keeps the cytosolic level of Ca^{2+} at the steady-state level characteristic of the dark state. Light closes the sodium channel and so the entry of Ca^{2+} from the extracellular medium is blocked. However, the antiporter continues to pump calcium out of the ROS until the calcium level in the cytosol becomes very low. The initial efflux rate is 10^7 Ca^{2+} per sec, which decays exponentially with a time constant of 0.4 sec (Yau & Nakatani 1985). Hence more than 4×10^6 Ca^{2+} are extruded following a saturating light pulse, and some 10^5 Ca^{2+} are extruded following formation of a single photoexcited rhodopsin molecule. These data strongly suggest that the net efflux of calcium after a light pulse is due to a diminished entry of calcium into the outer segment instead of an increased cytosolic level of calcium. Indeed, it now seems likely that the cytosolic Ca^{2+} level decreases rather than increases after a light pulse. A transient decrease in the cytosolic level of Ca^{2+} would serve to stimulate guanylate cyclase and inhibit the phosphodiesterase, which would increase the cGMP level and help restore the dark state.

These experiments demonstrate that there is a feedback relationship between cytosolic Ca^{2+} and cGMP levels. A high level of Ca^{2+} depresses the cGMP level, which leads to the closure of sodium channels. The entry of Ca^{2+} is then partially blocked but its efflux by the sodium-calcium exchanger continues, leading to a lower cytosolic Ca^{2+} level. Conversely, a low level of Ca^{2+} in the cytosol leads to an increase in the cGMP level, which opens sodium channels and thereby raises the cytosolic Ca^{2+} level. Likewise, the steady-state level of cGMP in the cytosol is feedback stabilized through its effect on the Ca^{2+} level. An increased cGMP level opens sodium channels and consequently accelerates the entry of Ca^{2+}, which in turn lowers the cGMP level. Thus, calcium plays a modulatory role in rods by regulating the cGMP level. It will be interesting to learn whether calcium has effects on the sodium channel that are not mediated through changes in the cGMP level.

Excitation Flows from Photoexcited Rhodopsin to Transducin to the Phosphodiesterase

The experiments described above provide strong evidence that the sodium channels in the plasma membrane are directly opened by cGMP and that the level of this nucleotide decreases rapidly on illumination, which leads to the closure of many channels. I turn now to the molecular mechanism of the light-triggered hydrolysis of cGMP and begin with an overview of this cascade.

It has been known for some time that ROS contain a light-activated phosphodiesterase that is specific for cGMP (Miki et al 1975). The high gain of this enzyme became evident when the rate of cGMP hydrolysis was directly monitored by the change in pH resulting from proton release. This assay showed that one photoexcited rhodopsin (R*) can trigger the hydrolysis of 10^6 cGMP in ROS suspensions (Yee & Liebman 1978). From the known turnover number of the phosphodiesterase (PDE), it was inferred that hundreds of enzyme molecules are activated by a single photon. This study also showed that GTP or a hydrolysis-resistant analogue such as GppNHp is required in addition to R* to activate the PDE. The need for GTP pointed to a guanyl-nucleotide binding protein as a likely link in the activation sequence. Support for this notion came from the finding that ROS contain a peripheral membrane protein that binds guanyl nucleotides and has GTPase activity in the presence of bleached membranes (Godchaux & Zimmerman 1979). It was then found that a single photoexcited rhodopsin catalyzes the exchange of 500 GTP for bound GDP in ROS (Fung & Stryer 1980). This experiment led to the proposal of a light-activated GTP-GDP amplification cycle involving a guanyl-nucleotide-binding protein with GTPase activity. The formation of the GTP form of this protein, now called transducin (T), was postulated to be the first stage of amplification in vision. It was proposed that GTP-transducin (T-GTP) is the activator of the phosphodiesterase and that the flow of information in the cascade is

R* → T-GTP → PDE*.

This scheme predicted that (a) T-GTP can be formed in vitro in the absence of phosphodiesterase and (b) the phosphodiesterase can be activated by T-GTP in the absence of photoexcited rhodopsin. Indeed, reconstituted membranes containing rhodopsin and transducin but devoid of phosphodiesterase exhibit amplified GTP-GDP exchange (Fung et al 1981). A single R* leads to the uptake of GppNHp by 71 molecules of transducin. Here is a direct amplified action of R* requiring just one other protein. The second prediction was then tested by the addition of activated transducin to unilluminated membranes. It was found that the phosphodiesterase can be fully stimulated in the absence of R* by the addition of T_α-GppNHp. These experiments established that transducin is the first amplified information-carrying intermediate in the cyclic nucleotide cascade of vision (Stryer et al 1981b, Stryer 1983).

The composition and roles of the excitatory and inhibitory proteins of the cGMP cascade are given in Table 1. The structure and interactions of rhodopsin are considered first, followed by a discussion of transducin and the phosphodiesterase. I then turn to rhodopsin kinase and arrestin, two inhibitory proteins of the cascade.

Table 1 Excitatory and inhibitory proteins of the cyclic GMP cascade

Protein[a]	Mass (kD)	Role
Rhodopsin (R)	40	Photoexcited rhodopsin (R*) activates transducin by catalyzing GTP–GDP exchange.
Transducin (T)	α 39	T_α-GTP is the amplified intermediate in the activation of the phosphodiesterase.
	β 36	Transducin interconverts between an inactive $T_{\alpha\beta\gamma}$-GDP form and an active
	γ 8	T_α-GTP form. $T_{\beta\gamma}$ is required for GTP–GDP exchange. T_α hydrolyzes bound GTP to GDP to return to the dark state.
Phosphodiesterase (PDE)	α 88	Hydrolyzes cyclic GMP when activated by T_α-GTP. The catalytic activity of $\alpha\beta$
	β 85	is blocked by the γ subunit of the PDE in the dark state.
	γ 11	
Rhodopsin kinase	68	Phosphorylates R* at multiple serine and threonine residues, enabling it to bind arrestin. Involved in the deactivation of R*.
Arrestin	48	Binds to phosphorylated R* and blocks its capacity to activate transducin.

[a] The molar amounts of these proteins in intact rod outer segments are estimated to be rhodopsin, 1000; transducin, 100; phosphodiesterase, 14; rhodopsin kinase, 1; arrestin, 100.

Structure and Evolutionary Conservation of Rhodopsin

Rhodopsin is a 40 kD transmembrane protein located in the disks and plasma membrane of ROS. The concentration of rhodopsin in ROS is very high, about 3 mM, which corresponds to a surface density of 27,000 per μm^{-2} of disk membrane. Indeed, rhodopsin comprises more than 90% of the integral membrane protein of disks. The determination of the amino acid sequence of bovine rhodopsin (Ovchinnikov 1982, Hargrave et al 1983) has led to a structural model (Figure 5) that brings together a wealth of chemical and physical information about this photoreceptor protein (reviewed by Dratz & Hargrave 1983). Rhodopsin is a single polypeptide chain consisting of 348 residues. The presence of seven predominantly hydrophobic stretches of 19 to 28 amino acids suggested that rhodopsin contains seven transmembrane helices. The proposed threading of the polypeptide chain (Figure 5) is based on the results of numerous studies of enzymatic labeling, proteolytic digestion, and lectin binding. The amino-terminus is on the intradiskal side and the carboxy-terminus is on the cytosolic (interdiskal) side of the disk membrane. About half of the mass of the protein is within the hydrophobic core of the membrane, one quarter is on the cytosolic side, and one quarter is on the intradiskal side. This postulated distribution of mass and orientation of α-helices agrees with the findings of spectroscopic, neutron-scattering, and magnetic-anisotropy studies (Chabre 1985). Rhodopsin contains two N-linked oligosaccharide units, attached to asparagine 2 and 15, which are located on the intradiskal side. The carboxy-terminal region on the cytosolic face is rich in serine and threonine residues, which become phosphorylated by rhodopsin kinase following illumination. Phosphorylation is a step in the deactivation of photoexcited rhodopsin, as is discussed below. Also noteworthy is the high degree of transmembrane charge asymmetry predicted by the model. The cytosolic face of the membrane has a net positive charge of about 4, whereas the intradiskal side has a net negative charge of about 7.

The 11-*cis* retinal chromophore of rhodopsin is bound to lysine 296 by a protonated Schiff base linkage. The long axis of retinal is nearly parallel to the plane of the membrane. The model places retinal near the center of the membrane, in agreement with fluorescence energy transfer studies (Thomas

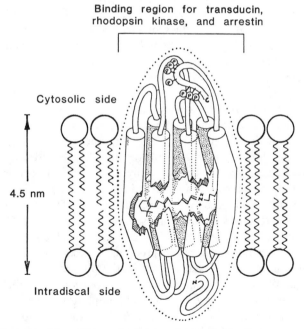

Binding region for transducin, rhodopsin kinase, and arrestin

Cytosolic side

4.5 nm

Intradiscal side

Figure 5   A model of the structure of vertebrate rhodopsin. (Based on Dratz & Hargrave 1983.)

& Stryer 1982). Some of the five charged residues within the membrane are likely to be critical in tuning the absorption maximum of rhodopsin, which is at 500 nm, compared with 440 nm for an unperturbed protonated Schiff base. Charged residues within the membrane may also be important in controlling the isomerization process.

The genes for bovine and human rhodopsin have recently been cloned and their nucleotide sequences have been determined (Nathans & Hogness 1983, 1984). The degree of evolutionary conservation is very high. The coding region of both genes is interrupted by four introns. Three of the four introns mark the boundary between the C-terminal end of a transmembrane helix and the start of the adjoining hydrophilic region. This arrangement supports the view that genes evolved by the coming together of exons that encode discrete structural units. Human and bovine rhodopsins have the same number of amino acid residues. Moreover, the degree of amino acid sequence identity is very high (94%). The retinal attachment site, the two glycosylation sites, and all but one of the seven phosphorylation sites are the same. Most striking, the three loops on the cytosolic face are perfectly conserved. These loops are probably involved in the binding of transducin, rhodopsin kinase, and arrestin.

Another important recent advance is the cloning of a rhodopsin gene from *Drosophila melanogaster* (O'Tousa et al 1985, Zuker et al 1985). Fruit fly rhodopsin contains 372 residues, of which 81 are the same as in bovine rhodopsin. The distribution of hydrophobic and hydrophilic residues suggests that *Drosophila* rhodopsin, like the mammalian ones, contains seven transmembrane helices. Several regions of *Drosophila* rhodopsin closely resemble mammalian rhodopsins. The lysine residue that forms a Schiff base with retinal is present in the same position, as are seven of 11 adjacent amino acids. Another region of homology is the first loop on the cytosolic face of rhodopsin, which may form part of the binding site for transducin or rhodopsin kinase. Moreover, the second loop on the intradiskal face of rhodopsin is highly conserved. The role of this intradiskal loop is unknown. Two charged residues inside the membrane, in addition to the Schiff base lysine, are identical in fly and mammalian rhodopsins. Diptera and mammals are separated by more than 500 million years of evolution. Hence the conservation of amino acid sequence points to the presence of functionally important regions of these molecules. The same fundamental architectural plan is used in the photoreceptor proteins of distant phyla. Where does rhodopsin have its evolutionary origin? Perhaps as far back as *Chlamydomonas*, a unicellular photosynthetic eucaryote. Phototaxis in a blind mutant was restored by the addition of retinal analogues (Foster et al 1984). The positions of the maxima of the action spectra for phototaxis strongly suggested that 11-*cis* retinal is the natural chromophore and that the protein environment of this chromophore is similar to that of mammalian rhodopsins. It now seems likely that rhodopsin is an ancient receptor protein, well over a billion years old.

Rhodopsin Is Photoisomerized to an Activated R* State

Light isomerizes the 11-*cis* retinal chromophore of rhodopsin to the all-*trans* form (reviewed in Honig et al 1979). Bathorhodopsin, the first photoproduct, is formed within picoseconds of the absorption of a photon. Bathorhodopsin converts to lumirhodopsin in nanoseconds, followed by metarhodopsin I in microseconds, and metarhodopsin II in about a millisecond. In metarhodopsin II, the Schiff base linkage is unprotonated and the chromophore is in the all-*trans* form. In times of minutes, metarhodopsin II becomes converted to metarhodopsin III, which is then hydrolyzed to opsin and all-*trans* retinal. Which of these intermediates triggers the cyclic nucleotide cascade? Spectroscopic studies have shown that metarhodopsin II serves as R*, the photoexcited rhodopsin species that catalyzes the activation of transducin (Emeis et al 1982, Bennett et al 1982). This conformationally activated state stores some 27 kcal/mol of the photon energy (Cooper 1981). The large activation barrier between rhodopsin and metarhodopsin II, more than 40 kcal/mol, contributes to the very low dark noise of the system. In the dark, a rod cell fires spontaneously about once a minute, which corresponds to a thermal isomerization rate of once per 3000 years for a single rhodopsin molecule (Baylor et al 1980). Thus, thermal isomerization in the dark proceeds 10^{23} times more slowly than does photoisomerization! Retinal combines two key properties: a very high degree of stability in the dark with an extremely rapid and efficient response to stimulation by light.

The photoisomerization of retinal leads to conformational changes that are transmitted from the center of the protein to its cytosolic face, a distance of at least 2 nm. The nature of these structural changes is not yet known but their functional consequences are evident. R*, but not R, interacts with transducin and induces release of its bound GDP (Kühn 1980, Fung & Stryer 1980). Furthermore, R*, but not R, binds rhodopsin kinase and becomes phosphorylated (Kühn 1984). Two-dimensional crystalline arrays of rhodopsin that diffract to a resolution of 2.2 nm have been obtained (Corless et al 1982), but this resolution does not suffice to reveal the folding of the protein. More highly ordered crystals are eagerly awaited.

Transducin Mediates a Light-activated Amplification Cycle

Transducin is a multisubunit peripheral membrane protein consisting of three polypeptide chains: α (39 kD), β (36 kD), and γ (8 kD) (Kühn 1980,

Stryer et al 1981b). Transducin has also been called G-protein, GTP-binding protein, and GTPase. I use the name transducin to emphasize its central role in visual transduction and to distinguish it from the hormonal G-protein, a related but distinct protein. The γ chain of transducin has recently been sequenced (Ovchinnikov et al 1985). The genes for the α chain (Lochrie et al 1985, Medynski et al 1985, Tanabe et al 1985, Yatsunami & Khorana 1985) and the γ chain (Hurley et al 1984a) of transducin have been cloned and their nucleotide sequences have been determined. These sequences show that the α and γ chains are synthesized individually rather than formed by cleavage of a polyprotein.

The functions of the subunits of transducin have been elucidated (Stryer et al 1981b, Fung 1983). The α subunit of transducin (T_α) contains the guanyl nucleotide binding site and the site for hydrolysis of GTP to GDP. T_α in the GTP form is the activator of the phosphodiesterase. The β and γ chains of transducin form a $T_{\beta\gamma}$ subunit. T_α and $T_{\beta\gamma}$ are together when GDP is bound, and separate when GTP is bound. The only known role of $T_{\beta\gamma}$ is to present T_α to photoexcited rhodopsin so that GTP can be exchanged for GDP. $T_{\beta\gamma}$ does not participate directly in the activation of the phosphodiesterase or in the hydrolysis of GTP bound to T.

A light-activated amplification cycle involving rhodopsin, transducin, and the phosphodiesterase is shown in Figure 6. The major features of this

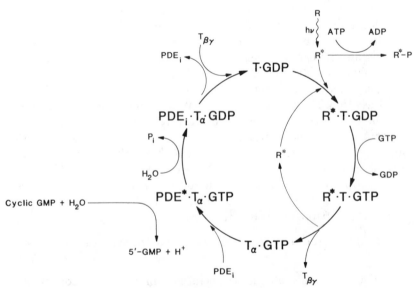

Figure 6   Light-activated transducin cycle. (Based on Stryer et al 1981a.)

cycle (Stryer 1983) are as follows:

1. In the dark, nearly all the transducin is in the T-GDP form, which does not activate the PDE.
2. R* encounters T-GDP by lateral diffusion in the plane of the disk membrane and forms an R*-T-GDP complex.
3. GTP exchanges for GDP in the R*-T-GDP complex. This exchange probably proceeds through an R*-T intermediate devoid of a bound nucleotide. In essence, the role of R* is to induce the release of GDP from transducin.
4. The exchange of GTP for GDP markedly diminishes the affinity of R* for transducin. In addition, the affinity of T_α for $T_{\beta\gamma}$ is very much decreased. These changes in binding affinities are crucial for the efficient operation of the cycle. The released R* is free to interact with another T-GDP, enabling R* to act catalytically. T_α-GTP is mobilized so that it can carry the excitation signal to the PDE. In this way, some 500 T_α-GTP are formed per R*.
5. T_α-GTP activates the PDE by overcoming an inhibitory constraint imposed by one of the subunits of the PDE, as discussed below. Activated PDE hydrolyzes cGMP at a very rapid rate.
6. The GTPase activity inherent in the α-subunit of transducin converts T_α-GTP to T_α-GDP in times of seconds to a minute; this in turn results in the deactivation of the PDE. T_α-GDP then rejoins $T_{\beta\gamma}$. Additional reactions needed to restore the dark state are discussed below.

Kinetics and Energetics of the Transducin Cycle

Information concerning the kinetics of activation of transducin comes from light-scattering studies. The near-infrared turbidity of ROS suspensions changes following excitation with a visible light pulse. Three types of scattering changes have been identified by using reconstituted membranes containing different amounts of transducin: an R* signal due to the bleaching of rhodopsin, a binding signal due to the association of R* with T-GDP, and an amplified dissociation signal due to the release of T-GTP from R* following GTP-GDP exchange (Kühn et al 1981, Kühn 1984, Michel-Villaz et al 1984). These studies showed that transducin is activated in less than 100 msec at moderate light levels. Subsequent light-scattering studies of magnetically oriented frog ROS have carried the analysis further in two respects (Vuong et al 1984). First, much of the lamellar organization of disks in native ROS was preserved in this preparation, and so the observed kinetics are highly pertinent to those occurring in vivo. Second, the high degree of orientation imposed by the magnetic field made it possible to distinguish between structural changes directed along the rod axis from those at right angles (Figure 7). A light-scattering signal arising from the release of activated transducin from the disk surface into the aqueous space between disks was observed. The dependence of the kinetics of this release signal on the mole fraction of R* revealed that a molecule of transducin is activated to the GTP form by a molecule of R* in 1 msec. This means that a single R* can activate 500 molecules of transducin in 0.5 sec, well within the 1 to 2 sec rise time of the change in membrane current. Thus, the activation of transducin occurs sufficiently rapidly to enable T_α-GTP to serve as an amplified intermediate in visual excitation. Indeed, the rate of activation of transducin by R* is close to the rate calculated for the diffusion-controlled encounter of R* and transducin (Liebman & Pugh 1981). The PDE is then rapidly activated by T_α-GTP (Bennett 1982).

The energetics of the transducin cycle also deserve comment. The cycle is powered by the hydrolysis of GTP and not by the energy of the absorbed photon. Light triggers the formation of the catalyst (R*) for GTP-GDP exchange but provides no input of energy for the cycle itself. Guanyl nucleotide exchange studies using hydrolysis-resistant analogues such as GTPγS have shown that the equilibrium for the reaction

$$T\text{-}GDP + GTP \rightleftharpoons T_\alpha\text{-}GTP + T_{\beta\gamma} + GDP$$

is displaced far to the right ($K > 100$) (Yamanaka & Stryer 1984). The hydrolysis of T_α-GTP is also essentially irreversible. The ratio of T-GDP to T_α-GTP in the dark is greater than 1000. Hence, there is a large thermodynamic driving force for both the formation of T_α-GTP and its subsequent hydrolysis to T_α-GDP. The phosphoryl potential of GTP is nicely partitioned to make both reactions irreversible so that the level of T_α-GTP, the activated form, is controlled kinetically. In the absence of R*, T-GDP is not converted appreciably to T_α-GTP because of a very large activation barrier. The GTP-GDP exchange time in the dark is at least

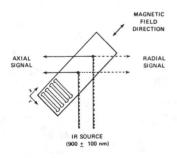
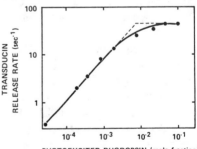

Figure 7   Kinetics of activation of transducin determined by light-scattering measurements of magnetically oriented fragments of ROS. (From Vuòng et al 1984.)

several hours. The role of R* is to lower this activation barrier by stabilizing the transition state for GTP-GDP exchange, which is most likely an R*-T species devoid of a bound nucleotide. The affinity of R* for T has also been optimized to assure the dissociation of R*-T-GTP even at high levels of R* (Stryer 1985). Thus, this amplification system is designed to operate over a very wide dynamic range in the concentration of R*, the catalyst.

The Phosphodiesterase is Inhibited by Its γ Subunit and Activated by T_α-GTP

The cGMP phosphodiesterase (PDE) in rod outer segments consists of three subunits: α (88 kD), β (84 kD), and γ (11 kD) (Baehr et al 1979). A key property of this enzyme is its very low catalytic activity in the dark state. The finding that limited tryptic digestion markedly increases its catalytic activity led to the proposal that the PDE is subject to an inhibitory constraint (Miki et al 1975). Three lines of evidence show that the PDE in the dark state is inhibited by its γ subunit (Hurley & Stryer 1982). First, trypsin rapidly degraded the γ subunit without markedly altering α and β. The kinetics of digestion of the γ subunit paralleled the activation of the enzyme. The α and β subunits of the PDE remained associated in the trypsin-activated enzyme. Second, the activity of trypsin-activated PDE was inhibited more than 99% by the addition of purified γ subunit, which was obtained from the holoenzyme by heat treatment or gel filtration in acidic media. Binding of γ to αβ was very tight. The dissociation constant for the complex of γ subunit and trypsin-activated PDE was 0.13 nM. Third, inhibitory activity copurified with the catalytic activity of the enzyme, indicating that the inhibitor is normally bound to the catalytic moiety of the PDE. These three lines of evidence show that the phosphodiesterase consists of distinct catalytic and regulatory subunits.

Transducin activates the phosphodiesterase by overcoming the inhibitory constraint imposed by its γ subunit (Fung et al 1981). The catalytic activity of phosphodiesterase bound to unilluminated disk membranes increased nearly linearly with the amount of T_α-GppNHp added, to a limiting value corresponding to a turnover number of $3700 \, s^{-1}$. In contrast, T-GDP does not activate the PDE. The maximal catalytic activity obtained by the addition of T_α-GppNHp is nearly the same as that achieved by tryptic activation. This identity strongly suggests that T_α-GTP activates the PDE physiologically by overcoming the inhibitory constraint imposed by its γ subunit. The kinetics of activation indicate that T_α-GTP binds first to the PDE. However, it is not yet known whether T_α-GTP displaces the γ subunit or carries it away from the PDE. The detailed mechanism of this important interaction merits further study.

The activated phosphodiesterase displays great catalytic prowess.

The ratio of the maximal catalytic rate of the enzyme to its Michaelis constant, k_{cat}/K_M, is 6×10^7 M^{-1} s^{-1}, a value close to the limit set by the diffusion-controlled encounter of enzyme and substrate. Some other enzymes with very high first-order rate constants are carbonic anhydrase (8×10^7 M^{-1} s^{-1}), catalase (4×10^7 M^{-1} s^{-1}), and acetylcholinesterase (16×10^7 M^{-1} s^{-1}). Thus, the PDE has come close to achieving kinetic perfection, which supports the notion that it plays a central role in visual excitation, one in which speed is of the essence.

The phosphodiesterase has binding sites for cGMP that are distinct from its catalytic site (Yamazaki et al 1980, 1982). These noncatalytic sites bind cGMP with high affinity ($K_d = 0.16 \mu M$ and $0.83 \mu M$). IBMX inhibits PDE activity and enhances the binding of cGMP to these sites, whereas tryptic digestion has opposite effects. The addition of GppNHp to illuminated disk membranes led to a reduction of noncatalytic binding. In contrast, the addition of a heat-stable factor (most likely, the γ inhibitory subunit of the PDE) enhanced the binding of cGMP to the noncatalytic sites. The physiological significance of these noncatalytic sites is not yet understood. However, these results raise the interesting possibility that much of the cGMP in ROS is bound in the dark to these high-affinity, noncatalytic sites on the PDE. The total cGMP content of ROS in the dark is about $60 \mu M$, comparable to the concentration of PDE, which is about $30 \mu M$. Might these noncatalytic sites serve as a storage depot for cGMP, which would be released from these sites a second or two after a light pulse to help restore the dark state?

The injection of activated transducin (T_α-GppNHP) into ROS results in a reversible hyperpolarization of the plasma membrane (Clack et al 1983). Different consequences of injecting phosphodiesterase have been observed. One laboratory finds that the injection of 300 molecules of trypsin-activated PDE has no effect in darkness but results in an augmented response to light (Shimoda et al 1984), whereas another observes a hyperpolarization similar to that obtained with illumination (Clack et al 1983). These studies demonstrate the feasibility of exploring transduction processes by the pressure injection of active proteins and point to the need for more work to resolve the apparent difference in the outcome of injecting PDE.

Rhodopsin Kinase and Arrestin Inhibit the Cyclic GMP Cascade

The hydrolysis of T_α-GTP to the T_α-GDP is a necessary step in the deactivation of the phosphodiesterase and the return to the dark state but it is not sufficient. R* must also be deactivated so that it does not continue to trigger the conversion of transducin to the GTP form. The release of all-

trans retinal and binding of 11-*cis* retinal to resynthesize rhodopsin occur in minutes, far slower than the loss of capacity of R* to catalyze the activation of transducin. An important clue concerning the inhibition of the cascade came from the finding that ATP rapidly quenches the photoactivation of the PDE in ROS (Liebman & Pugh 1980, Sitaramayya & Liebman 1983a). It has been known for many years that ROS contain a kinase that phosphorylates bleached rhodopsin but not unilluminated rhodopsin (Kühn & Dreyer 1972). Rhodopsin kinase is a 68 kD soluble enzyme present in the cytosolic space in a mole ratio of about one per 1000 rhodopsins (Kühn 1978). Following illumination, rhodopsin kinase binds to the cytosolic face of R* and phosphorylates up to nine serine and threonine residues in the C-terminal tail (Wilden & Kühn 1982). Phosphorylation of R* does not in itself diminish the capacity of R* to activate transducin. Yet another protein is needed to quench photoexcited rhodopsin. The partner is a 48 kD soluble protein that is highly abundant in the cytosol (Kühn 1978, Kühn et al 1984). This protein, now called "arrestin" (Zuckerman et al 1985), binds to phosphorylated R* but not to unphosphorylated R* or to phosphorylated rhodopsin (Kühn et al 1984). Moreover, arrestin and transducin compete in binding to phosphorylated R*. Thus, it seems likely that arrestin acts as an inhibitory cap on phosphorylated R*, blocking its capacity to catalyze the activation of transducin. The rate of phosphorylation of rhodopsin following a light pulse is sufficiently rapid to enable it to participate in the quenching of the cascade (Sitaramayya & Liebman 1983b). The extent of deactivation of R* is probably related to the number of phosphoryl groups incorporated into its C-terminal tail (Miller & Dratz 1984). A noteworthy feature of the kinetics of the cascade is that R* interacts first with transducin and then with the kinase because the concentration of T-GDP is about 100-fold higher than that of the kinase. R* becomes accessible to the kinase only after many transducins have been activated. It is evident that the amounts of the constituents of the cascade have been optimized so that activation precedes deactivation.

The capping of phosphorylated R* by arrestin is probably not the only inhibitory process. The light-activated PDE activity of ROS membranes incubated with GTP but not ATP has a decay time of about 30 sec, a decay time similar to that for the hydrolysis of T_α-GTP to T_α-GDP. However, the PDE activity is quenched in about 4 sec when ATP is present (Liebman & Pugh 1980, Sitaramayya & Liebman 1983b). This rapid inactivation suggests that R* is not the only site of deactivation. Recent studies suggest that arrestin undergoes ATP-ADP exchange and that the ATP form of arrestin directly inhibits activated PDE (Zuckerman et al 1982, 1985). The inhibitory effect of ATP on the light-activated PDE activity of intact ROS is most pronounced at low calcium levels (Kawamura & Bownds 1981). The dependence of PDE activity on light intensity shifts in the direction of decreased sensitivity to light by about a log unit when the Ca^{2+} concentration is decreased from 1 mM to 1 nM in the presence of ATP. These experiments have led to the proposal that calcium and ATP regulation of the phosphodiesterase play a role in light adaptation.

Transducin Belongs to a Family of Signal-coupling Proteins

Several lines of evidence indicate that transducin belongs to a family of signal-coupling proteins that includes the G proteins of the hormonally-regulated adenylate cyclase cascade. First, the mechanism of activation of the phosphodiesterase by light closely resembles that of adenylate cyclase by hormones. Second, transducin and the G proteins are specifically modified by cholera toxin and pertussis toxin. Third, transducin and the G proteins have the same subunit composition and are alike in structure. Fourth, components of the two systems can be interchanged with retention of function.

The activation of the phosphodiesterase by light resembles the activation of adenylate cyclase by hormones such as epinephrine and glucagon (Bitensky et al 1981, Stryer et al 1981a). These signal-amplifying cascades are compared in Figure 8. The G-protein of the hormonal system has a role like that of transducin. The hormone-receptor complex triggers GTP-GDP exchange in the G-protein (Schramm & Selinger 1984), just as R* triggers GTP-GDP exchange in transducin. G_α-GTP then separates from G and activates adenylate cyclase by relieving an inhibitory constraint. As with transducin, activation is terminated by the GTPase activity inherent in the α-subunits of the G protein. Recent studies have shown that the adenylate cyclase is under dual control by G_s, a stimulatory G protein, and G_i, an inhibitory G protein (Gilman 1984). It is not yet clear whether $G_{i\alpha}$ acts directly on adenylate cyclase or whether the inhibitory effect is mediated through $G_{\beta\alpha}$.

The effects of cholera toxin and pertussis toxin provide further evidence for the close relationship of transducin and the G-proteins. Cholera toxin ADP-ribosylates the stimulatory G protein and renders it persistently activated (Cassell & Pfeuffer 1977, Moss & Vaughan 1979), whereas pertussis toxin ADP-ribosylates the inhibitory G protein and prevents it from interacting with its receptor (Bokoch et al 1983). These toxins have very similar effects on transducin (Abood et al 1982, Van Dop et al 1984, Navon & Fung 1984). Light and GppNHp are needed for modification of transducin by cholera toxin, just as hormone and GppNHp promote labeling of G_s. The dark state of transducin, T-GDP, is the target of pertussis toxin, as in the hormone system, where G_i-GDP is modified. The consequences of ADP-ribosylation are also very similar: labeling of cholera toxin markedly inhibits the hydrolysis of T_α-GTP to T_α-GDP, whereas pertussis toxin modification prevents T-GDP from interacting with R*. A significant difference between the visual and hormonal systems is that transducin can be modified by both toxins. Different sites on the α-subunit of transducin are labeled by pertussis and cholera toxins (Manning et al 1984, Van Dop et al 1984a,b).

Transducin and the G-proteins have the same $\alpha\beta\gamma$ subunit structure. The 36 kD β subunits of T, G_s, and G_i are probably identical or nearly so (Manning & Gilman 1983). Polyclonal antibodies directed against the three subunits of transducin crossreact with the β subunit of G_s and G_i but not with their α or γ subunits (Gierschik et al 1985). The α subunits of these proteins differ in mass (T_α, 39 kD; $G_{s\alpha}$, 45 kD; $G_{i\alpha}$, 41 kD), but have homologous sequences. The proteolytic digestion pattern of the α-subunit of G_0, a G protein from bovine brain, is like that of transducin (Hurley et al 1984b). $G_{0\alpha}$ and T_α contain a stretch of 22 identical amino acids. An

(a) Vision   (b) Hormone Action

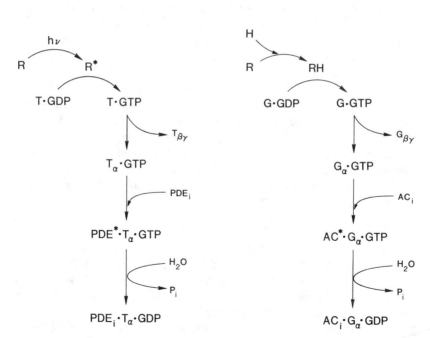

Figure 8 Homology between vision and hormone action.

additional intriguing finding is that the N-terminal portion of the *ras* protein has a similar sequence.

T_α — | V K L L L L G A G E S G K S T I V K Q M K I | —

$G_{0\alpha}$ — | V K L L L L G A G E S G K S T I V K Q M K I | —

c-N-*ras*   — Y| K L |V V V| G A G |G V| G K S |A L T I |Q| L I Q —

(A, alanine; E, glutamate; G, glycine; I, isoleucine; K, lysine; L, leucine; M, methionine; Q, glutamine; S, serine; T, threonine; V, valine; Y, tyrosine.)

The *ras* protein, which also cycles between a GTP and GDP form, has been implicated in the control of cell proliferation (reviewed by Cooper & Lane 1984). Oncogenic forms of the *ras* protein have diminished GTPase activity (Gibbs et al 1984, Sweet et al 1984); this would be expected to give rise to a persistently activated state of *ras*, like the effect of cholera toxin on the stimulatory G protein and transducin. It seems likely that the *ras* protein is a member of this family of signal-coupling proteins. Furthermore, *ras* and elongation factor Tu from *E. coli* show extensive sequence homology, suggesting that the GTP-binding site of these proteins is similar (Leberman & Egner 1984). Indeed, the transducin cycle is reminiscent of the elongation factor Tu-Ts cycle in bacterial protein synthesis, which is also driven by the hydrolysis of GTP (Kaziro 1978). The controlled uptake and release of proteins coupled to GTP-GDP exchange and hydrolysis was perfected early in evolution and retained over several billion years.

The kinship of transducin and the G proteins is also revealed by the interchangeability of components from the two systems. The activated GppNHp form of transducin stimulates adenylate cyclase in the absence of hormone (Bitensky et al 1982). Furthermore, R* triggers the activation of the adenylate cyclase cascade, showing that photoexcited rhodopsin can substitute for the hormone-receptor complex. R* can also interact with the inhibitory G protein (Kanaho et al 1984). The hydrolysis of GTP by G_i is stimulated by the addition of R* in the absence of an inhibitory hormone.

Recurring Motifs in Visual Excitation

The basic transduction mechanism of vertebrate cone cells is likely to be similar to that of rods. The sodium channels in the plasma membrane of cones are directly opened by cGMP (Haynes & Yau 1985). Furthermore, cones contain a light-activated cGMP phosphodiesterase that crossreacts immunologically with its rod counterpart (Ortez et al 1983, Hurwitz et al 1985). There has been little biochemical work on the proteins of cones because cone outer segments do not detach readily from the retina. Instead, molecular genetics is providing striking new information about the visual pigments of cone cells. The genes for the three color vision receptors have recently been cloned (J. Nathans and D. S. Hogness, personal communication). One of them (probably the blue absorbing pigment) is on an autosome, and the other two (probably the green and red absorbing pigments) are closely linked on the X-chromosome. Sequence studies now in progress are revealing that these receptor proteins for color vision have the same fundamental architecture as rhodopsin from rod cells.

The photoreceptor cells of vertebrates, molluscs, and arthropods are anatomically different and are believed to have evolved independently. However, it was shown years ago that their visual pigments use 11-*cis* retinal as the chromophore (Wald 1968). As discussed above, the structure of *Drosophila* rhodopsin is similar to that of mammalian rhodopsins. The initial events in visual excitation are likely to be the same in different phyla. Photoexcited rhodopsin from octopus or squid retinas can trigger the activation of bovine transducin (Ebrey et al 1980, Saibil & Michel-Villaz 1984). Deactivation too occurs similarly in different phyla. Photoexcited rhodopsin in flies and squid, as in vertebrates, becomes phosphorylated (Paulsen & Bentrop 1984, Vandenberg & Montal 1984b). Moreover, bovine R* can interact effectively with squid transducin. Further evidence for the presence of homologous proteins in the photoreceptors of these three phyla comes from the finding that cholera and pertussis toxins ADP-ribosylate transducin-like molecules in flies and squid (Vandenberg & Montal 1984a, Malbon et al 1984). Clearly, extensive regions of rhodopsin and transducin have been conserved over a long evolutionary period. The effector enzymes that are controlled by transducin in arthropods and molluscs are not yet known. Cyclic nucleotides appear not to be central in visual excitation in these phyla. However, it is evident that transducin plays an important role in the ventral eye of *Limulus*, the horseshoe crab. Nonhydrolyzable analogues of GTP, such as GTPγS, mimic the action of light (Fein & Corson 1981, Bolsover & Brown 1982).

The involvement of phosphoinositides in vision is likely to be another recurring motif. These compounds serve as messengers in many different cellular transduction processes (reviewed in Berridge & Irvine 1984). The injection of inositol trisphosphate into Limulus ventral photoreceptors both excites and adapts them in a manner similar to light (Fein et al 1984, Brown et al 1984b). Moreover, these cells contain phosphatidylinositol bisphosphate and hydrolyze this phospholipid on illumination. Likewise, vertebrate rods exhibit light-induced hydrolysis of phosphatidylinositol bisphosphate (Ghalayini & Anderson 1984) and are hyperpolarized by injection of inositol trisphosphate (Waloga & Anderson 1985). A new player has entered the game. An intriguing possibility is that phosphoinositide metabolism is linked to photoexcited rhodopsin by transducin.

Summary

Cyclic GMP is central to visual excitation in vertebrate retinal rod cells. Sodium channels in the plasma membrane of the outer segment are kept open in the dark by a high level of cGMP. Light closes these channels by activating an enzymatic cascade that leads to the rapid hydrolysis of cGMP. Photoexcited rhodopsin triggers transducin by catalyzing the exchange of GTP for bound GDP. The activated GTP-form of transducin then switches on the phosphodiesterase by overcoming an inhibitory constraint. The overall gain of this cascade is about 10^5. The cascade is turned off by the GTPase activity of transducin and by the action of rhodopsin kinase and arrestin. One of the challenges now is to delineate the interplay of cGMP, calcium ion, and phosphoinositides in excitation and adaptation. Transducin belongs to a family of signal-coupling proteins that includes the G proteins of the hormone-regulated adenylate cyclase cascade. The initial events in visual excitation in molluscs and arthropods are probably similar to those of vertebrates. The triggering of transducin by photoexcited rhodopsin is a recurring motif in visual transduction. The coming together of electrophysiology, biochemistry, and molecular genetics affords new opportunities in unraveling the molecular mechanism of visual transduction.

ACKNOWLEDGMENTS

I wish to thank Dr. Denis Baylor, Dr. Marc Chabre, Dr. Hermann Kühn, Dr. Paul Liebman, Dr. Jeremy Nathans, Dr. Gregory Yamanaka, and Dr. Anita Zimmerman for stimulating discussions. The research carried out in my laboratory was supported by grants from the National Eye Institute and the National Institute of General Medical Sciences.

Literature Cited

Abood, M. E., Hurley, J. B., Pappone, M. C., Bourne, H. R., Stryer, L. 1982. Functional homology between signal-coupling proteins: Cholera toxin inactivates the GTPase activity of transducin. *J. Biol. Chem.* 257:10540–43

Baehr, W., Devlin, M. J., Applebury, M. L. 1979. Isolation and characterization of cGMP phosphodiesterase from bovine rod outer segments. *J. Biol. Chem.* 254: 11669–77

Baylor, D. A., Lamb, T. D., Yau, K.-W. 1979a. Responses of retinal rods to single photons. *J. Physiol.* 288:613–34

Baylor, D. A., Lamb, T. D., Yau, K.-W. 1979b. The membrane current of single rod outer segments. *J. Physiol.* 288:589–611

Baylor, D. A., Matthews, G., Yau, K.-W. 1980. Two components of electrical dark noise in toad retinal rod outer segments. *J. Physiol.* 309:591–621

Bennett, N. 1982. Light-induced interactions between rhodopsin and the GTP-binding protein. Relation with phosphodiesterase activation. *Eur. J. Biochem.* 123:133–39

Bennett, N., Michel-Villaz, M., Kühn, H.

1982. Light-induced interaction between rhodopsin and the GTP-binding protein. *Eur. J. Biochem.* 127:97–103

Berridge M. J., Irvine, R. F. 1984. Inositol trisphosphate, a novel second messenger in cellular signal transduction. *Nature* 312:315–21

Bitensky, M. W., Gorman, R. E., Miller, W. H. 1971. Adenyl cyclase as a link between photon capture and changes in membrane permeability of frog photoreceptors. *Proc. Natl. Acad. Sci.* 68:561–62

Bitensky, M. W., Wheeler, G. L., Yamazaki, A., Rasenick, M. M., Stein, P. J. 1981. Cyclic nucleotide metabolism in vertebrate photoreceptors: A remarkable analogy and an unraveling enigma. *Curr. Top. Membr. Transp.* 15:237–71

Bitensky, M. W., Wheeler, M. A., Rasenick, M. M., Yamazaki, A., Stein, P. J., Halliday, K. R., Wheeler, G. L. 1982. Functional exchange of components between light-activated photoreceptor phosphodiesterase and hormone-activated adenylate cyclase systems. *Proc. Natl. Acad. Sci.* 79:3408–12

Bodoia, R. D., Detwiler, P. B. 1984. Patch clamp study of isolated frog rods. *Biophys. J.* 45:337a

Bokoch, G. M., Katada, T., Northup, J. K., Hewlett, E. L., Gilman, A. G. 1983. Identification of the predominant substrate for ADP-ribosylation by islet activating protein. *J. Biol. Chem.* 258:2072–75

Bolsover, S. R., Brown, J. E. 1982. Injection of guanosine and adenosine nucleotides into *Limulus* ventral photoreceptor cells. *J. Physiol.* 332:325–42

Brown, J. E., Kaupp, U. B., Malbon, C. C. 1984a. 3',5'-cyclic adenosine monophosphate and adenylate cyclase in phototransduction by Limulus ventral photoreceptors. *J. Physiol.* 353:523–39

Brown, J. E., Rubin, L. J., Ghalayini, A. J., Tarver, A. P., Irvine, R. F., Berridge, M. J., Anderson, R. E. 1984b. Myo-inositol polyphosphate may be a messenger for visual excitation in Limulus photoreceptors. *Nature* 311:160–63

Capovilla, M., Caretta, A., Cavaggioni, A., Cervetto, L., Sorbi, R. T. 1983. Metabolism and permeability in retinal rods. In *Progress in Retinal Research*, ed. N. Osborne, G. Chader, 1:233–47. New York: Pergamon

Caretta, A., Cavaggioni, A. 1983. Fast ionic flux activated by cyclic GMP in the membrane of cattle rod outer segments. *Eur. J. Biochem.* 132:1–8

Caretta, A., Cavaggioni, A., Sorbi, R. T. 1979. Cyclic GMP and the permeability of the disks of the frog photoreceptors. *J. Physiol.* 295:171–78

Cassel, D., Pfeuffer, T. 1978. Mechanism of cholera toxin action: Covalent modification of the guanyl nucleotide-binding protein of the adenylate cyclase system. *Proc. Natl. Acad. Sci. USA* 75:2669–73

Chabre, M. 1985. Trigger and amplification mechanisms in visual phototransduction. *Ann. Rev. Biophys. Biophys. Chem.* 14:331–60

Clack, J. W., Oakley, B. II, Stein, P. J. 1983. Injection of GTP-binding protein or cyclic GMP phosphodiesterase hyperpolarizes retinal rods. *Nature* 305:50–53

Cobbs, W. H., Pugh, E. N. Jr. 1985. Cyclic GMP can increase rod outer-segment light-sensitive current 10-fold without delay of excitation. *Nature* 313:585–87

Cooper, A. 1981. Rhodopsin photoenergetics: Lumirhodopsin and the complete energy profile. *FEBS Lett.* 123:324–26

Cooper, G. M., Lane, M. 1984. Cellular transforming genes and oncogenesis. *Biochim. Biophys. Acta* 738:9–20

Corless, J. M., McCaslin, D. R., Scott, B. L. 1982. Two-dimensional rhodopsin crystals from disk membranes of frog retinal rod outer segments. *Proc. Natl. Acad. Sci. USA* 79:1116–20

Cote, R. H., Biernbaum, M. S., Nicol, G. D., Bownds, M. D. 1984. Light-induced decreases in cGMP concentration precede changes in membrane permeability in frog rod photoreceptors. *J. Biol. Chem.* 259:9635–41

Cote, R. H., Burke, S. A., Nicol, B. D., Bownds, M. D. 1985. Involvement of cGMP in phototransduction by isolated frog rods in the absence of transmembrane calcium gradients. *Biophys. J.* 47:100a

Dahlem Conference. 1985. *The Molecular Mechanism of Photoreception.* Berlin/New York: Springer-Verlag

Detwiler, P. B., Conner, J. D., Bodoia, R. D. 1982. Gigaseal patch clamp recordings from outer segments of intact retinal rods. *Nature* 300:59–61

Dratz, E. A., Hargrave, P. A. 1983. The structure of rhodopsin and the rod outer segment disk membrane. *Trends Biochem. Sci.* 8:128

Ebrey, T. G., Kilbride, P., Hurley, J. B., Calhoon, R., Tsuda, M. 1980. Light control of cyclic nucleotide concentrations in the retina. *Curr. Top. Membr. Transp.* 15:133–56

Emeis, D., Kühn, H., Reichert, J., Hofmann, K. P. 1982. Complex formation between metarhodopsin II and GTP-binding protein in bovine photoreceptor membranes leads to a shift of the photoproduct equilibrium. *FEBS Lett.* 143:29–34

Fein, A., Corson, D. W. 1981. Excitation of *Limulus* photoreceptors by vanadate and by a hydrolysis-resistant analog of guanosine triphosphate. *Science* 212: 555–57

Fein, A., Payne, R., Corson, D. W., Berridge, M. J., Irvine, R. F. 1984. Photoreceptor excitation and adaptation by inositol 1,4,5-trisphosphate. *Nature* 311: 157–60

Fein, A., Szuts, E. Z. 1982. *Photoreceptors: Their Role in Vision*. Cambridge: Cambridge Univ. Press

Fesenko, E. E., Kolesnikov, S. S., Lyubarsky, A. L. 1985. Induction by cyclic GMP of cationic conductance in plasma membrane of retinal rod outer segment. *Nature* 313: 310–13

Fleischman, D., Denisevich, M. 1979. Guanylate cyclase of isolated bovine retinal rod axonemes. *Biochemistry* 18: 5060–66

Foster, K. W., Saranak, J., Patel, N., Zarilli, G., Okabe, M., Kline, T., Nakanishi, K. 1984. A rhodopsin is the functional photoreceptor for phototaxis in the unicellular eukaryote Chlamydomonas. *Nature* 311: 756–59

Fung, B. K.-K. 1983. Characterization of transducin from bovine retinal rod outer segments. Separation and reconstitution of the subunits. *J. Biol. Chem.* 258: 10495–10502

Fung, B. K.-K., Hurley, J. B., Stryer, L. 1981. Flow of information in the light-triggered cyclic nucleotide cascade of vision. *Proc. Natl. Acad. Sci.* 78: 152–56

Fung, B.K.-K., Stryer, L. 1980. Photolyzed rhodopsin catalyzes the exchange of GTP for GDP in retinal rod outer segment membranes. *Proc. Natl. Acad. Sci.* 77: 2500–4

Ghalayini, A., Anderson, R. E. 1984. Phosphatidylinositol 4,5-bisphosphate: Light-mediated breakdown in the vertebrate retina. *Biochem. Biophys. Res. Commun.* 124: 503–6

Gibbs, J. B., Sigal, I. S., Poe, M., Scolnick, E. M. 1984. Intrinsic GTPase activity distinguishes normal and oncogenic ras p21 molecules. *Proc. Natl. Acad. Sci.* 81: 5704–8

Gierschik, P., Codina, J., Simons, C., Birnbaumer, L., Spiegel, A. 1985. Antisera against a guanine nucleotide binding protein from retina cross-react with the beta subunit of the adenylyl cyclase-associated guanine nucleotide binding proteins, N_s and N_i. *Proc. Natl. Acad. Sci.* 82: 7727–31

Gilman, A. G. 1984. G proteins and dual control of adenylate cyclase. *Cell* 36: 577–79

Godchaux, W., Zimmerman, W. F. 1979. Soluble proteins of intact bovine rod cell outer segments. *Exp. Eye Res.* 28: 483–500

Gold, G. H. 1985. Plasma membrane Ca fluxes in intact rods are inconsistent with the "Ca hypothesis". *Biophys. J.* 47: 356a

Gold, G. H., Korenbrot, J. I. 1980. Light-induced calcium release by intact retinal rods. *Proc. Natl. Acad. Sci.* 77: 5557–61

Goldberg, N. D., Ames, A. III, Gander, J. E., Walseth, T. F. 1983. Magnitude of increase in retinal cGMP metabolic flux determined by ^{18}O incorporation into nucleotide alpha-phosphoryls corresponds with intensity of photic stimulation. *J. Biol. Chem.* 258: 9213–19

Goridis, C., Virmaux, N., Cailla, H. L., Delaage, M. A. 1974. Rapid, light-induced changes of retinal cyclic GMP levels. *FEBS Lett.* 49: 167–69

Gray, P., Attwell, D. 1985. Kinetics of light-sensitive channels in vertebrate photoreceptors. *Proc. R. Soc. Ser. B* 223: 379–88

Hagins, W. A. 1979. Excitation in vertebrate photoreceptors. In *The Neurosciences: Fourth Study Program*, ed. F. O. Schmitt, F. G. Worden, pp. 183–91. Cambridge, Mass.: MIT Press

Hagins, W. A., Penn, R. D., Yoshikami, S. 1970. Dark current and photocurrent in retinal rods. *Biophys. J.* 10: 380–412

Hargrave, P. A., McDowell, J. H., Curtis, D. R., Wang, J. K., Juszczak, E., Fong, S.-L., Mohana Rao, J. K., Argos, P. 1983. The structure of bovine rhodopsin. *Biophys. Struc. Mech.* 9: 235–44

Haynes, L., Yau, K.-W. 1985. Cyclic GMP-sensitive conductance in cell-free inside-out membrane patches from catfish cone outer segment. *J. Physiol.* In press

Hille, B. 1984. *Ionic Channels of Excitable Membranes*. Sunderland, Mass.: Sinauer Assoc.

Hodgkin, A. L., McNaughton, P. A., Nunn, B. J. 1985. The ionic selectivity and calcium dependence of the light-sensitive pathway in toad rods. *J. Physiol.* 358: 447–68

Honig, B., Ebrey, T., Callender, R. H., Dinur, U., Ottolenghi, M. 1979. Photoisomerization, energy storage, and charge separation: A model for light energy transduction in visual pigments and bacteriorhodopsin. *Proc. Natl. Acad. Sci.* 76: 2503–7

Hurley, J. B., Fong, H. K. W., Teplow, D. B., Dreyer, W. J., Simon, M. I. 1984a. Isolation and characterization of a cDNA clone for the gamma subunit of bovine retinal transducin. *Proc. Natl. Acad. Sci.* 81: 6948–52

Hurley, J. B., Simon, M. I., Teplow, D. B., Robishaw, J. D., Gilman, A. G. 1984b. Homologies between signal transducing G proteins and *ras* gene products. *Science* 226: 860–62

Hurley, J. B., Stryer, L. 1982. Purification and characterization of the gamma regulatory subunit of the cyclic GMP phosphodiesterase from retinal rod outer segments. *J. Biol. Chem.* 257: 11094–99

Hurwitz, R. L., Bunt-Milam, A. H., Chang,

M. L., Beavo, J. A. 1985. cGMP phosphodiesterase in rod and cone outer segments of the retina. *J. Biol. Chem.* 260:568–73

Kanaho, Y., Tsai, S.-C., Adamik, R., Hewlett, E. L., Moss, J., Vaughan, M. 1984. Rhodopsin-enhanced GTPase activity of the inhibitory GTP-binding protein of adenylate cyclase. *J. Biol. Chem.* 259:7378–81

Kawamura, S., Bownds, M. D. 1981. Light adaptation of the cyclic GMP phosphodiesterase of frog photoreceptor membranes mediated by ATP and calcium ions. *J. Gen. Physiol.* 77:571–91

Kaziro, Y. 1978. The role of guanosine 5'-triphosphate in polypeptide chain elongation. *Biochim. Biophys. Acta* 505:95

Kilbride, P., Ebery, T. G. 1979. Light-initiated changes of cyclic guanosine monophosphate levels in the frog retina measured with quick-freezing techniques. *J. Gen. Physiol.* 74:415–26

Kühn, H. 1978. Light-regulated binding of rhodopdin kinase and other proteins to cattle photoreceptor membranes. *Biochemistry* 17:4389–95

Kühn, H. 1980. Light- and GTP-regulated interaction of GTPase and other proteins with bovine photoreceptor membranes. *Nature* 283:587–89

Kühn, H. 1984. Interactions between photoexcited rhodopsin and light-activated enzymes in rods. In *Progress in Retinal Research*, ed. N. Osborne, J. Chader, 1:123–56. New York: Pergamon

Kühn, H., Bennett, N., Michel-Villaz, M., Chabre, M. 1981. Interactions between photoexcited rhodopsin and GTP-binding protein: Kinetics and stoichiometric analyses from light-scattering changes. *Proc. Natl. Acad. Sci. USA* 78:6873–77

Kühn, H., Dreyer, W. J. 1972. Light dependent phosphorylation of rhodopsin by ATP. *FEBS Lett.* 20:1

Kühn, H., Hall, S. W., Wilden, U. 1984. Light-induced binding of 48-kDa protein to photoreceptor membranes is highly enhanced by phosphorylation of rhodopsin. *FEBS Lett.* 176:473–78

Lamb, T. D. 1984. Electrical responses of photoreceptors. *Recent Adv. Physiol.* 10:29–65

Lamb, T. D., McNaughton, P. A., Yau, K.-W. 1981. Spatial spread of activation and background desensitization in toad rod outer segments. *J. Physiol.* 319:463–96

Leberman, R., Egner, U. 1984. Homologies in the primary structure of GTP-binding proteins: The nucleotide-binding site of EF-Tu and p21. *EMBO J.* 3:339–41

Liebman, P. A., Pugh, E. N. Jr. 1980. ATP mediates rapid reversal of cyclic GMP phosphodiesterase activation in visual receptor membranes. *Nature* 287:734–36

Liebman, P. A., Pugh, E. N. Jr. 1981. Control of rod disk membrane phosphodiesterase and a model for visual transduction. *Curr. Top. Membr. Transp.* 15:157–260

Lipton, S. A., Dowling, J. E. 1981. The relation between Ca^{2+} and cyclic GMP in rod photoreceptors. *Curr. Top. Membr. Transp.* 15:381–92

Lochrie, M. A., Hurley, J. B., Simon, M. I. 1985. Sequence of the alpha-subunit of photoreceptor G protein: Homologies between transducin, ras and elongation factors. *Science* 228:96–99

Lolley, R. N., Racz, E. 1982. Calcium modulation of cyclic GMP synthesis in rat visual cells. *Vision Res.* 22:1481–86

MacLeish, P. R., Schwartz, E. A., Tachibana, M. 1984. Control of the generator current in solitary rods of the *Ambystoma tigrinum* retina. *J. Physiol.* 348:645–64

Malbon, C. C., Kaupp, U. B., Brown, J. E. 1984. *Limulus* ventral photoreceptors contain a homologue of the alpha-subunit of mammalian N_s. *FEBS Lett.* 172:91–94

Manning, D. R., Gilman, A. G. 1983. The regulatory components of adenylate cyclase and transducin. A family of structurally homologous guanine nucleotide-binding proteins. *J. Biol. Chem.* 258:7059–63

Manning, D. R., Fraser, B. A., Kahn, R. A., Gilman, A. G. 1984. ADP-ribosylation of transducin by islet-activating protein. Identification of asparagine as the site of ADP-ribosylation. *J. Biol. Chem.* 259:749–56

Matthews, H. R., Torre, V., Lamb, T. D. 1985. Effects on the photoresponse of calcium buffers and cyclic GMP incorporated into the cytoplasm of retinal rods. *Nature* 313:582–85

Medynski, D. C., Sullivan, K., Smith, D., Van Dop, C., Chang, F.-H., Fung, B. K.-K., Seeburg, P. H., Bourne, H. R. 1985. Amino acid sequence of the alpha subunit of transducin deduced from the cDNA sequence. *Proc. Natl. Acad. Sci. USA* 82:4311–15

Michel-Villaz, M., Brisson, A., Chapron, Y., Saibil, H. 1984. Physical analysis of light-scattering changes in bovine photoreceptor membrane suspensions. *Biophys. J.* 46:655–62

Miki, N., Baraban, J. M., Keirns, J. J., Boyce, J. J., Bitensky, M. W. 1975. Purification and properties of the light-activated cyclic nucleotide phosphodiesterase of rod outer segments. *J. Biol. Chem.* 250:6320–27

Miller, J. L., Dratz, E. A. 1984. Phosphorylation at sites near rhodopsin's carboxylterminus regulates light initiated cGMP hydrolysis. *Vision Res.* 24:1509–21

Miller, W. H., ed. 1981. Molecular mechanisms of photoreceptor transduction. *Curr. Top. Membr. Transp.*, Vol. 15

Miller, W. H., Nicol, G. D. 1979. Evidence that cyclic GMP regulates membrane potential in rod photoreceptors. *Nature* 280: 64

Moss, J., Vaughan, M. 1979. Activation of adenylate cyclase by choleragen. *Ann. Rev. Biochem.* 48: 581–600

Mueller, P., Pugh, E. N. Jr. 1983. Protons block the dark current of isolated rod outer segments. *Proc. Natl. Acad. Sci. USA* 80: 1892–96

Nakatani, K., Yau, K.-W. 1985. cGMP opens the light-sensitive conductance in retinal rods. *Biophys. J.* 47: 356a

Nathans, J., Hogness, D. S. 1983. Isolation, sequence analysis, and intron-exon arrangement of the gene encoding bovine rhodopsin. *Cell* 34: 807–14

Nathans, J., Hogness, D. S. 1984. Isolation and nucleotide sequence of the gene encoding human rhodopsin. *Proc. Natl. Acad. Sci. USA* 81: 4851–55

Navon, S. E., Fung, B. K.-K. 1984. Characterization of transducin from bovine retinal rod outer segments. *J. Biol. Chem.* 259: 6686–93

O'Brien, D. F. 1982. The chemistry of vision. *Science* 218: 961–66

Ortez, R. A., Tamayo, A., Johnson, C., Sperling, H. G. 1983. Detection of a light-sensitive pool of cGMP in goldfish cone color receptors by immunocytochemistry. *J. Histochem. Cytochem.* 31: 1305–11

O'Tousa, J., Baehr, W., Martin, R., Hirsh, J., Pak, W. L., Applebury, M. L. 1985. The *Drosophila ninaE* gene encodes an opsin. *Cell* 40: 839–50

Ovchinnikov, Y. A. 1982. Rhodopsin and bacteriorhodopsin: Structure-function relationships. *FEBS Lett.* 148: 179–91

Ovchinnikov, Y. A., Lipkin, V. M., Shuvaeva, T. M., Bogachuk, A. P., Shemyakin, V. V. 1985. Complete amino acid sequence of gamma-subunit of the GTP-binding protein from cattle retina. *FEBS Lett.* 179: 107–10

Paulsen, R., Bentrop, L. 1984. Reversible phosphorylation of opsin induced by irradiation of blowfly retinae. *J. Comp. Physiol.* 155: 39–45

Saibil, H. R., Michel-Villaz, M. 1984. Squid rhodopsin and GTP-binding protein crossreact with vertebrate photoreceptor enzymes. *Proc. Natl. Acad. Sci. USA* 81: 5111–15

Schramm, M., Selinger, Z. 1984. Message transmission: Receptor controlled adenylate cyclase system. *Science* 225: 1350–56

Schroder, W. H., Fain, G. L. 1984. Light-dependent calcium release from photoreceptors measured by laser micro-mass analysis. *Nature* 309: 268–70

Schwartz, E. A. 1985. Phototransduction in vertebrate rods. *Ann. Rev. Neurosci.* 8: 339–67

Shimoda, Y., Hurley, J. B., Miller, W. H. 1984. Rod light response augmented by active phosphodiesterase. *Proc. Natl. Acad. Sci. USA* 81: 616–19

Sitaramayya, A., Liebman, P. A. 1983a. Mechanism of ATP quench of phosphodiesterase activation in rod disc membranes. *J. Biol. Chem.* 258: 1205

Sitaramayya, A., Liebman, P. A. 1983b. Phosphorylation of rhodopsin and quenching of cyclic GMP phosphodiesterase activation by ATP at weak bleaches. *J. Biol. Chem.* 258: 12106–9

Stryer, L. 1983. Transducin and the cyclic GMP phosphodiesterase: Amplifier proteins in vision. *Cold Spring Harbor Symp. Quant. Biol.* 48: 841–52

Stryer, L. 1985. Molecular design of an amplification cascade in vision. *Biopolymers* 24: 29–47

Stryer, L., Hurley, J. B., Fung, B. K.-K. 1981a. Transducin: An amplifier protein in vision. *Trends Biochem. Sci.* 6: 245–47

Stryer, L., Hurley, J. B., Fung, B. K.-K. 1981b. First stage of amplification in the cyclic nucleotide cascade of vision. *Curr. Top. Membr. Transp.* 15: 93–108

Sweet, R. W., Yokoyama, S., Kamata, T., Feramisco, J. R., Rosenberg, M., Gross, M. 1984. The product of ras is a GTPase and the T24 oncogenic mutant is deficient in this activity. *Nature* 311: 273–75

Tanabe, T., Nukada, T., Nishikawa, Y., Sugimoto, K., Suzuki, H., Takahashi, H., et al. 1985. Primary structure of the alpha-subunit of transducin and its relationship to *ras* proteins. *Nature* 315: 242–45

Thomas, D. D., Stryer, L. 1982. The transverse location of the retinal chromophore of rhodopsin in rod outer segment disc membranes. *J. Mol. Biol.* 154: 145–57

Tomita, T. 1970. Electrical activity of vertebrate photoreceptors. *Q. Rev. Biophys.* 3: 179–222

Vandenberg, C. A., Montal, M. 1984a. Light-regulated biochemical events in invertebrate photoreceptors. 1. Light-activated guanosinetriphosphatase, guanine nucleotide binding, and cholera toxin catalyzed labeling of squid photoreceptor membranes. *Biochemistry* 23: 2339–47

Vandenberg, C. A., Montal, M. 1984b. Light-regulated biochemical events in invertebrate photoreceptors. 2. Light-regulated phosphorylation of rhodopsin and phosphoinositides in squid photoreceptor membranes. *Biochemistry* 23: 2345–52

Van Dop, C., Tsubokawa, M., Bourne, H. R., Ramachandran, J. 1984a. Amino acid sequence of retinal transducin at the site ADP-ribosylated by cholera toxin. *J. Biol. Chem.* 259: 696–98

Van Dop, C., Yamanaka, G., Steinberg, F.,

Sekura, R. D., Manclark, C. R., Stryer, L., Bourne, H. R. 1984b. ADP-ribosylation of transducin by pertussis toxin blocks the light-stimulated hydrolysis of GTP and cGMP in retinal photoreceptors. *J. Biol. Chem.* 259:23–26

Vuong, T. M., Chabre, M., Stryer, L. 1984. Millisecond activation of transducin in the cyclic nucleotide cascade of vision. *Nature* 311:659–61

Wald, G. 1968. The molecular basis of visual excitation. *Nature* 219:800–7

Waloga, G., Anderson, R. E. 1985. Effects of inositol-1,4,5-trisphosphate injections into salamander rods. *Biochem. Biophys. Res. Commun.* 126:59–62

Wilden, U., Kühn, H. 1982. Light-dependent phosphorylation of rhodopsin: Number of phosphorylation sites. *Biochemistry* 21:3014

Woodruff, M. L., Bownds, M. D. 1979. Amplitude, kinetics, and reversibility of a light-induced decrease in guanosine 3',5'-cyclic monophosphate in frog photoreceptor membranes. *J. Gen. Physiol.* 73:629–53

Woodruff, M. L., Bownds, M. D., Green, S. H., Morrisey, J. L., Shadlovsky, A. 1977. Guanosine 3',5'-cyclic monophosphate and in vitro physiology of frog photoreceptor membranes. *J. Gen. Physiol.* 69:667–79

Yamanaka, G., Stryer, L. 1984. Interactions of transducin with two hydrolysis-resistant GTP analogs. *Invest. Ophthal. Vis. Sci. (Suppl.)* 25:157

Yamazaki, A., Bartucca, F., Ting, A., Bitensky, M. W. 1982. Reciprocal effects of an inhibitory factor on catalytic activity and noncatalytic cGMP binding sites of rod phosphodiesterase. *Proc. Natl. Acad. Sci. USA* 79:3702–6

Yamazaki, A., Sen, I., Bitensky, M. W. 1980. Cyclic GMP-specific, high affinity, non-catalytic binding sites on light-activated phosphodiesterase. *J. Biol. Chem.* 255:11619–24

Yatsunami, K., Khorana, G. K. 1985. GTPase of bovine rod outer segments: The amino acid sequence of the alpha-subunit as derived from the cDNA sequence. *Proc. Natl. Acad. Sci. USA* 82:4316–20

Yau, K.-W., Nakatani, K. 1984. Electrogenic Na-Ca exchange in retinal rod outer segment. *Nature* 311:661–63

Yau, K.-W., Nakatani, K. 1985. Light-induced reduction of cytoplasmic free calcium in retinal rod outer segment. *Nature* 313:579–82

Yee, R., Liebman, P. A. 1978. Light-activated phosphodiesterase of the rod outer segment. Kinetics and parameters of activation and deactivation. *J. Biol. Chem.* 253:8902–9

Yoshikami, S., George, J. S., Hagins, W. A. 1980. Light-induced calcium fluxes from outer segment layer of vertebrate retinas. *Nature* 287:395–98

Yoshikami, S., Hagins, W. A. 1973. Control of the dark current in vertebrate rods and cones. In *Biochemistry and Physiology of Visual Pigments*, ed. H. Langer, pp. 245–55. New York: Springer-Verlag

Yoshikami, S., Hagins, W. A. 1985. Cytoplasmic pH in rod outer segments and high-energy phosphate metabolism during phototransduction. *Biophys. J.* 47:101a

Zimmerman, A. L., Baylor, D. A. 1985. Electrical properties of the light-sensitive conductance of salamander retinal rods. *Biophys. J.* 47:357a

Zuckerman, R., Buzdygon, B., Philp, N., Liebman, P., Sitaramayya, A. 1985. Arrestin: An ATP/ADP exchange protein that regulates cGMP phosphodiesterase activity in retinal rod disk membranes (RDM). *Biophys. J.* 47:37a

Zuckerman, R., Schmidt, G. J., Dacko, S. M. 1982. Rhodopsin-to-metarhodopsin II transition triggers amplified changes in cytosol ATP and ADP in intact retinal rod outer segments. *Proc. Natl. Acad. Sci. USA* 79:6414–18

Zuker, C. S., Cowman, A. F., Rubin, G. M. 1985. Isolation and nucleotide sequence of a rhodopsin gene from *Drosophila melanogaster*. *Cell* 40:851–58

Reference added in proof:

Koch, K.-W., Kaupp, U. B. 1985. Cyclic GMP directly regulates a cation conductance in membranes of bovine rods by a cooperative mechanism. *J. Biol. Chem.* 260:6788–6800

Ann. Rev. Neurosci. 1986. 9:121–45

CHEMO-ELECTRICAL TRANSDUCTION IN INSECT OLFACTORY RECEPTORS

Karl-Ernst Kaissling

Max-Planck-Institut für Verhaltensphysiologie, Seewiesen, Federal Republic of Germany

INTRODUCTION

Insect behavior can be controlled by a single odor compound or by a blend of a few components in an exact ratio of concentrations. These odor compounds or odor mixtures are called *pheromones* if used for intraspecific communication. Each of the pheromone components is perceived by a particular type of receptor cell that is specifically tuned to its "key compound" and may respond to single odor molecules by firing single nerve impulses. The entire response range of a receptor cell may cover many decades of stimulus intensities. Each olfactory receptor cell is a bipolar neuron. Its dendrite innervates a specialized cuticular structure, usually a hollow cuticular hair that captures the stimulus molecules. The axon runs directly to the deutocerebrum—the first synaptic relay station of the central nervous system.

Chemo-electrical transduction in these organs comprises the processes leading from the adsorption of odor molecules on the cuticular surface via stimulus transport to the generation of receptor potentials and nerve impulses and, eventually, to the inactivation of the stimulus molecules. The interconnections of these processes and their influences on the ultimate impulse response of the receptor cells are largely unknown. This review covers electrophysiological investigations of cellular response characteristics as well as biophysical and biochemical work, much of which is based on radiolabeled pheromone compounds. These studies employed a few species of insects, mainly of moths with olfactory organs that are relatively large and accessible for experimental work. The olfactory hairs of

121

0147–006X/86/0301–0121$02.00

insects are well suited for recording with extracellular electrodes (Boeckh 1962, Kaissling 1974, van der Pers & den Otter 1978), and enable the study of single morphologically and functionally identified olfactory receptor cells. It is also possible to isolate the olfactory hairs, which contain no cellular material other than the odor-sensitive outer dendritic segments of the receptor cells. Isolated hairs are, therefore, an excellent preparation for isolating receptor cell membrane (Klein & Keil 1984) and for biochemical analysis of the reception mechanism (Vogt & Riddiford 1981b, Kaissling et al 1985).

In recent years many insect pheromone receptors and other olfactory cells have been examined with respect to their responses to systematically altered stimulus compounds. Some papers were concerned with the extraordinary specificity of molecular recognition, others elucidated the coding of odor blends in insects. These topics are not included here but have been partially covered in other reviews (Kaissling 1971, Priesner 1979a, 1985a,b, Payne 1974, Seabrook 1978, Hansen 1978, Schneider 1984, Mustaparta 1984). Several recent reviews deal with the morphology of insect sense organs (Zacharuk 1980, Altner & Prillinger 1980, Steinbrecht 1984, Keil & Steinbrecht 1984).

INSECT OLFACTORY ORGANS AND STIMULUS UPTAKE

The Olfactory Sensillum

Insect olfactory cells are characteristically grouped and closely associated with non-neuronal accessory or auxiliary cells. A group of four to six, seldom up to 20, cells together with the associated cuticular apparatus is called a *sensillum* (Figure 1). All of these cells may contribute to the electrical signals recorded from the sensillum. Each receptor cell of an olfactory sensillum is usually tuned to a different key compound, often to a pheromone component (Kaissling 1979, Priesner 1979b, 1985a,b). Typically three auxiliary cells envelope the soma region of the receptor neurons: the innermost is the *thecogen cell*, enclosed by the *trichogen* and the *tormogen cell* (Keil & Steinbrecht 1984). The latter two cells have a strongly folded apical membrane and may control the composition of the extracellular fluid within the hair (Thurm 1974, Phillips & Vande Berg 1976a,b, Wieczorek 1982), the *sensillum lymph*, also called *sensillum liquor* (Ernst 1969) or *receptor lymph*. This fluid provides the ionic milieu for the dendrites of the receptor cells (Thurm & Küppers 1980) and contains proteins that sequester the odor molecules within the hair (Vogt & Riddiford 1981a,b, Kaissling et al 1985). Size and shape of the cuticular parts of olfactory sensilla vary enormously and are used to categorize the sensilla (Slifer 1970, Altner & Prillinger 1980). The hair-like extensions of the sensilla trichodea in moths can reach 500 μm length and are a few micrometers thick (Sanes & Hildebrand 1976a, Keil 1984a).

Insect Antennae as Odor Filters

Olfactory sensilla occur in numbers up to many tens of thousands on the antennae, paired appendages of the insect head. The geometry of antennae and the array of olfactory sensilla in some insect groups are clearly adapted for the most efficient capture of odor molecules. For instance, the antennae of male saturniid moths have a feather-like structure with an "outline area" (F_{outl}) of up to 1 or 2 cm^2 (Boeckh et al 1960). About 30% of odor molecules in a free airstream passing a cross-sectional area equal to the antennal outline area were adsorbed on the antennae of male moths of *Antheraea polyphemus* (Kanaujia & Kaissling 1985). This fraction or "adsorption

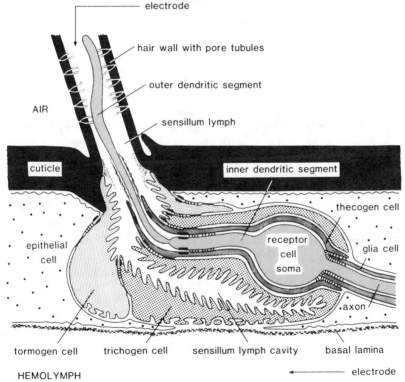

Figure 1 Schematic diagram of an olfactory sensillum trichodeum with one receptor cell and three auxiliary cells. The cuticular hair is 5 μm thick and 300 μm long (by courtesy of T. A. Keil).

quotient" (Q_{ads}) was determined using ^{3}H-labeled E-6, Z-11 hexadecadienyl acetate (HDA), a pheromone component of the female moth. A similar value (27%) was measured with ^{3}H-labeled bombykol (E-10, Z-12 hexadecadienol) in male moths of the silkworm, *Bombyx mori* (Kaissling & Priesner 1970, Kaissling 1971). These values are even slightly higher than those given for the "transmissivity" of an antenna, i.e. the fraction of a free airstream passing the antenna, as derived from hydrodynamic measurements (Vogel 1983). Of the adsorbed molecules, 80% ($= Q_{eff}$) were found on the long hairs of the sensilla trichodea in both species (Steinbrecht & Kasang 1972, Kanaujia & Kaissling 1985).

The above measurements in moths indicate a considerable accumulation of molecules near the sensitive dendrites of the receptor cells. The odor concentration in the hairs of *Antheraea polyphemus* increased by a factor $f_{hair/air}$ of about 10^5 over the concentration in air (c) within one second of stimulation time (t), when concentration in the hair was expressed as number of molecules adsorbed at the hairs (N_{effs}) per hair volume (v_h) of 2 pl and per 55,000 hairs (n_h) of one antenna. The factor ($f_{hair/air}$) was determined for a free airstream velocity (u) of 2.5 m/s by the formula (Kanaujia & Kaissling 1985)

$$f_{hair/air} = N_{effs}/v_h * n_h * c, \quad \text{with}$$

$$N_{effs} = c * u * F_{outl} * Q_{ads} * Q_{eff} * t \quad \text{(Kaissling 1971).}$$

In this calculation the minute desorption of molecules from the antennae was neglected. The remarkable "focusing" of the odor stimulus by the antenna can be understood on the basis of the laws of convective diffusion and was, in fact, predicted from these laws and from the geometry and arrangement of the sensilla on the antennae of moths (Adam & Delbrück 1968).

Values of Q_{ads} below 1% have been found in *Bombyx* antennae exposed to ^{3}H-labeled Z-7 dodecen-1-yl acetate, as can be recalculated from data given by Mankin & Mayer (1984). Their "deposition velocity" K corresponds to the product of Q_{ads} and the free airstream velocity u. This discrepancy may show a reduced adsorptivity of the *Bombyx* antenna for the acetate compared with bombykol ($Q_{ads} = 27\%$). In spite of this result, adsorption does not account for the high specificity of the cellular response. Thus, similar values of Q_{ads} were found for bombykol and hexadecanol, although the latter compound was about 10^6 times less effective as a stimulus (Kasang & Kaissling 1972).

Stimulus Transport

Odor molecules adsorbed on the hair surface probably penetrate the hair wall via the numerous pore tubules or other types of channels found in olfactory sensilla. The pore entrance and also the pore tubules have diameters in the range of 10 nm. The entire pore system is filled with electron-lucent material of unknown chemical composition which also covers the whole hair surface in a thickness of about 2.5 nm (Steinbrecht 1973). The pore densities are between less than 1 to 100 pores per μm^2 of hair surface (Keil & Steinbrecht 1984). The pore openings comprise a very small fraction of the hair surface. Therefore, most odor molecules hitting the surface between pores should either be reflected or migrate along the surface or within the outer surface layers until they reach the pore entrance. A molecule in an air stream passing the antenna would hit the antenna on average more than 100 times if it were never adsorbed (Boeckh et al 1965, Futrelle 1984). However, adsorption with first hit followed by surface diffusion would be 24 times more efficient—in terms of numbers of molecules reaching a pore—than reflection as calculated for *Bombyx mori* (Adam & Delbrück 1968).

The stimulus molecules are thought to interact with highly specific molecular receptors in the membrane of the receptor cell (Kafka 1970, Kafka & Neuwirth 1975). It is tempting to speculate that the lipophilic odor molecules do not enter the extracellular sensillum fluid on their way toward the receptor cell, but reach the dendritic membrane directly via the inner extensions of the pore tubules (Steinbrecht 1973, Keil 1982). However, in some electron-microscopic preparations, very few or no contacts between pore tubules and receptor cell membrane are visible (Keil 1984a,b).

The pore tubules clearly are not extensions of the receptor cell dendrite. The hair wall including the pore tubules is formed by the trichogen cell, one of the three accessory cells. Only after this cell withdraws from the hair lumen do the receptor cell dendrites invade the hair, where they may form contacts with the pore tubules (Ernst 1972, Sanes & Hildebrand 1976b). Cationic markers (cationized ferritin, ruthenium red) of the cell coat on the dendritic membrane also accumulate at the inner extensions of the pore tubules, indicating the presence of related material that may be involved in the formation of contacts (Keil 1984b).

Although it is still questionable whether the stimulus molecules reach the dendritic membrane directly via the tubules or through the sensillum lymph—possibly bound to protein (see below)—it is a fact that, eventually, a considerable amount of pheromone (or its metabolites) moves into the hair lumen. Sensillum lymph collected 1–2 min after stimulation of male *Antheraea polyphemus* with ^{3}H-labeled HDA contained 40% of the radioactivity associated with the hairs (Kanaujia & Kaissling 1985).

A stimulus molecule moving along the hair surface and through the pore tubules may reach the cell membrane within an average time of 5 ms. This figure was extrapolated from measurements of transport of ^{3}H-labeled pheromone from the hairs toward the antennal branches. The velocity of this transport corresponds to diffusion coefficients of 5×10^{-7} cm^2/s as estimated for bombykol in *Bombyx mori* (Steinbrecht & Kasang 1972) and 3×10^{-7} cm^2/s for HDA in *Antheraea polyphemus* (Kanaujia & Kaissling 1985). It seems questionable whether diffusion along the hairs can be used to calculate the time for transport toward the cell membrane. These diffusion coefficients may not apply for physiological stimuli because they were determined after extremely strong stimuli of about 10^8 molecules loaded on each hair within 10 s. High densities of stimulus molecules— about 10 molecules per nm^2 if all molecules diffuse on the hair surface— may alter the diffusion velocity. In fact, the average latencies of the receptor cell responses observed at low stimulus intensities (see below, and Figure 2) were several hundred ms (Kaissling & Priesner 1970, Kaissling & Thorson 1980) and would, consequently, allow 100 times lower diffusion coefficients. Such values are in the range of diffusion coefficients of molecules in lipid membranes (Träuble & Sackmann 1972) and may be not implausible for diffusional transport on the hairs under the certainly non-ideal conditions.

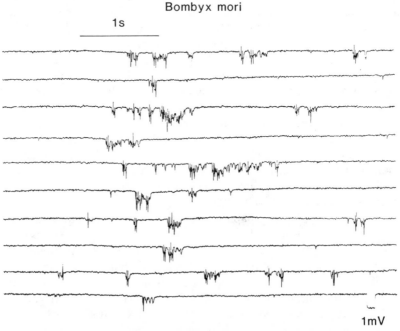

Figure 2 Elementary receptor potentials and nerve impulses as recorded extracellularly from a sensillum triochodeum. Each trace shows the response of a single receptor cell to one-second stimulation with the pheromone component E-10, Z-12 hexadecadienal (bombykal). The stimulus source was loaded with 10^{-3} μg bombykal.

ODOR INACTIVATION

After the stimulatory interaction with the presumed receptor molecules (see below), the odor molecules must be inactivated or removed from the cell membrane. This follows because, at physiological stimulus intensities, the cell response ceases immediately after termination of the stimulus although the stimulus molecules do not desorb significantly and do not migrate so quickly from the hairs. Two possible mechanisms of inactivation have been discussed for pheromone receptors of moths: (a) binding of the stimulus molecules to the abundant pheromone-binding protein found in the sensillum lymph (Vogt & Riddiford 1981b, Kaissling et al 1985), (b) enzymatic degradation of the pheromone molecules associated with the hairs (Kasang 1971, Ferkovich et al 1973a,b, Vogt et al 1985).

The Pheromone-Binding Protein

The most prominent soluble protein in the antennae of male *Antheraea polyphemus* binds radiolabeled phermone as found by fluorography after gel electrophoresis (Vogt & Riddiford 1981a,b). This protein has been called "pheromone-binding protein" and occurs in the sensillum lymph. The binding protein has a molecular weight in the range of 30,000 D, according to gel filtration under non-denaturing conditions (J. Hemberger, unpublished, Kaissling et al 1985). In SDS gels it shows a molecular weight of about 15,000 D (Vogt & Riddiford 1981b), suggesting the presence of two subunits. The purified binding protein shows saturable and specific binding of pheromone HDA with a dissociation constant of about 6×10^{-8} M. Saturation was reached with about 1 pheromone molecule per binding protein. In competition experiments with hexadecanyl acetate, or E-6, Z-11 hexadecadienol, the same amount of inhibition was reached at a 10- or 300-fold higher concentration, respectively, as compared to the unlabeled pheromone (J. Hemberger, unpublished, Kaissling et al 1985). A relatively broad binding specificity is demonstrated by the fact that the ^{3}H-labeled pheromone of *Antheraea polyphemus* binds to corresponding proteins in other species of moths, although these species are insensitive to this pheromone (Vogt & Riddiford 1981a). The concentration of the binding protein can be calculated as about 10 mM, on the assumption that all of the protein (up to 15 μg per antenna) is located in the sensillum lymph. This would mean a several–hundred-fold excess of binding protein molecules over the maximum number of receptors expected in the dendritic membrane.

Pheromone-Degrading Enzymes

Enzymatic degradation of the pheromone component, bombykol (E-10, Z-12 hexadecadienol), has been shown to occur on antennae of male and female *Bombyx mori* and also in other parts of the body. Two types of metabolic products, acids and esters, were eluted from intact antennae previously exposed to airborne pheromone stimuli. Therefore, at least two pheromone-degrading enzymes are associated with the antennae. Marked conversion of pheromone into acids but not into esters was found on the scales, "dry" cuticular extensions without cellular elements covering the whole body surface. One function of the pheromone degradation is, obviously, to avoid secondary stimulation by odor adsorbed on the insect body (Kasang 1971, Kasang & Kaissling 1972, Kasang & Weiss 1974, Kasang 1973).

Disparlure (*cis*-7,8-epoxy-2-methyloctadecane) was metabolized on antennae of the male gypsy moth *Lymantria dispar* (Kasang et al 1974). Z-7 dodecenyl acetate, a pheromone component of the cabbage looper, *Trichoplusia ni*, was converted into the corresponding alcohol in whole antennae and also in eluants of antennae after the tips of the sensilla had been fractured by sonication (Ferkovich et al 1973b). Gel filtration of such eluants revealed two peaks of enzymatic activity, one of which disappeared after treatment of the antennal sonicate with Triton X-100. This observation and the presence of cell membrane fragments in the fraction gave rise to the hypothesis that some of the esterase was membrane bound (Ferkovich & Mayer 1975, Mayer et al 1976). Gel electrophoresis demonstrated several bands stained for esterase activity by using naphthyl acetate as substrate (Ferkovich et al 1980). Three additional esterase bands appeared one to three days after eclosion of the moths when the males were maximally responsive to the pheromone (Taylor et al 1981). Esterase activity was found in other parts of the insect body and also in females (Ferkovich et al 1982a,b).

Electrophoresis of homogenized antennae of *Antheraea polyphemus* revealed "general" esterases not detected in isolated hairs and the so-called sensillar esterase found in male antennae and in isolated sensilla. General esterases were also obtained from wings and legs, but not from hemolymph. The sensillar esterase occurred in isolated sensillum lymph (Vogt & Riddiford 1981a,b). The molecular weight of the esterases ranged from 55,000 to 90,000 D, according to gradient microgel electrophoresis (U. Klein unpublished, Kaissling et al 1985). An esterase of 38,000 D was found in *Trichoplusia ni* (Mayer et al 1976). The enzymatic activity depended relatively little on the molecular structure of the stimulus molecule if compared to the respective dependence of the receptor cell response (Kasang & Kaissling 1972, Ferkovich & Mayer 1975).

Rapid Pheromone Inactivation

Determinations of the enzymatic decomposition of bombykol in intact *Bombyx* antennae revealed a half life of about 3 min over a more than 100-

fold range of stimulus loadings on the antenna (Kasang 1971, 1973). The half life of HDA adsorbed on intact antennae of *Antheraea polyphemus* was in the range of several minutes. It was even longer in the fraction of pheromone associated with the hairs (S. Kanaujia and G. Kasang, unpublished). In contrast to these results, the half life of pheromone expected from the decay of the receptor potential measured with similar stimulus loadings on the antennae was in the range of one or several seconds (Kaissling 1972, Zack 1979). Therefore a rapid nonenzymatic pheromone inactivation was postulated as opposed to the slower enzymatic degradation observed in intact antennae.

As a possible mechanism of inactivation, the reduction of free pheromone concentration (S_{free}) in the sensillum lymph due to pheromone binding can be calculated for *Antheraea polyphemus*. In all experiments the total initial concentration (S_{tot}) of pheromone in the hair was far below the total concentration of binding protein ($B_{tot} = 10^{-2}$ M). Therefore, B_{free} equals about B_{tot}. Using the dissociation constant ($K_D = 6 \times 10^{-8}$ M) of the pheromone binding protein and HDA, the mass action equation

$$S_{free}/S_{bound} = K_D/B_{free}$$

can be converted into

$$S_{free}/S_{bound} = K_D/B_{tot} = 6 \times 10^{-6}.$$

The formula shows a strong reduction of the free pheromone concentration under equilibrium conditions. Thus, binding could serve as a mechanism for rapid inactivation if the equilibrium is reached quickly enough.

The question of enzymatic versus nonenzymatic inactivation has recently been reopened (Vogt et al 1985). The electrophoretically isolated sensillar esterase of *Antheraea polyphemus* degraded one of the female pheromone components, E-6, Z-11 hexadecadienyl acetate (HDA), with an apparent K_M of about 2×10^{-6} M and a maximal velocity v_{max} of 5×10^{-9} M/s, for one antennal unit of enzyme in 1 ml. This value was adjusted to the volume in which the enzyme is present in the in vivo situation. Thus, v_{max} would be $F = \text{ml}/55,000 \text{ pl} = 1.82 \times 10^4$ times higher if the enzyme is located exclusively in the picoliter volume of sensillum lymph of each of the 55,000 sensilla trichodea per antenna. With $v_{max\,adj} = F \times v_{max}$ one calculates a half life (τ) of the pheromone compound of 16 ms under first order conditions, i.e. at free pheromone concentrations below K_M, from the equation $\tau = \ln 2 \times K_M/v_{max\,adj}$ (Vogt et al 1985). This value reduces to about 1.6 ms if one considers a 10% yield of the enzyme isolation (estimated from Klein & Keil 1984 and Vogt et al 1985). Therefore, the enzyme alone would be sufficiently fast for the rapid inactivation. However, the enzymatic degradation may proceed much more slowly if the concentration of free pheromone (S_{free}) in the hair is reduced due to the presence of the pheromone-binding protein. The half life of HDA within the hair could increase by the factor $S_{tot}/S_{free} = 1/6 \times 10^6$ to about 5 min if binding to the protein equilibrates faster than the enzymatic reaction. This effect could account for the (above-mentioned) pheromone half lives in the range of minutes reported for intact antennae and for isolated hairs.

One could ask whether, alternatively, such long half lives could be due to saturation of the enzyme by the high pheromone loadings on the hairs used in these experiments ($S_{tot} \approx 100 \times K_M$). This possibility can be excluded because—given no interference of the binding protein—the enzyme should degrade half of the initial pheromone load (S_{tot}) within about 0.1 s according to the formula $t_{1/2} = 0.5 \times S_{tot}/v_{max\ adj}$ for zero-order conditions and for a 10% yield of the enzyme isolation. Furthermore, $t_{1/2}$ in the alternative model should depend on the initial pheromone load (S_{tot}). No such dependence has been observed in Bombyx antennae over a more than hundred-fold range of stimulus concentrations (Kasang & Kaissling 1972). The fact that the fall time of the receptor-potential increases with the stimulus intensity at high stimulus strengths does not support the alternative model because this behavior is expected for stimuli saturating the receptors.

In conclusion, a consistent interpretation of all available data suggests that the activation of the pheromone is obtained by nonenzymatic binding that reduces the stimulus concentration at the receptors and also delays the enzymatic degradation of the pheromone. This model implies that the pheromone does not bind to the binding protein before arriving at the presumed receptors in the dendritic membrane and, thus, excludes a carrier function of the binding protein as suggested by Vogt et al (1985).

The discussed mechanisms of inactivation fail to explain how the fraction of 60% of stimulus molecules associated with the hair wall (Kanaujia & Kaissling 1985) is inactivated. It is also unclear how the inactivated molecules are transported along the hair. A better understanding of the rapid inactivation requires knowledge about the exact distribution of pheromone on and within the hair and about the dynamics of the processes involved.

GENERATION OF ELECTRICAL RESPONSES

Receptor Potentials and Nerve Impulses

In the resting state, a transepithelial potential of about $+30$ mV can be recorded from the opened tip (Figure 1) of an olfactory hair of male Antheraea polyphyemus with the reference electrode in the hemolymph space of the antenna (Kaissling & Thorson 1980). Odor stimulation causes a negative deflection of the transepithelial potential (Figures 2, 3). This change of potential is referred to here as a "receptor potential." The

Antheraea polyphemus

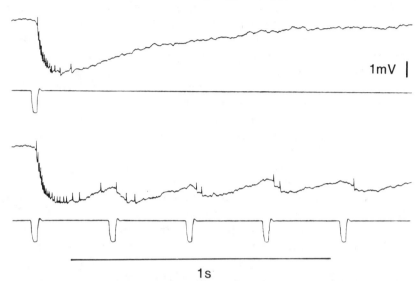

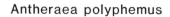

1 s

Figure 3   Receptor potentials and nerve impulses (*first and third traces*) of a single receptor cell elicited by 20-ms stimuli with the pheromone component E-6, Z-11 hexadecadienal. *Second and fourth traces:* downward deflections show the transient increase of air pressure in the stimulus line (from Kaissling 1985).

receptor potential can be as large as 20 or 30 mV; it rises within fractions of a second and falls more slowly, within seconds after the end of the stimulus.

In these recordings the receptor potential is usually accompanied by nerve impulses, with an initial positive phase lasting 1–2 ms and a longer negative phase. The peak-to-peak amplitude of the extracellularly recorded nerve impulses can reach 10 mV (De Kramer et al 1984) and is often characteristic of a particular receptor cell type. For example, the "acetate" cell of *Antheraea polyphemus* discharges larger impulses than the "aldehyde" cell and also has larger receptor potentials (Kaissling 1979). The differences in the extracellularly recorded response amplitudes were correlated with the diameters of the outer dendritic segments of the cells (Keil 1984a).

The Electrical Circuit of the Sensillum

Ideas about the generation of receptor potentials and nerve impulses in insect receptor cells were developed in experiments on taste (Morita & Yamashita 1959, Wolbarsht & Hanson 1965, Fujishiro et al 1984) and mechanoreceptive sensilla (Thurm 1963, 1972, 1974, Bernard & Guillet 1972, Guillet et al 1980, Erler & Thurm 1981). The receptor potential was thought to be caused by a stimulus-induced decrease of membrane resistance of the outer dendritic segment. The change in membrane potential would be electrotonically conducted along the dendrite and elicit nerve impulses near the soma region. This model explains the opposite polarity of the receptor potential and the first phase of the nerve impulse. The biphasic shape of the nerve impulses was thought to be due to an active antidromic impulse propagation from the cell soma into the hair.

Generation of electrical responses in olfactory receptors has been studied using equivalent circuit diagrams of the sensillum trichodeum of *Antheraea polyphemus* (Kaissling & Thorson 1980, de Kramer et al 1984, de Kramer 1985). As in earlier analyses (Thurm 1963, 1974, Rees 1968, Morita 1972), these diagrams have two main current paths from the hair tip toward the hemolymph space. One pathway leads from the sensillum lymph space across the folded apical membranes of the trichogen and tormogen cells into these accessory cells and continues through their basolateral membranes into the hemolymph. The other pathway enters the receptor cell through the dendritic membrane and leaves the cell in the soma region. Most probably, this pathway crosses the thecogen cell before entering the hemolymph. Further possible current paths through the clefts between the cells have been neglected because the presence of septate junctions (Figure 1) and the distribution of tracer compounds (Keil & Steinbrecht 1983; T. A. Keil, unpublished) suggested high extracellular resistance. The functional analyses were based on morphological measurements (Keil 1984a, 1984c, Steinbrecht & Gnatzy 1984, Gnatzy et al 1984), the ionic composition of the sensillum lymph, and the electrical responses to chemical and electrical stimuli.

Since it was not possible to measure the electrical parameters of each element of the equivalent circuit separately, a computer model of distributed resistances and voltage sources was used to simulate static responses of the sensillum and thereby to infer possible contributions of some sensillar elements (Kaissling & Thorson 1980, Kaissling 1980a). For instance, the transepithelial resistance of the sensillum was about $2 \times 10^8 \, \Omega$ and decreased by at most 20% during chemical stimulation in proportion to the receptor potential of maximally 30 mV. Based on the model circuit, these data are compatible with the idea that the stimulus does indeed affect the dendritic resistance. The "best guess" set of model parameters included a high resting resistance of the dendritic membrane ($\approx 10^4 \, \Omega \, cm^2$). This membrane resistance would be sufficiently high for an electrotonic spread of the receptor potential along the dendrite toward the soma region.

Dynamic Properties of the Sensillum

A recent analysis in *Antheraea polyphemus* (de Kramer et al 1984, de Kramer 1985) included dynamic properties of the sensillum up to a frequency of 10 kHz. The responses to current and voltage pulses under clamp conditions were studied in intact sensilla and after amputation of the hair or transepithelial perforation of accessory cells of a sensillum. The responses of the sensillum were interpreted in terms of a capacitance of about 30 pf in parallel with the main portion of the transepithelial resistance of $3 \times 10^8 \Omega$ (at 10°C). These values could be mainly ascribed to the apical membranes of the sensillar cells (including the receptor cells), which have a total area of about 3000 μm^2 (Gnatzy et al 1984). This area corresponds to the one expected from the measured capacitance if one assumes the usual specific capacitance of membranes (1 $\mu F/cm^2$). The specific resistance of the apical membranes would be in the range of $10^4 \Omega$ cm^2 ($3 \times 10^8 \Omega \times$ 3000 μm^2), which matches the value previously ascribed to the dendritic membrane.

Membrane capacitances in the sensillum could also be responsible for the biphasic shape of the nerve impulses (de Kramer et al 1984). This would be an explanation of the impulse shape that is alternative to the earlier explanation based on antidromic firing of impulses (Morita & Yamashita 1959). The earlier explanation was doubted because in sensilla trichodea of *Antheraea polyphemus*, simultaneous recordings from the hair tip and the side wall near the hair base did not show the delay expected for a backfired impulse at the hair tip (de Kramer et al 1984, de Kramer 1985).

A satisfactory understanding of the sensillum circuit is still lacking. Thus, a simple one-loop model with high resistance of the dendritic membrane would predict smaller nerve impulses recorded from shorter hairs. However, the extracellular size of nerve impulses was hardly affected if the hair was shortened experimentally. This result called into question the standard hypothesis of high dendritic membrane resistance and appreciable electrotonic spread of the receptor potential toward the soma region. Instead it was assumed that the nerve impulses initiate in the dendrite (de Kramer 1985). As a consequence of low dendritic resistance, the recorded receptor potential and the changes in sensillar resistance would no longer originate from the dendritic membrane. The new results cannot be simulated by the circuit model used so far and it remains open whether the conclusions would be supported by a more comprehensive model.

Voltage Sources in the Sensillum

A complete model of the sensillum circuit should include the sources of membrane potentials in the sensillum. The transepithelial potential may originate in an electrogenic potassium pump located in the folded apical membranes of the trichogen and tormogen cells. Evidence for this idea, first presented for mechanoreceptive sensilla (Thurm 1972, 1974), was the immediate drop of the potential after oxygen deprivation and the presence of membrane-associated particles (*portasomes*, Harvey 1980) in the folded membrane. The electrogenic pump may be partially responsible for the nonlinear voltage-current relationship measured across the sensillum (Kaissling & Thorson 1980). The transepithelial potential may serve to enhance the dendritic membrane potential and hence also the receptor potential. Experiments in zygaenid moths, however, showed no direct correlation of transepithelial and receptor potential after poisoning with HCN and during the recovery period (Levinson et al 1973).

The presence of ion pumps in the soma region of the receptor cell and the inner dendritic segment is suggested by the high densities of mitochondria (Thurm 1974, Steinbrecht & Gnatzy 1984). The outer dendritic segment is free of mitochondria and other intracellular organelles except for microtubules (Keil 1984a). Possibly, the outer dendrite has no Nernst potential at least for potassium, because no gradient of this ion is expected across the dendritic membrane (see below).

Ionic Conditions of the Sensillum Lymph

Any model of sensillum function has to take account of the unusual ionic conditions (Kaissling & Thorson 1980, Steinbrecht & Zierold 1982, 1984), especially around the outer dendritic segment: 200 mM of potassium were found in the sensillum lymph of the sensilla trichodea of *Antheraea polyphemus* and *Antheraea pernyi*, which had an osmolarity of about 475 mosmol. Similar potassium concentrations were also found in mechanoreceptive sensilla and may be due to the action of the electrogenic potassium pump (Küppers 1974, Thurm & Küppers 1980, Wieczorek 1982). The hemolymph of these species of moths showed much lower ion concentrations, 40 mM of K and 12 mM of Na.

It is not known which ion carries the receptor current in these sensilla, i.e. the additional current entering the receptor cell dendrite due to chemical stimulation. The sodium concentration in the sensillum lymph was 25 mM. The low Cl/K ratio suggests a lack of anions in the sensillum lymph, a lack that may possibly be compensated by sulfur-containing groups. The extraordinary sulfur concentration, above 100 mM, may indicate the presence of acidic mucopolysaccharides. Strong pheromone stimulation further reduced the Cl/K ratio, whereas the S/K ratio increased (Kaissling & Thorson 1980, Kaissling unpublished). These effects still cannot be explained.

Responses to Single Pheromone Molecules

A combined electrophysiological, behavioral, and radiometric study of *Bombyx mori* led to the conclusion that one molecule of bombykol is sufficient to trigger a nerve impulse (Kaissling & Priesner 1970, Kaissling

1971). The first electrical response to a single pheromone molecule is the so-called elementary receptor potential preceding the nerve impulse (Kaissling 1974, 1977a). These extracellularly recorded events are negatively directed potential fluctuations of an amplitude up to 0.5 mV and a duration of 10–50 ms or more (Figure 2). Sometimes these potentials are rectangular in shape, but ordinarily they are rounded, possibly due to capacitive coupling in the sensillar circuit. According to the static circuit analysis, the elementary potentials reflect a change of conductance in the dendritic membrane in the range of 30 pS. Therefore, these events may indicate opening of single ionic channels in the cell membrane (Kaissling & Thorson 1980).

This model of chemo-electrical transduction does not require amplification mechanisms other than ion gating in a charged cell membrane such as has been found, for instance, in muscle cells sensitive to acetylcholine (Neher & Sakmann 1976). A few investigators have discussed a possible role of adenylcyclase for transduction in insect sensilla (Felt & Vande Berg 1977, Villet 1978).

The coupling between receptor potential and impulse firing may be more complicated than in axonal impulse propagation. Clearly, the elementary receptor potentials are not necessarily connected with nerve impulses; they often occur without being followed by an impulse (Figure 2). Nerve impulses can be selectively blocked without affecting the elementary receptor potentials—for example, by application of the insecticide, permethrin (Kaissling 1980b). On the other hand, impulses can be elicited electrically without preceding elementary potentials (K.-E. Kaissling, unpublished).

At weak stimulus intensities the elementary receptor potentials are irregularly distributed as expected for random molecular stimuli (Figure 2). The reaction time of the responses is several hundred milliseconds on average and varies between several tens of milliseconds and several seconds. This reaction time may reflect the transport of the molecule from the adsorption site at the hair surface through the pore tubule toward the cell membrane, as discussed above. Alternatively, the transport could be quicker, and other stochastic processes, such as the assumed interaction between stimulus and receptor molecule, may be responsible for the delay of the responses.

With increased stimulus intensities, the elementary responses form the usual fluctuating receptor potential. At very high stimulus intensities there is a typical decrease in the amplitude of fluctuation, which could have a number of causes (Kaissling & Thorson 1980): (a) saturation in the number of open channels, (b) a decrease in the duration and/or width of opening of the individual channels, and (c) a nonlinear relation between membrane conductance and voltage.

Strongly reduced fluctuations were also observed in receptor potentials elicited by certain pheromone derivatives (Kaissling 1974, 1977a). This suggests that their lower stimulatory effectiveness was due to smaller ion flow per opened ionic channel. This result is reminiscent of the effects of ligand structure on the lifetime of open ionic channels in acetylcholine receptors (Neher & Sakmann 1976). Curiously, reduced fluctuations of the receptor potential were correlated with faster rise and fall of the receptor potential (Kaissling 1977a). Other pheromone derivatives produced fluctuations similar to those observed with pheromone stimulation but at higher stimulus intensities. Conceivably, the probability of the hypothetical channel activation was reduced but the manner of channel opening was not altered. These results show that consecutive molecular steps of interaction may have different chemical specificities. This may be the basis for the extreme specificity of olfactory cells that sometimes respond 1000-fold less sensitively to little altered derivatives of the key compound.

CELLULAR RESPONSE CHARACTERISTICS

Intensity Coding at the Receptor-Potential Level

Olfactory receptor cells of insects usually respond to a wide range of stimulus intensities covering at least two to three (Kafka 1970, Vareschi 1971, Kaib 1974, O'Connell 1975, Sass 1976, den Otter 1977, Dickens et al 1984, Mustaparta et al 1984), sometimes many more, orders of magnitude (Schneider et al 1967, Dumpert 1972, Zack 1979, Wadhams et al 1982). Often the steady state amplitude of the receptor potential increases proportionally to the logarithm of stimulus intensity until it approaches saturation. Less effective stimulus compounds produce dose response curves shifted toward higher stimulus intensities and often saturating at lower amplitudes of the receptor potential. In addition, they may have a steeper slope in a double log plot (Kaissling 1972, 1974). Shifted dose response curves and steeper curves with lower saturation levels are also observed after adaptation or inhibition of olfactory cells (see below).

Some of the features of dose response curves have been accounted for by a model that includes binding between stimulus compound and receptor molecule, an activation of the receptor molecule and a change of membrane conductance proportional to the number of activated receptors (Kaissling 1977a). This model reveals a relationship between the steady receptor potential amplitude and the stimulus concentration that appears as a typical S-shaped curve in a semilog plot, with a slope of 1 for lower stimulus concentrations in the double-log plot. Flatter dose response curves, with slopes of 0.5, may be produced by electrical cable properties of long and relatively thin dendrites (Kaissling 1971). Formal explanations have been discussed for slopes below 0.5 (Thorson & Biederman-Thorson 1974), which are often found in olfactory cells.

The above model demonstrates that the relative height of the receptor potential is not directly proportional to the fraction of activated receptor molecules. It also shows that shift and lower saturation level of a dose response curve may have quite different causes—a lower ratio of activated/occupied receptors or a lower conductance increase per activated receptor. In this minimal model, processes such as transport of stimulus molecules toward the cell membrane or odor inactivation are neglected or considered as irrelevant for the kinetics of the receptor potential.

As expected from the model, the rise time of the receptor potential decreases with increasing stimulus intensity (Kaissling 1971, Zack 1979). The fall time after end of odor stimulation would be expected to remain constant, indicating the dissociation of the ligand receptor complex, but actually it increases markedly at high stimulus strengths. These deviations evidently occur because processes not included in the model dominate the time course of the response, at least at high stimulus intensities. Thus a greatly prolonged falling phase of the receptor potential may be due to the presence of active odor molecules at the receptor sites after end of stimulation and may reflect processes of odor inactivation. The prolongation is not correlated with the amplitude of the receptor potential because it also occurs with less effective stimulus compounds that produce saturation of the receptor potential at lower amplitudes (Kaissling 1972, 1977a,b).

Intensity Coding and Temporal Resolution at the Nerve Impulse Level

At very low stimulus intensities the distribution of nerve impulses in time corresponds about to that of elementary receptor potentials (Figure 2). At higher stimulus intensities and with stimulus durations above 1 s the impulse response shows a phasic-tonic behavior: that is, an initial burst of impulses followed by a lower level of firing that may remain constant for minutes (Lacher 1964, Kaissling 1985). Immediately after the end of stimulation, a transient break in impulse firing can occur. The subsequent after-response may last for minutes, depending on stimulus strength and duration. Very strong stimuli may cause a brief initial burst of impulses followed by a silent period and a rebound of impulse activity (Zack 1979, Boeckh & Boeckh 1979, Rumbo 1981).

The mechanism for generation of the complicated pattern of impulse activity depending on stimulus strength and duration is not understood. The impulse response seems to be especially adapted to signal the onset rather than the end of a stimulus. Furthermore, permanent levels of low and medium stimulus intensities are well represented by the impulse frequency.

The dynamics of the olfactory response vary among different cell types (Rumbo 1983, Hansen 1984), even within the same sensillum. Thus, the receptor cell for HDA in *Antheraea polyphemus* has a slower rise and fall of the receptor potential and of the impulse frequency than the cell sensitive to the corresponding aldehyde. However, the third cell type present in 30% of the sensilla responds more slowly than the HDA cell (Kaissling 1979). These properties are not correlated with the diameter of the dendrites. They may have a behavioral significance.

Within a natural odor plume insects are exposed to frequent and extensive changes of odor concentration (Murliss & Jones 1981). Correspondingly, it has been found that repetitive odor pulses are more effective stimuli for the orientation behavior of male *Bombyx mori* than constant odor concentrations (Kramer 1985). The resolution of stimulus frequencies in the impulse response of the olfactory cell may be best at high stimulus intensities, at which the latency between adsorption of the stimulus molecules and the nerve impulses is shortest (Boeckh 1962, Kaissling & Priesner 1970). Minimal latencies of about 15 ms for the receptor potential and about 25 ms for the first nerve impulse have been observed in the receptor cell type responding to E-6, Z-11 hexadecadienal of *Antheraea polyphemus* (Kaissling 1985). These cells responded to pulsed stimuli of 20 ms duration at least up to three stimuli per second (Figure 3). At this repetition rate, one or two nerve impulses followed each stimulus onset with a delay of 40 to 110 ms. At repetition rates above 4 stimuli per second not every stimulus elicited a nerve impulse. However, the observed nerve impulses remained time-locked to the stimulus onset. Therefore, averaging over many such cell responses would enable the animal to detect each odor pulse.

Olfactory Adaptation

Sensory adaptation is here understood as the reversible decrease in sensitivity of a receptor cell due to previous adequate stimulation. Adaptation can occur at the level of the receptor potential, the nerve impulse response, and possibly at other levels. Thus, the sensitivity of the impulse response can be significantly more reduced than expected from the reduction of the receptor potential, indicating adaptation of the nerve impulse generator (Zack 1979). The typical changes of the receptor potential observed with adaptation—shift of the dose response curve toward higher stimulus intensities and lower level of saturation (Kaissling 1972, Zack 1979)—may be interpreted by the minimal model as discussed above.

Adaptation of the receptor potential is accompanied by a decrease in the stimulus-induced resistance change (Kaissling & Thorson 1980). This result implies that during adaptation, desensitization of the mechanisms controlling membrane permeability may occur. The role of membrane-bound effects in adaptation is also indicated by the local adaptation observed after local application of pheromone to the hair (Zack 1979). Other possible mechanisms of adaptation may depend upon general changes in the sensillum—for example, altered ionic conditions in the sensillum lymph. In this case one would expect reciprocal cross-adaptation of receptor cells within the same sensillum. In *Antheraea polyphemus* and *Antheraea pernyi*, stimulation with the acetate component of the pheromone (HDA) reduced the sensitivity of the aldehyde receptor cell, but there was a much smaller reciprocal effect of the aldehyde on the actate cell. This result may indicate a direct inhibitory effect of the acetate (HDA) on the aldehyde cell (Zack 1979).

The fact that the excited cell always adapts more than the other cells in the sensillum allows one to distinguish the cells functionally. This method of selective adaptation has been used in cases where cells cannot be distinguished by the size of their nerve impulses (Boeckh et al 1965, Priesner 1979a, Vareschi 1971, Kaissling 1979). Selective adaptation, may, in such cases, indicate that when one type of molecular receptor is desensitized the same cell can still be stimulated via a second receptor type, sensitive to a different stimulus compound. Several receptor types per cell have not been convincingly demonstrated, but may be difficult to prove or to disprove by this method.

Recovery after adaptation, or deadaptation, requires minutes to hours, depending on strength and duration of the conditioning stimulus (Kaissling 1972, Zack 1979). Deadaptation proceeds more slowly in isolated antennae than in whole animal preparations (Zack 1979). Evidently the recovery depends on factors supplied via the hemolymph.

Inhibition of Receptor Cell Response

In many cases spontaneous firing of nerve impulses is suppressed by certain olfactory stimuli, whereas other stimulus compounds excite the same cell. Inhibitory compounds can be excitatory for another receptor cell in the same sensillum. These results suggest different molecular receptors and effector mechanisms in the cell membrane for inhibitory and excitatory effects (Kafka 1970). They also show that inhibition may be a natural way of coding odor stimuli (Boeckh 1962, 1967, Lacher 1964, Schneider et al 1964, Den Otter et al 1978).

In other cases the inhibitory compound has little or no excitatory effect but it inhibits the response to an excitatory stimulus. For example, geraniol inhibits the HDA-receptor cell in *Antheraea pernyi* (Schneider et al 1964), and linalool inhibits the bombykol receptor of *Bombyx* males (Kaissling 1977b). Linalool is a very potent stimulus for a receptor cell type of female *Bombyx mori* that does not respond to bombykol (E. Priesner, unpublished, cited in Boeckh et al 1965). It is not known whether such inhibitory effects are biologically significant. They require relatively high inhibitor concentrations and the inhibition wanes after a few minutes.

Octylamine and other amines inhibit the response to excitatory stimuli in many cells and by themselves produce a small positive receptor potential, indicating hyperpolarization of the cell. In high stimulus concentrations these compounds depolarize the cell and elicit nerve impulses. However, the cell becomes irreversibly damaged. Such compounds may interfere with the lipid portion of the cell membrane and indirectly influence the transducer mechanism in the cell membrane (Kaissling 1972, 1977b).

FINAL REMARKS

Various biochemical inhibitors, including sulfhydryl and amino reagents, have been used to modify the reactivity of antennae (Frazier & Heitz 1975, Villet 1974, Norris et al 1971, 1977, Kasang 1971, Ma 1981). These experiments were undertaken to study the roles of proteins in transduction. However, the sites of action of these drugs have not been determined, nor was it tested whether they are specific for olfactory responses. The involvement of membrane lipids in transduction was suggested by temperature-dependent hysteresis of olfactory responsiveness (Bestmann & Dippold 1983). Most recently it became possible to exchange the fluid in the hair during electrophysiological recording and to apply watersoluble agents and also pheromone stimuli directly to the dendritic membrane of the receptor cell (K.-E. Kaissling, J. Thorson, G. Adamek, unpublished). More refined methods in biochemistry and genetics (Venard & Pichon 1984) together with careful fine structural, and electrophysiological studies may help to compensate for the small size of inset olfactory organs and may make it possible to unravel chemo-electrical transduction in these unique chemoreceptors.

ACKNOWLEDGMENT

The author wishes to thank U. Klein, J. J. de Kramer, T. A. Keil, E. Priesner, and R. A. Steinbrecht for helpful discussions and valuable suggestions and A. Günzel and U. Lauterfeld for technical help.

Literature Cited

Adam, G., Delbrück, M. 1968. Reduction of dimensionality in biological diffusion processes. In *Structural Chemistry and Molecular Biology*, ed. A. Rich, N. Davidson, pp. 198–215. San Francisco: Freeman

Altner, H., Prillinger, L. 1980. Ultrastructure of invertebrate chemo-, thermo-, and hygroreceptors and its functional significance. *Int. Rev. Cytol.* 67:69–139

Bernard, J., Guillet, J. C. 1972. Changes in the

receptor potential under polarizing currents in two insect receptors. *J. Insect Physiol.* 18:2173–87

Bestmann, H. J., Dippold, K. 1983. Temperaturabhängigkeit von Elektroantennogrammen bei Lepidopteren. Pheromone 43 (1). *Naturwissenschaften* 70:47–48

Boeckh, J. 1962. Elektrophysiologische Untersuchungen an einzelnen Geruchsrezeptoren auf den Antennen des Totengräbers (*Necrophorus*, Coleoptera). *Z. Vergl. Physiol.* 46:212–48

Boeckh, J. 1967. Inhibition and excitation of single insect olfactory receptors, and their role as a primary sensory code. *Int. Symp. Olfaction and Taste II*, ed. Hayashi, pp. 721–35. Oxford: Pergamon

Boeckh, J., Boeckh, V. 1979. Threshold and odor specificity of pheromone-sensitive neurons in the deutocerebrum of *Antheraea pernyi* and *A. polyphemus* (Saturnidae). *J. Comp. Physiol.* 132:235–42

Boeckh, J., Kaissling, K.-E., Schneider, D. 1960. Sensillen und Bau der Antennengeißel von *Telea polyphemus*. *Zool. Anat.* 78:559–84

Boeckh, J., Kaissling, K.-E., Schneider, D. 1965. Insect olfactory receptors. *Cold Spring Harbor Symp. Quant. Biol.* 30:263–80

De Kramer, J. J. 1985. The electrical circuitry of an olfactory sensillum in *Antheraea polyphemus*. *J. Neurosci.* 5: In press

De Kramer, J. J., Kaissling, K.-E., Keil, T. 1984. Passive electrical properties of insect olfactory sensilla may produce the biphasic shape of spikes. *Chem. Senses* 8:289–95

Den Otter, C. J. 1977. Single sensillum responses in the male moth *Adoxophyes orana* (F.v.R.) to female sex pheromone components and their geometrical isomers. *J. Comp. Physiol.* 121:205–22

Den Otter, C. J., Schuil, H. A., Sander-Van Oosten, A. 1978. Reception of host-plant odours and female sex pheromone in *Adoxophyes orana* (Lepidoptera: Tortricidae): electrophysiology and morphology. *Ent. Exp. Appl.* 24:370–78

Dickens, J. C., Payne, T. L., Ryker, L. C., Rudinsky, J. A. 1984. Single cell responses of the Douglas-fir beetle, *Dendroctonus pseudotsugae* Hopkins (Coleoptera: Scolytidae) to pheromones and host odors. *J. Chem. Ecol.* 10:583–600

Dumpert, K. 1972. Alarmstoffrezeptoren auf der Antenne von *Lasius fuliginosus* (Latr.) (Hymenoptera, Formicidae). *Z. Vergl. Physiol.* 76:403–25

Erler, G., Thurm, U. 1981. Dendritic impulse initiation in an epithelial sensory neuron. *J. Comp. Physiol.* 142:237–49

Ernst, K.-D. 1969. Die Feinstruktur von Reichsensillen auf der Antenne des Aaskäfers *Necrophorus* (Coleoptera). *Z. Zellforsch. Mikrosk. Anat.* 94:72–102

Ernst, K.-D. 1972. Die Ontogenie der basiconischen Reichsensillen auf der Antenne von *Necrophorus* (Coleoptera). *Z. Zellforsch. Mikrosk. Anta.* 129:217–36

Felt, B. T., Vande Berg, J. S. 1977. Localization of adenylate cyclase in the blowfly labellar chemoreceptors. *J. Insect Physiol.* 23:543–48

Ferkovich, S. M., Mayer, M. S., Rutter, R. R. 1973a. Conversion of the sex pheromone of the cabbage looper. *Nature* 242:53–55

Ferkovich, S. M., Mayer, M. S., Rutter, R. R. 1973b. Sex pheromone of the cabbage looper: Reactions with antennal proteins in vitro. *J. Insect Physiol.* 19:2231–43

Ferkovich, S. M., Mayer, M. S. 1975. Localization and specificity of pheromone degrading enzyme(s) from antennae of *Trichoplusia ni*. *Int. Symp. Olfaction and Taste V*, ed. D. A. Denton, J. P. Coghlan, pp. 337–42. New York: Academic

Ferkovich, S. M., Oliver, J. E., Dillard, C. 1982a. Pheromone hydrolysis by cuticular and interior esterases of the antennae, legs, and wings of the cabbage looper moth, *Trichoplusia ni* (Hübner). *J. Chem. Ecol.* 8:859–66

Ferkovich, S. M., Oliver, J., Dillard, C. 1982b. Comparison of pheromone hydrolysis by the antennae with other tissues after adult eclosion in the cabbage looper moth, *Trichoplusia ni*. *Ent. Exp. Appl.* 31:327–28

Ferkovich, S. M., van Essen, F., Taylor, T. R. 1980. Hydrolysis of sex pheromone by antennal esterases of the cabbage looper, *Trichoplusia ni*. *Chem. Senses* 5:33–46

Frazier, J. L., Heitz, J. R. 1975. Inhibition of olfaction in the moth *Heliothis virescens* by the sulfhydryl reagent fluorescein mercuric acetate. *Chem. Senses Flavor* 1:271–81

Fujishiro, N., Kijima, H., Morita, H. 1984. Impulse frequency and action potential amplitude in labellar chemosensory neurones of *Drosophila melanogaster*. *J. Insect Physiol.* 30:317–25

Futrelle, R. P. 1984. How molecules get to their detectors. The physics of diffusion of insect pheromones. *TINS* 7:116–20

Gnatzy, W., Mohren, W., Steinbrecht, R. A. 1984. Pheromone receptors of *Bombyx mori* and *Antheraea pernyi*. II. Morphometric data. *Cell Tissue Res.* 235:35–42

Guillet, J. C., Bernard, J., Coillot, J. P., Callec, J. S. 1980. Electrical properties of the dendrite in an insect mechanoreceptor: Effects of antidromic or direct electrical stimulation. *J. Insect Physiol.* 26:755–62

Hansen, K. 1978. Insect chemoreception. In *Taxis and Behavior*, ed. G. L. Hazelbauer, pp. 233–92. London: Chapman & Hall

142   KAISSLING

Hansen, K. 1984. Discrimination and production of disparlure enantiomers by the gypsy moth and the nun moth. *Physiol. Entom.* 9:9–18

Harvey, W. R. 1980. Water and ions in the gut. In *Insect Biology in the Future*, ed. M. Locke, D. S. Smith, pp. 105–24. New York: Academic

Kafka, W. A. 1970. Molekulare Wechselwirkungen bei der Erregung einzelner Reichzellen. *Z. Vergl. Physiol.* 70:105–43

Kafka, W. A., Neuwirth, J. 1975. A model of pheromone molecule-acceptor interaction. *Z. Naturforsch. Teil C* 30:278–82

Kaib, M. 1974. Die Fleisch- und Blumenduftrezeptoren auf der Antenne der Schmeißfliege *Calliphora vicina*. *J. Comp. Physiol.* 95:105–21

Kaissling, K.-E. 1971. Insect olfaction. In *Handbook of Sensory Physiology*, ed. L. M. Beidler, 4(1):351–431. Berlin: Springer-Verlag

Kaissling, K.-E. 1972. Kinetic studies of transduction in olfactory receptors of *Bombyx mori*. In *Int. Symp. Olfaction and Taste IV*, ed. D. Schneider, pp. 207–13. Stuttgart: Wiss. Verlagsges.

Kaissling, K.-E. 1974. Sensory transduction in insect olfactory receptors. In *Biochemistry of Sensory Functions*, ed. L. Jaenicke, pp. 243–73. Berlin: Springer-Verlag

Kaissling, K.-E. 1977a. Structures of odour molecules and multiple activities of receptor cells. In *Int. Symp. Olfaction and Taste VI*, ed. J. Le Magnen, P. MacLeod, pp. 9–16. London: Information Retrieval

Kaissling, K.-E. 1977b. Control of insect behavior via chemoreceptor organs. In *Chemical Control of Insect Behavior: Theory and Application*, ed. H. H. Shorey, J. J. McKelvey, Jr., pp. 45–65. New York: Wiley

Kaissling, K.-E. 1979. Recognition of pheromones by moths, especially in Saturniids and *Bombyx mori*. In *Chem. Ecology: Odour Communications in Animals*, ed. F. J. Ritter, pp. 43–56. Amsterdam: Elsevier/North-Holland/Biomed. Press

Kaissling, K.-E. 1980a. Studies on the functional organisation of insect olfactory sensilla (*Antheraea polyphemus* and *A. pernyi*). In *Int. Symp. Olfaction and Taste VII*, ed. H. van der Starre, p. 81. London: Information Retrieval

Kaissling, K.-E. 1980b. Action of chemicals, including (+)trans-Permethrin and DDT, on insect olfactory receptors. In *Insect Neurobiology and Pesticide Action (Neurotox 79)*, pp. 351–58. London: Soc. Chem. Industry

Kaissling, K.-E. 1985. Temporal characteristics of pheromone receptor cell responses in relation to orientation behaviour of moths. In *Mechanisms in Insect Olfaction*, ed. T. L. Payne. Oxford: Univ. Press. In press

Kaissling, K.-E., Klein, U., de Kramer, J. J., Keil, T. A., Kanaujia, S., Hemberger, J. 1985. Insect olfactory cells: Electrophysiological and biochemical studies. In *Molecular Basis of Nerve Activity. Proc. Int. Symp. in Memory of D. Nachmansohn*, ed. J.-P. Changeux, F. Hucho, A. Maelicke, E. Neumann, pp. 173–83. Berlin/New York: de Gruyter

Kaissling, K.-E., Priesner, E. 1970. Die Riechschwelle des Seidenspinners. *Naturwissenschaften* 57:23–28

Kaissling, K.-E., Thorson, J. 1980. Insect olfactory sensilla: Structural, chemical and electrical aspects of the functional organization. In *Receptors for Neurotransmitters, Hormones and Pheromones in Insects*, ed. D. B. Sattelle, L. M. Hall, J. G. Hildebrand, pp. 261–82. Amsterdam: Elsevier/North-Holland

Kanaujia, S., Kaissling, K.-E. 1985. Interactions of pheromone with moth antennae: Adsorption, desorption and transport. *J. Insect Physiol.* 31:71–81

Kasang, G. 1971. Bombykol reception and metabolism on the antennae of the silkmoth *Bombyx mori*. In *Gustation and Olfaction*, ed. G. Ohloff, A. F. Thomas, pp. 245–50. London/New York: Academic

Kasang, G. 1973. Physikochemische Vorgänge beim Riechen des Seidenspinners. *Naturwissenschaften* 60:95–101

Kasang, G., Kaissling, K.-E. 1972. Specificity of primary and seondary olfactory processes in *Bombyx* antennae. In *Int. Symp. Olfaction and Taste IV*, ed. D. Schneider, pp. 200–6. Stuttgart: Wiss. Verlagsges.

Kasang, G., Knauer, B., Beroza, M. 1974. Uptake of the sex attractant ^{3}H-disparlure by male gypsy moth antennae (*Lymantria dispar*) (= *Porthetria dispar*). *Experientia* 30:147–48

Kasang, G., Weiss, N. 1974. Dünnschichtchromatographische Analyse radioaktiv markierter Insektenpheromone. Metaboliten des (^{3}H)Bombykols. *J. Chromatography* 92:401–17

Keil, T. 1982. Contacts of pore tubules and sensory dendrites in antennal chemosensilla of a silkmoth: Demonstration of a possible pathway for olfactory molecules. *Tissue Cell* 14:451–62

Keil, T. A. 1984a. Reconstruction and morphometry of silkmoth olfactory hairs: A comparative study of sensilla trichodea on the antennae of male *Antheraea polyphemus* and *Antheraea pernyi* (Insecta, Lepidoptera). *Zoomorphology* 104:147–56

Keil, T. A. 1984b. Surface coats of pore tubules and olfactory sensory dendrites of a silkmoth revealed by cationic markers. *Tissue Cell* 16:705-17

Keil, T. A. 1984c. Very tight contact of tormogen cell membrane and sensillum cuticle: Ultrastructural basis for high electrical resistance between receptorlymph and subcuticular spaces in silkmoth olfactory hairs. *Tissue Cell* 16:131-35

Keil, T., Steinbrecht, R. A. 1983. Beziehungen zwischen Sinnes-, Hüll- und Gliazellen in epidermalen Mechano- und Chemorezeptoren von Insekten. *Verh. Dtsch. Zool. Ges.* 76:294

Keil, T. A., Steinbrecht, R. A. 1984. Mechanosensitive and olfactory sensilla of insects. In *Insect Ultrastructure*, ed. R. C. King, H. Akai, 2:477-516. New York: Plenum

Klein, U., Keil, T. A. 1984. Dendritic membrane from insect olfactory hairs: Isolation method and electron microscopical observations. *Cell. Molec. Neurobiol.* 4: 385-96

Kramer, E. 1985. Turbulent diffusion and pheromone triggered anemotaxis. In *Mechanisms in Insect Olfaction*, ed. T. L. Payne. London: Oxford Univ. Press. In press

Küppers, J. 1974. Measurements on the ionic milieu of the receptor terminal in mechanoreceptive sensilla of insects. In *Abh. Rheinisch-Westfälische Akad. d. Wiss.*, Vol. 53. *Symp. Mechanoreception*, ed. J. Schwartzkopff, pp. 387-94. Opladen: Westdeutscher Verl.

Lacher, V. 1964. Elektrophysiologische Untersuchungen an einzelnen Receptoren für Geruch, Kohlendioxyd, Luftfeuchtigkeit und Temperatur auf den Antennen der Arbeitsbiene und der Drohne. (*Apis mellifica L*). *Z. Vergl. Physiol.* 48:587-623

Levinson, H. Z., Kaissling, K.-E., Levinson, A. R. 1973. Olfaction and cyanide sensitivity in the six-spot burnet moth *Zygaena filipendulae* and the silkmoth *Bombyx mori*. *J. Comp. Physiol.* 86:209-14

Ma, W. C. 1981. Receptor membrane function in olfaction and gustation: Implications from modification by reagents and drugs. In *Perception of Behavioral Chemicals*, ed. D. M. Norris, pp. 267-87. Amsterdam: Elsevier/North-Holland Biomed. Press

Mankin, R. W., Mayer, M. S. 1984. The insect antenna is not a molecular sieve. *Experientia* 40:1251-52

Mayer, M. S., Ferkovich, S. M., Rutter, R. R. 1976. Localization and reactions of a pheromone degradative enzyme isolated from an insect antenna. *Chem. Senses Flavor* 2:51-61

Morita, H. 1972. Primary processes of insect chemoreception. *Adv. Biophys.* 3:161-98

Morita, H., Yamashita, S. 1959. The back-

firing of impulses in a labellar chemosensory hair of the fly. *Mem. Fac. Sci. Kyushu Univ., Ser. E (Biol.)* 3:81-87

Murliss, J., Jones, C. D. 1981. Fine-scale structure of odour plumes in relation to insect orientation to distant pheromone and other attractant sources. *Physiol. Entomol.* 6:71-86

Mustaparta, H. 1984. Olfaction. In *Chemical Ecology of Insects*, ed. W. J. Bell, R. T. Cardé, pp. 37-70. London: Chapman & Hall. 524 pp.

Mustaparta, H., Tommeras, B. A., Baeckström, P., Bakke, J. M., Ohloff, G. 1984. Ipsdienol-specific receptor cells in bark beetles: Structure-activity relationships of various analogues and of deuteriumlabelled ipsdienol. *J. Comp. Physiol. A* 154: 591-95

Neher, E., Sakmann, B. 1976. Single-channel currents recorded from membrane of denervated frog muscle fibers. *Nature* 260:779-802

Norris, D. M., Ferkovich, S. M., Baker, J. E., Rozental, J. M., Borg, T. K. 1971. Energy transduction in quinone inhibition of insect feeding. *J. Insect Physiol.* 17:85-97

Norris, D. M., Rozental, J. M., Samberg, G., Singer, G. 1977. Protein-sulfur dependent differences in the nerve receptors for repellent 1,4-naphthoquinones in two strains of *Periplaneta americana*. *Comp. Biochem. Physiol. C* 57:55-59

O'Connell, R. J. 1975. Olfactory receptor responses to sex pheromone components in the redbanded leafroller moth. *J. Gen. Physiol.* 65:179-205

Payne, T. L. 1974. Pheromone perception. In *Pheromones*, ed. M. C. Birch, pp. 35-61. Amsterdam/London: North-Holland

Phillips, C. E., Vande Berg, J. S. 1976a. Directional flow of sensillum liquor in blowfly (*Phormia regina*) labellar chemoreceptors. *J. Insect Physiol.* 22:425-29

Phillips, C. E., Vande Berg, J. S. 1976b. Mechanism for sensillum fluid flow in trichogen and tormogen cells of *Phormia regina* (Meigen). (Diptera: Calliphoridae). *Int. J. Insect Morphol. Embryol.* 5(6):423-31

Priesner, E. 1979a. Progress in the analysis of pheromone receptor systems. *Ann. Zool. Ecol. Anim.* 11(4):533-46

Priesner, E. 1979b. Specificity studies on pheromone receptors of noctuid and tortricid Lepidoptera. In *Chemical Ecology: Odour Communication in Animals*, ed. F. J. Ritter, pp. 57-71. Amsterdam: Elsevier/North-Holland/Biomedical Press

Priesner, E. 1985a. Pheromone als Sinnesreize. Verh. 113. *Vers. Ges. Deutsch. Naturf, u. Ärzte*, ed. P. Karlson et al, pp. 207-26. Stuttgart: Wiss. Verlagsges

144   KAISSLING

Priesner, E. 1985b. Correlating sensory and behavioural responses in multi-chemical pheromone systems of Lepidoptera. In *Mechanisms in Insect Olfaction*, ed. M. Birch, C. Kennedy, T. L. Payne. London: Oxford Univ. Press. In press

Rees, C. J. C. 1968. The effect of aqueous solutions of some 1:1 electrolytes on the electrical response of the type 1 ("salt") chemoreceptor cell in the labella of *Phormia*. *J. Insect Physiol.* 14:1331–64

Rumbo, E. R. 1981. Study of single sensillum responses to pheromone in the light-brown apple moth, *Epiphyas postvittane*, using an averaging technique. *Physiol. Entomol.* 6:87–98

Rumbo, E. R. 1983. Differences between single cell responses to different components of the sex pheromone in males of the light brown apple moth (*Epiphyas postvittana*). *Physiol. Entomol.* 8:195–201

Sanes, J. R., Hildebrand, J. G. 1976a. Structure and development of antennae in a moth, *Manduca sexta*. *Dev. Biol.* 51:282–99

Sanes, J. R., Hildebrand, J. G. 1976b. Origin and morphogenesis of sensory neurons in an insect antenna. *Dev. Biol.* 51:300–19

Sass, H. 1976. Zur nervösen Codierung von Geruchsreizen bei *Periplaneta americana*. *J. Comp. Physiol.* 107:49–65

Schneider, D. 1984. Insect olfaction—our research endeavor. *Found. Sens. Science*, ed. W. W. Dawson, J. M. Enoch, pp. 381–418. Berlin/Heidelberg: Springer-Verlag

Schneider, D., Lacher, V., Kaissling, K.-E. 1964. Die Reaktionsweise und das Reaktionsspektrum von Riechzellen bei *Antheraea pernyi* (Lepidoptera, Saturniidae). *Z. Vergl. Physiol.* 48:632–62

Schneider, D., Block, B. C., Boeckh, J., Priesner, E. 1967. Die Reaktion der männlichen Seidenspinner auf Bombykol und seine Isomeren: Elektroantennogramme und Verhalten. *Z. Vergl. Physiol.* 54:192–209

Seabrook, W. D. 1978. Neurobiological contributions to understanding insect pheromone systems. *Ann. Rev. Entomol.* 23:471–85

Slifer, E. H. 1970. The structure of arthropod cheomoreceptors. *Ann. Rev. Entomol.* 15:121–42

Steinbrecht, R. A. 1973. Der Feinbau olfaktorischer Sensillen des Seidendpinners (Insecta, Lepidoptera): Rezeptorfortsätze und reizleitender Apparat. *Z. Zellforsch. Mikrosk. Anat.* 139:533–65

Steinbrecht, R. A. 1984. Arthropoda: Chemo-, thermo-, and hygroreceptors. In *Biology in the Integument*, ed. J. Bereiter-Hahn, A. G. Matoltsy, K. S. Richards, 1:523–53. Berlin: Springer-Verlag

Steinbrecht, R. A., Gnatzy, W. 1984. Pheromone receptors of *Bombyx mori* and *Antheraea pernyi*. I. Reconstruction of the cellular organization of the sensilla trichodea. *Cell Tissue Res.* 235:25–34

Steinbrecht, R. A., Kasang, G. 1972. Capture and conveyance of odour molecules in an insect olfactory receptor. In *Olfaction and Taste IV*, ed. D. Schneider, pp. 193–99. Stuttgart: Wiss. Verlagsges.

Steinbrecht, R. A., Zierold, K. 1982. Cryo-embedding of small frozen specimens for cryo-ultramicrotomy. *Proc. 10th Int. Congr. Electron Microscopy, Hamburg*, 3:183–84. Frankfurt/Main: Dtsch. Ges. Elektronenmikrosk.

Steinbrecht, R. A., Zierold, K. 1984. A cryo-embedding method for cutting ultrathin cryosections from small frozen specimens. *J. Microscopy* 136:69–75

Taylor, T. R., Ferkovich, S. M., Van Essen, F. 1981. Increased pheromone catabolism by antennal esterases after adult eclosion of the cabbage looper moth. *Experientia* 37:729–31

Thorson, J., Biederman-Thorson, M. 1974. Distributed relaxation processes in sensory adaptation. *Science* 183:161–72

Thurm, U. 1963. Die Beziehungen zwischen mechanischen Reizgrößen und stationären Erregungszuständen bei Borstenfeld-Sensillen von Bienen. *J. Comp. Physiol.* 46:351–82

Thurm, U. 1972. The generation of receptor potentials in epithelial receptors. In *Int. Symp. Olfaction Taste IV*, ed. D. Schneider, pp. 95–101. Stuttgart: Wiss. Verlagsges.

Thurm, U. 1974. Basics of the generation of receptor potentials in epidermal mechanoreceptors of insects. In *Abh. Rheinisch-Westf. Akad. der Wiss. Symp. Mechanoreception*, ed. J. Schwartzkopff, 53:355–85. Opladen: Westdeutscher Verlag

Thurm, U., Küppers, J. 1980. Epithelial physiology of insect sensilla. In *Insect Biology in the Future*, ed. M. Locke, D. S. Smith, pp. 735–63. New York: Academic

Träuble, H., Sackmann, E. 1972. Studies of the crystalline-liquid crystalline phase transition of lipid model membranes. III. Structure of a steroid-lecithin system below and above the lipid-phase transition. *Am. Chem. Soc.* 94:4499–4510

Van der Pers, J. N. C., Den Otter, C. J. 1978. Single cell responses from olfactory receptors of small ermine moths to sex-attractants. *J. Insect Physiol.* 24:337–43

Vareschi, E. 1971. Duftunterscheidung bei der Honigbiene—Einzelzell-Ableitungen und Verhaltensreaktionen. *Z. Vergl. Physiol.* 75:143–73

Venard, R., Pichon, Y. 1984. Electrophysiological analysis of the peripheral response to odours in wild type and smell-deficient olf C mutant of *Drosophila melanogaster*. *J. Insect Physiol*. 30:1–5

Villet, R. H. 1974. Involvement of amino and sulphydryl groups in olfactory transduction in silk moths. *Nature* 248:707–9

Villet, R. H. 1978. Mechanism of insect sex-pheromone sensory transduction: Role of adenyl cyclase. *Comp. Biochem. Physiol. C* 61:389–94

Vogel, S. 1983. How much air passes through a silkmoth's antenna? *J. Insect Physiol*. 29:597–602

Vogt, R. G., Riddiford, L. M. 1981a. Pheromone deactivation by antennal proteins of lepidoptera. In *Regulation of Insect Development and Behavioir*, ed. F. Sehnal, A. Zabza, J. Menn, B. Cymborowski, pp. 955–67. Wrocklaw: Polytechnical Univ. of Wrocklaw Press

Vogt, R. G., Riddiford, L. M. 1981b. Pheromone binding and inactivation by moth antennae. *Nature* 293:161–63

Vogt, R. G., Riddiford, L. M., Prestwich, G. D. 1985. Kinetic properties of a phero-

mone degrading enzyme: The sensillar esterase of *Antheraea polyphemus*. *Proc. Natl. Acad. Sci. USA*. In press

Wadhams, L. J., Angst, M. E., Blight, M. M. 1982. Responses of the olfactory receptors of *Scolytus scolytus* (F.) (Coleoptera: Scolytidae) to the stereoisomers of 4-methyl-3-heptanol. *J. Chem. Ecol*. 8:477–92

Wieczorek, H. 1982. A biochemical approach to the electrogenic potassium pump of insect sensilla: Potassium sensitive ATPases in the labellum of the fly. *J. Comp. Physiol*. 148:303–11

Wolbarsht, M. L., Hanson, F. E. 1965. Electrical activity in the chemoreceptors of the blowfly III. Dendritic action potentials. *J. Gen. Physiol*. 48:673–83

Zacharuk, R. Y. 1980. Ultrastructure and function of insect chemosensilla. *Ann. Rev. Entomol*. 25:27–47

Zack, C. 1979. *Sensory adaptation in the sex pheromone receptor cells of saturniid moths*. Dissertation Fak. Biol. 1–99. Ludwig-Maximilians-Universität, München

Ann. Rev. Neurosci. 1986. 9 : 147–70

ON REACHING

Apostolos P. Georgopoulos

The Philip Bard Laboratories of Neurophysiology, Department of
Neuroscience, The Johns Hopkins University, School of Medicine,
Baltimore, Maryland 21205

I deal in this review with behavioral and developmental aspects of arm
movements aimed at targets in immediate extrapersonal space. In addition,
neural mechanisms that may subserve this function are discussed on the
basis of neuronal recording and brain lesion studies. However, this is not a
comprehensive review of each subject; thus, no exhaustive references are
given, nor is a detailed tracing of the history of ideas underlying modern
studies attempted. These subjects can be found in specialized articles within
each of the separate subfields considered above. Moreover, anatomical
findings are not discussed. These as well as other topics related to certain
aspects of cerebrocortical mechanisms subserving reaching movements
were reviewed previously by Humphrey (1979).

THE AIMED MOVEMENT DISSECTED

Arm movements aimed at targets in space typically involve motion at the
shoulder and elbow joints. If the movement is directed to a graspable object,
additional motion at the wrist and finger joints may occur to preshape the
hand and prepare it for grasping the object. In contrast, if the subject directs
a tool to an object, the sequence may be reversed. The concomitant motion
about the shoulder and elbow joints is basic in both cases for the transport
of the hand and/or tool to the desired point in space. These motions result
from the application of torques generated about each joint by the
contraction of muscles. The muscles activated and the temporal course of
the intensity of their contraction will depend on the movement trajectory in
space, the velocity of the movement, and the magnitude and direction of
fixed or changing external loads. Regarding the torques, one has to consider
gravitational, inertial, and Coriolis forces, especially at higher velocities
(Soechting & Lacquaniti 1981, Hollerbach & Flash 1982). The relations

147

between the temporal pattern of these torques and the trajectory of the hand are complicated even for two-dimensional hand trajectories (Hollerbach & Flash 1982). This is true both for deriving the trajectory given the torques (*integral kinematics*) or the torques at each joint given the trajectory (*inverse kinematics*). Inverse kinematics for unrestrained three-dimensional (3-D) arm movements are almost intractable (Saltzman 1979).

The muscle patterns underlying the performance of aimed movements differ according to the movement path and dynamics. Forward reaching movements involve primarily activation of the anterior deltoid to implement the protraction at the shoulder, and various degrees of activation of the biceps to brake the action of gravity on the extending forearm (Soechting & Lacquaniti 1981, Murphy et al 1982). The extension at the elbow is aided variably, or may even be produced by gravity, depending on the height of the target; accordingly, the magnitude of activation of the biceps will also vary widely (Soechting & Lacquaniti 1981). These patterns of muscle activation pertain to forward reaching movements made at (Soechting & Lacquaniti 1981) or near (Murphy et al 1982) the sagittal plane. When the initial position of the hand and the location of the target require aimed movements in different directions, the muscles involved and the patterns of their activation are much more complicated. In an electromyographic (EMG) analysis of movements made by monkeys in eight different directions on a plane tilted 15° from the horizontal toward the animal, it was found that at least ten muscles, acting on the shoulder joint and girdle, were involved (Georgopoulos et al 1984). The magnitude of EMG activation was correlated significantly among several muscles, a finding that suggests that the movements were subserved by muscle synergies. These synergies differed in an orderly fashion with the direction of movement (Georgopoulos et al 1984). We still lack a complete description of the simultaneous muscle patterns during unconstrained arm movements in 3-D space.

The complexity of the muscle patterns of activation underlying aimed and other multijoint movements arises from several factors. First, these movements are implemented by the concomitant contraction of several muscles. Therefore, a population of muscles has to be considered rather than an agonist-antagonist pair, as is usually the case for movements around joints possessing a single degree of freedom (e.g. elbow). Second, the exact contribution of each of these muscles to the torques generated at each joint in the course of the movement is often difficult to measure because the direction along which a muscle will exert its mechanical action may change during the movement. Third, a complete analysis of the muscular patterns of activity subserving these movements has to take into account the relative timing differences between the evolving contractions of the participating muscles, for such differences may affect the movement trajectory. This problem is further complicated by methodological considerations. For example, muscle activation is usually assessed using EMG methods, yet there is no general agreement concerning the relations between the EMG activity and the force exerted by the muscle. This is an especially difficult problem in the case of pinnate muscles, whose fibers do not run along the axis between the insertion points of the muscle: for an approximate assessment of the relation between the EMG activity and the force developed along that axis, and within different compartments of these muscles, the geometrical arrangement of the fibers must be considered.

CHARACTERISTICS OF AIMED MOVEMENTS

Paths and Trajectories

I shall follow Hollerbach & Flash (1982) in distinguishing between path and trajectory. *Path* is the sequence of positions that the hand follows in space; *trajectory* is the time sequence of these successive positions. Several factors can influence the shape of the paths of aimed movements, including the location of the movement path and its direction in extrapersonal space, the attachment of the hand to a manipulandum, or the restriction of motion to particular planes at the elbow and shoulder joints. These effects have not been evaluated carefully. Handpaths of unconstrained reaching movements made on the sagittal plane in the 3-D space are usually straight or slightly curved (Soechting & Lacquaniti 1981). In general, they can be described as curvilinear, with segments of different degrees of curvature. It is significant that these paths are not affected by the speed of the movement (Soechting & Lacquaniti 1981), by large, constant loads, or by changes of the effective length of the arm (Lacquaniti et al 1982).

A different invariance holds for the shape of the velocity profile of the aimed movement. It is dome-shaped and, if no emphasis on accuracy is placed, single-peaked. The curve may become slightly asymmetric with practice, with the ascending, accelerating slope steeper than the descending, decelerating slope (Beggs & Howarth 1972b; see also Soechting 1984). The shape of the velocity profile seems to be unaffected by the location of the movement vector in space (Georgopoulos et al 1981, Morasso 1981), although this has not been investigated systematically in 3-D space. In general, peak velocity increases with movement amplitude, so that the movement time tends to remain constant.

What determines the shape of the velocity profile? Probably the strategy for making the movement and the compromises in performance characteristics that this strategy entails (Nelson 1983). First of all, a basic strategy that human and monkey subjects follow in making aimed movements is to arrive near the target with a single movement that is large, relative to the distance to be covered. Moreover, the acceleration of the movement changes continuously, so that there is no portion of the movement in which the velocity is constant. We do not know why this strategy is adopted. However, we do know that it is acquired gradually during infancy (see below) and that it breaks down in patients with Parkinson's disease. The initial reaching response of infants is composed of a series of smaller movements, and it is only later that the large amplitude, initial component develops (von Hofsten 1979). In Parkinson's disease, the first movement component is too small, and a series of smaller movements is again employed by the patient to get to the target (Flowers 1975).

Nelson (1983) investigated systematically the effects of different strategies on the shape of the velocity profile. This profile is usually dome-shaped: its exact shape depends on the variable being optimized. This is illustrated in Figure 1 for a movement of a fixed duration and distance. It should be kept in mind that the variable being optimized depends on the instructions given to the subject, or, more generally, on the behavioral goal of a particular movement. It is reasonable to suppose that in everyday life there is not a *single* movement variable to be optimized, for the behavioral goals change, depending on, for example, whether one wants to catch a fly, to play the violin, or to shake hands. It seems that several criteria will be involved in each of these cases and it could be supposed that a particular movement will

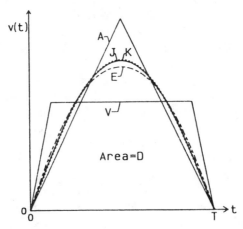

Figure 1 Velocity patterns for the same movement time and distance that are optimum with respect to five different objectives: *A*, minimum peak acceleration (*solid line*); *E*, minimum energy (*dashed line*); *J*, minimum jerk (*solid*); *K*, constant stiffness (*dotted*); and *V*, minimum peak velocity, or impulse (*solid line*). (From Nelson 1983; reproduced by permission from the author and Springer Verlag.)

reflect these different criteria with different weights, according to the particular circumstances. These weights reflect the fact that no single factor is optimized, and, therefore, represent a compromise. Nelson (1983) called these solutions of the velocity shape problem "performance trade-offs." It will be important for future research to describe these trade-offs for different classes of movements and associate them with the behavioral goals that these movements subserve.

Accuracy

The accuracy of an aimed movement is usually referred to the target at which it is aimed. The trajectory itself is not commonly constrained, although accuracy boundaries could be placed on it as well. Fitts (1947) and Fitts & Crannell (1950) investigated the accuracy of blind reaching at targets in different locations in extrapersonal space. In the first series of experiments (Fitts 1947), pilots memorized the letter-coded locations of 20 targets and reached with a pencil at a requested target while fixating a red light in front of them. The subjects did not see the target areas, which were constructed like a bull's eye, so that the accuracy of reaching could be determined directly (Figure 2). It was found that the accuracy of reaching was not uniform throughout extrapersonal space. Most accurate were the movements directed straight ahead, whereas least accurate were those directed on either side, and slightly behind (Figure 3). In subsequent experiments (Fitts & Crannell 1950) these findings were confirmed, and further experiments were performed in which the initial position of the hand was varied; movements started with the hand placed on the sides rather than in front of the body. Although the movements made to targets located directly ahead were still the most accurate, reaching accuracy to other locations differed appreciably from that observed when the starting point of the movement was in front of the body. Finally, movements of smaller distance were also more accurate. These findings indicate that the starting point, the movement amplitude, and the movement endpoint all influence the accuracy of blind reaching.

The relation among accuracy, direction, and extent of aimed movements in the dark were studied by Brown et al (1948). With respect to direction, it was found that aimed movements in the horizontal plane were more accurate when directed away from than toward the body. With respect to the length of the movement, two kinds of errors were distinguished: constant and variable. The former concern the location of the errors relative to the target, whereas the latter pertain to the variance of the error distribution. A range effect was observed for the constant errors. The subjects tended to overshoot the target in small movements (2.5 cm) and to undershoot in large movements (40 cm), whereas in movements of intermediate length no systematic effect was observed. The variable errors increased with increasing movement amplitude. This was probably due to increasing variability in the motor output that initiates larger movements (Schmidt et al 1979). In the case of visual aiming, errors in initial reaching are corrected by additional, smaller movements. The whole aiming response is then composed of at least two parts: an initial, large amplitude movement that brings the hand near the target, and subsequent one or more smaller amplitude movements that bring the hand to its final position. During the first movement the motions at the shoulder and elbow joints are tightly coupled and relate to the position of the target in space (Soechting 1984). In contrast, the smaller, corrective movements affect the motion at the wrist and finger joints, and seem to relate to the angular orientation of

Figure 2 Experimental arrangement used by Fitts (1947) to study the accuracy of blind reaching. Red goggles allowed the subject to look at a light straight ahead but prevented him from seeing the target areas. (From Fitts 1947. Reproduced by permission of AFAMRL, May 1985.)

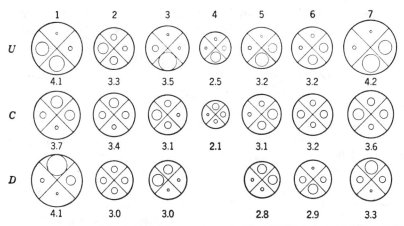

Figure 3 Summary of the results of the experiment by Fitts (1947) illustrated in Figure 2. *U*, *C*, and *D*: rows of targets (7, 7, and 6, respectively) corresponding to the three rows shown in Figure 2. The size of each *circle* is proportional to the average error in reaching (indicated by the *number*, in inches, below each circle) at the corresponding target location: the smaller the circle, the better the accuracy. The *little circles* in the quadrants of the large circles indicate the relative number of errors in each quadrant. (After Fitts 1947. Reproduced by permission of AFAMRL, May 1985.)

the target in space (Soechting 1984). These corrections are probably guided visually (Beggs & Howarth 1972a).

Movement Time and the Speed-accuracy Trade-off

The duration of the aimed movement is a behaviorally important variable, for it specifies how quickly the target will be reached. With respect to dynamics, changing the time of moving between two points involves a substantial recomputation of the joint motion dynamics. Hollerbach & Flash (1982) found that these computations could be simplified by scaling the joint torques according to the speed desired. For example, "If the velocity profile is scaled by a factor r, then simply by scaling the time-dependent portion of the torque program by a factor r^2 and then adding in the gravity contribution without amplitude change, the same path will be followed but at the new speed" (Hollerbach 1982, p. 192). It was found that human subjects seemed to adopt a strategy compatible with this scaling procedure (Hollerbach & Flash 1982). These results suggest that the overall speed of movement could be specified at a behavioral level and implemented without an increased computational load. This would be a behavioral advantage for the subject, especially during motor learning. For example, "One might conceive of a practice strategy beginning with slow movements to learn the basic torque profiles, then simply scaling these profiles to increase the speed of movement" (Hollerbach 1982, p. 192).

What determines the duration of the aimed movement? Behavioral considerations pertaining partly to the duration of the movement itself and partly to the accuracy of the movement appear important. In the absence of appreciable accuracy constraints, the duration of the movement tends to remain constant as the amplitude of the movement increases because, under these conditions, the speed of the movement increases as well. However, when accuracy constraints are present, there is a trade-off between the speed and the accuracy of the aimed movement: more accurate movements are performed more slowly, and conversely, faster movements are less accurate. This phenomenon has been studied extensively (see Keele 1981 for a review). It seems that the increase in duration of movement under conditions that require increased accuracy results from an increase in the number of corrective movements that bring the hand on the target following the large-amplitude first aiming component.

Information Transmission

Fitts (1954) applied an information-theoretical approach to the relations linking movement time, accuracy, and distance. He calculated the rate of information transmission during an aimed movement as follows:

$$I_p = -1/t \log_2 W/2A$$

where I_p, the index of performance, is the information transmitted in bits/s, t is the movement time, W is the target width, and A is the movement amplitude. The term $-\log_2 W/2A$ is the index of task difficulty. It is a composite measure of informational load, relating to both accuracy and distance. The upper limits of information transmitted by human subjects vary with the task (Fitts 1954, Fitts & Peterson 1964) and decrease with increasing age (Welford et al 1969).

Two points are worth mentioning here. First, the considerations above refer to the calculation of information transmission regarding the amplitude of the movement in one dimension. It is important that this approach be extended to include the direction of movement in 3-D space as well. And second, the traditional approach treats the movement in a discrete fashion, by disregarding the time-varying movement trajectory. Yet, if it is assumed that the motor system plans the ongoing movement trajectory, an extension of this approach to the continuous case will be needed.

DEVELOPMENT OF REACHING IN INFANCY

The capacity to reach to objects of interest develops gradually over a period of several months after birth. However, a rudimentary form of eye-hand coordination is present in the newborn, and it is on this background that visual reaching is established. Von Hofsten has performed recently a series of quantitative studies in infants by recording the movements of their arms using two television cameras and following the development of reaching to moving and stationary objects of interest during the first several months of life (von Hofsten 1979, 1980, 1982, 1984, von Hofsten & Lindhagen 1979). The object used in those experiments was a spherical tuft made of bright red, blue, and yellow yarn. It hung down from the end of a 70 cm long rod attached to an electric motor. The distance of the object from the infant's eyes was 12 or 16 cm, depending on the infant's age. The object could be moved at different speeds (3.4, 15, or 30 cm/s) and in either direction along a horizontal circular path of 140 cm diameter, with the center straight behind the infant. In the newborn (von Hofsten 1982), the slowest speed was used. In the presence of the object, there were more forward extensions than other movements, and the movements made while the neonate was looking at the object were aimed closer at the object than were other movements. Moreover, the hand slowed down near the object in the best aimed of these so-called "prereaching" movements. The frequency and form of these movements were studied longitudinally in 23 infants from the first to the sixteenth–nineteenth week of age (von Hofsten 1984). The number of prereaching movements during the presentation of the object decreased gradually, reaching a minimum at seven weeks of age, and then increased dramatically several-fold during the subsequent weeks. A concomitant change was observed in the hand posture accompanying the forward extension of the arm, from an open hand in the beginning to a fisted hand by the seventh week, and to an open hand again in the following weeks.

A detailed analysis of the forward extensions of the arm aimed at the target was performed in infants from the fifteenth–sixteenth to the thirty-sixth week of age (von Hofsten 1979). Remarkable changes in the reaching skill were observed, so that at the end of the observation period it closely resembled the adult pattern. These changes affected mostly the pattern of the reaching response. At the age of 12–16 weeks, this response consisted of a series of movements, as judged by the zero crossings in the acceleration record. About 80% of reachings consisted of three or more successive movements. This resulted in long, fragmented movement paths and in first approaches that covered no more than 40% of the total distance. In contrast, at 36 weeks of age the total movement path had decreased by several times and the first approach now covered more than 70% of the total distance (Figure 4). The fragmentation of the response decreased. In fact, the essence of the development lay in the gradual buildup of a dominance of the first-step movement toward the target: at the end of the observation period the reaching response consisted predominantly of one forceful movement that brought the hand near the target, a pattern that resembles

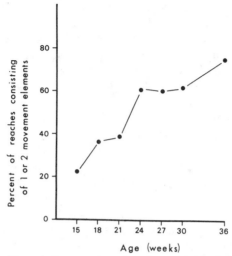

Figure 4   Increase with age in the percentage of reaches consisting of one or two movement elements. (Graph was constructed from data of Table 2 in von Hofsten 1979.)

that observed in adults. Two additional observations are noteworthy. First, in most instances a preference for the use of the right arm was observed; and second, the reaching performance was better to the fast-moving target (at speeds of 15 and 30 cm/s), than to the stationary or slowly-moving (at 3.4 cm/s) target. In fact, there were clear indications that some of the reaching responses were predictive in nature (von Hofsten 1980).

VISUAL GUIDANCE OF REACHING

The behavioral repertoire of reaching to a target does not consist of the arm movement alone. It is made of concomitant movements of the eyes and head toward the target. These responses are generated in parallel, for the latency of *muscle* (EMG) activation of the head and arm muscles and of changes in the electrooculogram (EOG) are almost identical when subjects are instructed to track a visual target, when it appears, by eye, head, and hand as quickly as possible, without specifying a sequence in those movements (Biguer et al 1982). In contrast, the latency of the onset of the *movement* of the eyes, head, and arm differ—that of the saccade is the shortest and that of the arm is the longest (Biguer et al 1982)—due to the different inertial loads that have to be overcome before the movement starts. These observations concerning the simultaneous muscle activation and the rank ordering of the onset times of the eye, head, and arm movements in reaching seem to reflect a central pattern, for they were unaffected when vision of the arm was eliminated (Biguer et al 1982). The sequencing of movement onsets can be of particular significance, because the eye movement may be completed before the arm movement has even started, due to the high velocity of the saccade (Biguer et al 1982; A. P. Georgopoulos, J. F. Kalaska, J. T. Massey, unpublished observation in the rhesus monkey). In fact, if eye or head movements are not allowed, reaching to eccentric targets is very inaccurate (Prablanc et al 1979a). The contribution of the eye and head movements to the accuracy of reaching is probably due to the foveation of the target and the visual monitoring of the hand movement, for allowing the saccade to occur in the absence of vision of the hand did not improve reaching accuracy at eccentric targets (Prablanc et al 1979a).

This introduces the question of visual guidance of reaching. There are at least three aspects of this problem: (a) visual localization of the target in extrapersonal space and suitable coding of that information for use by the arm motor system; (b) visual monitoring of the hand before and during its movement through space; and (c) visual adjustment of the final position of the hand to touch, grasp, or retrieve successfully the object of interest. Since this review deals with reaching, I do not discuss the continuous tracking by the arm of a moving object.

The coding of absolute target position in space is a large subject that is usually treated in the context of perception rather than movement. It will suffice to mention that the perceived location of the target need not coincide with the location used by the motor system to direct the arm. This was shown, for example, in experiments in which subjects were asked to hit with a hammer targets illuminated at different positions in space during the onset of a saccade (Skavenski & Hansen 1978). Remarkably, the subjects hit the targets accurately, despite the fact that during saccades visual perception mechanisms are suppressed (Volkmann et al 1968) and visual localization of objects in space is poor (Matin 1972).

The monitoring of the arm and hand motion throughout the reaching movement is apparently important for accurate performance. It has been shown unequivocally that reaching is more accurate in the presence than in the absence of vision of the arm just before (Prablanc et al 1979b) and during the movement (Conti & Beaubaton 1976, Prablanc et al 1979a, 1979b). Since this improvement was observed even for movements that were completed within 200 ms, it was proposed (Paillard 1982) that visual cues from arm motion are being processed at higher speeds than the times (190–260 ms) assumed necessary to utilize external visual feedback (Keele & Posner 1968).

Two visual systems have been identified that process information related to arm/hand movement during reaching (Paillard 1982, Jeannerod & Biguer 1982). Their contribution to the visual guidance of movement and the cues they use have been studied in experiments that allowed separate control of target and hand vision through a colored filter in normal and split-brain monkeys (Beaubaton et al 1978). In other experiments, continuous or stroboscopic illumination was used to dissociate position cues from motion cues during the course of prismatic adaptation in human subjects (Paillard et al 1981). It was found that one system utilizes central vision (8°), is facilitated by the presence of a foveated target, and analyzes positional (displacement) cues, for it is unaffected under conditions of stroboscopic illumination (Paillard et al 1981). Presumably this system subserves the accurate placement of the hand on the target near the end of the reaching movement. The other system employs peripheral vision and analyzes motion cues, as evidenced by its impairment under conditions of stroboscopic illumination. More importantly, the motion cues that seem to be meaningful to this system are those arising from the motion of the arm when *actively* moved by the subject but not when passively moved by the experimenter (Paillard et al 1981). These findings are consistent with earlier results that suggested that the development of visually guided reaching depends on "self-produced movement with its concurrent visual feedback" (Held & Hein 1963). In general, the two visual systems mentioned above resemble the distinction made previously by Trevarthen (1968).

Paillard (1982) has summarized the implications of these findings as follows. Three aspects of visual information are utilized in reaching. The first concerns the visual localization of the target in space. Although this information can be restricted to one hemisphere, it can be effectively used to trigger reaching by either arm. The second piece of visual information concerns the relative position of hand and target; and the third comes from the motion of the limb across the visual fields. The last two cues are processed most efficiently by the hemisphere contralateral to the moving arm (Beaubaton et al 1978).

PLANNING OF REACHING

Reaching as a behavioral act is the result of a complex sensorimotor coordination. Reaching itself is a complicated multijoint movement directed to a defined point in space and performed under behavioral and biomechanical constraints. In self-paced, cyclical tasks in which alternate reaching movements are performed between two targets, the planning, initiation, and execution of successive movements will overlap partially in time. In contrast, in reaction tasks in which, for example, a fast aiming movement has to be made in response to a stimulus, it can be assumed that a good part of the planning process happens during the reaction time (RT). Several aspects of this process have been studied successfully using RT paradigms. The basic assumption is that the duration of the RT reflects the difficulty in generating a response, given a certain stimulus. The least complicated case is that of a simple RT task, in which a response (e.g. movement of the hand) is required as soon as a stimulus is presented. In contrast, in choice RT tasks the response is contingent on a decision concerning specific attributes of the stimulus. For example, given a set of two responses (e.g. movement of the left or right hand) and two stimuli (e.g. a blue and a red light), a choice RT task could be as follows: "Move the right hand in response to the blue light, and the left hand in response to the red light." Choice RT increases with the number of alternatives (N) involved in decision making; choice RT is a linear function of stimulus uncertainty ($\log_2 N$) (Hick 1952).

Is the RT of a movement aimed at a target a simple or a choice RT? It is not a simple RT because the response is constrained by the location of the stimulus (see also Prablanc et al 1979a), but if it is a choice RT, what are the alternative choices? Assuming that the movement starts from the same point in space and is of constant amplitude and accuracy, it is reasonable to suppose that the alternatives will be determined by the number of targets presented in a task, for each target specifies a different movement. Yet, remarkably, the RT under these conditions may increase only slightly or not at all with stimulus uncertainty (Sanders 1967, Georgopoulos et al 1981), probably because reactions toward the source of stimulation are fast (Simon 1969), and because aimed movements are usually highly practiced and possess a high degree of spatial compatibility with the target. Both of the latter factors have been shown to reduce the effect of stimulus uncertainty on choice RT (Mowbray & Rhoades 1959, Brainard et al 1962). Therefore, aimed movements seem to be generated very efficiently, as if there were a direct link between the target and the movement directed to it. This link is probably based on the strong similarity between the ways in which the target and the aimed movement are coded spatially, according to theories of stimulus-response compatibility in the spatial domain (Nicoletti et al 1982, Duncan 1977, Wallace 1971).

Another interesting feature in the planning of the aimed movements is that they can be elicited in quick succession as responses to targets changing during the RT, without delays usually observed in other tasks under similar conditions; that is, without an appreciable psychological refractory period (PRP). The PRP is a delay beyond the normal RT that is observed in the response to the second of two stimuli presented in quick succession. The occurrence of a PRP has been well documented (see Bertelson 1966 for review). The most widely accepted explanation is the single-channel theory, which postulates that there exists a gated channel of limited capacity that cannot handle the stimulus-response requirements of both stimuli simultaneously; therefore, the generation of responses to the two stimuli are treated sequentially and without overlap. When the second stimulus is presented during the RT to the first stimulus, this successive processing of information results in a delayed response to the second stimulus (i.e. in addition to its own RT); this additional delay equals approximately $RT_1 - ISI$ (where RT_1 is the RT to the first stimulus, and ISI is the interstimulus interval). Indeed, a PRP of the predicted magnitude has been observed in several studies (see Bertelson 1966 for review). However, a PRP is not observed when the required responses are movements of the same band aimed to different targets. This has been documented in experiments involving both monkeys (Georgopoulos et al 1981, 1983b) and human subjects (Soechting & Lacquaniti 1983, Massey et al 1985). This result probably reflects the highly efficient information processing of aimed movements, an efficiency probably arising from the high spatial stimulus-response compatibility and extensive practice, as discussed in the preceding paragraph. In fact, it was found that the information transmitted during the second of two aimed movements emitted in quick succession was much more than that of a single aimed movement (Massey et al 1985).

The considerations above refer to the spatial aspects of the aimed movement. A different question concerns the planning of its temporal characteristics and dynamics. For example, the desired total movement time can be specified on behavioral grounds for making a slow, intermediate, or fast movement, in relation, of course, to the accuracy desired. Similarly, other factors will be incorporated to complete the final planning of the movement trajectory; for example, the energy to be spent, the jerk to be avoided, etc. All of these general considerations come under Nelson's (1983) term "performance trade-offs" (see above) and permeate the decision making concerning the dynamic shaping of the trajectory. Given these constraints, the specific dynamics can then be calculated; namely, the torques to be applied to the appropriate joints to generate the desired trajectory of the aimed movement. These torques could be recalculated efficiently using scaling procedures (Hollerbach & Flash 1982) to implement changes in speed for the same trajectory. Finally, the appropriate combination of muscles will be activated in the correct sequence and strength. These processes have been discussed succinctly by Hollerbach (1982). The possibility has also been discussed that a solution to this problem might be achieved by a process that requires specification of only the endpoint of the movement; the shortcomings of this idea when applied to multijoint movements have been pointed out (Hollerbach & Flash 1982, Hollerbach 1982).

Hollerbach (1982) considered "a three-level hierarchical movement plan which converts a movement command to muscle activations by first planning the movement at the object level, then translating the object trajectory into coordinated joint movement, and finally converting from joint movement to muscle activations" (Hollerbach 1982, p. 190). The question then is, at which level is the movement planned? It seems that the answer will depend on the instructions given to the subject and on the particular peripheral motor conditions. Assuming that we are dealing with nonisometric muscle contractions of sufficient strength, there will always be some movement. If now the instruction to a subject is, "extend your elbow," there is no *a priori* reason to assume planning at the object level; and if, for example, under conditions of biofeedback a subject is instructed to increase the EMG activity of the biceps, the unconstrained forearm will probably move, yet there is no need to assume that that motion was planned at all, although it did occur. If, however, a subject is asked to describe a circle in the air with the index finger, it is reasonable to assume that the trajectory will be planned at the object level, the more so the less familiar the required trajectory will be. Now, since the motion of the hand will be effected by changes in the angles of joints produced by the action of muscles, it is obvious that planning and/or transformations from one level to the other will take place. The real problem is to identify the level at which a neuron or a neuronal population participates in the generation of the movement. The problem is further complicated by the fact that this relation may change as a function of time. For example, the pattern of discharge of a particular cell during the early part of the RT may reflect input from cells operating at a higher hierarchical level, and, conversely, its discharge later on may reflect feedback from a lower level. These are important questions that await rigorous treatment at the theoretical and practical levels.

EFFECTS OF BRAIN LESIONS

Reaching to targets will be adversely affected by general motor disorders that affect the initiation, performance, or braking of the movement. These disorders include, for example, paralysis or paresis, akinesia, hypotonia, ataxia, and others. In some cases it is difficult to distinguish between such a fundamental disorder and a particular defect in aiming. In contrast, the accuracy of reaching can be affected in the presence of normal motor function. This is mostly observed with lesions of the posterior parietal areas of the cerebral cortex. Defects in reaching have been observed both in human subjects and in subhuman primates. Parietal lobe syndromes with misreaching are heterogeneous and comprise, mainly, the syndromes of optic ataxia (Balint 1909, Hecaen & Ajuriaguerra 1954, Rondot et al 1977) and visual disorientation (Holmes 1918, Brain 1941). Misreaching cannot be accounted for by motor disability of the limb or a visual defect. Thus, other movements are performed well, and visual functions in the field where misreaching occurs are carried out normally, despite occasional presence of amblyopia. In cases with visual disorientation there is usually a disturbance of perceived spatial relations regarding the absolute distance of objects from the patient's body, especially in the sagittal plane. In addition, the distance between objects and the motion of objects may be misjudged. Stereopsis is usually intact. Infrequently, a disturbance with orientation in extrapersonal space and route finding may be present. These perceptual disturbances could account for a substantial part of the inaccurate reaching.

Patients with optic ataxia, however, show a severe impairment in visually guided reaching in the absence of perceptual disturbance in estimating distance. In several cases an oculomotor disorder is also present (Balint 1909), but in others eye movements are normal (Rondot et al 1977). The disorder can be complicated, for it can be localized in homonymous half-fields or it can affect only one hand (Rondot et al 1977). Thus, neural mechanisms related to the visual localization of the target or to the guidance of the hand toward the target can apparently be differentially affected. The performance of patients with optic ataxia following unilateral posterior parietal lobe lesions was studied quantitatively by Perenin & Vighetto (1983) in a reaching task similar to that employed by Prablanc et al (1979a,b). The accuracy of pointing (with the head immobilized) was normal when concomitant, orienting movements of the eyes as well as view of the moving arm were allowed. Accuracy fell when either of these conditions was not fulfilled, especially when vision of the moving arm was obstructed. Most inaccurate performance was observed when the contra-lateral hand reached to targets in the contralateral visual field, relative to the side of the cortical lesion.

Misreaching has been observed following lesions of the posterior parietal cortex of monkeys (Ettlinger & Kalsbeck 1962, Hartje & Ettlinger 1973, LaMotte & Acuna 1975, Faugier-Grimaud et al 1978). Constant errors are usually toward the medial side of the target. Impaired reaching is present in the light or dark, and affects predominantly the contralateral, and to a lesser extent the ipsilateral, hand.

The reaching sequence is composed of an initial, large amplitude movement about the shoulder and elbow joints that brings the hand near the object of interest, and of subsequent, smaller movements about the wrist and finger joints that orient the hand and the fingers and prepare them for touching accurately, grasping, or retrieving the object (see above). These two parts seem to be subserved by different neuronal systems: the first by the medial, the second by the lateral system of Kuypers (1964). This was suggested by the differential effects of lesions interrupting selectively the medial (Lawrence & Kuypers 1968a) or the lateral descending systems (Lawrence & Kuypers 1968b). It appears that the proximal reaching movement can be controlled effectively by either cerebral hemisphere, but the relatively independent finger movements seem to be controlled predominantly by the contralateral hemisphere. The visual guidance of the latter, distal movements seems to depend on intrahemispheric pathways arising probably in the occipital lobe (Haaxma & Kuypers 1975). It is noteworthy that patients with optic ataxia show a disturbance in the visual guidance of both proximal and distal movements associated with reaching (Perenin & Vighetto 1983).

NEURONAL RECORDING STUDIES

Mountcastle et al (1975) described a class of cells found in the posterior parietal cortex (areas 5 and 7) of the behaving monkey that were activated only with active projection of the arm in extrapersonal space ("reaching neurons"). These neurons were subsequently found to be activated with reaching in complete darkness (Mountcastle et al 1980). Classes of neurons with a rich variety of functional response properties, however, have been observed in area 7 of the monkey cortex, including neurons with responses to stationary or moving visual stimuli (see Lynch 1980 for a review), to the fixation of gaze in 3-D space (Sakata et al 1980), and to the attentional state (Mountcastle et al 1984). How these neuronal subsets interact and ultimately contribute to the successful reaching remains to be elucidated.

Murphy et al (1982) recorded the activity of cells in the motor cortex during forward reaching by monkeys. Intracortical microstimulation at the site of single cell recording produced, for every cell studied, a simple joint rotation. There were three salient findings of this study, among others. First, no simple relation was observed between the single cell activity and the EMG, even when the muscles from which the EMG was recorded were activated by the microstimulation. Second, single cells related to motion about the shoulder or elbow joints behaved similarly in the task, although the motions about these joints were produced quite differently. Finally, the discharge of shoulder-related cells varied systematically with the movement trajectory. These workers concluded that "the production of any movement, however complex or discrete it may seem peripherally, engages a complex population of precentral neurons, such that any one neuron may behave similarly for overtly different movements" (Murphy et al 1982, p.

144). It seems, then, that the unique trajectory of the movement might be represented at the level of the neuronal population and not of the single neuron.

This problem was investigated by Georgopoulos et al (1982, 1983a), who studied neuronal discharge in relation to the direction of 2-D aimed movements in the motor and posterior parietal (area 5) cortex (Kalaska et al 1983) of behaving monkeys. The discharge of single cells was broadly tuned to the direction of the 2-D movement, frequently in a sinusoidal fashion. We have recently extended these studies in 3-D space and observed that the discharge of motor cortical cells is indeed tuned to the 3-D movement direction (A. B. Schwartz, R. Kettner, and A. P. Georgopoulos, unpublished). An example is shown in Figure 5. Monkeys made movements (1–8, Figure 5, top) from the center to the corners of a 15 cm edge cube. The center of the cube was in the middle, straight ahead, at shoulder level, and at a distance of 22 cm from the animal's body. Impulse activity of a motor cortical cell whose activity varied in an orderly fashion with movement direction is shown in the middle of Figure 5; the cell's preferred direction is plotted on the bottom.

Although motor cortical cells possess a preferred direction, their broad tuning suggests that the coding of movement direction may be, instead, a function of a neuronal population. Indeed, this was found to be the case, given certain assumptions ("vector hypothesis," Georgopoulos et al 1983a, 1984). Briefly, the relations between the direction of movement and the population discharge were formulated within a spatial vector context, in which a population "vector" is derived on the basis of the preferred direction and the change in activity of individual constituent cells. Experimental results showed a good correspondence between the direction of the population vector and the direction of the upcoming movement (Georgopoulos et al 1983a, 1984). Now that the directional tuning can be generalized to the 3-D case (see above), it will be interesting to test the hypothesis under these general conditions.

The discussion above referred to the neural coding of spatial attributes of the aimed movement. Unfortunately, we lack knowledge of how more abstract characteristics of the aimed movement, e.g. its accuracy, are represented in the brain, and the same is true for its visual guidance. It is remarkable, however, that studies in brain areas related to motor control of the arm have not revealed invariances relating to the endpoint of the aimed movement; instead, a clear relationship to movement parameters (direction and amplitude) has been observed (Georgopoulos et al 1982, 1983c, Kalaska et al 1983). Indeed, when monkeys made movements from several points on a circle to the same final endpoint, neuronal discharge in the motor and parietal cortex was related to the movement direction but not to

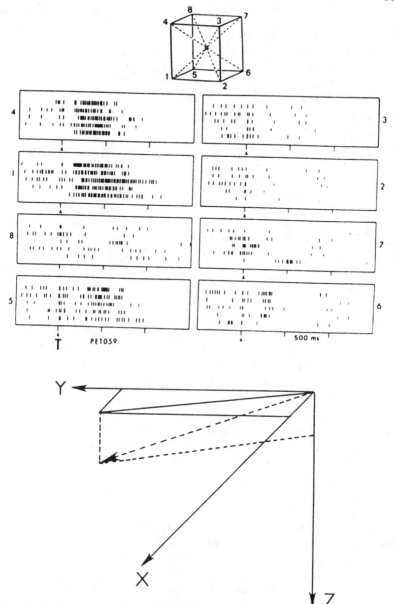

Figure 5 Three-dimensional tuning of a motor cortical cell (*dashed line with arrow, bottom*). *Boxes* (*middle*) show five repetitions of impulse activity (*T* = target onset); *numbers* on left of boxes correspond to numbered movements made from the center to the corners of the cube shown on top. See text for explanation. (A. B. Schwartz, R. Kettner, and A. P. Georgopoulos, unpublished observations.)

the final position of the hand (Georgopoulos et al 1985). These results indicate that, if the endpoint control hypothesis (Polit & Bizzi 1979) is true, the motor and parietal areas studied are involved after the process that specifies the endpoint.

The neuronal events subserving the aimed movement at the level of the spinal cord are largely unknown. Ultimately, of course, the appropriate motoneuronal pools will be excited and/or inhibited to implement the reaching movement. These pools could be addressed *directly* at their segments by supraspinal signals, but the possibility exists that this action is *indirect*, via propriospinal neuronal networks. In fact, a series of studies by Lundberg and his colleagues in the cat have shown that this may be the case. A population of propriospinal neurons at the C3–C4 level has been identified that receives monosynaptic, convergent input from almost all the major descending motor tracts (cortico-, rubro-, tecto, reticulospinal tracts) and, in turn, excites monosynaptically or inhibits disynaptically motoneurons innervating forelimb muscles (Illert et al 1978). Surgical interruption of the projection from these propriospinal neurons to the segmental forelimb pools resulted in a severe impairment of reaching but not of grasping (Alstermark et al 1981). These workers suggested that "the precise aiming in the normal target-reaching movement depends exclusively or very largely on the C3–C4 propriospinal neurons and that the segmental neuronal circuitry alone cannot appropriately substitute for them" (Alstermark et al 1981, p. 315). The findings about this propriospinal system indicate that it may subserve the synergistic activation of the motoneuronal pools involved in reaching. Moreover, the rich convergence on these neurons suggests that several supraspinal systems possess a direct access to the reaching synergy for its on-line modification, if needed, according to changes in the environment or in the behavioral goal.

CONCLUDING REMARKS

The temporal patterns of activity in the muscle populations subserving aimed movements reflect spatiotemporal patterns of activity within the motoneuronal pools innervating those muscles, and these, in turn, are the result of more central influences. What picture, then, might one expect to encounter under these conditions at levels of the central nervous system (CNS) successively more remote from the muscles and their motoneuronal pools and progressively closer to the sensory system in which the target was presented? The answer to this question depends upon the theoretical formulation of the problem (that is, of what was termed "picture" above) and upon the underlying hypotheses concerning the CNS processes that lead ultimately to the patterned activation at the spinal cord level.

However, some general statements can be made. First, these processes involve populations of neurons that may ultimately exert their action either on motoneuronal pools at the segmental level or on integrative propriospinal systems. And second, they proceed in parallel in separate structures. Both of these statements are supported by studies of the single cell activity in several brain areas during motor performance by monkeys. Thus, it has been shown that participation of a structure in the control of movement involves the activation of a population of cells; and that changes in neuronal activity occur in overlapping times in different structures (see, for example, Georgopoulos et al 1983c, Kalaska et al 1983, Thach 1978).

It is important to realize that the information processing relating to aimed movements, and also to other movements, is performed by neuronal ensembles and that this processing can be properly understood only at the population level. The relationship between the ensemble activated and the movement would be reflected both in the time course of changes in activity of individual neurons as well as in the relative timing of these changes among the constituent neurons of the ensemble. The latter cannot be derived from one-at-a-time recordings of single cell activity. In any case, assuming that the determination of activity in neuronal ensembles is feasible, the crucial problem is to define and measure the information processing involved. This remains one of the most fundamental and urgent problems of modern CNS neurophysiology. It should be noted in this context that active neuronal ensembles are not homogeneous, for they are commonly composed of cells possessing different attributes. This differentiation may be important in understanding the function of a structure. A common approach has been the parcellation of the population into subsets of cells with similar functional properties. Another pervasive principle is based on the origin of the anatomical input and the destination of the output of cells in a population. Since the information processing in a subset can be thought of as a recoding of the input into the output, knowledge of the anatomical connectivity could give valuable clues regarding this recoding operation. Let us assume then, for the sake of argument, that not only can we decipher the operations of a given neuronal subset but, even more importantly, we can understand the rules of communication between subsets belonging to different structures. This would lead to the identification and functional characterization of a distributed system (Mountcastle 1978) and its contribution to the control of a given function—in this case, of the aimed movement.

ACKNOWLEDGMENT

This work was supported by USPHS Grant NS17413.

Literature Cited

Alstermark, B., Lundberg, A., Norrsell, U., Sybirska, E. 1981. Integration in descending motor pathways controlling the forelimb in the cat. 9. Differential behavioral defects after spinal cord lesions interrupting defined pathways from higher centres to motoneurones. *Exp. Brain Res.* 42 : 299–318

Balint, R. 1909. Seelenlahmung des "Schauens", Optische Ataxie, raumliche Storung der Aufmarksamkeit. *Mon. Psychiat. Neurol.* 25 : 51–81

Beaubaton, D., Grangetto, A., Paillard, J. 1978. Contribution of positional and movement cues to visuo-motor reaching in split-brain monkey. In *Structure and Function of Cerebral Commissures*, ed. I. Russell, M. W. van Hoff, G. Berlicchi, pp. 371–84. Baltimore, MD : University Park Press

Beggs, W. D. A., Howarth, C. I. 1972a. The accuracy of aiming at a target. *Acta Psychol.* 36 : 171–77

Beggs, W. D. A., Howarth, C. I. 1972b. The movement of the hand towards a target. *Q. J. Exp. Psychol.* 24 : 448–53

Bertelson, P. 1966. Central intermittency twenty years later. *Q. J. Exp. Psychol.* 18 : 153–63

Biguer, B., Jeannerod, M., Prablanc, C. 1982. The coordination of eye, head, and arm movements during reaching at a single visual target. *Exp. Brain Res.* 46 : 301–4

Brain, W. R. 1941. Visual disorientation with special reference to lesions of the right cerebral hemisphere. *Brain* 64 : 244–72

Brainard, R. W., Irby, T. S., Fitts, P. M., Alluisi, E. A. 1962. Some variables influencing the rate of gain of information. *J. Exp. Psychol.* 63 : 105–10

Brown, J. S., Knauft, E. B., Rosenbaum, G. 1948. The accuracy of positioning reactions as a function of their direction and extent. *Am. J. Psychol.* 61 : 167–82

Conti, P., Beaubaton, D. 1976. Utilization des informations visuelles dans le controle du movement : Etude de la precision des pontages chez l'homme. *Travail Humain* 39 : 19–32

Duncan, J. 1977. Response selection rules in spatial choice reaction tasks. In *Attention and Performance VI*, ed. S. Dornic, pp. 49–61. Hillsdale, NJ : Erlbaum

Ettlinger, G., Kalsbeck, J. E. 1962. Changes in tactile discrimination and in visual reaching after successive and simultaneous bilateral posterior ablations in the monkey. *J. Neurol. Neurosurg. Psychiat.* 25 : 256–68

Faugier-Grimaud, S., Frenois, C., Stein, D. G. 1978. Effects of posterior parietal lesions on visually guided behavior in monkeys. *Neuropsychologia* 16 : 151–68

Fitts, P. M. 1947. A study of location discrimination ability. In *Psychological Research on Equipment Design*, ed. P. M. Fitts, pp. 207–17. Washington, DC : USGPO

Fitts, P. M. 1954. The information capacity of the human motor system in controlling the amplitude of the movement. *J. Exp. Psychol.* 47 : 381–91

Fitts, P. M., Crannell, C. 1950. Location discrimination. II. Accuracy of reaching movements to twenty-four different areas. *USAF Air Materiel Command Tech. Rep.* 5833

Fitts, P. M., Peterson, J. R. 1964. Information capacity of discrete motor responses. *J. Exp. Psychol.* 67 : 103–12

Flowers, L. A. 1975. Ballistic and corrective movements on an aiming task : Intention tremor and parkinsonian movement disorders compared. *Neurology* 25 : 413–21

Georgopoulos, A. P., Kalaska, J. F., Massey, J. T. 1981. Spatial trajectories and reaction times of aimed movements : Effects of practice, uncertainty, and change in target location. *J. Neurophysiol.* 46 : 725–43

Georgopoulos, A. P., Kalaska, J. F., Caminit, R., Massey, J. T. 1982. On the relations between the direction of two-dimensional arm movements and cell discharge in primate motor cortex. *J. Neurosci.* 2 : 1527–37

Georgopoulos, A. P., Caminiti, R., Kalaska, J. F., Massey, J. T. 1983a. Spatial coding of movement : A hypothesis concerning the coding of movement direction by motor cortical populations. *Exp. Brain Res.* (*Suppl.*) 7 : 327–36

Georgopoulos, A. P., Kalaska, J. F., Caminiti, R., Massey, J. T. 1983b. Interruption of motor cortical discharge subserving aimed arm movements. *Exp. Brain Res.* 49 : 327–40

Georgopoulos, A. P., DeLong, M. R., Crutcher, M. D. 1983c. Relations between parameters of step-tracking movements and single cell discharge in the globus pallidus and subthalamic nucleus of the behaving monkey. *J. Neurosci.* 3 : 1586–98

Georgopoulos, A. P., Kalaska, J. F., Crutcher, M. D., Caminiti, R., Massey, J. T. 1984. The representation of movement direction in the motor cortex : Single cell and population studies. In *Dynamic Aspects of Neocortical Function*, ed. G. M. Edelman, W. E. Gall, W. M. Cowan, pp. 501–24. New York : Wiley

Georgopoulos, A. P., Kalaska, J. F., Caminiti, R. 1985. Relations between two-dimensional arm movements and single cell discharge in motor cortex and area 5: Movement direction versus movement endpoint. *Exp. Brain Res. (Suppl.)* 10:175–83

Haaxma, R., Kuypers, H. G. J. M. 1975. Intrahemispheric cortical connexions and visual guidance of hand and finger movements in the rhesus monkey. *Brain* 98:239–60

Hartje, W., Ettlinger, G. 1973. Reaching in light and dark after unilateral posterior parietal ablations in the monkey. *Cortex* 9:346–54

Hecaen, H., de Ajuriaguerra, J. 1954. Balint's syndrome (psychic paralysis of visual fixation) and its minor forms. *Brain* 77:373–400

Held, R., Hein, A. 1963. Movement-produced stimulation in the development of visually guided behavior. *J. Comp. Physiol. Psychol.* 56:872–76

Hick, W. E. 1952. On the rate of gain of information. *Q. J. Exp. Psychol.* 4:11–26

Hollerbach, J. M. 1982. Computers, brains and the control of movement. *Trends Neurosci.* 5:189–92

Hollerbach, J. M., Flash, T. 1982. Dynamic interactions between limb segments during planar arm movement. *Biol. Cybernet.* 44:67–77

Holmes, G. 1918. Disturbances of visual orientation. *Br. J. Ophthalmol.* 2:449–68; 506–16

Humphrey, D. R. 1979. On the cortical control of visually directed reaching: Contributions by nonprecentral motor areas. In *Posture and Movement*, ed. R. E. Talbott, D. R. Humphrey, pp. 51–112. New York: Raven

Illert, M., Lundberg, A., Padel, Y., Tanaka, R. 1978. Integration in descending motor pathways controlling the forelimb in the cat. 5. Properties of and monosynaptic excitatory convergence on C3–C4 propriospinal neurones. *Exp. Brain Res.* 33:101–34

Jeannerod, M., Biguer, B. 1982. Visuomotor mechanisms in reaching within extrapersonal space. In *Analysis of Visual Behavior*, ed. D. J. Ingle, M. A. Goodale, R. J. W. Mansfield, pp. 387–409. Cambridge, MA: MIT

Kalaska, J. F., Caminiti, R., Georgopoulos, A. P. 1983. Cortical mechanisms related to the direction of two-dimensional arm movements: Relations in area 5 and comparison with motor cortex. *Exp. Brain Res.* 51:247–60

Keele, S. W. 1981. Behavioral analysis of movement. In *Handbook of Physiology. The Nervous System*. Sect. 1, 2(2):1391–1414. Bethesda, MD: Am. Physiol. Soc.

Keele, S. W., Posner, M. I. 1968. Processing of visual feedback in rapid movements. *J. Exp. Psychol.* 77:155–58

Kuypers, H. G. J. M. 1964. The descending pathways to the spinal cord, their anatomy and function. In *Organization of the Spinal Cord*, ed. J. C. Eccles, J. C. Shade, pp. 178–200. Amsterdam: Elsevier

Lacquaniti, F., Soechting, J. F., Terzuolo, C. A. 1982. Some factors pertinent to the organization and control of arm movements. *Brain Res.* 252:394–97

LaMotte, R. H., Acuna, C. 1975. Defects in accuracy of reaching after removal of posterior parietal cortex in monkeys. *Brain Res.* 139:309–26

Lawrence, D. G., Kuypers, H. G. J. M. 1968a. The functional organization of the motor system in the monkey. I. The effects of bilateral pyramidal lesions. *Brain* 91:1–14

Lawrence, D. G., Kuypers, H. G. J. M. 1968b. The functional organization of the motor system in the monkey. II. The effects of lesions of the descending brain stem pathways. *Brain* 91:15–36

Lynch, J. C. 1980. The functional organization of posterior parietal association cortex. *Behav. Brain Sci.* 3:485–99

Massey, J. T., Schwartz, A. B., Georgopoulos, A. P. 1985. On information processing and performing a movement sequence. *Exp. Brain Res. (Suppl)* In press

Matin, L. 1972. Eye movements and perceived visual direction. In *Handbook of Sensory Physiology*, ed. D. Jameson, L. Hurvicz, Vol. 7, Part 4: *Visual Psychophysics*. New York: Academic

Morasso, P. 1981. Spatial control of arm movements. *Exp. Brain Res.* 42:223–27

Mountcastle, V. B. 1978. *The Mindful Brain*, ed. G. M. Edelman, V. B. Mountcastle, p. 39. Cambridge, MA: MIT

Mountcastle, V. B., Lynch, J. C., Georgopoulos, A. P., Sakata, H., Acuna, C. 1975. Posterior parietal association cortex of the monkey: Command functions for operations within extrapersonal space. *J. Neurophysiol.* 38:871–908

Mountcastle, V. B., Motter, B. C., Andersen, R. A. 1980. Some further observations on the functional properties of neurons in the parietal lobe of the waking monkey. *Behav. Brain Sci.* 3:520–22

Mountcastle, V. B., Motter, B. C., Steinmetz, M. A., Duffy, C. J. 1984. Looking and seeing: The visual functions of the parietal lobe. See Georgopoulos et al 1984, pp. 159–93

Mowbray, G. H., Rhoades, M. V. 1959. On the reduction of choice reaction times with practice. *Q. J. Exp. Psychol.* 11:16–23

Murphy, J. T., Kwan, H. H., MacKay, W. A., Wong, Y. C. 1982. Precentral unit activity correlated with angular components of a compound arm movement. *Brain Res.* 246:141–45

Nelson, W. L. 1983. Physical principles for economies of skilled movements. *Biol. Cybernet.* 46:135–47

Nicoletti, R., Anzola, G. P., Luppino, G., Rizzolatti, G., Umilta, C. 1982. Spatial compatibility effects on the same side of the body midline. *J. Exp. Psychol. Hum. Perc. Perf.* 8:664–73

Paillard, J. 1982. The contribution of peripheral and central vision to visually guided reaching. See Jeannerod & Biguer 1982, pp. 367–85

Paillard, J., Jordan, P., Brouchon, M. 1981. Visual motion cues in prismatic adaptation: Evidence of two separate and additive processes. *Acta Psychol.* 48:253–70

Perenin, M. T., Vighetto, A. 1983. Optic ataxia: A specific disorder in visuomotor coordination. In *Spatially Oriented Behavior*, ed. A. Hein, M. Jeannerod, pp. 305–26. New York: Springer

Polit, A., Bizzi, E. 1979. Characteristics of motor programs underlying arm movements in monkeys. *J. Neurophysiol.* 42:183–94

Prablanc, C., Echalier, J. F., Komilis, E., Jeannerod, M. 1979a. Optimal response of eye and hand motor systems in pointing at a visual target. I. Spatio-temporal characteristics of eye and hand movements and their relationships when varying the amount of visual information. *Biol. Cybernet.* 35:113–24

Prablanc, C., Echalier, J. F., Jeannerod, M., Komilis, E. 1979b. Optimal response of eye and hand motor systems in pointing at a visual target. II. Static and dynamic visual cues in the control of hand movement. *Biol. Cybernet.* 35:183–87

Rondot, P., de Recondo, J., Dumas, J. L. R. 1977. Visuomotor ataxia. *Brain* 100:355–76

Sakata, H., Shibutani, H., Kawano, K. 1980. Spatial properties of visual fixation neurons in posterior parietal association cortex of the monkey. *J. Neurophysiol.* 43:1654–72

Saltzman, E. 1979. Levels of sensorimotor representation. *J. Math. Biol.* 20:91–163

Sanders, A. F. 1967. Some aspects of reaction processes. *Acta Psychol.* 27:115–30

Schmidt, R. A., Zelaznik, H., Hawkins, B., Frank, J. S., Quinn, J. T. 1979. Motor output variability: A theory for the accuracy of rapid motor acts. *Psychol. Rev.* 86:415–51

Simon, J. R. 1969. Reactions toward the source of stimulation. *J. Exp. Psychol.* 81:174–76

Skavenski, A. A., Hansen, R. M. 1978. Role of eye position information in visual space perception. In *Eye Movements and the Higher Psychological Functions*, ed. J. W. Senders, D. F. Fisher, R. A. Monty, pp. 15–34. Hillsdale, NJ: Erlbaum

Soechting, J. F. 1984. Effect of target size on spatial and temporal characteristics of a pointing movement in man. *Exp. Brain Res.* 54:121–32

Soechting, J. F., Lacquaniti, F. 1981. Invariant characteristics of a pointing movement in man. *J. Neurosci.* 1:710–20

Soechting, J. F., Lacquaniti, F. 1983. Modification of trajectory of a pointing movement in response to a change in target location. *J. Neurophysiol.* 49:548–64

Thach, W. T. 1978. Correlation of neural discharge with pattern and force of muscular activity, joint position, and direction of intended next movement in motor cortex and cerebellum. *J. Neurophysiol.* 41:654–76

Trevarthen, C. B. 1968. Two mechanisms of vision in primates. *Psychol. Forsch.* 31:299–337

Volkmann, F. C., Schick, A. M., Riggs, L. A. 1968. Time course of visual inhibition during voluntary saccades. *J. Opt. Soc. Am.* 58:562–69

von Hofsten, C. 1979. Development of visually directed reaching: The approach phase. *J. Human Mov. Studies* 5:160–78

von Hofsten, C. 1980. Predictive reaching for moving objects by human infants. *J. Exp. Child Psychol.* 30:369–82

von Hofsten, C. 1982. Eye-hand coordination in the newborn. *Dev. Psychol.* 18:450–61

von Hofstein, C. 1984. Developmental changes in the organization of prereaching movements. *Dev. Psychol.* 20:378–88

von Hofsten, C., Lindhagen, K. 1979. Observations on the development of reaching for moving objects. *J. Exp. Child Psychol.* 28:158–73

Wallace, R. J. 1971. S-R compatibility and the idea of a response code. *J. Exp. Psychol.* 88:354–60

Welford, A. T., Norris, A. H., Shock, N. W. 1969. Speed and accuracy of movement and their changes with age. *Acta Psychol.* 30:3–15

Ann. Rev. Neurosci. 1986. 9:171–207

INTERACTIONS BETWEEN RETINAL GANGLION CELLS DURING THE DEVELOPMENT OF THE MAMMALIAN VISUAL SYSTEM

Carla J. Shatz and David W. Sretavan

Department of Neurobiology, Stanford University School of Medicine, Stanford, California 94305

INTRODUCTION

The development of connections in the mammalian visual system requires the solution of several problems to achieve the adult pattern of connectivity. In the visual pathways from the retina to the lateral geniculate nucleus (LGN) and superior colliculus (SC), and in layer 4 of the primary visual cortex, axons representing the two eyes must terminate in separate and largely non-overlapping territories: ocular dominance patches in cortical layer 4 (Hubel & Wiesel 1969, 1972, Wiesel et al 1974, Shatz et al 1977); layers in the LGN (Hayhow 1958, Guillery 1970, Hickey & Guillery 1974); bands in the superior colliculus (Harting & Guillery 1976, Graybiel 1975, Hubel et al 1975). Within these target structures, axons representing each eye must also establish a topographically ordered map of the retina (Sanderson 1971, Tusa et al 1978, Berman & Cynader 1972). Finally, these maps must be aligned and interdigitated with each other in such a way that axons from the two eyes representing topographically similar coordinates in the visual field project to adjacent locations within each target structure.

In the past few years, much attention has been focused on how these problems are solved during the prenatal and postnatal development of the mammalian visual system. The retinogeniculate system, particularly of higher mammals, is an especially suitable choice for studying many of these

171

issues and their solutions during development because, in the adult, the retinal ganglion cell axons from each eye terminate in an easily recognizable and stereotyped pattern of completely separate layers within the LGN. In addition, the ganglion cells and their axonal projections can be labeled and manipulated with relative ease even in fetal animals. As a consequence, a good deal is now known at the descriptive level about the sequence of steps leading to the development of eye-specific layers in the LGN in a variety of mammalian species. Our review therefore centers on a consideration of the development of the retinogeniculate pathway.

In every species studied, the adult pattern of segregated eye input is not present initially and only emerges following an early developmental period during which inputs from the two eyes share common territory within the LGN (monkey: Rakic 1977; opposum: Calvacante & Rocha-Miranda 1978; hamster: So et al 1978, Frost et al 1979, So et al 1984; rat: Bunt et al 1983; ferret: Card-Linden et al 1981; cat: Shatz 1983). Several hypotheses concerning the mechanisms underlying this transition from a mixed to segregated state have been formulated, all of which involve the notion that competitive interactions between the retinal ganglion cell axons of the two eyes are responsible for eliminating inappropriate connections and retaining and stabilizing appropriate ones (Rakic 1981, Rakic & Riley 1983b, Williams et al 1983b, Shatz & Kirkwood 1984). Our purpose in this article is to review current information about the events associated with the development of eye-specific layers in the mammalian LGN in order to assess the role of binocular competition in the establishment of layers. Evidence we present here suggests that binocular interactions must play an important role in this process. However, binocular interactions alone cannot account for all aspects of the development of the retinogeniculate system, particularly those having to do with the establishment of topographic order. We suggest here that interactions between the retinal ganglion cell axons of one eye alone must also be considered.

ADULT ORGANIZATION OF THE RETINOGENICULATE SYSTEM

Before considering the developmental issues raised above, we provide a brief review of organization of the projection pattern of retinal ganglion cells in the adult cat, with particular emphasis on the retinogeniculate pathway. Retinal ganglion cells project to a number of different targets, including the lateral geniculate nucleus in the thalamus and the superior colliculus in the midbrain. In the cat, retinal ganglion cells have been subdivided into a number of separate classes on the basis of their morphology, physiological response properties, and pattern of central projection. Three main classes have been identified physiologically (X, Y, and W: Enroth-Cugell & Robson 1966, Stone & Hoffmann 1972, Cleland & Levick 1974) and much evidence suggests that these functional classes correspond to the three main classes defined on morphological grounds (beta, alpha, and gamma, respectively: Boycott & Wässle 1974, Wässle 1982, Saito 1983). Each target of retinofugal projections receives a characteristic mixture of inputs from the different ganglion cell classes. For example, the superior colliculus receives its major inputs from Y- and W-cells, while the LGN receives inputs from X-cells as well (for reviews see Rodieck 1979, Wässle 1982).

The class identity of each retinal ganglion cell, along with its location in the retina, correlates with whether or not its axon will cross in the optic chiasm. In general, retinal ganglion cells of all classes located in the nasal half of each retina cross to innervate the contralateral LGN and/or SC. However, the projection from the temporal retina is considerably more variable, depending on ganglion cell class. For instance, the majority of X-cells situated in the temporal retina project ipsilaterally, whereas only about 50% of the W-cells do (the other 50% project contralaterally) (see Wässle, 1982). As a result of the chiasmatic routing of ganglion cell axons, the LGN and the SC receive inputs from both eyes, but in each target, the proportion of input from contralateral nasal and ipsilateral temporal retinae varies. In the LGN, the projection from all three ganglion cell classes arises from the nasal retina of the contralateral eye and the temporal retina of the ipsilateral eye, thereby conferring a representation of the contralateral visual field only. However, the SC receives a significant input from the contralateral temporal retina (mainly from W-cells), and, as a consequence, the SC contains a representation not only of the contralateral visual field but also of part of the ipsilateral visual field (see Wässle 1982 for review).

Within the LGN, ganglion cell axons from the two eyes terminate in several separate and distinct layers that are arranged parallel to the optic tract. The innermost two layers (i.e. those furthest from the optic tract) are designated layers A and A1 (Guillery 1970) (see also Figure 4 here) and occupy roughly two-thirds of the nucleus; the remaining third consists of the C-complex and is subdivided into layers C, C1, C2, and C3 (closest to the optic tract). The pattern of projection from one eye to the various LGN layers has been shown by injecting radioactive tracers such as tritiated leucine intraocularly, allowing time for axonal transport of the tracer to the LGN, and then processing the tissue for autoradiography (Hickey & Guillery 1974). After an injection into one eye, layers A, C, and C2 in the contralateral LGN are labeled, while label is found in layers A1 and C1 of the ipsilateral LGN. (Layer C3 does not receive retinal input.) A comparison of the labeling pattern on the two sides shows that afferents from the two eyes occupy mutually exclusive layers within the LGN. Thus, the projection from each eye to the LGN forms several layers alternating with layers from the other eye.

The number of layers representing the two eyes is variable, depending on the species. Input from the two eyes may simply be segregated into a single ipsilateral zone and a single contralateral zone, as is the case in rodents (Bunt et al 1983, So et al 1978, Frost et al 1979), while up to ten separate layers of afferent input are seen in some marsupials (Sanderson et al 1984). One reason suggested for the presence of multiple layers in the LGN in some animals is that each pair of contralateral and ipsilateral layers receives a different functional class of retinal input. In the cat, layers A, A1, and C receive a mixture of X and Y cell input, while W cell input is found only in layers C1 and C2 (see Rodieck 1979 and Wässle 1982 for reviews). Therefore, it has been proposed that multiple layers serve as separate channels for visual information conveyed by unique mixtures of retinal ganglion cell classes. This explanation is not entirely satisfactory, since the extent to which individual classes of retinal ganglion cells are sorted out into different layers varies with the species. For example, retinal ganglion cells, which are classified according to their center surround organization into ON- or OFF-center cells, project to separate layers in monkey (Schiller & Malpeli 1978), mink (LeVay & McConnell 1982), and ferret (Stryker & Zahs 1983), whereas in the cat, ON- and OFF-cell axons project within the same layers (Bowling & Michael 1984). (Recently, however, there is some evidence (Bowling 1984) that ON- and OFF-LGN cells within a layer may be located preferentially in different dorsal-ventral positions.)

The retinal projection into a given layer is organized further according to topographic rules. Central retina projects medially in the nucleus, whereas more peripheral parts of the retina project laterally. As a result, the retinal surface is systematically represented within each of the five layers such that adjacent layers are topographically in register with each other (Sanderson 1971). It is worth noting here that in the cat, as in other species, the area within the LGN devoted to the representation of the central visual field is expanded relative to that for the peripheral visual field. This arrangement exists in large part because of the central-to-peripheral gradient in retinal ganglion cell density: in the adult cat, the density of retinal ganglion cells in central retina is over 80 times that in the periphery (Stone 1978).

This characteristic segregation of afferents into alternating layers according to eye preference, and the systematic mapping of the retina in each layer of the LGN, are the result of the precise axonal projections of individual retinogeniculate axons. Therefore, as expected, individual retinogeniculate axons arborize in a manner consistent with the organization described above. The terminal arborizations of single retinal afferents have been studied by filling optic axons intracellularly with HRP and completely reconstructing their terminal arbors (Bowling & Michael 1980, 1984, Sur & Sherman 1982). Retinal afferents run into the LGN from the optic tract below and elaborate terminal arbors only in the layers appropriate for their eye of origin. Axons, as they course through the entire nucleus, are completely smooth; preterminal branches and boutons are not given off in the territory occupied by the other eye. In addition, not only are the terminal arbors of these afferents localized exclusively to appropriate layers, but within these layers they are also restricted in width. It is very likely that this restriction in width forms the morphological basis for the fine-grain retinotopic map in the LGN.

The general features discussed above characterize the organization of the LGN in every mammalian species studied. Common to all is the presence of a set, or sets, of eye-specific layers arranged in topographic register with each other.

DEVELOPMENT OF RETINOGENICULATE CONNECTIONS

In the main part of this article, we first review what is currently known about the sequence of developmental events leading to the final adult pattern of connections between retina and LGN. The events we discuss include genesis and maturation of retinal ganglion cells, the development of the adult pattern of ganglion cell projection from nasal and temporal retinae, the development of topographic order, and the segregation of eye input. Finally, we consider some of the possible mechanisms underlying these events.

Genesis of Retinal Ganglion Cells

Ganglion cells of the mammalian retina, unlike those in lower vertebrates, are generated over a restricted period of time during prenatal life. This finding is based on studies in which ^{3}H-thymidine was administered to fetuses at known gestational ages in order to identify the set of neurons generated at that time. (Most cells undergoing their final round of cell division will remain heavily labeled at all later ages, indicating that they were generated on the day of ^{3}H-thymidine administration.) In the cat, ^{3}H-thymidine studies have shown that the majority of retinal ganglion cells are generated during the two-week period between embryonic day 20 (E20) and E35 (gestation is 65 days in the cat) (Kliot & Shatz 1982, Walsh et al 1983, Walsh & Polley 1985). Administration of ^{3}H-thymidine at E19 does not result in ganglion-cell labeling, even when the retinae are examined prenatally (as early as E30), thus indicating that a population of earlier-generated ganglion cells that is only transiently present during embryonic life is unlikely to exist (Kliot & Shatz 1982 and in preparation).

Ganglion cells are generated in a central retina-first, peripheral retina-last spatio-temporal sequence in all species studied (rat: Sidman 1961; monkey: Rakic 1977; cat: Walsh et al 1983, Walsh & Polley 1985, Kliot & Shatz 1982). In the cat, where more information is available, the gradient is extremely coarse and, although at any given time more ganglion cells may be generated in one part of the retina as compared to another, ganglion cells can also be generated simultaneously within central and peripheral retinal regions. This point is important for any consideration of mechanisms responsible for ordering ganglion cell axons within the optic nerve or tract: plainly, the establishment of topographically orderly projections cannot be due simply to timing of axon outgrowth, as is likely to be the case in some lower vertebrates.

In the cat, ganglion cells belonging to the different classes tend to be generated at slightly different times. The onset of production of medium-sized retinal ganglion cells (β-cells) starts several days before significant production of large-sized retinal ganglion cells (α-cells), whereas small cells (primarily γ-cells) are produced throughout the entire period between E20–E35. As expected, each cell class is produced in a roughly central-to-peripheral sequence (Walsh et al 1983, Kliot & Shatz 1982, Walsh & Polley 1985). Thus, the majority of medium-sized ganglion cells generated on any given day will be situated at slightly greater retinal eccentricities than the majority of large-sized ganglion cells generated on the same day.

Development of the Central-to-Peripheral Gradient

The histological appearance of the retinal ganglion cell layer during the fetal period immediately following the end of ganglion cell genesis is strikingly different from that seen in the adult: cell density is higher and is initially uniform across the central and peripheral retina (Stone et al 1982, 1984, Rapaport & Stone 1983), whereas in the adult there is a markedly central-to-peripheral gradient in ganglion cell density, as described above. In the cat, a difference in density of cells in central verus peripheral retina can be detected by histological inspection by E52 (Rapaport & Stone 1983), although the adult-like gradient is not present until well after birth (Mastronarde et al 1984, Rapaport & Stone 1983). The interpretation of the fetal histology is complicated by difficulties associated with unambiguously identifying ganglion cells, since at early ages the majority of cells in the ganglion cell layers are of roughly similar size and shape. In order to distinguish ganglion cells from other cell types known to reside in the ganglion cell layer (e.g. displaced amacrine cells or glial cells), it is desirable to demonstrate, by using retrograde labeling techniques, that cells supply an axon to the optic nerve and/or tracts. Indeed, when massive horseradish peroxidase (HRP) injections are made bilaterally into the thalamus and midbrain of cat fetuses as young as E34, a gradient in the density of retrogradely labeled cells in the ganglion cell layer is already apparent (Lia et al 1983), thus suggesting that at least some of the cells in the ganglion cell layer are not retinal ganglion cells. Alternatively, since neurogenesis in peripheral retina lags that in central retina, outgrowth of peripheral retinal ganglion cell axons must also lag, and therefore it may not be possible to label retrogradely all the peripherally located ganglion cells with injections placed within the central targets of the retinofugal projection at the earliest ages (e.g. E34). Thus, although there is general agreement that the density distribution of cells in the ganglion cell layer is initially uniform, it is not yet clear whether all are ganglion cells and consequently whether the initial distribution of ganglion cells follows a central-to-peripheral gradient of cell density. What is clear, however, is that such a gradient exists by E57 (Stone et al 1982).

A number of suggestions have been made concerning the way in which the adult gradient of ganglion cell density may be produced from an initially uniform distribution. As mentioned above, one suggestion is that although the initial distribution is uniform, it is not homogeneous, and a disproportionately large number of cells located in the peripheral retina differentiate into amacrine rather than ganglion cells, as has been suggested for rodent retina (Hinds & Hinds 1983). Recently, this issue was addressed directly by Perry et al (1983). Cells in the ganglion cell layer of the rat retina were retrogradely labeled by HRP injections into the thalamus. Twenty-four to 48 hours later their axons were severed by an intraretinal incision, under the assumption that cells with their axon cut would die, while cells without axons would not be affected. Therefore, if amacrine cells are derived from ganglion cells by the simple loss of an axon, some should be HRP-labeled but immune to retrograde degeneration subsequent to the intraretinal incision. No remaining HRP-labeled cells were found; this suggested that amacrine cells are not derived from ganglion cells that retract an axon, at least in rodents. Unfortunately, this experiment is open to several criticisms, including the possibility that amacrine cells themselves may have undergone retrograde transneuronal degeneration due to the removal of their synaptic targets, the ganglion cells. Thus the possibility that the gradient in retinal ganglion cell density arises in part as a consequence of the preferential differentiation of peripheral retinal cells into amacrine cells cannot yet be eliminated.

More likely is the possibility that the central-to-peripheral gradient in ganglion cell density arises from a combination of ganglion cell death and retinal growth occurring preferentially in the peripheral retina. Several lines of evidence indicate that during prenatal and postnatal development a significant amount of cell death occurs in the ganglion cell layer of the mammalian retina. In the cat retina, cell counts have shown that there is about an eight-fold decrease in total number of cells in the ganglion cell layer between E47 and birth (Stone et al 1982), and in hamster retina it has been possible to estimate by means of extrapolating from counts of pyknotic nuclei that a 50% reduction in cell number in the ganglion cell layer occurs between birth and 10 postnatal days (Sengelaub & Finlay 1982). Both of these approaches have the disadvantage that it is not possible to prove that ganglion cells rather than other cell types are lost. Direct evidence for loss of ganglion cells comes from the study of Potts et al (1982), in which massive injections of HRP were used to retrogradely label ganglion cells at different postnatal times in the rat: a progressive decrease in the number of labeled cells was observed. More recently, Insausti et al (1984) labeled ganglion cells retrogradely by injecting long-lasting fluorescent dyes at birth and examined the number of retrogradely labeled ganglion cells at subsequent times: Again a progressive decrease in the number of labeled cells was found, indicating that ganglion cell death indeed occurs.

Another line of evidence that provides some indication of the magnitude of loss comes from estimates of the total number of axons in the optic nerve at different times during development. Results in all species studied (monkey: Rakic & Riley 1983a; cat: Ng & Stone 1982, Williams et al 1983a; rat: Lam et al 1982, Perry et al 1983, Crespo et al 1984) indicate that there is an initial overproduction of axons, followed by a large decrease (two- to four-fold) to adult numbers. For example, in the cat, as shown in Figure 1, the number of optic nerve axons reaches a maximum at about E48, with 600,000 or so axons, then declines until the adult level of 150,000 axons is achieved. The question naturally arises as to whether optic nerve counts accurately reflect ganglion cell number, and therefore give reliable quantitative information concerning ganglion cell death. In instances where counts of cells in the ganglion cell layer and axons in the optic nerve have been correlated, the numbers are in reasonably good agreement (rat: Perry et al 1983; cat: B. Lia and L. M. Chalupa, personal communication), and suggest that virtually all cells in the ganglion cell layer supply an axon to the optic nerve. An alternative, that some cells do not have axons in the nerve while many others have axons that branch in the nerve, seems unlikely in view of previous Golgi studies (Morest 1970, Cajal 1972) and recent experiments in which HRP has been used to fill axons in the optic nerve of fetal cats (M. Kliot, D. W. Sretavan, and C. J. Shatz, in preparation) and mice (P. Bovalenta and C. A. Mason, in preparation): Thus far, no axons have been found to branch within the optic nerve. This observation receives further support from an electron microscopic study in which axon counts were made both at the distal and proximal ends of the fetal cat optic nerve: More axons were found nearer the eye than nearer the chiasm, arguing against bifurcation (Lia et al 1985). However, Murakami et al (1982) have observed one ganglion cell whose axon bifurcated within the *retina* in an adult Siamese cat. Nevertheless, the accumulated evidence suggests that optic nerve counts are quite likely to reflect reliably the loss of retinal ganglion cells during development.

It is worth noting that the time course of axon loss shown in Figure 1 can be correlated with the time of appearance of the central-to-peripheral gradient in ganglion cell density: the major axon loss occurs by E57—a time when the gradient is first apparent. If we assume that axon loss is a reliable reflection of ganglion cell death (see above), this correlation suggests that cell death, particularly if it occurred preferentially in peripheral retina, could contribute significantly to the establishment of the gradient in ganglion cell density. Two sets of experiments in fact support this suggestion. In the hamster retina, Sengelaub & Finlay (1982) have reported that the number of pyknotic nuclei is consistently greater in peripheral versus central retina. And in the cat, injections of ^{3}H-thymidine given on a day selected to label cells of the ganglion cell layer result in a labeling index that is far greater initially than at later fetal ages in the peripheral but not central retina (Kliot & Shatz 1982). Since the labeling index is based on the number of ^{3}H-thymidine–labeled cells per 100 unlabeled cells, the age-related decrease in peripheral retina cannot be due exclusively to differential retinal growth, but rather indicates that a real and differential loss of cells has occurred in peripheral relative to central retina.

Nevertheless, differential retinal growth must also contribute to the establishment of the adult central-to-peripheral gradient in ganglion cell density. It is well known that during the cat's fetal and postnatal development, the retina undergoes a large increase in area (Rapaport & Stone 1983, Mastronarde et al 1984) that is not accompanied by the addition of neurons to the ganglion cell layer, since the major period of retinal growth begins at around E40, well after the final genesis of cells belonging to the retinal ganglion cell layer. Furthermore, a recent study (Mastronarde et al 1984) has suggested that retinal growth, at least between three weeks after birth and adulthood, is nonuniform, with peripheral retinal regions growing more than central regions: such a growth pattern would certainly contribute to the production of the central-to-peripheral gradient in ganglion cell density, although the exact extent to which it does so is currently unknown. However, since the major phase of ganglion cell death occurs before birth, it is quite possible that cell death is largely responsible for creating the initial central-to-peripheral gradient observed

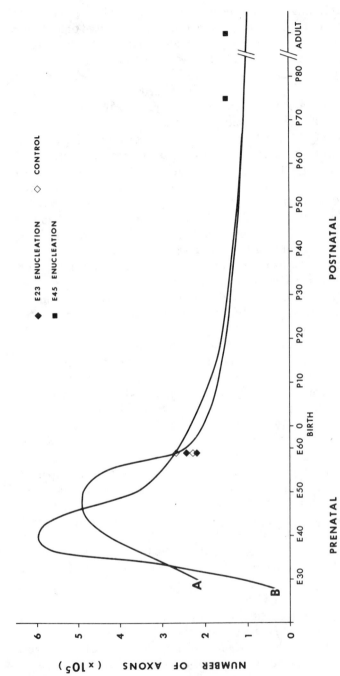

Figure 1   To show the initial addition and subsequent loss of optic nerve axons during the development of the cat's visual system. *Curve A* : taken from the results of Ng & Stone (1982). *Curve B* : taken from the results of Williams et al (1983a, 1986). Symbols indicate results of studies designed to examine the effects of unilateral enucleation on the elimination of optic nerve axons. *Squares* represent results of enucleations performed at E45 and examined at or near adulthood (from Williams et al 1983b). *Diamonds* represent results of enucleations performed at E23 and examined at E59 (*filled diamonds*), or of control normal animals examined at E59 (*open diamonds*) (from Sretavan & Shatz 1986). In all cases data was compiled using similar quantitative sampling methods at the electron microscopic level.

prenatally, which is further steepened postnatally by differential retinal growth.

Maturation of Retinal Ganglion Cells

Little is known about the sequence of morphological and physiological steps that accompany the maturation of the different classes of retinal ganglion cells. An analysis of the distribution of soma sizes in the ganglion cell layer of the cat retina indicates that a small but distinct population of larger size somas is evident by E57 (Rapaport & Stone 1983). Unfortunately, it is not possible to identify this larger-sized population on the basis of soma size alone, since it could represent either α-cells or β-cells (or both) beginning a period of growth and differentiation.

The conclusive identification of retinal ganglion cells according to morphological and physiological criteria is possible by three weeks postnatally in the cat. In studies of Golgi-impregnated retinas at this age (Rusoff & Dubin 1978, Rusoff 1979), soma size and dendritic branching patterns are sufficiently distinct that it is possible to identify some α-, β-, and γ-cells according to the criteria of Boycott & Wässle (1974), as shown in Figure 2A. Prior to birth, very little is known about the morphological

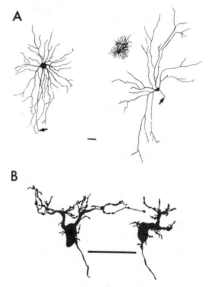

Figure 2   To show the morphology of Golgi-impregnated ganglion cells in the cat's retina at several different developmental ages. *A :* From Rusoff (1979). *Left,* adult α-cell ; *right,* three-week old kitten α and β (small) cells viewed in wholemount. *B :* From C. J. Shatz and S. J. Eng (in preparation). Two ganglion cells from a fetal (E36) cat retina viewed in cross section. Scale bars are 50 μm.

development of ganglion cells in any mammalian species due to the difficulties in obtaining good Golgi impregnations or adequate dye-filling of ganglion cells. However, the Golgi studies of Morest (1970) indicate that in the rat retina, by birth (about two days after the last ganglion cells have been generated) the ganglion cells have significant dendritic trees and an axon within the optic fiber layer. Similarly, in the cat, Golgi studies in progress have shown that as early as E36 (again, several days after the end of neurogenesis) some ganglion cells have remarkably extensive dendritic arbors (C. J. Shatz and S. J. Eng, in preparation), as shown in Figure 2B.

A good deal of maturation of retinal ganglion cell morphology continues well into postnatal life, since soma size increases up to two to three months postnatally (Rapaport & Stone 1983), and considerable growth of dendrites occurs as well (Rusoff & Dubin 1978, Mastronarde et al 1984). As might be expected in view of the central-to-peripheral neurogenetic gradient, a gradient in the morphological maturation of ganglion cells has also been reported (Rusoff 1979): for example, Golgi studies have shown that β-cells in the central retina achieve their adult size and dendritic extent by about three weeks after birth, whereas those in the peripheral retina are still immature.

Physiologically, cat retinal ganglion cells are also immature at three postnatal weeks. However, at this age it is possible to distinguish between X- and Y-like properties, at least in some instances, since microelectrode recordings from single optic tract axons indicate that some axons can be categorized as having linear (X) versus nonlinear (Y) characteristics of spatial and temporal summation (Sur et al 1984, Friedlander et al 1985) or, alternatively, as responding to visual stimulation in a brisk-sustained (X) versus a brisk-transient (Y) fashion (Rusoff & Dubin 1977). All studies have also indicated that receptive field size at these young ages is much larger than in the adult (Hamasaki & Flynn 1977, Sur et al 1984, Friedlander et al 1985) and that center-surround antagonism is weaker (Rusoff & Dubin 1977). It is worth mentioning that the interpretation of changes in the receptive field properties of ganglion cells at this early age is complicated by the fact that the optical quality of the kitten eye is poor, and improves substantially throughout early postnatal life until adult-like optics are achieved, by about five weeks of age (Thorn et al 1976, Bonds & Freeman 1978). (Indeed, very little is known about the functional development of retinal ganglion cells during the period between birth and three postnatal weeks precisely because of poor optics.) Even after five weeks, however, the receptive-field properties of ganglion cells are not yet adult-like: receptive-field size (both center and surround) is still larger than normal (Rusoff & Dubin 1977, Sur et al 1984) and the nonlinear component of the Y-cell response does not reach adult strength until about three months postnatally. These observations indicate that optics alone cannot be responsible for producing the immature receptive field properties of retinal ganglion cells throughout all of early postnatal life.

Another aspect of the immaturity of retinal ganglion cells at three postnatal weeks is that their terminal arborizations within the LGN are not adult-like in morphological appearance. By examining physiologically identified optic tract axons that have been filled with HRP, Sur et al (1984) and Friedlander et al (1985) have shown that from three weeks onwards X- and Y-axons undergo striking and different morphological changes. At three weeks (the earliest time studied), X-axons have extensive terminal arbors that are subsequently pruned in width until the adult size is achieved by three months, as illustrated in Figure 3 (right); there is also the indication that the number of boutons per X-axon may decrease concurrently (Sur et al 1984). In contrast, as shown in Figure 3, Y-axons at three weeks have arbors that are substantially smaller than at adulthood, and arbor width and bouton number increase over an extended period of postnatal life, well beyond three months (Sur et al 1984, Friedlander et al 1985).

Prior to three postnatal weeks, it has not proved possible to distinguish the terminal arborizations of different classes of retinal ganglion cell axons either morphologically or physiologically. Physiologically, experiments are complicated by poor optics, as discussed above, and the fact that photoreceptors even in central retina do not appear histologically mature until about four weeks postnatally (Donovan 1966). Anatomically, however, it has been possible to study the morphological development of retinal ganglion cell axon arbors within the cat's LGN during the first few postnatal weeks (Mason 1982) and during the final third of gestation (Sretavan & Shatz 1984, 1985), by filling optic tract axons with HRP. Results show that although axons grow extensively during prenatal and early postnatal life (see below: *Eye-Specific Segregation of Retinal Ganglion Cell Axons*), it is not possible to distinguish the different classes on morphological grounds. For example, at each fetal age studied, all axons had similar arbor widths, as shown in Figure 3 (*left*). This observation suggests that all classes of axon have similar morphologies during prenatal life. The alternative, that only one of the various classes (e.g. X-axons) was HRP-filled in these studies, is unlikely in view of the fact that, as in the adult, in the fetal animals it was possible to fill not only axons that arborize only within the LGN (e.g. likely to be X-axons) but also those supplying collaterals to the medial intralaminar nucleus (MIN) and toward SC as well (e.g. Y-axons). However, the possibility that some X-axons may extend extra collaterals during prenatal development, cannot be discounted at present, and further work is required to resolve this point, as well as to learn about axon arbor development during the first three weeks of postnatal life.

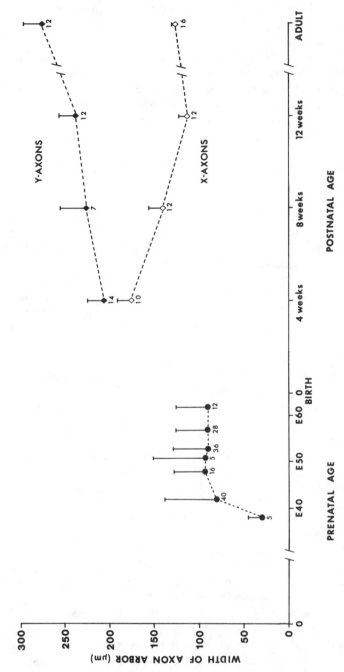

Figure 3 To show the progressive change in width of retinogeniculate axon arbors during prenatal and postnatal development of the cat's visual system. Data from prenatal ages are taken from Sretavan & Shatz (1985); data from postnatal ages are taken from Sur et al (1984). Currently, no information is available for the period between birth and four weeks.

To sum up so far, it is clear that although a good deal of the development of retinal ganglion cells occurs prenatally, functional and morphological maturity is not achieved until at least three months postnatally. Thus, maturation extends well into the cortical critical period for the effects of visual deprivation (Hubel & Wiesel 1970).

Topographic Ordering of Retinal Ganglion Cell Projections

In the adult mammalian visual system, the projection of each retina onto target structures is highly ordered topographically, and the projections from the two eyes are topographically in register. In order to achieve this organization during development, retinal ganglion cell axons from the two eyes must first make the appropriate decision at the optic chiasm such that ganglion cell axons from the corresponding regions of the nasal retina of one eye and the temporal retina of the other project to the same side of the brain. Although little is known about the prenatal development of this chiasmatic routing pattern, a substantial amount of evidence indicates that at birth there is precision in the routing of ganglion cell axons, although some refinement occurs even during postnatal development. For example, in the cat's visual system, Jacobs et al (1984) have shown that at birth about 600 ganglion cells in the nasal retina project ipsilaterally—about six times as many as are present in the adult. Precision in the chiasmatic routing of retinal ganglion cell axons is also suggested by the results of several studies showing that ganglion cell axons rarely bifurcate to project to both sides of the brain, either in the adult (Jeffrey et al 1981, Hsiao et al 1984) or during early postnatal development (Insausti et al 1984). In view of the observed postnatal refinement in the ipsilateral projection from the nasal retina in the cat, it seems likely that this process is a continuation of one that begins during prenatal life. However, without direct information concerning the chiasmatic routing of axons early in development, we can only speculate about the degree of specificity in the initial chiasmatic choice made by axons from the nasal and temporal retinae.

Once the chiasmatic decision has been made, axons from the two eyes must establish topographically orderly connections within their targets. One way in which this may occur is that, as in lower vertebrates, the neighbor relationship of retinal ganglion cells is preserved in a systematic manner throughout the pathway from retina to target (Scholes 1979, Easter et al 1981, Bunt & Horder 1983, Reh et al 1983). In lower vertebrates, this orderly arrangement arises as a direct consequence of the orderly central-to-peripheral gradient of retinal neurogenesis; thus, the axons of ganglion cells generated at the same time tend to travel together in the optic nerve (Easter et al 1981, Scholes 1979, Johns 1977, Straznicky & Gaze 1971). In contrast, in mammals the neurogenetic gradient is very coarse at best (see above) and even if axons from concurrently-generated retinal ganglion cells were to join together as they enter the optic nerve, topographic order also would be coarse—exactly what is found in physiological and anatomical studies of adult optic nerve (Horton et al 1979, Torrealba et al 1982; for review see Guillery 1982). Furthermore, a recent electron microscopic analysis of cat optic nerve has shown that axons of different diameters are intermixed with each other in the optic nerve (Williams & Chalupa 1983), thus indicating that different classes of retinal ganglion cell axons are not clearly segregated from each other. This observation, again, is consistent with the fact that the periods of neurogenesis for the various classes of retinal ganglion cells overlap substantially.

The mammalian optic tract, in contrast, contains a much more orderly representation of the visual field. For example, in the cat, recent studies using degeneration and anterograde tracing techniques (Torrealba et al 1982) or microelectrode recordings (Aebersold et al 1981, Torrealba et al 1981, Mastronarde 1984) have shown that axons are arranged clearly according to topographic coordinates in the optic tract. Moreover, axons are also partially segregated according to class within the tract, with the larger diameter Y-axons placed ventral to the medium-diameter X-axons (Guillery et al 1982, Mastronarde et al 1984). Thus, Guillery and his colleagues have argued that each optic tract contains multiple, rough maps in partial register, each map representing one axon class (for review see Walsh & Guillery 1984).

These observations suggest that during development, a good deal of reorganization of axon trajectories must occur en route from optic nerve head to optic tract in order to achieve the coarse topography and class segregation seen in the mammalian optic tract. Although very little is currently known about the path followed by individual growing axons, support for the above suggestion comes from a recent study in which the nearest neighbor relationships between axons in the optic nerve of fetal monkeys were examined at the ultrastructural level (Williams & Rakic 1985). Over a remarkably short distance (less than 1 mm), axons lost up to 90% of their original neighbors and moved freely between fascicles, indicating that indeed, substantial rearrangements of axon trajectories can occur in the optic nerve during development. Thus, it is conceivable that during development, these rearrangements within the optic nerve give rise to the rough order seen in the optic tract. If so, as suggested by Walsh & Guillery (1984), retinal ganglion cell axons may be in a good position to establish many of the basic features of organization seen in the adult when they initially grow into their targets.

From the above considerations, it is plain that during development much of the topographic ordering of axons can occur en route to, rather than within, target structures. However, because topography in the optic tract is coarse, at best, a good deal of topographic fine-tuning must also take place within the target, particularly in order to achieve the precise alignment of topographic maps according to eye of origin and ganglion cell class. Perhaps the best evidence in favor of this idea comes from developmental studies of the ipsilateral retino-collicular projection. For example, when the ipsilateral retino-collicular pathway in carnivores and rodents is labeled by the anterograde transport of tracers, the projection seen early on occupies a much larger proportion of the superior colliculus than in adult life; over subsequent days, labeling within caudal collicular regions disappears, thereby producing the restricted ipsilateral projection present in the adult (cat: Williams & Chalupa 1982; rodent: Bunt et al 1983, Frost et al 1979, Cusick & Kaas 1982, Godemont et al 1984). This sequence of events suggests that the fine retinotopic order present in the adult colliculus arises from a rougher order present initially.

Although it is clear that the restriction of the ipsilateral projection results in the precise topographic alignment of the projections from ipsilateral and contralateral retinae, the details of this process are not well understood. A number of events may contribute to the restriction of the ipsilateral projection seen during development. One possibility suggested by experiments in the rodent visual system is that axons initially projecting to the caudal superior colliculus arise from retinal ganglion cells situated at topographically inappropriate retinal locations, and these cells are then eliminated by cell death. In the adult, the ipsilateral retinocollicular projection arises from ganglion cells located in the temporal crescent of the retina. Cells in this region of retina project to the rostral portion of the superior colliculus, which also receives input from ganglion cells in the nasal retina of the contralateral eye that subserve the same portion of the visual field; hence, this is the binocular region of the superior colliculus (Drager & Olsen 1980, Tiao & Blakemore 1976). During development, however, experiments in which large injections of retrograde tracers were made into the superior colliculus have shown that the ipsilateral projection arises not only from the temporal crescent, but also from numerous ganglion cells scattered throughout the retina (Insausti et al 1984, Jeffrey & Perry 1982), thus indicating that the superior colliculus indeed receives topographically inappropriate input from the ipsilateral eye. Recently, it has been shown that at least some of the inappropriate ipsilateral input is removed by cell death. Insausti et al (1984) employed the method of retrograde labeling of retinal ganglion cells with long-lasting fluorescent dyes injected into the superior colliculus in hamsters to show that, following injections at birth, there is a progressive loss in the number of retrogradely labeled ganglion cells with time: such a loss can only occur due to removal of labeled ganglion cells by cell death.

Although this result demonstrates that cell death is directly correlated with the restriction of the ipsilateral projection, it does not prove that cell death selectively eliminates topographically inappropriate connections arising from regions of the retina outside the temporal crescent. To do so requires local injections of tracers either into the retina or the colliculus in order to demonstrate directly that the cells that are eliminated indeed project to the caudal, topographically inappropriate part of the colliculus. Another possibility that deserves consideration here is that the refinement of topography is accomplished not only by cell death, but also by the reorganization of axons and their terminals. For example, it is conceivable that during development, retinal ganglion cell axons destined to innervate the rostral colliculus also extend terminal branches into the caudal colliculus. These branches could then be removed by simple retraction rather than cell death. This suggestion can be verified by examining the morphological changes undergone by individual axons in conjunction with localized injections of fluorescent long-term tracers into the caudal colliculus.

Much less is known about the development of topographic order in the retinogeniculate projection, but here too it is reasonable to suppose that similar events lead to the precise retinotopy present in the adult. It is now known in the ferret that the projection from the retina to the LGN is already roughly topographic early in development, well before the eye-specific layers are formed (Jeffery 1985). Our own observations, based on filling individual retinal ganglion cell axons with HRP in order to examine their shapes during the development of the cat's retinogeniculate projection, are consistent with this result (Sretavan & Shatz 1984, 1985). Throughout development, axons have a restricted pattern of arborization (see Figure 3) and their trajectories are oriented in a direction that will later coincide with the lines of projection of the visuotopic map, suggesting that there exists initially a reasonable amount of topographic order. Of course this suggestion also must be verified by making small localized injections in the LGN in order to examine directly the precision of topography at different times during development. In addition, the extent to which ganglion cell death contributes to the establishment of topographic order in the retinogeniculate projection remains to be determined.

Eye-Specific Segregation of Retinal Ganglion Cell Axons

The question to be considered in this section is how the adult segregated pattern of eye input is established during development. In every mammalian species studied to date, the eye-specific segregation of retinal ganglion cell axons characteristic of adult retinofugal projections is not present during the initial development of the visual system. Rather, segregated inputs only emerge gradually during development from an initially diffuse pattern in which inputs from the two eyes are intermixed. One example of this phenomenon is in the development of the retino-collicular projection in which the adult patches of ipsilateral eye input to the superior colliculus are not present initially (Williams & Chalupa 1982, Godement et al 1984, Bunt et al 1983, Frost et al 1979, Cusick & Kaas 1982).

Perhaps the best example of segregation from an initially intermixed state is in the development of the retinogeniculate projection. Intraocular injections of tritiated amino acids into one eye and HRP into the other during prenatal development in the cat have shown that retinal axons from the contralateral eye invade the LGN by E32, while those from the ipsilateral eye arrive several days later. During the next four and a half weeks inputs from the two eyes first intermix and then gradually segregate from each other, so that by birth (E65) the eye-specific layers are present (Shatz 1983). Figure 4 illustrates this sequence of events. In each drawing, LGN territory shared by inputs from the two eyes is shown in black, while the territory belonging to the right eye is shown in light gray, and that

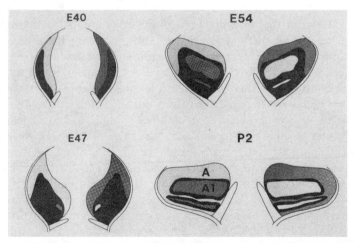

Figure 4 To show the initial intermixing and subsequent segregation of retinogeniculate axons into eye-specific layers during prenatal development in the cat. Modified from Shatz (1983). *Black* indicates regions in which inputs from the two eyes are intermixed. *Light gray* indicates input from the right eye; *darker gray* indicates input from the left eye. By postnatal day 2, inputs to the contralaterally-innervated layer A and ipsilaterally-innervated layer A1 and the layers of the C complex (not labeled here) are almost completely segregated from each other, except at laminar borders.

belonging to the left eye is shown in dark gray. Thus, by E40, inputs from the two eyes are intermixed within LGN territory immediately adjacent to the optic tract, while contralateral fibers occupy exclusively the innermost portion of the LGN. Over the next several days, afferents from both eyes grow further into the nucleus and begin to sort out from each other (E47). As segregation continues, the territory shared by the two eyes is diminished and the eye-specific layers emerge (E54); by birth (P2) segregation is almost complete.

The pattern of development of retinal projections to the LGN in many mammalian species is similar in broad outline to that found in the cat (ferret: Card-Linden et al 1981; hamster: Frost et al 1979, So et al 1978, 1984; marsupial: Cavalcante & Rocha-Miranda 1978, Sanderson et al 1984; monkey: Rakic 1977; mouse: Godement et al 1984; rat: Bunt et al 1983). For example, in every species in which it has been possible to perform early studies, the projection from the contralateral eye has been found to invade the LGN before that from the ipsilateral eye. Also common to all is the intermixing and subsequent segregation of eye input. However, the extent to which the inputs from the two eyes are intermixed with each other varies depending on the species studied. In general the maximum amount of shared territory found during development is related to the size of the ipsilateral projection and therefore to the degree of binocular vision found in the adult.

It is also clear that measurements of the exact amount of intermixing depend on several methodological variables including the anterograde tracer or tracers used, choice of post-injection survival time, and criteria for distinguishing labeled from unlabeled territory. Because of these variables it is possible that the virtually complete intermixing seen in primates (Rakic 1977) has been overestimated due to long post-injection survival times and the use of tritiated proline, both of which would tend to increase the background labeling due to spillover (LeVay et al 1978). However, the maximum amount of intermixing seen in the cat (40%; Shatz 1983) could be an underestimate due to the choice of criteria used to distinguish labeled from unlabeled territories. Nevertheless, it is worth emphasizing that conclusions concerning the maximum amount of LGN territory shared by afferents from the two eyes in the cat were based on examination of double-labeled material in which one eye received an injection of tritiated leucine and the other an injection of HRP. This approach allows for a more accurate assessment of the area of intermixing. More important, it has shown that a region of the LGN destined to become the innermost rim of contralaterally-innervated layer A in the adult is almost exclusively innervated by the contralateral eye throughout prenatal development (see

Figure 4), an observation relevant for subsequent considerations of the mechanisms responsible for segregation (for discussion, see Shatz 1983).

In order to learn more about the cellular interactions underlying the formation of eye-specific layers in the cat's lateral geniculate nucleus, the development of individual retinal ganglion cell axons has been studied by filling axons with HRP at different times during the periods of intermixing and segregation (Sretavan & Shatz 1984, 1985). The appearance of axons at a time when segregation of eye input is almost complete (near birth) resembles the adult morphology in almost every respect: axons have smooth main trunks with extensive terminal arborizations located in LGN layers appropriate to the eye of origin. The major difference is one of size. At birth, the width of axon terminal arborizations is narrower (see Figure 3), and the density of the arbor less, than that seen in the adult. The morphological appearance of axons at these ages is perfectly consistent with the finding, based on intraocular injections of label, that the development of eye-specific layers is almost complete (Shatz 1983).

In contrast, axons during the period when the amount of territory shared by the two eyes is at a maximum (E46–E54; Shatz 1983) are distinguished by the presence of many short processes (side branches) that stud the main axon trunk along its course through the nucleus. The presence of side branches at these ages and their subsequent loss suggests that side branches may play an important role in the segregation of eye input. This suggestion is supported by the finding that during prenatal development the number of side branches per axon first increases and then decreases with a time course similar to that observed for the initial intermixing of inputs and subsequent reduction of territory shared by the two eyes (compare Sretavan & Shatz 1985, Figure 11, with Shatz 1983, Figure 15). Further support comes from the observation that the location of side branches is highly correlated with territories of intermixing of eye input (Sretavan & Shatz 1985). This point is illustrated in the diagrams of Figure 5, in which the relationship between the morphological appearance of retinogeniculate axons and LGN territories occupied by both eyes (*shaded areas*) is shown at different prenatal ages. Figure 5 also shows that eye-specific layers emerge as individual axons lose their side branches (or, in the case of the ipsilateral eye, their distal tips) and elaborate complex terminal arbors within territories appropriate to their eye of origin. Thus, it is highly likely that this sequence of morphological change undergone by individual axons contributes significantly to the segregation of eye input within the LGN. In view of the fact that development of the retinogeniculate projection in many mammals undergoes a similar sequence of intermixing followed by segregation, it is reasonable to suppose that similar changes at the level of individual axons contribute to the segregated pattern of eye input present in other species as well.

The morphological changes in retinogeniculate axons associated with the segregation of eye input are remarkably conservative in the sense that extensive terminal arborizations are apparently laid down only in LGN territories destined in the adult to belong to the eye of axon origin. In territories destined to belong to the other eye, axons contribute only short side branches. These considerations suggest that the development of eye-specific layers involves a large net increase in the amount of retinal axon arborization, a suggestion substantiated by direct measurements of total linear length of axon arbor with age (Sretavan & Shatz 1985). The conservative nature of the growth pattern of retinogeniculate axons contrasts with that of geniculocortical axons, which are thought to undergo extensive pruning to give rise to the segregated state within the visual cortex (LeVay & Stryker 1978).

In view of the intermixing of inputs from the two eyes and the transient nature and location of side branches, the question naturally arises as to whether the process of segregation involves the formation and elimination of synapses, as has been demonstrated elsewhere during development of the mammalian nervous system (for review see Purves & Lichtman 1980).

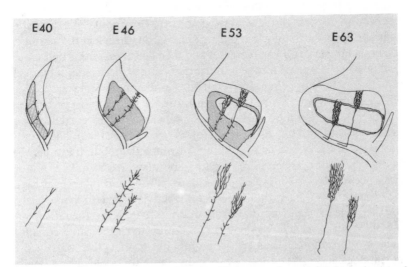

Figure 5 To summarize the proposed relationship between the pattern of arborization of retinal ganglion cell axons and the regions in which inputs from the two eyes are intermixed at different times during the prenatal development of the cat's retinogeniculate projection. At each age, zones of intermixing are shown by the *shaded regions*, and a representative axon from the ipsilateral (shorter axon in each case) and the contralateral eye is drawn (from Sretavan & Shatz 1985).

Electron microscopic studies of the developing cat (Shatz et al 1982) and hamster (Campbell et al 1984) retinogeniculate projection have directly demonstrated, by means of EM-HRP, that throughout the period of intermixing and segregation retinal axons synapse onto LGN neurons. Ultrastructural studies of the development of the primate retinogeniculate projection (Hendrickson & Rakic 1977) have also indicated that synapses are present during the relevant period; however, conclusive identification of retinal origin was not made. Moreover, Campbell et al (1984) have shown that some of the retinal synapses in the hamster's LGN are indeed located in territory ultimately destined to belong exclusively to the other eye. Thus, there is good evidence that synapse formation and elimination accompany the segregation of eye input within the mammalian LGN.

Not only are synapses present during the development of the retino-geniculate projection, but recent physiological studies in cat have shown that they are also capable of functioning (Shatz & Kirkwood 1984). Microelectrode recordings from fetal cat LGN maintained in vitro have shown that synaptic transmission between ganglion cell axons and LGN neurons can be elicited by electrical stimulation of the optic nerves as early as E40. Furthermore, during the period of intermixing, LGN neurons receive convergent excitatory inputs from both optic nerves, whereas by birth many neurons receive an adult-like mixture of excitatory input from one nerve and inhibition from the other. Thus, the inhibitory circuitry within the LGN appears relatively late during development, at a time when segregation is almost complete. This observation is based not only on the physiological evidence mentioned above but also on the basis of staining for glutamic acid decarboxylase immunoreactivity at different times during prenatal and postnatal life (Shotwell et al 1984). It is noteworthy that the early presence and subsequent loss of convergent excitation occurs concomitantly with the appearance and withdrawal of side branches from retinogeniculate axons. This correlation suggests that the side branches themselves may be at least partly responsible for the synaptic interactions observed during the period of segregation.

These observations permit a more complete understanding of the cellular interactions underlying the development of eye-specific layers in the mammalian LGN. If retinal ganglion cells axons from the two eyes are to interact with each other, the most likely site for this interaction is the set of postsynaptic LGN neurons that receive synaptic input from both eyes. Then, the transition from the intermixed to the segregated state involves the remodeling of synaptic connections, some or all of which are capable of function as assessed in vitro. It is now of great importance to determine whether synaptic function is also present in vivo. If so, such function would have to be driven by spontaneous activity in the ganglion cell population, because, at least in the cat, photoreceptor outer segments are not present until after the formation of layers is complete (Donovan 1966).

The Role of Binocular Competition in Segregation of Eye Input

It is commonly thought that the segregation of retinal ganglion cell axons from the two eyes into separate layers during development is achieved through a process of binocular competition. Such a process is assumed to involve some kind of competitive interaction between the two sets of ganglion cell axons for postsynaptic territory within the LGN. In this section, we consider the evidence supporting the hypothesis that binocular competition is responsible for segregation of eye input and discuss the possible roles played by activity and cell death in the competitive process.

Many experiments designed to prove that binocular competition is responsible for segregation are based on the premise that removal of one of the two competing elements should prevent segregation. In these experiments one eye is removed at a time when inputs from the two sets of ganglion cells are intermixed within the LGN. Then the consequences for the development of the retinogeniculate projection from the remaining eye are examined at later times when, under normal circumstances, inputs from the two eyes have segregated from each other. Results from studies of several different species show that the projection from the remaining eye fails to become restricted to the appropriate eye-specific layers (Godement et al 1980, Rakic 1981, Chalupa & Williams 1984, So et al 1984). While at first glance these results would seem to provide direct confirmation of the hypothesis, unfortunately another interpretation of the results is possible. Since these experiments are performed relatively late in development, eye removal causes axonal degeneration and denervation of LGN neurons. Either or both of these events could be responsible for the failure of the remaining eye to segregate. Therefore this type of experiment by itself cannot distinguish between the effects of denervation on the remaining retinal projection versus the true absence of binocular competition.

To circumvent this particular drawback, the development of the retinogeniculate projection from one eye has been examined under conditions in which axons from the remaining eye never have the opportunity to interact with those from the enucleated eye. This approach has been used in the cat's retinogeniculate system, where the enucleations were performed at E23, when retinal ganglion cell axons have not yet even reached the optic chiasm (J. Silver, M. Kliot, C. J. Shatz, personal communication; Williams et al 1986). The resulting pattern of projection was studied close to birth (Sretavan et al 1984, Sretavan & Shatz 1986). As described above, here too the projection from the remaining eye filled both the ipsilateral and the contralateral LGN, including regions normally occupied by afferents from the other eye. This observation suggests that binocular competition is indeed involved in formation of the segregated pattern of eye input, since such a pattern fails to form when one of the competitors is absent.

Experiments in which one eye is removed early, although they do avoid some difficulties in interpretation, are still subject to a number of criticisms. The most important one is that the manipulation itself might have not only eliminated competitive interactions but also perturbed development in other unknown ways, perhaps by altering the chiasmatic routing of ganglion cell axons or the extent of ganglion cell death in the remaining eye. For instance, in cat fetuses enucleated at E23 and examined at E59, the size of the ipsilateral projection is somewhat smaller than normal (Sretavan & Shatz 1986), suggesting that chiasmatic routing might be altered. [However, it should be noted that enucleations performed at analogous times during the development of the mouse retinogeniculate system (E12–E13) do not substantially alter the size or distribution of the ipsilaterally projecting ganglion cell population (Godement 1984).] Furthermore, early enucleation in the cat does not increase the number of ganglion cell axons, when examined at E59, in the remaining optic nerve (see Figure 1: *diamond symbols*), suggesting that the amount of cell death is within the usual range (Sretavan et al 1984, Sretavan & Shatz 1986). Nevertheless, even though chiasmatic routing and cell death may appear essentially normal, the possibility remains that the enucleations have caused other perturbations that could confound the interpretation of the results.

Any true test for the presence of competitive interactions should involve conditions in which competition is selectively removed or blocked without necessarily physically eliminating one of the competitors. Consequently, the experiments described above for the retinogeniculate system are informative but not definitive. Indeed, if binocular competition is responsible for the development of eye-specific layers, then it should be possible to perturb the formation of layers by altering the competition without removing one of the two sets of eye inputs. To design an appropriate experiment of this sort requires an understanding of the mechanisms underlying the competitive process. In fact, on the basis of many studies, it is reasonable to assume that neuronal activity is an important driving force for competitive interactions that lead to the segregated state (Stent 1973; for reviews see Purves & Lichtman 1980, Fawcett & O'Leary 1985). For instance, at the mammalian neuromuscular junction, it has been possible to alter the outcome of competition between motoneurons for muscle fibers either by blocking activity or neuromuscular transmission, or by altering the pattern of activity by electrical stimulation (Thompson et al 1979, Thompson 1983, Van Essen 1982).

In the vertebrate visual system, there are now several examples in which activity is known to play a role in the segregation of eye input from an initially intermixed state. These include the development of the system of ocular dominance columns in layer 4 of the cat's visual cortex (Stryker & Strickland 1984, Stryker & Harris 1985) and the experimentally-induced formation of eye-specific stripes in the optic tecta of frogs (for review see Constantine-Paton 1983) and goldfish (Meyer 1982, Boss & Schmidt 1984). In each instance, blocking retinal ganglion cell activity by intraocular injections of tetrodotoxin (TTX) prevents the segregation of eye input. These observations strongly suggest that activity is also the driving force for the formation of eye-specific layers in the mammalian retinogeniculate system. Indirect support for this suggestion comes from several different lines of evidence. For instance, Archer et al (1982) have shown that intraocular injections of TTX during early postnatal life can alter the normal development of receptive field properties within the cat's LGN. And, although it is currently unknown whether the formation of eye-specific layers during the prenatal development of the mammalian retinogeniculate projection can be prevented by TTX application, it is quite likely in view of the findings, mentioned above, that functional synaptic connections are present between retinal ganglion cells and their target LGN neurons throughout the relevant fetal period.

While it is plain from the above considerations that binocular competitive interactions such as those between retinal ganglion cells can be mediated through some kind of activity-dependent process, how such a process gives rise to the segregated state is currently unknown. In this context, it is worth noting that at least in the visual system, activity plays a role not only in the eye-specific segregation of inputs, but also in refining retinotopic maps. For example, fine-tuning of topography during regeneration of goldfish retinotectal connections is prevented by intraocular application of TTX (Meyer 1983, Schmidt & Edwards 1983). In addition, the normal restriction of the ipsilateral retinocollicular projection during postnatal development in rodents is also prevented if ganglion cell activity is blocked with bilateral injections of TTX (Fawcett et al 1984). A common theme to all of these cases is that activity is involved in the remodeling of connections. However, exactly how activity exerts its effects is currently unknown, although it is generally thought to involve synaptic interactions between the competing inputs and their postsynaptic targets (Stent 1973, Changeux & Danchin 1976, Purves & Lichtman 1980).

To summarize so far, available evidence suggests that the formation of eye-specific layers in the retinogeniculate system involves an activity-mediated binocular competition for postsynaptic territory within the LGN. The question naturally arises as to the consequences of the competition for both sets of inputs. The outcome of competition is manifested within the LGN by the retention of inputs from a given eye within certain LGN territories while those from the other eye are removed. Removal of inputs might be accomplished either by ganglion cell death, collateral retraction, or even a combination of these. Thus, it has been argued that competitive interactions lead to the segregated pattern of eye input by controlling the amount of cell death and/or collateral retraction during development of the retinogeniculate system. If so, then manipulations that alter or remove competition should also affect cell death and/or ganglion cell axon morphology.

Many experiments have been designed to examine the role of binocular interactions in the control of cell death, all of which involve the enucleation of one eye during development. Surprisingly, the results of these experiments show that a massive amount of cell death still occurs even in the absence of binocular competition. For example, in one set of experiments, one eye is removed during development at a time when retinal ganglion cell axons from both eyes are intermixed with each other within the LGN, and then the number of axons in the remaining optic nerve is examined in adult animals. Results in the cat show that of the 400,000 or so axons lost under normal circumstances, almost 90% are still lost despite the enucleation (see Figure 1, *square symbols*, and Williams et al 1983b); in the monkey, over 60% are still lost (Rakic & Riley 1983b). These findings indicate that binocular interactions, at best, control only 10–30% of optic axon loss (and presumably a corresponding amount of ganglion cell death) during development, while other factors must be responsible for the major share of ganglion cell death.

It is conceivable that the loss of axons resulting from binocular interactions, small though it is, nevertheless is sufficient to account for the segregation of eye input in the LGN. Cell death might operate in the following way to give rise to segregated layers of eye input: During the time when inputs are intermixed, axons from one eye might be divided into those that are correctly located in regions of the nucleus destined to belong to that eye, and those that project incorrectly to territory destined to be occupied by the other eye. Axons projecting to inappropriate territories might then be eliminated by cell death. However, in view of the results described above showing that virtually all axons initially have side branches that are subsequently retracted (Sretavan & Shatz 1984, 1985), we consider it more likely that segregation of retinal input into eye-specific layers during prenatal development is largely due to the remodeling of the branching pattern by elimination of inappropriately placed branches. Furthermore, one might expect that the number of optic nerve axons should decline during normal development with a time course that parallels the time course of the formation of eye-specific layers within the LGN. Although this is generally true, in both the cat (Williams et al 1983a, 1986, Figure 1) and the monkey (Rakic & Riley 1983a), axon loss also continues well into postnatal life—far beyond the time when segregation of eye input is complete. This observation suggests that if cell death is indeed involved in the segregation of retinogeniculate axons into eye-specific layers, then the effects of enucleation on axon number in the remaining optic nerve should be evident as soon as segregation is complete. Unfortunately, the results of several experiments designed to investigate this suggestion do not lend it much support. For example, when enucleations are performed, either as above, at a time when retinogeniculate axons from the two eyes are intermixed with each other within the LGN (rodents, Lam et al 1982, Crespo et al 1984), or even earlier, prior to their arrival at the optic chiasm (cat, Sretavan et al 1984, Sretavan & Shatz 1986), but the number of surviving optic nerve axons is assessed just after the completion of segregation, rather than in the adult, virtually no effect can be seen: axon loss is indistinguishable from that occurring during normal development. This result is shown in Figure 1 (*diamond symbols*) for the cat's optic nerve, where it is evident that the decline in axon number that is coincident with the period of segregation of retinogeniculate axons is not prevented by unilateral enucleation. Thus, at least in the cat, it is possible that binocular competition may not be the primary interaction responsible for the massive loss of axons occurring during this prenatal period and consequently may not contribute significantly to the formation of eye-specific layers within the LGN. However, the axon loss that continues beyond this prenatal period is likely to be controlled directly by binocular interactions, since it can be prevented by unilateral enucleation.

If cell death does not play a major role in eliminating retinal ganglion cell axons from inappropriate regions of the LGN, then the question arises as to what is the function of the massive elimination of optic nerve axons during prenatal life? There are many possibilities including, as discussed above in preceding sections, the correction of topographic errors (*Topographic Ordering of Retinal Ganglion Cell Projections*) and the establishment of the central-to-peripheral gradient of ganglion cell density in the retina (*Development of the Central-to-Peripheral Gradient*). In this context, it is noteworthy that in the cat, the central-to-peripheral gradient develops normally in animals even after early (Sretavan et al 1984, Sretavan & Shatz 1986) or late (Chalupa et al 1984) unilateral enucleation, an observation consistent with the suggestion that cell death independent of binocular interactions may mediate this developmental process. However, before any reasonable decision can be made concerning the role of cell death in this or any other aspect of the development of retinal ganglion cells and their projections, it is essential to resolve one critical issue: It is currently not even known whether the optic nerve axons targeted for elimination ever reach the optic chiasm, much less innervate the LGN.

In view of the above considerations, it seems unlikely that cell death is exclusively or even principally responsible for the elimination of axons from inappropriate LGN territories during the formation of eye-specific layers. Rather, a much more likely alternative is that segregation of eye inputs is mediated in large part by the retraction of ganglion cell axon branches from inappropriate LGN territories, concurrently with the elaboration of terminal arborizations within territories appropriate to the eye of axon origin. This suggestion receives strong support from the results of studies, described in the preceding section (*Eye-specific Segregation of Retinal Ganglion Cell Axons*), of the morphological changes undergone by individual retinogeniculate axons during the development of the cat's visual system: The changes observed are perfectly consistent with the notion that local retraction and specific growth of axon arbors can account for the segregation of ganglion cell axons from the two eyes within the LGN. Thus, the outcome of binocular competition is probably largely manifested in retraction rather than in cell death for the set of inappropriate inputs. However, it is worth emphasizing that this scenario by no means eliminates the possibility that some ganglion cell axons may be removed by cell death; but if so, cell death may only make a minor contribution to the formation of eye-specific layers.

Interactions Between the Ganglion Cell Axons from One Eye

If binocular competition between retinal ganglion cell axons from the two eyes indeed leads to the retraction of inappropriate branches and the retention and further elaboration of appropriate ones, then enucleation of one eye during development should affect the morphological appearance of axons by permitting the retention and growth of branches otherwise destined to be eliminated through competitive interactions with inputs from the other eye. This suggestion was tested directly in the development of the cat's visual system by removing one eye at an early stage (E23) when retinal ganglion cell axons have not yet reached the optic chiasm, and then examining the morphological appearance of individual retinogeniculate axons by filling them with HRP at a later time (E59) when, under normal circumstances, the segregation of axons from the two eyes would have been nearly complete. The results were totally unexpected, and showed that even in the absence of binocular interactions, retinogeniculate axons are capable of acquiring restricted terminal arborizations almost indistinguishable in size, shape, and complexity from those seen in normal animals at the same age (Sretavan et al 1984, Sretavan & Shatz 1986). Moreover, as shown in Figure 6, in the enucleated animals, the terminal arborizations of axons appear to be separated into several distinct tiers within the LGN in a fashion that is remarkably reminiscent of the eye-specific layers present in the normal animals.

The observation that unilateral enucleation does not prevent the normal developmental restriction of retinal ganglion cell terminal arbors within the

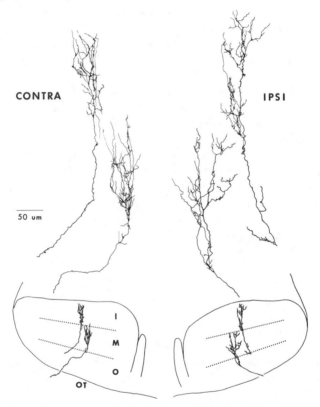

Figure 6 To show the pattern of arborization and location of four HRP-filled retino-geniculate axons reconstructed from a E59 cat fetus in which one eye was removed at E23. One pair of axons was located in the LGN contralateral to the remaining eye; the other pair was ipsilateral. Axons are indistinguishable from normal in the shape and complexity of their terminal arborizations. Moreover, axons appear to be distributed in tiers with respect to each other: one axon in each pair here is located in an innermost tier (I), while the other is located in a middle (M). (From Sretavan & Shatz 1986.)

LGN might at first seem to imply that this feature of axonal development is rigidly predetermined and therefore simply immune to the effects of competitive interactions of any sort. This is not the case, because axon morphology can be altered in the expected direction (e.g. excessive branching along the full length of the axon as it traverses the LGN) if the enucleations are made later on during development, at a time of maximal intermixing of axons from the two eyes within the LGN (Sretavan et al 1985). Thus, the results of the early enucleation experiments described above suggest that interactions between ganglion cell axons *from the same eye*, as well as those between ganglion cell axons from the two separate eyes, are capable of influencing the final morphological appearance of retinogeniculate axons. This suggestion should not, perhaps, be too surprising in view of the recent report by Perry & Linden (1982) that ganglion cell dendrites may compete with each other during development for territory: When a small group of ganglion cells in the retina of newborn rats is destroyed, the dendrites of nearby ganglion cells move into the vacated space. There is also evidence from studies of the effects of monocular visual deprivation in the cat that during postnatal development, retinal ganglion cell axons from the same eye compete with each other for LGN territory: In animals raised with one eye closed from birth, the terminal arborizations of one class of retinal ganglion cell axon originating in the deprived eye, the X-axons, are abnormally large, while those of the Y-axons are abnormally small (Sur et al 1982), suggesting that deprivation somehow perturbs the normal competitive interactions that control the arbor size of X-axons and Y-axons from the same eye.

If similar interactions were to occur between ganglion cell axons from the same eye during prenatal development, then such interactions might simply be accentuated when binocular competition is absent due to early unilateral enucleation. For example, during normal development of the cat's retinogeniculate projection, axons from the same eye come to be only mildly segregated from each other according to axon class (X versus Y) or other functional properties (on-center versus off-center). (See above: *Adult Organization of the Retinogeniculate System.*) Consequently, the tier-like arrangement of retinogeniculate axons seen in the cat's LGN following enucleation (Figure 6) could reflect the complete segregation of afferents from the remaining eye according to these or perhaps even other functional distinctions. Obviously, future experiments are required to resolve this point, but regardless of their outcome, the fact remains that interactions between the ganglion cell axons of one eye alone are capable of producing many normal features of the adult retinogeniculate projection. This does not imply that binocular interactions are not important in the development of eye-specific layers. Rather, it serves to emphasize that axons can interact with each other not only on the basis of their eye of origin, but also based on other differences as well.

CONCLUDING REMARKS

Here we have considered the sequence of developmental events that lead to the adult pattern of connectivity in the retinogeniculate system. In order to achieve the adult pattern, retinal ganglion cell axons must sort out from each other in an eye-specific manner and must also establish a topographically ordered map of the retina. Furthermore, the maps representing the visual fields in the two eyes must be aligned with each other. With regard to the development of topographic order, evidence reviewed here suggests that by the time ganglion cell axons reach the LGN, a coarse topographic order is already present within the optic tract, and this order is further fine-tuned within the target structures by eliminating topographically inappropriate connections through cell death, probably also combined with collateral retraction. With regard to the development of eye-specific layers within the LGN, it is thought that an activity-mediated binocular competition between ganglion cell axons for postsynaptic territory leads to the segregated state. In this case, however, recent evidence suggests that a major consequence of binocular competitive interactions within the LGN is the remodeling of axon arborizations to give rise to the segregated pattern of eye input. Competitive interactions can occur between the ganglion cells of one eye, and these are also likely to be involved in producing the adult pattern of connectivity in the retinogeniculate system.

ACKNOWLEDGMENTS

We wish to thank Drs. A. Rusoff, J. Stone, M. Sur, and R. W. Williams and their colleagues for kindly allowing us to use results from their studies in assembling some of the figures and Dr. M. Sherman for his insightful criticisms. The authors' original research cited herein was supported by grants from the National Institutes of Health (EY 02858), the National Science Foundation (BNS 8317228), the McKnight Foundation, and the March of Dimes.

Literature Cited

Aebersold, H., Creutzfeld, O. D., Kuhnt, U., Sanides, D. 1981. Representation of the visual field in the optic tract and the optic chiasma of the cat. *Exp. Brain Res.* 42:127–45

Archer, S. M., Dubin, M. W., Stark, L. A. 1982. Abnormal retinogeniculate connectivity in the absence of action potentials. *Science* 217:743–45

Berman, N., Cynader, M. 1972. Comparison

of receptive-field organization of the superior colliculus in Siamese and normal cats. *J. Physiol.* 224:363–89

Bonds, A. B., Freeman, R. D. 1978. Development of optical quality in the kitten eye. *Vision Res.* 18:391–98

Boss, V. C., Schmidt, J. T. 1984. Activity and the formation of ocular dominance patches in dually innervated tectum of goldfish. *J. Neurosci.* 4:2891–2905

Bowling, D. B. 1984. Differences in the timing and sensitivity of responses from cells across single layers of the lateral geniculate nucleus in the cat. *Soc. Neurosci. Abstr.* 10:296

Bowling, D. B., Michael, C. R. 1980. Projection patterns of single physiologically characterized optic tract fibers in the cat. *Nature* 286:899–902

Bowling, D. B., Michael, C. R. 1984. Terminal patterns of single physiologically characterized optic tract fibers in the cat's lateral geniculate nucleus. *J. Neurosci.* 4:198–216

Boycott, B. B., Wässle, H. 1974. The morphological types of ganglion cells of the domestic cat's retina. *J. Physiol.* 240:397–419

Bunt, S. M., Horder, T. J. 1983. Evidence for an orderly arrangement of optic axons within the optic nerves of the major non-mammalian vertebrate classes. *J. Comp. Neurol.* 213:94–114

Bunt, S. M., Lund, R. D., Land, P. W. 1983. Prenatal development of the optic projections in albino and hooded rats. *Dev. Brain Res.* 6:149–68

Cajal, S. R. 1972. The structure of the retina. In *The Development of Retinal Cells*, Ch. 9. Springfield, Ill.: Thomas. (Orig. publ. in *La Cellule*, 1892)

Calvacante, L. A., Rocha-Miranda, C. E. 1978. Postnatal development of retinogeniculate, retinopretectal and retinotectal projections in the opossum. *Brain Res.* 146:231–48

Campbell, G., So, K.-F., Lieberman, A. R. 1984. Normal postnatal development of retinogeniculate axons and terminals and identification of inappropriately located transient synapses: Electron microscope studies of HRP-labeled retinal axons in the hamster. *Neuroscience* 13:743–59

Card-Linden, D., Guillery, R. W., Cucchiaro, J. 1981. The dorsal lateral geniculate nucleus of the normal ferret and its postnatal development. *J. Comp. Neurol.* 203:189–211

Chalupa, L., Williams, R. W. 1984. Organization of the cat's lateral geniculate nucleus following interruption of prenatal binocular competition. *Human Neurobiol.* 3:103–7

Changeux, J. P., Danchin, A. 1976. Selective stablization of developing synapses, a mechanism for the specification of neuronal networks. *Nature* 264:705–12

Cleland, B. G., Levick, W. R. 1974. Brisk and sluggish concentrically organized ganglion cells in the cat's retina. *J. Physiol.* 240:421–56

Constantine-Paton, M. 1983. Position proximity in the development of maps and stripes. *Trends Neurosci.* 6:32–36

Crespo, D., O'Leary, D. D. M., Cowan, W. M. 1984. Reduction of optic nerve axons during the postnatal development of the albino rat. *Soc. Neurosci. Abstr.* 10:464

Cusick, C. G., Kaas, J. H. 1982. Retinal projections in adult and newborn grey squirrels. *Dev. Brain Res.* 4:275–84

Donovan, A. 1966. Postnatal development of the cat retina. *Exp. Eye Res.* 5:249–54

Drager, U. C., Olson, J. F. 1980. Origins of crossed and uncrossed retinal projections in pigmented and albino mice. *J. Comp. Neurol.* 191:383–412

Easter, S. S., Rusoff, A. C., Kish, P. E. 1981. The growth and organization of the optic nerve and tract in juvenile and adult goldfish. *J. Neurosci.* 1:793–811

Enroth-Cugell, C., Robson, J. G. 1966. The contrast sensitivity of retinal ganglion cells of the cat. *J. Physiol.* 187:517

Fawcett, J. W., O'Leary, D. D. M. 1985. The role of electrical activity in the formation of topographic maps in the nervous system. *Trends Neurosci.* 8:201–6

Fawcett, J. W., O'Leary, D. D. M., Cowan, W. M. 1984. Activity and the control of ganglion cell death in the rat retina. *Proc. Natl. Acad. Sci. USA* 81:5589–93

Friedlander, M. J., Martin, K. A. C., Vahle-Hinz, C. 1985. The structure of the terminal arborizations of physiologically identified retinal ganglion cell Y-axons in the kitten. *J. Physiol.* 359:293–314

Frost, D. O., So, K. F., Schneider, G. E. 1979. Postnatal development of retinal projections in Syrian hamsters: A study using autoradiographic and anterograde degeneration techniques. *Neuroscience* 4:1649–77

Godement, P. 1984. Development of retinal projections in the mouse. In *Development of Visual Pathways in Mammals*, ed. J. Stone, B. Dreher, D. H. Rapaport, pp. 127–43. New York: Liss

Godement, P., Saillour, P., Imbert, M. 1980. The ipsilateral optic pathway to the dorsal lateral geniculate nucleus and superior colliculus in mice with prenatal or postnatal loss of one eye. *J. Comp. Neurol.* 190:611–26

Godement, P., Salaün, J., Imbert, M. 1984.

Prenatal and postnatal development of retinogeniculate and retinocollicular projections in the mouse. *J. Comp. Neurol.* 230: 552–75

Guillery, R. W. 1970. The laminar distribution of retinal fibers in the dorsal lateral geniculate nucleus of the cat: A new interpretation. *J. Comp. Neurol.* 138: 339–68

Guillery, R. W. 1982. The optic chiasm of the vertebrate brain. In *Contributions to Sensory Physiology*, ed. W. D. Neff, 7: 38–73. New York: Academic

Guillery, R. W., Polley, E. H., Torrealba, F. 1982. The arrangement of axons according to fiber diameter in the optic tract of the cat. *J. Neurosci.* 2: 714–21

Graybiel, A. M. 1975. Anatomical organization of retinotectal afferents in the cat: An autoradiographic study. *Brain Res.* 96: 1–24

Hamasaki, D. I., Flynn, J. T. 1977. Physiological properties of retinal ganglion cells of 3-week-old kittens. *Vision Res.* 17: 275–84

Harting, J. K., Guillery, R. W. 1976. Organization of retinocollicular pathways in the cat. *J. Comp. Neurol.* 166: 133–44

Hayhow, W. R. 1958. The cytoarchitecture of the lateral geniculate body in relation to the distribution of crossed and uncrossed fibers. *J. Comp. Neurol.* 110: 1–51

Hendrickson, A., Rakic, P. 1977. Histogenesis and synaptogenesis in the dorsal lateral geniculate nucleus (LGd) of the fetal monkey brain. *Anat. Rec.* 187: 602

Hickey, T. L., Guillery, R. W. 1974. An autoradiographic study of the retinogeniculate pathways in the cat and the fox. *J. Comp. Neurol.* 156: 239–54

Hinds, J. W., Hinds, P.-L. 1983. Development of retinal amacrine cells in the mouse embryo: Evidence for two modes of formation. *J. Comp. Neurol.* 213: 1–23

Horton, J. C., Greenwood, M. M., Hubel, D. H. 1979. Non-retinotopic arrangement of fibres in cat optic nerve. *Nature* 282: 720–22

Hsiao, K., Sachs, G. M., Schneider, G. E. 1984. A minute fraction of syrian golden hamster retinal ganglion cells project bilaterally. *J. Neurosci.* 4: 359–67

Hubel, D. H., LeVay, S., Wiesel, T. N. 1975. Mode of termination of retinotectal fibers in Macaque monkey: An autoradiographic study. *Brain Res.* 96: 25–40

Hubel, D. H., Wiesel, T. N. 1969. Anatomical demonstration of columns in the monkey striate cortex. *Nature* 221: 747–50

Hubel, D. H., Wiesel, T. N. 1970. The period of susceptibility to the physiological effects of unilateral eye closure in kittens. *J. Physiol.* 206: 419–36

Hubel, D. H., Wiesel, T. N. 1972. Laminar and columnar distribution of geniculocortical fibers in the Macaque monkey. *J. Comp. Neurol.* 146: 421–50

Insausti, R., Blakemore, C., Cowan, W. M. 1984. Ganglion cell death during development of ipsilateral retino-collicular projection in golden hamster. *Nature* 308: 362–65

Jacobs, D. S., Perry, V. H., Hawken, M. J. 1984. The postnatal reduction from the uncrossed projection from the nasal retina in the cat. *J. Neurosci.* 4: 2425–33

Jeffrey, G. 1985. Retinotopic order appears before ocular segregation in developing pathways. *Nature* 313: 575–76

Jeffrey, G., Cowey, A., Kuypers, H. G. J. M. 1981. Bifurcating retinal ganglion cell axons in the rat, demonstrated by retrograde double labelling. *Exp. Brain Res.* 44: 34–40

Jeffrey, G., Perry, V. H. 1982. Evidence for ganglion cell death during development of the ipsilateral retinal projection in the rat. *Dev. Brain Res.* 2: 176–80

Johns, P. R. 1977. Growth of the adult goldfish eye. III. Source of the new retinal cells. *J. Comp. Neurol.* 176: 343–57

Kliot, M., Shatz, C. J. 1982. Genesis of different retinal ganglion cell types in the cat. *Soc. Neurosci. Abstr.* 8: 815

Lam, K., Sefton, J., Bennett, M. R. 1982. Loss of axons from the optic nerve of the rat during early postnatal development. *Dev. Brain Res.* 3: 487–91

LeVay, S., McConnell, S. K. 1982. On and Off layers in the lateral geniculate nucleus of the mink. *Nature* 300: 350–51

LeVay, S., Stryker, M. P. 1978. The development of ocular dominance columns in the cat. In *Soc. Neurosci. Symp. 4: Aspects of Developmental Neurobiology*, ed. J. Ferendelli, pp. 83–98. Bethesda: Soc. Neurosci.

LeVay, S., Stryker, M. P., Shatz, C. J. 1978. Ocular dominance columns and their development in layer IV of the cat's visual cortex: A quantitative study. *J. Comp. Neurol.* 179: 223–44

Lia, B., Williams, R. W., Chalupa, L. M. 1983. Early development of retinal specialization: The distribution and decussation patterns of ganglion cells in the prenatal cat demonstrated by retrograde peroxidase labeling. *Soc. Neurosci. Abstr.* 9: 702

Lia, B., Williams, R. W., Chalupa, L. M. 1985. Overproduction of optic nerve fibers in the fetal cat does not reflect axonal branching within the nerve. *Suppl. Invest. Ophthalmol. Vis. Sci.* 26: 286

Mason, C. A. 1982. Development of terminal arbors of retinogeniculate axons in the kitten—I. Light microscopial observations. *Neuroscience* 7: 541–559

Mastronarde, D. N. 1984. Organization of the cat's optic tract as assessed by single axon recordings. *J. Comp. Neurol.* 227: 14–22

Mastronarde, D. N., Thibeault, M. A., Dubin, M. W. 1984. Non-uniform postnatal growth of the cat retina. *J. Comp. Neurol.* 228: 598–608

Morest, K. D. 1970. The pattern of neurogenesis in the retina of the rat. *Z. Anat. Entwickl.-Gesch.* 131: 45–67

Meyer, R. L. 1982. Tetrodotoxin blocks the formation of ocular dominance columns in goldfish. *Science* 218: 589–91

Meyer, R. L. 1983. Tetrodotoxin inhibits the formation of refined retinotopography in goldfish. *Dev. Brain Res.* 6: 293–98

Murakami, D., Sesma, M., Rowe, M. H. 1982. Characteristics of nasal and temporal retina in Siamese and normally pigmented cats: Ganglion cell composition, axon trajectory and laterality of projection. *Brain Behav. Evol.* 21: 67–113

Ng, A., Stone, J. 1982. The optic nerve of the cat: Appearance and loss of axons during normal development. *Dev. Brain Res.* 5: 263–71

Perry, V. H., Henderson, Z., Linden, R. 1983. Postnatal changes in retinal ganglion cell and optic axon populations in the pigmented rat. *J. Comp. Neurol.* 219: 356–68

Perry, V. H., Linden, R. 1982. Evidence for dendritic competition in the developing retina. *Nature* 297: 683–85

Potts, R. A., Dreher, B., Bennett, M. R. 1982. The loss of ganglion cells in the developing retina of rat. *Dev. Brain Res.* 3: 481–86

Purves, D., Lichtman, J. W. 1980. Elimination of synapses in the developing nervous system. *Science* 210: 153–57

Rakic, P. 1977. Prenatal development of the visual system in rhesus monkey. *Philos. Trans. R. Soc. London Ser. B* 278: 245–60

Rakic, P. 1981. Development of visual centers in the primate brain depends on binocular competition before birth. *Science* 214: 928–31

Rakic, P., Riley, K. P. 1983a. Overproduction and elimination of retinal axons in the fetal rhesus monkey. *Science* 219: 1441–44

Rakic, P., Riley, K. 1983b. Regulation of axon number in primate optic nerve by prenatal binocular competition. *Nature* 305: 135–37

Rapaport, D. H., Stone, J. 1983. Time course of the morphological differentiation of cat retinal ganglion cells: Influences on cell size. *J. Comp. Neurol.* 221: 42–52

Reh, T. A., Pitts, E. C., Constantine-Paton, M. 1983. The organization of the fibers in the optic nerve of normal and tectum-less *Rana pipiens. J. Comp. Neurol.* 218: 282–96

Rodiek, R. W. 1979. Visual pathways. *Ann.*

Rev. Neurosci. 2: 193–225

Rusoff, A. C. 1979. Development of ganglion cells in the retina of the cat. In *Developmental Neurobiology of Vision*, ed. R. Freeman, pp. 19–30. New York: Plenum

Rusoff, A. C., Dubin, M. W. 1977. Development of receptive-field properties of retinal ganglion cells in kittens. *J. Neurophysiol.* 40: 1188–98

Rusoff, A. C., Dubin, M. W. 1978. Kitten ganglion cells: Dendritic field size at 3 weeks of age and correlation with receptive field size. *Invest. Ophthalmol. Visual Sci.* 17: 819–21

Saito, H. A. 1983. Morphology of physiologically identified X-, Y-, and W-type retinal ganglion cells of the cat. *J. Comp. Neurol.* 221: 279–88

Sanderson, K. J. 1971. The projection of the visual field to the lateral geniculate and medial interlaminar nuclei in the cat. *J. Comp. Neurol.* 143: 101–18

Sanderson, K. J., Haight, J. R., Pettigrew, J. D. 1984. The dorsal lateral geniculate nucleus of macropodial marsupials: Cytoarchitecture and retinal projections. *J. Comp. Neurol.* 224: 85–106

Schiller, P. H., Malpeli, J. G. 1978. Functional specificity of lateral geniculate nucleus laminae of the Rhesus monkey. *J. Neurophysiol.* 41: 788–97

Schmidt, J. T., Edwards, D. L. 1983. Activity sharpens the map during the regeneration of the retinotectal projection in goldfish. *Brain Res.* 269: 29–39

Scholes, J. H. 1979. Nerve fibre topography in the retinal projection to the tectum. *Nature* 278: 620–24

Sengelaub, D. R., Finlay, B. L. 1982. Cell death in the mammalian visual system during normal development: I. Retinal ganglion cells. *J. Comp. Neurol.* 204: 311–17

Shatz, C. J. 1983. The prenatal development of the cat's retinogeniculate pathway. *J. Neurosci.* 3: 482–99

Shatz, C. J., Kirkwood, P. A. 1984. Prenatal development of functional connections in the cat's retinogeniculate pathway. *J. Neurosci.* 4: 1378–97

Shatz, C. J., Kirkwood, P. A., Siegel, M. W. 1982. Functional retinogeniculate synapses in fetal cats. *Soc. Neurosci. Abstr.* 8: 815

Shatz, C. J., Lindström, S., Wiesel, T. N. 1977. The distribution of afferents representing the right and left eyes in the cat's visual cortex. *Brain Res.* 131: 103–16

Shotwell, S. L., Luskin, M. B., Shatz, C. J. 1984. Development of GAD immunoreactivity correlates with onset of inhibition in cat lateral geniculate nucleus. *Soc. Neurosci. Abstr.* 10: 142

Sidman, R. L. 1961. Histogenesis of mouse

retina studied with thymidine-H[3]. From *The Structure of the Eye*, pp. 487–506. New York: Academic

Sretavan, D. W., Kliot, M., Shatz, C. J. 1984. Observations on the development of the cat's retinogeniculate pathway in the absence of binocular interactions. *Soc. Neurosci. Abstr.* 10:670

Sretavan, D. W., Shatz, C. J. 1984. Prenatal development of individual retinogeniculate axons during the period of segregation. *Nature* 308:845–48

Sretavan, D. W., Garraghty, P. E., Sur, M., Shatz, C. J. 1985. Development of retinogeniculate axon arbors following prenatal unilateral enucleation. *Soc. Neurosci. Abstr.* 11: In press

Sretavan, D. W., Shatz, C. J. 1985. Prenatal development of retinal ganglion cell axons: Segregation into eye-specific layers. *J. Neurosci.* In press

Sretavan, D. W., Shatz, C. J. 1986. Prenatal development of retinogeniculate axon arbors in the absence of binocular interactions. *J. Neurosci.* In press

So, K.-F., Schneider, G. E., Frost, D. O. 1978. Postnatal development of retinal projections to the lateral geniculate body in Syrian hamsters. *Brain Res.* 142:343–52

So, K.-F., Woo, H. H., Jen, L. S. 1984. The normal and abnormal postnatal development of retinogeniculate projections in golden hamsters: An anterograde horseradish peroxidase tracing study. *Dev. Brain Res.* 12:191–205

Stent, G. S. 1973. Physiological mechanism for Hebb's postulate of learning. *Proc. Natl. Acad. Sci. USA* 70:997–1001

Stone, J. 1978. The number and distribution of ganglion cells in the cat's retina. *J. Comp. Neurol.* 180:735–72

Stone, J., Hoffmann, K. P. 1972. Very slow-conducting ganglion cells in the cat's retina: A major, new functional type? *Brain Res.* 43:610–16

Stone, J., Rapaport, D. H., Williams, R. W., Chalupa, L. 1982. Uniformity of cell distribution in the ganglion cell layer of prenatal cat retina: Implications for mechanisms of retinal development. *Dev. Brain Res.* 2:231–42

Stone, J., Maslim, J., Rapaport, D. H. 1984. The development of the topographical organization of the cat's retina. In *Development of Visual Pathways in Mammals*, ed. J. Stone, B. Dreher, D. H. Rapaport, pp. 3–21. New York: Liss

Straznicky, K., Gaze, R. M. 1971. The growth of the retina in xenopus laevis: An autoradiographic study. *J. Embryol. Exp. Morphol.* 26:67–79

Stryker, M. P., Harris, W. 1985. Binocular impulse blockade prevents the formation of ocular dominance columns in cat visual cortex. *J. Neurosci.* In press

Stryker, M. P., Strickland, S. L. 1984. Physiological segregation of ocular dominance columns depends on the pattern of afferent electrical activity. *Invest. Ophthalmol. Vis. Sci.* (Suppl.) 25:278

Stryker, M. P., Zahs, K. 1983. On and Off sublaminae in the lateral geniculate nucleus of the ferret. *J. Neurosci.* 3:1943–51

Sur, M., Sherman, S. M. 1982. Retinogeniculate terminations in cats. Morphological differences between X- and Y-cell axons. *Science* 218:389–91

Sur, M., Humphrey, A. L., Sherman, S. M. 1982. Monocular deprivation affects X- and Y-cell retinogeniculate terminations in cats. *Nature* 300:183–85

Sur, M., Weller, R. E., Sherman, S. M. 1984. Development of X- and Y-cell retinogeniculate terminations in kittens. *Nature* 310:246–49

Torrealba, F., Guillery, R. W., Polley, E. H., Mason, C. A. 1981. A demonstration of several independent, partially overlapping, retinotopic maps in the optic tract of the cat. *Brain Res.* 219:428–32

Torrealba, F., Guillery, R. W., Eysel, U., Polley, E. H., Mason, C. A. 1982. Studies of retinal representations within the cat's optic tract. *J. Comp. Neurol.* 211:377–96

Thompson, W. 1983. Synapse elimination in neonatal rat muscle is sensitive to pattern of muscle use. *Nature* 302:614–16

Thompson, W., Kuffler, D. P., Jansen, J. K. S. 1979. The effect of prolonged, reversible block of nerve impulses on the elimination of polyneuronal innervation of new-born rat skeletal muscle fibers. *Neuroscience* 4:271–81

Thorn, F., Gollender, M., Erickson, P. 1976. The development of the kitten's visual optics. *Vision Res.* 16:1145–49

Tiao, Y.-C., Blakemore, C. 1976. Regional specialization in the golden hamster's retina. *J. Comp. Neurol.* 168:439–58

Tusa, R. J., Palmer, L. A., Rosenquist, A. C. 1978. The retinotopic organization of area 17 (striate cortex) in the cat. *J. Comp. Neurol.* 177:213–36

Van Essen, D. C. 1982. In *Neuronal Development*, ed. N. C. Spitzer, pp. 333–76. New York: Plenum

Wässle, H. 1982. Morphological types and central projections of ganglion cells in the cat retina. *Progress in Retinal Research*, ed. N. Osborne, G. Chandler, pp. 125–52. Oxford: Pergamon

Walsh, C., Guillery, R. W. 1984. Fibre order in the pathways from the eye to the brain. *Trends Neurosci.* 7:208–11

Walsh, C., Polley, E. H. 1985. The topography of ganglion cell production in the cat's retina. *J. Neurosci.* 5:741-50

Walsh, C., Polley, E. H., Hickey, T. L., Guillery, R. W. 1983. Generation of cat retinal ganglion cells in relation to central pathways. *Nature* 302:611-14

Wiesel, T. N., Hubel, D. H., Lam, D. M. K. 1974. Autoradiographic demonstration of ocular dominance columns in the monkey striate cortex by means of transneuronal transport. *Brain Res.* 79:273-79

Williams, R. W., Bastiani, M. J., Chalupa, L. M. 1983a. Addition and attrition of axons within the optic nerve during fetal development: Appearance of growth cones and necrotic axons. *Invest. Ophthal. Vis. Sci. Suppl.* 24:8

Williams, R. W., Bastiani, M., Chalupa, L. M. 1983b. Loss of axons in the cat optic nerve following fetal unilateral enucleation: An electron microscopic analysis. *J. Neurosci.* 3:133-44

Williams, R. W., Chalupa, L. M. 1982. Prenatal development of retinocollicular projection in the cat: An autoradiographic tracer transport study. *J. Neurosci.* 2:604-22

Williams, R. W., Chalupa, L. M. 1983. An analysis of axon caliber within the optic nerve of the cat: Evidence of size groupings and regional organization. *J. Neurosci.* 3:1554-64

Williams, R. W., Rakic, P. 1985. Dispersion of growing axons within the optic nerve of the embryonic monkey. *Proc. Natl. Acad. Sci. USA* 82:3906-10

References added in proof:

Chalupa, L. M., Williams, R. W., Henderson, Z. 1984. Binocular interaction in the fetal cat regulates the size of the ganglion cell population. *Neuroscience* 12:1139-46

Williams, R. W., Bastiani, M., Lia, B., Chalupa, L. M. 1986. Growth cones, dying axons, and developmental fluctuations in the fiber population of the cat's optic nerve. *J. Comp. Neurol.* In press

Ann. Rev. Neurosci. 1986. 9: 209–54

HISTAMINE AS A NEUROREGULATOR

George D. Prell and Jack Peter Green

Department of Pharmacology, Mount Sinai School of Medicine of the City University of New York, New York, New York 10029

INTRODUCTION

Strong evidence now supports the view that histamine is a neuroregulator. This assertion is not easily refuted unless one invokes demands that have not been applied to analogous assertions that have been made for other biogenic amines, e.g. 5-hydroxytryptamine. Histamine has a nonuniform regional distribution in the brain of all species examined, the phylogenetically older regions of vertebrate brain showing higher levels (see reviews by Green 1970, Taylor 1975, Schwartz et al 1979a, Hough & Green 1984). It sediments with synaptosomes (Carlini & Green 1963, Kataoka & DeRobertis 1967, Dismukes et al 1974, Picatoste et al 1977, Schwartz et al 1979a, Sperk et al 1981b, Almeida & Beaven 1981). The histamine-synthesizing enzyme, histidine decarboxylase, has high specificity for histidine and is distinct from aromatic L-amino acid decarboxylase (Schwartz et al 1970, Palacios et al 1976). Histamine methyltransferase, the enzyme that metabolizes histamine in mammalian brain, has no other known endogenous substrate (Brown et al 1959). The regional distribution in brain of the synthesizing enzyme is similar to the distribution of histamine (Schwartz et al 1979a), as are the distributions of the metabolites—both *tele*-methylhistamine (i.e. N^τ-methylhistamine), the immediate metabolite (Hough & Domino 1979a, Hough et al 1981), and *tele*-methylimidazoleacetic acid (i.e. N^τ-methylimidazoleacetic acid), the metabolite of *tele*-methylhistamine (Khandelwal et al 1984). Histamine turns over in brain, and the turnover rates vary over 50-fold in different brain regions, correlating with the regional concentrations of histamine (Pollard et al 1974, Hough et al 1984a, Oishi et al 1984). Immunohistochemistry shows histaminergic fibers in many parts of the brain (Tran & Snyder 1981,

209

0147–006X/86/0301–0209$02.00

Wilcox & Seybold 1982, Watanabe et al 1983, 1984, Panula et al 1984, Steinbusch & Mulder 1985) and in the spinal cord (Wahlestedt et al 1985), originating from cell bodies in the hypothalamus. Lesions of the medial forebrain bundle produce a fall in histidine decarboxylase activity in brain regions distal to the lesions (Garbarg et al 1974), and transection of fibers to the hippocampus nearly eliminates histidine decarboxylase activity in the hippocampus (Barbin et al 1976). Histamine is released from brain slices by K^+ (Taylor & Snyder 1973, Atack & Carlsson 1972, Verdiere et al 1975, Subramanian & Mulder 1976) and by electrical stimulation (Biggs & Johnson 1980) by a Ca^{2+}-dependent process. Histamine receptors, homologous to those found in peripheral tissues, are present in brain tissue (for reviews see Schwartz 1979, Schwartz et al 1982, Green 1983). Neurons respond to histamine, and the electrical responses are reduced by antagonists of histamine. Stimulation of fibers produced electrophysiological responses at postsynaptic sites that were mimicked by histamine and attenuated by a histamine antagonist (Sastry & Phillis 1976a,b, Haas & Wolf 1977; see below). In the marine gastropod, *Aplysia californica*, histamine has been shown to be a neurotransmitter (see Weinreich 1978, 1985; see below).

This evidence has accumulated slowly (Green 1970, Green et al 1978, Schwartz et al 1979a, 1982, Hough & Green 1984) despite the fact that histamine was one of the first biogenic amines to be found in brain (Abel & Kubota 1919, Kwaitkowski 1943) and despite observations, accumulated over 20 years ago, suggesting that histamine has a function in the brain (Green 1964). One reason for the delay has been the problem of measuring histamine in the brain, a problem solved by a radioenzymatic method (Snyder et al 1966). The histaminergic pathways in brain eluded histochemical visualization and were finally revealed by immunohistochemical methods (Tran & Snyder 1981, Wilcox & Seybold 1982, Watanabe et al 1983, Panula et al 1984, Steinbusch & Mulder 1985). Measurements of histamine turnover in brain without disturbing steady-state levels were not practical before convenient methods to measure the main (perhaps only) histamine metabolite in mammalian brain, *tele*-methylhistamine (Hough & Domino 1979a, Hough et al 1981, Oishi et al 1984) had been developed; an early tedious procedure to measure *tele*-methylhistamine (Fram & Green 1965) was specific and yielded concentrations in brain (Fram & Green 1968) that were confirmed by advanced methods, but this early procedure, more an ordeal than a method, invited little use (Green et al 1964, Fram & Green 1968). Probably more determinant than methodology for the relatively sparse studies of histamine in the nervous system has been the historic and overwhelming concern with the role of histamine in disease, in contrast with the interest in other biogenic amines which was initially provoked by their association with the nervous system. Despite this relative neglect, histamine adventitiously initiated the development of drugs to treat psychoses, and concomitantly, the development of biological psychiatry: phenothiazines were designed as antihistaminic drugs and noticed by French neurosurgeons to produce a different kind of sedation, a "euphoric quietude," an observation that led to their use in schizophrenia (Swazey 1974, Kety 1978). From the phenothiazines evolved the dibenzazepines and other drugs that proved effective in treating depression (Kuhn 1970). Drugs of both these classes react with both histamine H_1 and H_2 receptors (Green 1983, Hough & Green 1984).

This review is mainly concerned with evidence that histamine is a neuroregulator. As detailed reviews have recently appeared on formation, metabolism, and receptors of histamine (Schwartz et al 1979a, 1982, Green 1983, Hough & Green 1984, Green & Khandelwal 1985), these subjects are summarily discussed here. Other subjects that have been recently reviewed are not discussed: the effects of neurally active drugs on histamine and its receptors (Schwartz et al 1982, Green 1983, Hough & Green 1984), and the neural functions that have been proposed for histamine, which have been summarized (Green et al 1978, 1980, Schwartz et al 1979a, 1980, 1981, 1982, Green & Hough 1980, Hough & Green 1980, 1984, Gross 1982, Green 1983, Mazurkiewicz-Kwilecki 1984). Its role in specific processes has been discussed: thermal regulation (Clark & Clark 1980, Lomax & Green 1981), cerebral circulation (Edvinsson & MacKenzie 1977, Gross 1981, 1982), cardiovascular regulation (Roberts & Calcutt 1983), drinking and feeding behavior (Leibowitz 1979, 1980), emesis (Bhargava et al 1982), self-stimulation, conditioning, motor activity (Schwartz et al 1979a, Schwartz et al 1982, Hough & Green 1984), analgesia (Rumore & Schlichting 1985, Hough et al 1985a), sleep and wakefulness (Monnier et al 1970, Hough & Green 1984, Roberts & Calcutt 1983), and neuroendocrine regulation (Roberts & Calcutt 1983, Donoso & Alvarez 1984).

TURNOVER OF HISTAMINE IN MAMMALIAN BRAIN: THE EXISTENCE OF DIFFERENT POOLS

In mammalian brain, histamine is methylated postsynaptically (Garbarg et al 1973, Haas et al 1978, Bischoff & Korf 1978, Sperk et al 1981a) at the *tele*-position, i.e. the ring-nitrogen farther from the side chain, to *tele*-methylhistamine, most of which, both in mature guinea pigs (Fram & Green 1968) and very young rats (Hough et al 1982), is bound to particulate material. *Tele*-Methylhistamine is oxidatively deaminated, mainly by monoamine oxidase B (Hough & Domino 1979b), to *tele*-

methylimidazoleacetic acid. Although histamine is also oxidatively de-aminated to imidazoleacetic acid in mammalian peripheral tissues and by the brain of lower vertebrates (Almeida & Beaven 1981), mammalian brain lacks diamine oxidase (Schwartz et al 1971, Almeida & Beaven 1981), and no imidazoleacetic acid could be detected in rat brain (Khandelwal et al 1982). Because a high-affinity uptake system for histamine has not been demonstrated in mammalian brain (Snyder & Taylor 1972, Schwartz et al 1973) and because tele-methylhistamine appears to be the only metabolite of histamine in mammalian brain, the accumulation of tele-methylhistamine after inhibition of its metabolism by treating animals with an irreversible inhibitor of monoamine oxidase, e.g. pargyline, should indicate histamine turnover. The turnover rates in regions of brain of rat, mouse, and guinea pig were similar (Hough et al 1984a, Oishi et al 1984, Nishibori et al 1984). The hypothalamus showed the highest turnover rate but not the highest rate constant (Hough et al 1984a), a divergence implying that a relatively small portion of the histamine there is turning over, probably because a portion of the hypothalamic histamine is in cell bodies (see below). The highest rate constants and hence shortest half lives of histamine in rat brain were in the caudate nucleus and cortex, where the half lives of histamine (Hough et al 1984a) were shorter than those reported for dopamine. The kinetic values were based on a one-compartment model.

If the measurements of tele-methylhistamine also embrace a non-neuronal compartment, that compartment is not likely to include mast cells if brain mast cells have properties even grossly similar to those of peritoneal mast cells. Early experiments with radioactive histamine failed to show metabolism of histamine by peritoneal mast cells of the rat (Furano & Green 1964a), which may, like mouse neoplastic mast cells, elaborate histamine into the surrounding medium (Green & Day 1963, Furano & Green 1964b, Green 1966a,b). Similarly, recent work (Goldschmidt et al 1984a) failed to show histamine methyltransferase activity in rat peritoneal mast cells; and the very low levels of tele-methylhistamine that could be detected by mass spectrometry were not elevated by incubation of the cells with pargyline. The small amounts of tele-methylhistamine that were found could have been sequestered from plasma, for peritoneal mast cells take up as well as synthesize biogenic amines (Furano & Green 1964a). Since turnover rates of histamine in brain were derived from measurements of tele-methylhistamine (to which mast cells contribute insignificantly) after treatment with pargyline (which does not raise the low levels of tele-methylhistamine in mast cells), brain mast cells are most unlikely to contribute to the turnover measurements of histamine in regions of rat brain. Moreover, even if mast cells in rat brain differ from those in the peritoneum and vigorously methylate histamine, the histamine turnover measurements can be confounded by mast cells only in thalamus, where this cell type is almost exclusively localized in rat brain (Dropp 1972, Persinger 1979, Goldschmidt et al 1984b).

Additional experiments fail to support the idea that mast cells are a major source of histamine in the brain. In the brain of newborn rat, where a larger portion of the histamine is found in the nonsynaptosomal P-1 fraction than it is in adult brain, the P-1 fraction lacked metachromasia (Young et al 1971), which is an essential property of a mast cell (Padawer 1963). In newborn rat brain, as in the adult, in contrast with mast cells, the contents of histamine and *tele*-methylhistamine are similar, as are the ontogenies of histamine, *tele*-methylhistamine, and histamine methyltransferase (Hough et al 1982): once formed in neonatal brain, histamine turns over. In mouse brain, more direct evidence suggests that mast cells make little contribution to brain histamine. Mutant W/W^v mice deficient in mast cells have about the same concentrations of histamine and *tele*-methylhistamine as +/+ (controls), W/+, and W^v/+ heterozygote mice, not only in whole brain but in brain regions and subcellular fractions (Orr & Pace 1984, Hough et al 1984b). Other workers found lower levels of histamine in the mutant mice (Yamatodani et al 1982, Maeyama et al 1983), which were attributed (Orr 1984) to immersion of the decapitated head in liquid nitrogen with resultant degranulation of mast cells in the dura.

Another argument against mast cells accounting for a significant pool of histamine is that mammals vary in the numbers and regional distribution of mast cells in brain (Dropp 1976, Persinger 1979), yet the histamine levels in whole brains and in brain regions are similar (see tabulation by Green 1970). Within the thalamus, the numbers of mast cells vary greatly among rats (Persinger 1977, Goldschmidt et al 1985, Hough et al 1985b), but the thalamic histamine levels do not (see Hough & Green 1984). There may be a cell in brain similar to a mast cell but lacking metachromasia. In adult rat median eminence, a region where histamine levels were unchanged after deafferentation (Brownstein et al 1976), histamine was reduced by the mast cell degranulator, compound 48/80 (Pollard et al 1976). Compound 48/80 released endogenous histamine from adult rat hypothalamic slices (Schwartz et al 1976b, Verdiere et al 1975). A neurolipomastocytoid cell has been described in brain (Ibrahim et al 1979, Sturrock 1980), and it is degranulated by compound 48/80 (Ibrahim 1970). The same particulates from rat brain that contain histamine have a sulfomucopolysaccharide distinct from heparin, including a lipid soluble sulfomucopolysaccharide not found in other organs (Robinson & Green 1962). Brain sulfomucopolysaccharides may sequester histamine in analogy with the sulfomucopolysaccharides in mast cells (Green & Day 1963, Robinson et al 1965, Green 1966a,b).

More than one pool of histamine is probably present in brain. On subcellular fractionation, a portion, 15–40%, of histamine in rat and guinea pig brain is found in the P-1 fraction, which has few synaptosomes (Carlini & Green 1963, Michaelson & Coffman 1967, Kataoka & DeRobertis 1967, Kuhar et al 1971, Young et al 1971, Snyder & Taylor 1972, Snyder et al 1974, Dismukes et al 1974, Martres et al 1975, Barbin et al 1976, Picatoste et al 1977, Schwartz et al 1979a, Sperk et al 1981b). In neonatal rat brains, histamine is found almost entirely in the P-1 fraction, and with development the portion in the P-2 fraction increases along with histidine decarboxylase activity (Martres et al 1975, Picatoste et al 1977, Hough et al 1982). Lesions of putative histaminergic fibers (Garbarg et al 1976, Barbin et al 1976) strikingly reduce histidine decarboxylase activity and histamine content in the P-2 fraction containing synaptotosomes, without altering histamine content in the P-1 fraction. α-Fluoromethylhistidine, an irreversible inhibitor of histidine decarboxylase (Kollonitsch et al 1978), nearly abolishes histidine decarboxylase activity but reduces histamine content only to about half (Garbarg et al 1980b, Sperk et al 1981b, Maeyama et al 1982, Bouclier et al 1983, Oishi et al 1984, Duggan et al 1984). After chronic treatment with α-fluoromethylhistidine, this pool too is nearly abolished (Bouclier et al 1983, Duggan et al 1984). These findings imply that some of the histamine in brain is localized to another, perhaps nonneuronal, compartment(s) that slowly forms and slowly dissipates histamine.

Histamine is present in brain blood vessels (El-Ackad & Brody 1975a, Jarrott et al 1979, Head et al 1980, Karnushina et al 1980, Robinson-White & Beaven 1982). The microvessels from rat and guinea pig brain had low activities of histidine decarboxylase and histamine methyltransferase (Karnushina et al 1979, 1980, Robinson-White & Beaven 1982). Histological studies on these fractions from several species failed to show significant numbers of mast cells (El-Ackad & Brody 1975b, Karnushina et al 1980, Head et al 1980, Robinson-White & Beaven 1982). Compound 48/80 had no effect on the microvessel histamine levels (Karnushina et al 1980, Robinson-White & Beaven 1982).

In peripheral nerve bundles, mast cells are abundant (see Olsson 1965, 1968, Green 1970, Hough & Green 1984). The histamine concentration is not changed by denervation (Ryan & Brody 1972) but is depleted by compound 48/80 (Howland & Spector 1972). Mast-cell–like granules were observed by electron microscopy near catecholaminergic terminals in walls of bovine splenic vein (Dzielak et al 1983). Early evidence suggested that histamine in some peripheral nerves mediates reflex-induced, active vasodilation (Tuttle 1967, Ryan & Brody 1972).

USE OF AGONISTS AND ANTAGONISTS TO CLASSIFY HISTAMINE RECEPTORS AND TO IMPUTE FUNCTIONS TO HISTAMINE

The principles for classifying receptors rest on receptor theory and apply equally to the characterization of receptors by high-affinity binding studies or by the response of tissues. Histamine receptors have been classified by both techniques. Histamine receptors have also been misclassified by both techniques because of innocence about receptor theory and a penchant to attribute inordinate specificity to agonists and antagonists (Green et al 1978, 1979, Green & Hough 1980, Green 1983, Hough & Green 1984). The method of binding is rapid and simple; it is, however, subject to artifacts, and there are criteria that must be met before a binding site can be postulated to be a receptor. The substances that compete for binding of the labeled ligand must, for a known receptor (e.g. the H_1 receptor labeled with the H_1-antagonist, [^{3}H]pyrilamine), have relevance to their pharmacology; and their affinities (e.g. K_D values) for the binding site should reflect affinities for the receptor mediating a response. The IC-50 of the agonist (the concentration to inhibit binding by 50%) in binding experiments may not reflect the effectiveness of the agonist in producing a response because affinity of the agonist for the receptor site is not the sole determinant of agonist effect. An inference that a binding site is a receptor must await demonstration of homologies with a functioning receptor (i.e. one associated with a biological effect), e.g. affinities of the antagonists for the receptor should correlate with affinities for the binding sites. Saturable and reversible binding sites have been described with [^{3}H]histamine that show far higher affinity in binding to brain membranes than in evoking a response on any isolated tissue or in competing for sites bound by antagonists (e.g. [^{3}H]pyrilamine). These [^{3}H]histamine binding sites do not reflect the presence of a functional histamine receptor (see Laduron 1984). Other deficiencies presented by studies of agonist binding are that partial agonists are not distinguished from full agonists and that desensitization of the receptor, which could alter the binding, is not apparent from equilibrium experiments (although kinetic experiments may reveal this).

The use of labeled antagonists offers fewer problems that those of agonists, but they too may show high enough nonspecific binding to create confusing results. The H_1 antagonist, astemizole, has high affinity for glassware as well as for nonspecific sites on membranes, and its affinity for membranes was influenced by the dilution of the incubation medium; its K_D value for the H_1 receptor could not be estimated (Laduron et al 1982). Claims that the H_2 receptor in brain had been labeled in binding studies with [^{3}H]cimetidine had to be incorrect because, as noted (Green 1983,

Hough & Green 1984), the ligand was used at a concentration (6.26 nM) about one-hundredth of its K_D value, a concentration that could occupy only a fraction of a percent of the H_2 receptors. The inability of H_2-active substances to inhibit binding and additional experiments further showed that sites other than H_2 receptors were being labeled (Smith et al 1980, Rising et al 1980, Bristow et al 1981, Laduron 1984). [^{3}H]Tiotidine has been shown to label H_2 receptors in the guinea pig cerebral cortex with a K_D not different from that on functioning H_2 receptors; other H_2 antagonists inhibited the binding, showing K_D values consonant with those for the H_2 receptor (Gajtkowski et al 1983). The high nonspecific binding of tiotidine may explain the inability to show specific binding to the H_2 receptor in the rat cerebral cortex, rat uterus, guinea pig atrium, guinea pig gastric mucosa (Gajtkowski et al 1983) and guinea pig hippocampus (Maayani et al 1982).

In classification of receptors by the response of tissues (O'Brien 1979, Kenakin 1984), agonists and antagonists are essential. Usually, a series of agonists are examined for their relative potencies in producing an effect, e.g. contraction of the guinea pig ileum to measure H_1 activity. Complete dose-response curves must be determined to obtain the EC-50 values, i.e. the concentration of each agonist to produce 50% of its maximal effect (Emax). For the agonists to be assumed to be acting on only one receptor, the slopes of their dose-response curves must not differ and the rates of onset and offset of the effects must be similar.

The EC-50 values of a series of agonists relative to histamine give their rank order of potency on the receptor. The EC-50 values are often converted to percentage of activities relative to histamine. For example, the EC-50 values of histamine, 4-methylhistamine, and 2-methylhistamine in increasing the rate of the guinea pig atrium are 1.1, 2.6, and 25 μM, respectively; the respective activities relative to the activity of histamine, derived from these EC-50 values, are 100, 43.0, and 4.4% (Black et al 1972). If the rank order is the same as that on a known receptor, then the functions in the two tissues are postulated to be subserved by the same receptor. For example, the rank order of potency of histamine, 4-methylhistamine, 2-methylhistamine, and other agonists were the same in increasing the rate of the guinea pig atrium, inhibiting contraction of the rat uterus (Black et al 1972, Durant et al 1975), and stimulating adenylate cyclase activity in brain homogenates of the guinea pig (Hegstrand et al 1976, Green et al 1977). The congruence suggests that the histamine receptors associated with these specific responses in these different tissues comprise a homologous population, in these examples, the H_2 receptor. However, discrepancies in the rank order of potencies do not necessarily imply that the receptors are different. The relative potencies of a series of agonists that act on the same receptor may vary among tissues—sometimes for reasons that have been discerned, but sometimes for unknown reasons (for examples, see Green 1983). For example, both 2-pyridylethylamine and impromidine, relatively selective agonists for the H_1 and H_2 receptors, respectively, enhanced norepinephrine release from the electrically-stimulated guinea pig atria, but the use of antagonists failed to show that a histamine receptor was associated with the release (Rand et al 1982).

A common error in the design and interpretation of experiments is to attribute absolute *specificity* rather than *relative selectivity* to an agonist. Most histamine agonists stimulate both histamine receptors: the discrimination is quantitative, not qualitative. 2-Methylhistamine, as noted above, has higher affinity for the H_1 receptor than for the H_2 receptor; the respective affinities are 16.5% and 4.4% of the affinities of histamine for the respective receptors (Black et al 1972). The percentage differences, which derive from the EC-50 values of 2-methylhistamine and histamine for the two receptors, have led to the imputation of H_1 *specificity* to 2-methylhistamine. But 2-methylhistamine is a full agonist at both receptors (Black et al 1972). From the EC-50 values, activation of the receptor by any concentration of agonist can be calculated from the equation: Percent of Emax $= e[A] \times 100/(EC\text{-}50 + [A])$, where e is the intrinsic activity (here, 1, because 2-methylhistamine is a full agonist) and $[A]$ is the concentration of agonist. At 0.1 mM, a concentration commonly used, 2-methylhistamine produces 97% of the maximum effect on the H_1 receptor and 80% of the maximum effect on the H_2 receptor. No specificity can be confidently affirmed at this concentration, as has been emphasized (Green & Hough 1980, Hough et al 1980, Green 1983, Hough & Green 1984).

The use of competitive antagonists has provided less ambiguity in the classification of histamine (and other) receptors, for ideally a discriminative antagonist annuls only the effect of the agonist on the receptor. Parallel displacement of the log concentration-response curve of the agonist in the presence of the antagonist suggests that the agonist and antagonist are acting on the same receptor. From the EC-50 values in the presence and absence of the antagonist can be obtained the apparent dissociation constant, K_D, of the antagonist. Several concentrations of an antagonist, B, over several log units, should be used to obtain a secure K_D value. A convenient way to test for competitive antagonism and to obtain the K_D value is to plot log (dose ratio-1) against log B, the Schild plot (Schild 1947, Arunlakshana & Schild 1959). A line with slope not different from unity implies competitive antagonism, and the intercept with the abscissa is $-\log K_D$ or pA_2. More than one antagonist should be used to classify a receptor. Additional assurance is gained by estimating the pA_2 of an antagonist with more than one agonist; agonists with very different potencies yield indistinguishable pA_2 values for the antagonist, since the affinity of an antagonist is independent of that of an agonist. Histamine-stimulated adenylate cyclase activity in homogenates of guinea pig brain was shown to be linked to the H_2 receptor by measurements of the pA_2 values of seven H_2 antagonists and, for some antagonists, with both histamine and dimaprit as agonists (Green et al 1977, Maayani et al 1982). The use of an antagonist with more than one agonist can help to avoid false inferences due to adventitious effects of an agonist. When the relative activities of an agonist differ on two preparations, the temptation to herald a new receptor is alleviated by the observation that a series of antagonists give nearly identical K_D values on the two preparations: the discords of agonists can be turned to consonances with antagonists.

Even on isolated preparations in vitro, interaction of an antagonist with a receptor may be obscured by other actions of the antagonist. Both amitriptyline (Green & Maayani 1977, Kanof & Greengard 1978) and propantheline derivatives (Hough & Barker 1981) competitively block the H_2 receptor linked to adenylate cyclase in brain homogenates and the H_2 receptor in the guinea pig papillary muscle, but blockade of the H_2 receptor in the guinea pig atrium could not be demonstrated (Angus & Black 1980, Hough & Barker 1981). This paradox almost certainly rests on other actions of these drugs on the atrium that obscured the interaction with the H_2 receptor. In a system simpler than a contracting muscle, i.e. brain homogenates containing histamine-stimulated adenylate cyclase activity, the heirarchy of EC-50 values of agonists and the pA_2 values of seven H_2 antagonists clearly showed the H_2 receptor to be the mediator of increased cyclic AMP formation (Maayani et al 1982). But modeling the system revealed that as much as 15% of the response could be mediated by an H_1 receptor, and no combination of agonists or antagonists could prove or disprove the hypothesis that the H_1 receptor contributes to cyclic AMP formation in these homogenates (Hough et al 1980).

The difficulties in making sure attribution of an event to a receptor are compounded in complex systems, especially in living systems, where functions of histamine have been inferred from observations on the effects of histidine, histamine agonists, and histamine antagonists, all of which have actions in addition to those on the histaminergic system. Since the K_M of histidine for histidine decarboxylase is much higher than the usual concentrations of histidine in brain and blood, administration of L-histidine raises brain levels of histamine (Schwartz et al 1970, Taylor & Snyder 1972, Schwartz et al 1972, Abou et al 1973). But it also raises the brain levels of carnosine and homocarnosine (Chung-Hwang et al 1976) and lowers the level of 5-hydroxytryptamine (Taylor & Snyder 1972); furthermore, since histidine can serve as substrate for L-amino acid decarboxylase, some of the resultant histamine may be localized in nerve endings not normally containing histamine. Many histamine agonists and antagonists inhibit histamine methyltransferase (Taylor & Snyder 1972, Barth & Lorenz 1978, Beaven & Roderick 1980), thereby increasing the concentrations of endogenous histamine (Boissier et al 1970). Histamine itself is catabolized to products that have pharmacological effects unrelated to those of histamine (see Green 1970).

Just as agonists must be used at concentrations related to their EC-50 values, antagonists must be used at concentrations related to their K_D values whenever the experimental system permits. One source of confusion is that H_1 and H_2 antagonists are applied at arbitrary, often equal, concentrations despite large differences in their affinities for their respective receptors. For example, the affinity of pyrilamine for the H_1 receptor ($K_D = 0.4$ nM) is about 1000-fold the affinity of cimetidine for the H_2 receptor ($K_D = 0.6$ μM). Inferences about functions associated with each receptor are muddled if, as commonly occurs, both are tested at the same concentrations—e.g. 1 μM, which is for pyrilamine equivalent to about 2000 K_D units for the H_1 receptor and for cimetidine about 2 K_D units for the H_2 receptor. Inferences are especially risky at high concentrations, where, as noted below, the antagonists block other receptors—for some compounds (e.g. pyrilamine) both histamine receptors. In experimental systems where the concentrations of antagonists (and agonists) cannot be precisely controlled (e.g. iontophoresis or in whole animals) the attribution of an event to a specific receptor must be vested with less confidence. In the whole animal, the interpretation of drug effects is confounded by pharmacokinetic considerations and the unknown concentration of the drug at the receptor.

Although the commonly used H_1 antagonists have high affinity for the H_1 receptor, at higher concentrations they are local anesthetic, and they block the muscarinic receptor (see van den Brink & Lein 1978), the H_2 receptor (Green et al 1977, Kanof & Greengard 1978, Hough et al 1980), the binding sites for benzodiazepines, γ-aminobutyrate, dihydromorphine, and a β-adrenergic antagonist (Spreeg et al 1981), and they block the effects of glutamate and aspartate on lobster muscle (Constanti & Nistri 1976). Astemizole, an H_1 antagonist, lacks affinity for the muscarinic receptor, but it has affinities for the α_1-adrenergic receptor and a 5-hydroxytryptamine receptor (Laduron et al 1982). Many H_1 antagonists also block the sites of uptake for 5-hydroxytryptamine, norepinephrine, and dopamine (Carlsson & Lindquist 1969, Coyle & Snyder 1969, Lidbrink et al 1971, Brown & Vernikos 1980).

In vivo, many H_1 antagonists block the effects of apomorphine, tremorine, tryptamine, and, because of their affinities for muscarinic sites, the effects of physostigmine and pilocarpine (Wauquier et al 1981). In blocking isolation-induced fighting in mice (Barnett et al 1971), the potencies of H_1 antagonists more nearly paralleled antimuscarinic potency than that for the H_1 receptor (Barnett et al 1971). Most H_1 antagonists in most species produce sedation, but there is no clear relationship between affinities for the H_1 receptor and effectiveness in producing sedation in mice and rats (Heinrich 1953); and in adult man, no consistent relationship was found, in concomitant measurements, between peripheral response to H_1 blockers and central effects, including sedation (Peck et al 1975, Carruthers et al 1978). Similarly, potency of H_1 antagonists in preventing motion sickness in man was not correlated with affinities for the H_1 or muscarinic receptors (Brand 1968). The anticonvulsant effect of some H_1 antagonists is independent of both their potency as H_1 antagonists and their effects in producing sedation (Dashputra et al 1966). The effect of H_1 antagonists in producing sedation is also independent of the effect in impairing psychomotor performance (Seppala et al 1981). Further evidence that H_1 antagonists have pharmacological effects in vivo that cannot be ascribed to any single receptor are their disparate effects. Some, but not all, reverse reserpine-induced hypothermia, and some potentiate whereas others suppress hypermotility in mice induced by caffeine or methamphetamine. Some, but not all, increase cortical synchrony in cat brain (Faingold & Berry 1972b); this effect is almost certainly unrelated to H_1 receptor blockade because the effect produced by intravenous infusion of racemic chlorpheniramine was similar to that produced by an equal dose of the l-isomer, despite differences in their affinities for the H_1 receptor (Faingold & Berry 1972a).

Cimetidine blocks the H_2 receptor with a K_D value of 0.6 μM, but at higher concentrations it blocks the H_1, β-adrenergic, and muscarinic receptors (Parsons 1977, Barker 1981) and the binding sites for dihydromorphine, benzodiazepines, and γ-aminobutyrate (Spreeg et al 1981). Like ranitidine (Bertaccini & Coruzzi 1981), which is another H_2 antagonist, cimetidine releases acetylcholine (Vyas & Verma 1981). Several H_2 antagonists have been shown to release norepinephrine (Brimblecombe et al 1975, McCulloch et al 1979).

An adverse reaction to cimetidine are dose-related, reversible mental symptoms, which range from restlessness to disorientation and visual and auditory hallucinations (see Schentag et al 1979). In two of the patients with mental symptoms in Schentag and co-workers' study, cerebrospinal fluid levels of cimetidine were 0.87 and 1.4 μg/ml, more than three times those of three patients without these symptoms (Schentag et al 1979). These values could suggest that in the symptomatic patients 82 and 88% of the H_2 receptors were occupied by cimetidine, whereas in the asymptomatic patients 48–57% were occupied (Green 1983). These are crude estimates, for they are based on studies at equilibrium that may not at all reflect the circumstances in a living animal; the estimates also assume that the drug in spinal fluid is in equilibrium with the histamine receptors in brain, and that not enough histamine is present to surmount the effect of cimetidine. Furthermore, by the same reasoning, these concentrations of cimetidine are enough to reduce β-adrenergic activity slightly and to increase noradrenaline release. Despite these caveats, it is worth considering that the mental symptoms are due to H_2 blockade. Ranitidine, another H_2 antagonist in clinical use, has produced similar symptoms in patients (Hughes et al 1983, Silverstone 1984, Epstein 1984, Mani et al 1984); its cerebrospinal fluid levels in these patients were not measured. Maybe relevant to this adverse effect of the H_2 blockers is that LSD competitively antagonizes histamine at the H_2 receptor (Green et al 1977).

Attribution of an effect of an antagonist to a specific receptor is made with greater confidence if agonists are used in tandem with the antagonists, and if a variety of agonists and antagonists are used at concentrations that reflect their affinities for the receptor. For example, the use of agonists and antagonists in tandem revealed electrophysiological effects associated with histamine receptors (see below); the analeptic response that followed the injection of dimaprit into the brain ventricle was not attributed to H_2 receptor stimulation because impromidine, another H_2 agonist, failed to elicit this response and an H_2 antagonist failed to block it (Pakkari & Karppanen 1982).

RECEPTORS AND SECOND MESSENGERS

The density of H_1 binding sites in mammalian brain is higher than in any other organ (Chang et al 1979a). Binding with labeled pyrilamine or doxepin also showed a low affinity site (Chang et al 1979b, Hill & Young 1980, Tran et al 1981, Taylor & Richelson 1982, Hadfield et al 1983). Similar observations were made on other organs, including lung, where the high nonspecific binding of [^{3}H]pyrilamine (which could be reduced by the addition of sodium) was shown to account for the apparent heterogeneity of H_1 binding sites (Carswell & Nahorski 1982). A variable influencing K_D values was the concentration of the membranes used in the assays (Taylor & Richelson 1980). The K_D values of [^{3}H]pyrilamine and of some other H_1 antagonists appear to vary among brains of different species (Chang et al 1979b, Hill & Young 1980). The affinities of H_1 antagonists for the H_1 receptor linked to contraction of the guinea pig ileum correlated better with their affinities for rat brain membranes than for guinea pig brain membranes (Chang et al 1979b).

More striking than these relatively slight differences in affinities are the differences among species in the density of the H_1 binding sites in brain regions (Chang et al 1979b, Hill & Young 1980). For example, in rat the cerebellum was lowest, the hypothalamus highest; in guinea pig, the cerebellum was highest, the basal ganglia lowest; in man, the frontal, parietal, cingulate, and temporal cortices were highest, the cerebellum lowest (Chang et al 1979b). It is not clear that all the H_1 receptors are associated with neural tissue (Chang et al 1980, Palacios et al 1981b). Radioautography of guinea pig brain (Palacios et al 1979, 1981a) showed high densities of the H_1 receptor in the molecular layer of the cerebellum; few densities were seen in this layer in rat cerebellum. In hippocampus of the guinea pig, the dentate gyrus was especially rich in H_1 receptors, and CA4 had greater densities than CA1, CA2, or CA3 (Palacios et al 1979). High densities in structures of the rat brain that are associated with the auditory system (Palacios et al 1981a) may relate to the impairment of auditory vigilance occurring with H_1 blockers in man (Peck et al 1975). Rats showed high densities in CA3 of the hippocampus and in the subiculum (Palacios et al 1981a); the hypothalamus showed high density in the supraoptic and suprachiasmatic nuclei, ventromedial nucleus, and the nucleus premammilaris ventralis. Some of these may be the sites where intraventricularly injected histamine acts to alter food and water intake, body temperature, hormone secretion, and other processes, as summarized by Palacios et al (1981a). Prolonged treatment of guinea pig with pyrilamine did not alter the B_{max} (or K_D) of pyrilamine binding by membranes from different parts of the brain or from intestinal smooth muscle (Hill et al 1981). Tolerance to an H_1 antagonist (Heinrich 1953) may therefore not be related to an effect on the H_1 receptor.

The H_1 receptor is associated with stimulation of formation of both cyclic AMP and cyclic GMP and with turnover of inositol phospholipids. Stimulation of the H_1 receptor in brain slices of different mammalian species increases cyclic AMP formation (see Daly 1977). In conventionally prepared homogenates, stimulation of the H_1 receptor did not increase cyclic AMP formation (Chasin et al 1974, Hegstrand et al 1976, Green et al 1977, Maayani et al 1982), but in a cell-free preparation containing large vesicular sacs (Chasin et al 1974, Psychoyos 1978), stimulation of the H_1 receptor increased cyclic AMP formation. Evidence that both the H_1 and H_2 receptors are linked to adenylate cyclase in guinea pig brain slices was provided by the use of antagonists in appropriate concentrations and over a range sufficient to do a Schild plot (on the H_1 antagonist) that showed competitive antagonism (Palacios et al 1978). Others (Daum et al 1982) failed to show a response to the H_2 agonist dimaprit in slices. In slices, the increased cyclic AMP formation on stimulation of the H_1 receptor depends either on stimulation of the H_2 receptor (Palacios et al 1978) and/or on the presence of adenosine (Daly et al 1980); treatment of brain slices with adenosine deaminase abolished the cyclic AMP response to H_1 stimulation (Schwabe et al 1978, Hill et al 1981). H_1 agonists but not dimaprit potentiated adenosine-stimulated cyclic AMP formation in slices (Dismukes et al 1976, Daum et al 1982), an effect that was blocked by pyrilamine (Daum et al 1982).

Histamine-stimulated cyclic GMP formation in mouse neuroblastoma cells is linked to the H_1 receptor, as was revealed by the effects of agonists and antagonists (Taylor & Richelson 1979). Desensitization of the receptor by agonists was prevented by an H_1 antagonist (Taylor & Richelson 1979). Histamine increases the cyclic GMP content of cerebral cortical slices of rabbit (Kuo et al 1972) and guinea pig (Schwabe et al 1978) but not cerebellar slices of rabbit (Kuo et al 1972), guinea pig (Ohga & Daly 1977), or mouse (Ferrendelli et al 1975). Since this effect of histamine requires Ca^{2+} (Schwabe et al 1978), it is pertinent that histamine appears to increase Ca^{2+} influx in membranes of smooth muscle containing the H_1 receptor, not in membranes of smooth muscle lacking this receptor (Uchida 1980). Ca^{2+} may then activate formation of cyclic GMP (see Schwabe et al 1978, Study & Greengard 1978). H_1 receptor stimulation also increases glycogen breakdown in mouse cortex minces, which could also be mediated by increased cytosolic Ca^{2+} (Quach et al 1980). Both the increased cyclic GMP formation and glycogenolysis could rest on H_1 stimulation of the turnover of the inositol lipids with a resultant increase in cytosolic Ca^{2+}.

Stimulation of many receptors including the H_1 receptor increases turnover of inositol-containing phospholipids in brain and in numerous other tissues and cells, a subject that has been recently reviewed (Berridge 1984, Berridge & Irvine 1984, Hokin 1985). The first indication that histamine influenced phospholipid turnover was in measurements of phosphate incorporation into phospholipids in rat brain in vivo after intracisternal injections of histamine, an effect that was blocked by H_1 antagonists (Friedel & Schanberg 1975, Subramanian et al 1980). The ontogeny of H_1 receptors in the rat brain as revealed by [^{3}H]pyrilamine binding was similar to the ontogeny of the phospholipid response produced by intracisternal injections of histamine (Subramanian et al 1981). Evidence that the histamine response was due to breakdown of inositol phospholipids was obtained (by labeling with phosphate) in the guinea pig ileum, prompting the inference that the effect was essential to stimulus-response coupling at the H_1 receptor (Jones et al 1979). In slices of rat cerebral cortex, histamine stimulated the turnover of [^{3}H]inositol-labeled phospholipid (Berridge et al 1982). In slices of cerebral cortex of both rat (Brown et al 1984) and guinea pig (Daum et al 1984), the effects of agonists and antagonists unequivocally showed that the H_1 receptor is linked to inositol phospholipid turnover. This response to histamine in different regions of the guinea pig brain reflects the density of [³H]pyrilamine binding sites (Daum et al 1982). The inositol lipids that are sensitive to receptor stimulation may be in a discrete pool, consisting of phosphatidylinositol, phosphatidylinositol-4,5-bisphosphate and phosphatidylinositol 4-phosphate (Berridge 1984, Berridge & Irvine 1984, Hokin 1985). Receptor-stimulated hydrolysis produces both inositol-1,4,5-triphosphate and a 1,2-diacylglycerol, mainly 1-stearoyl-2-arachidonyl-sn-glycerol. In many cells, inositol-1,4,5-triphosphate has been shown to elevate cytosolic Ca^{2+}, which activates many processes, among which is activation of calmodulin-dependent phosporylation of proteins; the Ca^{2+} and diacylglycerol (along with phospholipids, e.g. phosphatidylserine) are required for maximum activation of protein kinase C, diacylglycerol increasing the affinity of Ca^{2+} for the kinase (Berridge 1984, Hokin 1985). The physiological consequences of these effects have been indicated by injection of inositol-1,4,5-triphosphate into frog oocytes, which produced an effect that mimicked the Ca^{2+}-dependent muscarinic depolarizing chloride current (Oron et al 1985); injection into bag cell neurons of *Aplysia* of either protein kinase C or a phorbol ester, which activates protein kinase C, enhanced the voltage-sensitive Ca^{2+} current (De Reimer et al 1985). There has been much discussion (Cockcroft 1981, Hawthorne & Azila 1982, Irvine et al 1982) of the hypothesis (Michell 1983) that inositol-1,4,5-triphosphate not only fosters Ca^{2+} mobilization but also opens the Ca^{2+} gate on the surface membrane.

Ca^{2+}-independent and Ca^{2+}-dependent changes in inositol phospholipid metabolism may coexist in the same cell (Berridge 1984). Perhaps supporting this suggestion are observations in the rat cerebral cortical slice showing that the presence of low Ca^{2+} decreased histamine-stimulated breakdown of [³H]inositol-labeled lipids as measured by the accumulation of [³H]inositolphosphates in the presence of Li^+, which inhibits inositol-1-monophosphatase (Allison et al 1976, Hallcher & Sherman 1980); but low Ca^{2+} did not affect the accumulation produced by stimulation of the muscarinic receptor or by potassium- or veratradine-depolarization. Thus the Ca^{2+} requirement for turnover of the inositol lipid(s) varies among receptors, and different pools of the lipid(s) may be associated with different receptors and perhaps in different cells (Downes 1982, Kendall & Nahorski 1984).

That agonist-stimulated cyclic GMP formation and agonist-stimulated turnover of inositol phospholipids are related (Berridge 1984, Berridge & Irvine 1984, Hokin 1985) has received additional support from studies of the effects of histamine and other agonists on neuroblastoma cells (Snider et al 1984). Histamine-stimulated increase of cyclic GMP in these cells is

Ca^{2+}-dependent, but aequorin failed to show an increase in intracellular Ca^{2+} after stimulation (Snider et al 1984). Histamine increased arachidonate release under the same conditions that it increased cyclic GMP formation. Both stimulation of cyclic GMP formation and stimulation of the release of arachidonate by histamine were antagonized by pyrilamine and also by quinacrine, which inhibits phospholipase A_2, one of the enzymes that releases arachidonate from lipids.

Two inhibitors of lipoxygenase, which converts arachidonate to hydroxyicosatetraneoic acids, blocked histamine-stimulated cyclic GMP formation. An inhibitor of cyclooxygenase, which converts arachidonate to prostaglandins, did not reduce stimulated cyclic GMP formation and instead seemed to increase it, probably by shunting arachidonate to the lipoxygenase pathway. The authors (Snider et al 1984) suggest that the products of arachidonate stimulate guanylate cyclase, and that the role of extracellular Ca^{2+} for histamine-stimulated cyclic GMP formation (after stimulation of turnover of the inositol phospholipids) may be to activate phospholipase A_2 and/or diacylglycerol lipase to release arachidonate. Maybe relevant to the failure to find increased cytosolic Ca^{2+} in the stimulated neuroblastoma cells (Snider et al 1984) is that arachidonate releases and extrudes sequestered calcium from cells (see Hokin 1985, e.g. Kolesnick & Gershengorn 1985).

Li^+ has effects that appear to distinguish H_1 activation from activation of other receptors. In rat cerebral cortical slices, the presence of Li^+ did not increase the accumulation of total inositol phosphates produced by a histamine stimulation, whereas Li^+ greatly increased the accumulation produced by stimulating the muscarinic, α_1-adrenergic and 5-hydroxytryptamine$_2$ receptors (Brown et al 1984). The effect of Li^+ also varies with species. Li^+ enhanced histamine-stimulated accumulation of inositol-1-phosphate in guinea pig cerebral cortical slices (Daum et al 1984) but not in rat cerebral cortical slices (Brown et al 1984). In guinea pig, hippocampal slices were also responsive to Li^+ but cerebellar slices were not (Daum et al 1984). The reasons for the different responses to Li^+ have not been explored.

The effects of agonists on phospholipid metabolism, first observed over 30 years ago (Hokin & Hokin 1953), have thus been shown to initiate a series of products that almost certainly function as second messengers (see Berridge 1984, Takai et al 1984, Hokin 1985). Inositol-1,4,5-triphosphate functions to release intracellular Ca^{2+} which, among other effects, produces glycogenolysis and stimulates calmodulin-dependent phosphorylation of proteins. Both Ca^{2+} and diacylglycerol are needed to activate protein kinase C to phosphorylate specific proteins. Arachidonate releases and extrudes calcium from cells and may be metabolized to other substances, one or more of which may stimulate guanylate cyclase to form cyclic GMP. The

H_1 receptor is one of the receptors that mediates these activities. Stimulation of turnover of inositol phospholipids may have yet additional import to histamine activity. Histamine has been implicated in cell growth and development (Kahlson & Rosengren 1971, e.g. Hough et al 1982, Bartholeyns & Bouclier 1984). Diacylglycerol may function in cell growth by activating protein kinase C (Davis et al 1985). Hydrolysis of inositol phospholipids may produce in some cells another compound, inositol-1,3,4-triphosphate, which may regulate the proliferation of cells (see Berridge & Irvine 1984).

The H_2 receptor in guinea pig brain homogenates, as in many other tissues (see Green 1983), is directly linked to stimulation of adenylate cyclase activity as shown by the relative EC-50 values of agonists and the K_D values of a series of H_2 antagonists (Hegstrand et al 1976, Green et al 1977, Maayani et al 1982). In guinea pig brain, activity was highest in homogenates of the hippocampus and neocortex; other regions of the guinea pig brain (and whole rat brain) showed little or nondetectable responses (Hegstrand et al 1976, Green et al 1978, Hough & Green 1981), even where electrophysiological studies show evidence of the presence of the H_2 receptor (see Green 1983 and below).

H_2 activation of adenylate cyclase and H_1 activation of inositol phospholipid turnover may converge in the GTP-binding protein that controls both cyclic AMP formation and the hydrolysis of phosphatidylinositol-4,5-bisphosphate. Cyclic AMP in turn inhibits the breakdown of inositol phospholipid (Berridge & Irvine 1984, Takai et al 1984). Other convergences of the effects of cyclic AMP with the second messengers derived from inositol phospholipid hydrolysis are (a) cyclic AMP also regulates glycogenolysis in brain cells (Cummins et al 1983) and (b) a cyclic AMP–dependent protein kinase may activate a calcium pump in membranes (see Johnson 1982a,b). H_1 stimulation releases arachidonate, which is metabolized by lipoxygenase to a product(s) that stimulates cyclic GMP formation (Snider et al 1984); H_2 stimulation (of the myenteric plexus of the guinea pig ileum) releases arachidonate, which is metabolized by cyclooxygenase to a product(s) that contracts the muscle (Barker & Ebersole 1982). These relationships are important to histaminergic activities, since H_1 and H_2 receptors may be found on the same cells, as implied in the studies of brain slices that suggest that the H_2 receptor needs to be activated before the H_1 receptor can increase cyclic AMP formation (Palacios et al 1978, cf Daum et al 1982).

Evidence has been presented (Arrang et al 1983) for a third histamine receptor that controls histamine release from presynaptic sites. Slices of cerebral cortex of the rat were incubated with labeled histidine, and the effects of histamine agonists and antagonists in inhibiting the potassium-evoked release of labeled histamine were assessed. Histamine inhibited release, and the inhibition was blocked by impromidine (an H_2 agonist) and burimamide (an H_2 antagonist) in a competitive manner, as shown by Schild plots. The potencies of compounds in this system differed markedly from the potencies on the H_1 and H_2 receptors, including, as the authors discuss, the EC-50 of histamine itself, which was 40 nM, lower than its EC-50 values for the H_1 and H_2 receptors (Arrang et al 1983).

HISTAMINERGIC PATHWAYS IN MAMMALIAN BRAIN

Studies with neurotoxins suggested that histaminergic neurons in some brain regions are neurochemically distinct, and therefore probably anatomically distinct from other aminergic neurons. Cortical and striatal histidine decarboxylase activities remained unchanged or were elevated after administration of 6-hydroxydopamine or 5,6-dihydroxytryptamine, while cortical levels of norepinephrine and 5-hydroxytryptamine and activities of DOPA and 5-hydroxytryptophan decarboxylase were reduced (Garbarg et al 1976).

Histaminergic fiber pathways and cell bodies were visualized in rat brain by immunohistochemical methods with antibodies reactive to histidine decarboxylase (Fukui et al 1980, Tran & Snyder 1981, Watanabe et al 1983, 1984, Takeda et al 1984a,b, Wahlestedt et al 1985, Pollard et al 1985) and to histamine (Wilcox & Seybold 1982, Panula et al 1984, Steinbusch & Mulder 1985, Leslie et al 1984, Dirks et al 1984). The antibody had no influence on rat brain DOPA decarboxylase activity (Tran & Snyder 1981, Watanabe et al 1983, 1984); the antibody to one of the histamine-conjugates was shown to react with *tele*-methylhistamine (Panula et al 1984). Histidine decarboxylase-like immunoreactivity was widely distributed in rat brain, localized in neuronal fibers and cell bodies but not in glial elements. Immunoreactive cell bodies were observed (Watanabe et al 1984) in the posterior hypothalamic area but not in any other region, even in colchicine-treated rats. Most of these cells were large (20–30 μm diameter), polymorphic or spindle shaped, with numerous well-developed dendrites (Watanabe et al 1984, Takeda et al 1984b). Several immunoreactive cell clusters were apparent in the tuberal magnocellular, caudal magnocellular, posterior hypothalamic and dorsomedial hypothalamic nuclei, and the lateral hypothalamus at the level of the posterior hypothalamic nucleus.

Most histidine decarboxylase–containing fibers that projected to the rat cortex originated from cell bodies in the caudal, tuberal, and postmamillary caudal magnocellular nuclei of the posterior hypothelamus (Takeda et al 1984b). Total transection of tissue caudal to the posterior hypothalamus failed to alter cortical histidine decarboxylase immunohistochemical staining; the latter completely disappeared five to seven days after total transection was made rostral to this region. Hemisection rostral to the posterior hypothalamus caused histidine decarboxylase–immunopositive intensity to be reduced bilaterally, but this intensity was not abolished in the forebrain or cortex, although regions ipsilateral to the lesion exhibited the greatest qualitative reduction. Therefore, a significant portion of the histidine decarboxylase–containing fibers were inferred to cross the midline and project to rostral contralateral regions (Takeda et al 1984b).

After injection of horseradish peroxidase into the frontal, somatosensory, or occipital cortices of rats, labeled cells were observed mainly in the three magnocellular nuclei; this was reflective of retrograde transport. Cells labeled with horseradish peroxidase were apparent in these nuclei irrespective of the cortical region injected, and they were distributed bilaterally with ipsilateral predominance. Injection of horseradish peroxidase into the somatosensory cortex and use of the histidine decarboxylase antibody in a combined study showed a bilateral distribution, with ipsilateral predominance, of numerous immunopositive cells in the three magnocellular nuclei that also contained the horseradish peroxidase label (Takeda et al 1984b). Histidine decarboxylase immunoreactive fibers projecting to the spinal cord also appeared to originate from the caudal portion of the posterior hypothalamus near the ventral surface (Wahlestedt et al 1985). Rat brain histaminergic cell bodies were also demonstrated with guinea pig antiserum raised against histamine-methylated bovine serum albumin (Wilcox & Seybold 1982) or other conjugates (Panula et al 1984, Steinbusch & Mulder 1985). Histamine immunoreactive cell bodies, many bipolar, were also observed in regions lateral and ventral to the posterior hypothalamic nucleus; those caudal to the posterior hypothalamus in the ventral tegmentum were partially intermingled with dopaminergic cells (Steinbusch & Mulder 1985, Dirks et al 1984). Large cell bodies were also observed in the ventral horn of the spinal cord (Leslie et al 1984).

The distribution of histaminergic fibers by immunohistochemical techniques for histidine decarboxylase and for the histamine conjugates was similar. Fiber projections were widespread, with a thin but varicose appearance. The highest density of fibers was in the hypothalamus; intermediate density was observed in the mammillary body, amygdala, hippocampal dentate gyrus, and bed nucleus of the stria terminalis; few fibers appeared in the thalamus (Tran & Snyder 1981, Watanabe et al 1983, 1984). Hypothalamic fibers projected to the lateral geniculate and periventricular thalamic nucleus. All subnuclei of the olfactory nuclei received histaminergic fiber input; fibers going to the olfactory nuclei sent collaterals to the stria terminalis. Fibers were evident in ventral segments of the medial forebrain bundle, from which they dissociated into smaller branches at the level of the anterior hypothalamic nucleus and projected to the amygdaloid complex, piriform cortex, and capsula interna. Fiber plexuses were observed in the central gray matter of the midbrain and pons; possibly they were a continuation of fibers from the mammillary body. Projections to the inferior colliculus, lateral lemniscus, and corpus trapeziodeum, dorsal cochlear, and ventral cochlear nuclei as well as to the medial vestibular nucleus were apparent, but few, if any, impinged upon other cranial nuclei.

In the cerebral cortex no regional predominance of fibers was observed (Watanabe et al 1984). Two pathways have been proposed, a major pathway that ascends through the median eminence, and a minor pathway that descends dorsocaudally and projects to the substantia griseum centralis and further caudally in the griseum centralis (Watanabe et al 1984, Steinbusch & Mulder 1985). Fibers have also been found in the cerebellar cortex and spinal cord, in the latter, predominantly in the dorsal regions, arising from posterior hypothalamic cells (Wahlestedt et al 1985).

The distribution of histidine decarboxylase–immunoreactive fibers differs from the distribution of H_1 receptors. The immunoreactive fibers are evenly distributed throughout the regions and layers of the cerebral cortex and show low density in these brain stem nuclei (Watanabe et al 1984). Radioautography shows especially high concentrations of H_1 receptors in layer IV of the cortex and brain stem (Palacios et al 1981a). Discordance between the distribution of receptors and nerve terminals of other transmitters is well known (Snyder & Bennett 1976, e.g. Meibach et al 1980).

Many of the pathways revealed by immunohistochemistry were anticipated by observations on the neurochemical and electrophysiological changes produced by lesions (Schwartz 1975, Schwartz et al 1975, 1976a, 1979a,b, 1981, Barbin et al 1976, Ben-Ari et al 1977, Bischoff & Korf 1978, Chronister et al 1982, Dismukes et al 1974, 1975, Garbarg et al 1973, 1974, 1976, 1980a, Haas et al 1978, Pollard et al 1978, Sperk et al 1981b). The reduction of histamine content and histidine decarboxylase activity in the hippocampus after transection of the ventral and, especially, the dorsal input to the hippocampus indicated that histaminergic fibers project to the hippocampus (Barbin et al 1976). These findings may complement those from immunohistochemistry (Watanabe et al 1984), in which only a few scattered fibers to the hippocampus were observed. Lesion studies suggested the presence of cell bodies in the rostral mesencephalon (Pollard et al 1978, Schwartz et al 1979b), but these were not observed by immunohistochemistry of histidine decarboxylase (Watanabe et al 1984).

The immunohistochemical finding that histaminergic fibers cross the midline to innervate the contralateral side rostral to the posterior hypothalamus (Takeda et al 1984b) was not supported by other work after hemisection of the rat medial forebrain bundle. Histidine decarboxylase activity on the contralateral side was not significantly different from the activity in the same regions of unoperated animals (Garbarg et al 1974). Perhaps the histaminergic fibers that project to contralateral forebrain and cortical regions, though visible by immunohistochemistry, make little contribution to the quantitation of histidine decarboxylase activity. It is also possible that all or a portion of the fibers cross the midline between the lateral and posterior hypothalamus, since these were the respective lesion sites in the neurochemical (Garbarg et al 1974) and histochemical (Takeda et al 1984b) studies. Another consideration in the apparently discrepant findings is that young rats, i.e. those weighing 50–100 g, were used for immunohistochemistry (Watanabe et al 1984, Takeda et al 1984b) and mature rats were used for lesion studies (e.g. Garbarg et al 1974). It is important to know whether or not fibers cross to the contralateral side, for contralateral regions have been used as controls in neurochemical and electrophysiological studies on histamine (e.g. Garbarg et al 1974, Barbin et al 1976, Ben-Ari et al 1977, Haas et al 1978, Pollard et al 1978).

Immunohistochemistry suggests that in the hypothalamus, histidine decarboxylase and glutamate dehydrogenase may coexist in the same cell bodies. Antibodies to both these enzymes, each applied to alternate consecutive rat brain sections, labeled numerous neurons in the three magnocellular nuclei (see above) of the posterior hypothalamus (Takeda et al 1984a). Application of both antibodies to single tissue sections showed the coexistence of both enzymes. These observations imply that γ-aminobutyrate and histamine may coexist in these neurons.

ELECTROPHYSIOLOGICAL EFFECTS ON MAMMALIAN NEURONS

The electrophysiological effects of histamine in invertebrates are discussed below. The relationship of second messengers to some electrical events has been noted above.

Histamine has both excitatory and inhibitory effects, the latter defined as decreased firing rates, hyperpolarizations, decreased excitatory postsynaptic potentials (PSPs), or reduction of glutamate-evoked and electrically-evoked excitation. Iontophoresis of histamine produces inhibitory responses in many parts of the central nervous system, e.g. spinal cord (Phillis et al 1968b, Engberg et al 1976), brain-stem medullary reticular formation (Bradley 1968, Hosli & Haas 1971, Anderson et al 1973, Haas et al 1973), cerebellum (Siggins et al 1971), thalamus and hippocampus (Haas & Wolf 1977), and cerebral cortex (Krnjevic & Phillis 1963, Phillis et al 1968a,b,

1975, Haas & Bucher 1975). Histamine increases the spontaneous firing rate of many hypothalamic neurons: in culture (Geller 1976, 1981) and in slices in vitro (Haas & Geller 1982) and in vivo in cats, rats, and guinea pigs (Haas 1974, Haas et al 1975, Haas & Wolf 1977, Carette 1978). In some hypothalamic nuclei, both excitations and inhibitions are observed, but the predominant action of histamine appears to be inhibitory; e.g. on rat tuberal nuclei in culture (Geller 1979, 1981), rat ventromedial nuclei in vivo (Renaud 1976), cat rostromedial nuclei in vivo (Sweatman & Jell 1977), rat suprachiasmatic nuclei in hypothalamic slices (Liou et al 1983), and guinea pig arcuate-median eminence in vivo (Carette 1978). Experiments designed to classify the histamine receptors that mediate the responses have sometimes presented equivocal data because the agonists and antagonists were sometimes used in perfusion experiments without concern of their EC-50 and K_D values.

Iontophoresis of histamine onto cultured rat tuberal hypothalamic neurons evoked excitatory and inhibitory effects on spontaneous firing rates (Geller 1981). Histamine-induced excitations were blocked by the H_1-antagonists, diphenhydramine (10 μM) and promethazine (1–10 μM), but not by the H_2-antagonist, metiamide (20 μM). Histamine and the H_2 agonist, dimaprit, caused dose-related inhibitions, with similar onset and termination times, that were blocked by H_2-antagonists, metiamide and cimetidine (each up to 10 μM), but not by promethazine (up to 10 μM) (Geller 1981). Iontophoretic application of histamine caused mainly an inhibition of the spontaneous rate of firing of neurons in cat medial and lateral vestibular nuclei in vivo, which was blocked by local application of metiamide but not diphenhydramine (Kirsten & Sharma 1976). The few excitatory responses that occurred could not be blocked by either antagonist, and, curiously, cimetidine had no effect on the inhibitory response (Satayavivad & Kirsten 1977). In rat suprachiasmatic nuclei, iontophoretically applied histamine mainly inhibited spontaneous firing, and the effect was attenuated by cimetidine, not by pyrilamine, the H_1-antagonist; the few excitatory responses that occurred were blocked by pyrilamine (Liou et al 1983).

Histamine perfusion of antidromically identified, supraoptic neurons of rat hypothalamo-neurohypophyseal explants evoked excitatory responses characterized by single bursts in slow/silent neurons or increased numbers of bursts or continuous firing in phasic bursting cells (Armstrong & Sladek 1984). Concentration-dependent excitation could be observed with histamine and with the H_1 agonists, 2-pyridylethylamine and 2-thiazolyl-ethylamine, but not with the H_2 agonists, impromidine or dimaprit. Histamine-induced excitations were blocked by promethazine (0.01–1 μM) but not cimetidine (10 μM), thus suggesting that histamine stimulation of vasopressin-containing cells is mediated by the H_1 receptor (Armstrong & Sladek 1984), an inference supported by experiments in the living goat (Tuomisto et al 1984). In anesthetized cats, microiontophoresis of histamine increased the firing rate of neurosecretory neurons in the supraoptic nuclei (Haas & Wolf 1977). Some osmosensitive neurons were not spontaneously active, and they did not respond to histamine applied at high currents; however, after injection of hypertonic NaCl, these cells became spontaneously active and were further excited by histamine (Haas et al 1975). This example of state-dependent responsiveness to histamine recalls other observations. α-Fluoromethylhistidine, an irreversible inhibitor of histidine decarboxylase, did not alter ACTH secretion in normal rats but prevented the 11-fold surge of ACTH secretion that followed bilateral adrenalectomy (Wada et al 1985).

Microdrop application of histamine onto hippocampal CA3 cells caused slow depolarizations, elevated excitatory PSPs, and increased firing rates— actions that were attributed to stimulation of the H_1 receptor, since they were sensitive to promethazine but not cimetidine. These effects may, in part, be attributed to presynaptic actions, since they were markedly reduced in media containing low Ca^{2+}, high Mg^{2+}, or in the presence of tetrodotoxin (Segal 1980, 1981). Iontophoresis or pressure injection in vivo or in vitro of histamine or impromidine onto hippocampal CA1, pyramidal or dentate granule cells induced a hyperpolarization that was blocked by metiamide (Haas & Wolf 1977, Haas 1981a,b). Histamine and impromidine also reduced spontaneous firing and the firing induced by homocysteic acid and depolarizing current injection (Haas 1981a,b). The hyperpolarizations were probably direct (i.e. postsynaptic), because they were unaffected by synaptic isolation (Haas 1981a). The H_2 receptor in the isolated superior cervical ganglion of the rabbit (Brimble & Wallis 1973) and rat (Lindl 1979) was also associated with an inhibitory effect, as was shown by the reduction of the compound action potential. Stimulation of the H_1 receptor in the guinea pig ileum resulted in membrane depolarization and an increase in action potential frequency by opening ion channels similar to those linked to muscarinic receptors (Bolton et al 1981).

In contrast to the predominantly inhibitory effects observed on iontophoretic or pressure injection of histamine onto hippocampal cells, bath perfusion of histamine produced excitatory responses in hippocampal slices of rats (Haas & Konnerth 1983, Haas et al 1984, Haas 1984) and guinea pigs (Olianas et al 1982, 1984, Tagami et al 1984, Pellmar 1984). Intracellular recording of CA1 pyramidal cells revealed that histamine reduced the magnitude of the late afterhyperpolarization and its associated Ca^{2+}-dependent K^+-conductance, whether elicited by a train of action potentials or by calcium spikes in tetrodotoxin-poisoned cells (Haas & Konnerth

1983, Haas 1985). Histamine also elevated spontaneous firing in tetrodotoxin-free media. As with local microapplication techniques, the time of onset of the action was slow and its duration long. Impromidine (1 μM) but not thiazolylethylamine (100 μM) mimicked histamine; and metiamide and cimetidine (each 2 μM) but not pyrilamine (10 μM) blocked these effects (Haas & Konnerth 1983). H_2-mediated excitations were also observed in guinea pig CA3 pyramidal cells. Perfusion with histamine (0.1–10 μM) or dimaprit (1 μM), but not 2-pyridylethylamine (10 μM), increased the amplitudes of population spikes and excitatory postsynaptic potentials, and they depolarized resting membrane potentials (Tagami et al 1984). These actions were reduced by cimetidine but not pyrilamine (each 10 μM).

A postsynaptic locus for H_2-mediated excitatory effects was demonstrated in rat hippocampal CA1 pyramidal cells synaptically isolated with low Ca^{2+}, high Mg^{2+} (Haas et al 1984). Histamine, perfused over the slice, caused a concentration-related increase in field-burst firing frequency. The H_2 agonists, impromidine and 4-methylhistamine, produced these effects, which were blocked by cimetidine and metiamide (each up to 10 μM). Whereas perfusion with H_2 agonists caused excitatory responses, perfusion with thiazolylethylamine (1 μM) depressed CA1 bursting activity; this latter effect occurred in the presence of cimetidine, but was blocked by pyrilamine (1 μM) (Haas et al 1984). On the perfused myenteric plexus of the isolated guinea pig ileum, both histamine and dimaprit produced excitatory effects such as membrane depolarization, reduced afterhyperpolarizations, augmented repetitive spike discharges, and membrane depolarization—effects that were blocked by cimetidine but unaffected by pyrilamine or diphenhydramine (each at 10 μM) (Nemeth et al 1984). It is interesting that H_2 stimulation of the myenteric plexus releases acetylcholine, 5-hydroxytryptamine, a prostaglandin-like substance, and a peptide (Barker & Ebersole 1982).

These paradoxes—that both histamine receptors, when stimulated by local application of histamine, evoke an electrophysiological response remarkably different from that evoked on perfusion of histamine—may have precedent. The depressant effects of histamine on cat precruciate cortical neurons in vivo were converted to excitatory responses when the applied histamine current was doubled (Phillis et al 1968a). Intermittent and variable responses were observed when histamine was iontophoretically applied to neonatal mouse brain neurons in culture (Bonkowski & Dryden 1977), whereas bath application (0.1–10 μM) produced reproducible depolarizations (Bonkowski & Dryden 1976). To account for the inhibitory responses observed on local application of histamine and the excitatory responses on perfusion of histamine, it was proposed (Haas 1984) that the inhibitory responses are overcome by excitatory actions that occur when a larger population of H_2 receptors on hippocampal neurons are stimulated.

After electrical stimulation of the rat medial forebrain bundle, reduced neuronal firing rate was recorded in the ipsilateral cortex (Sastry & Phillis 1976b). Iontophoresis of metiamide attenuated the duration of the stimulation-induced inhibition. Application of histamine to those neurons responsive to metiamide inhibited the spontaneous firing rate, an effect that was reversibly antagonized by metiamide. Haas & Wolf (1977) showed that metiamide reversibly reduced both the cortical inhibition induced by medial forebrain bundle stimulation in the rat and the inhibition of hippocampal pyramidal neurons that occurred after stimulation of the fornix. These observations are consistent with the hypothesis that histamine, released from axons of an ascending pathway(s), may act on H_2 receptors to elicit inhibitory responses.

A relationship between the electrophysiological responses produced by H_2 receptor stimulation and increased cyclic AMP formation was suggested in rat tuberal hypothalamic neurons in culture (Geller 1979). The electrophysiological effects of histamine and dimaprit, which occurred in the absence of Ca^{2+}, were enhanced in the presence of phosphodiesterase inhibitors. The EC-50 of histamine in elevating CA3 firing rate in the guinea pig hippocampal slice was very similar to the EC-50 for stimulation of adenylate cyclase in the hippocampal homogenates (Olianas et al 1984). Perfusion of hippocampal CA1 cells in low Ca^{2+} with a cyclic AMP analogue or a phosphodiesterase inhibitor produced effects like those of histamine (Haas et al 1984). Also like histamine, cyclic AMP analogues were shown to increase population spikes (Mueller et al 1981) and to reduce late afterhyperpolarizations (Madison & Nicoll 1982). Histamine produces electrophysiological and biochemical responses similar to those of norepinephrine (Geller & Hoffer 1977, Siggins et al 1971, Haas & Konnerth 1983).

Not all evidence supports the idea that the electrophysiological effects of H_2 receptor stimulation are associated with H_2-linked adenylate cyclase. Electrolytic lesions of the guinea pig medial forebrain bundle reduced histidine decarboxylase activity and histamine concentration in the ipsilateral cerebral cortex. Supersensitivity of the histamine-induced electrophysiological response was recorded with no change in EC-50 or Emax for H_2-stimulated adenylate cyclase activity in the cortical homogenates (Haas et al 1978); the latter observation was in agreement with that of others (Dismukes et al 1976). Chronic administration of imipramine and mianserin, two drugs that block the H_2 receptor, produced supersensitivity to histamine-induced firing rate in guinea pig hippocampal CA3 neurons but failed to alter the maximal response of histamine-stimulated adenylate cyclase activity (Olianas et al 1982).

The effects of histamine on hippocampal cells are slow in onset and of long duration, like the effects of norepinephrine and 5-hydroxytryptamine, and unlike the rapid and short-lived effects of glutamate and γ-aminobutyrate (Haas 1981a). It has been inferred that this action of histamine "allows modulation of the quick information transfer but no rapid communication" (Haas 1981a). On iontophoresis, histamine, without itself exciting the neurons, potentiated the excitatory effects of homocysteate and glutamate on quiescent (but not spontaneously active) rat medullary neurons; the effect, probably mediated by the H_2 receptor (Jones et al 1983), lasted up to 30 min (Bradley et al 1984). In the rabbit nucleus accumbens, histamine potentiates the activity of γ-aminobutyrate (Chronister et al 1982). A modulatory role of histamine was also observed in rat cerebral cortex in situ and in rat hippocampal slice (Geller et al 1984). Similar modulatory roles of norepinephrine (Segal 1982) and 5-hydroxytryptamine have been shown. For example, iontophoretically applied 5-hydroxytryptamine failed to excite facial motoneurons of rat, but it greatly facilitated the excitatory effects of glutamate and, as well, the effects of excitatory synaptic inputs (McCall & Aghajanrian 1979).

HISTAMINE IN THE INVERTEBRATE NERVOUS SYSTEM

Histamine has been identified in numerous invertebrate species (see reviews of Reite 1972, Kehoe & Marder 1976, Leake & Walker 1980, Martin & Spencer 1983) and studied systematically in marine (*Aplysia*), freshwater (*Lymnaea*), and terrestial (*Helix*) gastropods, most extensively in *Aplysia californica*. A strikingly heterogeneous distribution of histamine characterizes the major ganglia of the *Aplysia* nervous system (Weinreich et al 1975). The pedal and pleural ganglia have highest and lowest levels, respectively; intermediate histamine levels are found in the cerebral, abdominal, and buccal ganglia. Of several peripheral nerves examined, only three nerve trunks, all emanating from the cerebral ganglion, contained histamine. The E clusters of the cerebral ganglion contain the highest levels of histamine in the ganglion. Few cell bodies contain histamine but the C2 neurons are particularly rich; they contain about 400 μM, about 2–3 orders of magnitude higher in concentration than their neighbors (Weinreich et al 1975)—a concentration similar to that of 5-hydroxytryptamine and acetylcholine in C1 and R2 neurons, respectively (Brownstein et al 1974). The C2 neurons are smaller than most *Aplysia* neurons, and are almost transparent, with a pigmented axon hillock region and an invariant cluster position. The finding that histamine coexists with 5-hydroxytryptamine and acetylcholine in neurons such as C1 and R2 (Brownstein et al 1974) has been challenged and attributed instead to exogenous material adhering to hand-dissected neurons (Osborne 1977); the levels of histamine in C1 and R2 cells were at or below 1% of those in C2 cells (Weinreich et al 1975).

Using a different technique for cell isolation, Ono & McCaman (1979) found additional histamine-rich cells, designated C3, in the ventral L clusters of the cerebral ganglion (Ono & McCaman 1980). These cells are smaller than C2 cells but contain nearly the same levels of histamine, and they share some of the same follower cells, including the serotonergic metacerebral cell. Ono & McCaman (1980) found much higher histamine concentrations in C2 cells (2.9 mM) and C3 cells (3.4 mM) than Weinreich et al (1975) found in the C2 cells. Weak electronic coupling was observed between C2 and C3 histamine-containing neurons and between many E and L cluster-follower cells, characterized by attenuated signals lacking a synaptic delay (Ono & McCaman 1980). The histaminergic C2 cell projects to and excites the serotonergic C1 cell, a cell that potentiates feeding behavior by release of 5-hydroxytryptamine (Weiss et al 1978a,b). C2 cells fire when stimuli are applied to the lips of *Aplysia* and continue to fire as long as rhythmic mouth movements persist. Since mechanical stimuli to the mouth can generate spikes, even in media containing high Mg^{2+} and low Ca^{2+}, C2 is probably a primary mechanoafferent neuron (Weiss et al 1983). C2 synaptic output occurs only when a full spike successfully passes the axon trifurcation, a large-diameter segment of low safety (see Gotoh & Schwartz 1982). The C2 cell therefore requires a high rate of axon input for passage of proprioceptive information to the C1 cell (Weiss et al 1983) and to other cells that modulate motor patterns of feeding (Chiel et al 1983). It has been suggested that the C2 cell "is normally in a state in which it fails to pass information into the nervous system. If its axon input occurs at a high rate, it switches into a transmitting mode and passes the proprioceptive information to an arousal system" (Weiss et al 1983). On this basis, histamine performs a modulating rather than an obligatory role in feeding.

Histidine has a uniform regional distribution in *Aplysia* (Weinreich & Weinreich 1977); accumulation of labeled histidine by neurons maintained in culture was not specific to histamine-containing cells but reflective of cell size (Weinreich & Yu 1977). Histamine is formed by histidine decarboxylase, which has properties similar to those of the enzyme in mammalian brain (Weinreich & Yu 1977, Weinreich & Rubin 1981). In C2 neurons the concentration of histidine is nearly ten-fold greater than the K_M of histidine decarboxylase, thus saturating the enzyme (Weinreich & Yu 1977), in contrast with the circumstance in mammalian brain (Schwartz et al 1980). Unlike mammalian brain, histamine is not metabolized in the gastropod nervous system by methylation (Gustafsson & Plym Forshell 1964. Huggins & Woodruff 1968, Weinreich 1979; see review by Reite

1972). Incubation of *Aplysia* ganglia with labeled histamine resulted in up to 90% recovery of a compound that had chromatographic and electrophoretic mobilities indistinguishable from those of γ-glutamylhistamine (Weinreich 1979). Its formation is dependent on a single soluble enzyme, γ-glutamylhistamine synthetase, that requires L-glutamate and ATP (Stein & Weinreich 1982). Postsynaptic follower cells of C2 have higher enzymatic activity than other cells; the highest activity is present in the connective tissue surrounding the ganglia containing the follower cells (Stein & Weinreich 1982, 1983). Bath application of γ-glutamylhistamine (up to 1 mM) produced no electrophysiological effects on *Aplysia* neurons (Weinreich 1985).

After injection of labeled histamine into the C2 cell body, radioactivity along a C2 axon showed a rate of movement of 40–50 mm per day. Selective accumulation of labeled histamine, but not its metabolite, occurred proximal to a ligature on the posterior lip nerve axon; little was measured distal to the ligature (Gotoh & Schwartz 1982). Rapid transport was slowed but not abolished by colchicine or by temperature reduction, a finding that suggests that substantial diffusion occurred. Subcellular fractionation showed one-third of the labeled histamine in particulate fractions as early as 3 hr after injection. The C2 cell failed to transport serotonin or choline rapidly (Gotoh & Schwartz 1982). In contrast, the serotonergic C1 cell transported histamine, dopamine, octopamine and 5-hydroxytryptamine at about the same rates (Goldberg & Schwartz 1980).

After C2 cells were injected with horseradish peroxidase, two populations of vesicles were observed in the axon varicosities by electron microscopy (Bailey et al 1982). Some were large with electron-dense cores that nearly filled the entire vesicle, and others were small and electronlucent. Both types were dissimilar from mast cell vesicles. Synapses characteristic of *Aplysia* were apparent (Bailey et al 1982). The vesicles in the axon terminal were qualitatively similar to those observed in the C2 soma of *Aplysia* (Gotoh & Schwartz 1982).

In *Aplysia*, two distinct postsynaptic responses to iontophoretically applied histamine have been observed in unidentified neurons (Carpenter & Gaubatz 1975). Histamine produced a marked depolarization with short latency mediated by elevated Na^+ conductance in the pleural ganglion, which was antagonized by pyrilamine (1 mM) but not by burimamide (1 mM), the H_2 antagonist. In a cerebral ganglion cell, histamine elicited a slow hyperpolarization due to elevated K^+ conductance, which was reversibly blocked by burimamide (1 mM) but not by pyrilamine (1 mM). The discovery of histamine in C2 and C3 cells, their axonal projections, and their monosynaptically connected follower cells in *Aplysia* provided an opportunity to compare the postsynaptic responses elicited by stimulation of C2, an electrically silent cell (Weinreich 1977), with those generated by application of histamine to which only a few *Aplysia* neurons respond (Carpenter & Gaubatz 1975, Weinreich 1977). Several of the classes of follower monosynaptic neurons exhibit spontaneous firing. C2 stimulation produces depression of the endogenous spiking of these cells. The depression of each spike was qualitatively mimicked by application of histamine (Weinreich 1977). In response to C2 firing, another class of followers demonstrated a rise in firing rate characterized by a short duration fast-excitatory PSP and a long duration slow-excitatory PSP; both were mimicked by histamine (Weinreich 1977). Additional responses to C2 firing were also mimicked by histamine when recordings were made near the axon processes (McCaman & McKenna 1978, McCaman & Weinreich 1982, 1985). The C2 neurons and followers were postulated (McCaman & McKenna 1978) to comprise an "elementary divergent aggregate" system in which C2 neurons have a functional role as "multiaction interneurons." Besides the chemical synapses between the C2 cells and follower cells, these cells also interact at nonrectifying electrical synapses; the follower cells are reciprocally connected to each other through similar, nonchemical synapses (McCaman & Weinreich 1985).

Iontophoretic application of histamine on identified A cluster neurons of the cerebral ganglion of *Aplysia* resulted in two distinct inhibitory responses: a monophasic slow hyperpolarization of long duration, and a biphasic hyperpolarization with a fast hyperpolarization superimposed upon a slow hyperpolarization (Gruol & Weinreich 1979a,b). Changes in membrane potential due to altering ions in the medium and the assessment of reversal potential values indicated that the fast and slow responses were mediated by conductance elevations in Cl^- and K^+, respectively. Cimetidine (10 μM) completely and reversibly blocked the histamine-induced slow inhibitory (K^+-mediated) response, but not the responses produced by acetylcholine or dopamine; chlorpheniramine and diphenhydramine (up to 0.5 mM) had no effect on this response (Gruol & Weinreich 1979a). Further analysis revealed an increase in amplitude of this response with increasing histamine concentrations. Cimetidine (1 μM) produced a parallel, dextral shift of the concentration-response curve to histamine, but the slope (1.7) of a modified Schild plot implied that the competition was not simple (Gruol & Weinreich 1979b). Failure of two other H_2 antagonists, metiamide and burimamide (up to 0.5 mM), to abolish the histamine response is puzzling. Also provocative is the observation that a slow, K^+-mediated hyperpolarization is elicited by at least three agonists (histamine, acetycholine, and dopamine) on these A cells (Gruol & Weinreich 1979a,b, Ascher & Chesnoy-Marchais 1982), and the response to all three agonists exhibited both cross-desensitization and cross-potentiation (Ascher & Chesnoy-Marchais 1982, Chesnoy-Marchais & Ascher 1983).

A histamine-rich (275 μM) giant visceral ganglion cell in the pond snail, *Lymnaea stagnalis* L, has an enzyme that synthesizes histamine (Turner & Cottrell 1977, Turner et al 1980). The presence of histamine was confirmed by immunohistochemistry (Leslie et al 1984). After injection of labeled histamine, numerous labeled axons and processes were observed to project to several other ganglia, which then projected to the periphery. Labeled histamine accumulates against a concentration gradient in this neuron (Turner & Cottrell 1977, Turner et al 1980) as well as in the subesophageal ganglion of the snail, *Helix pomatia* (Osborne et al 1979). In the latter, histamine appeared to be actively transported. Labeled material was released by K^+ in a Ca^{2+}-dependent process; the released material was not identified (Osborne et al 1979).

Within the nervous systems of the locust (*Schistocerca americana gregaria*), cockroach (*Periplaneta americana*), and moth (*Manduca sexta*), histamine levels are highest in the cerebral ganglion, optic lobe and retina and lowest within the abdominal ganglion (Elias & Evans 1983). Endogenous histamine levels appear to reflect the capacity to synthesize histamine from histidine (Maxwell et al 1978, Elias & Evans 1983). In contrast to its high concentration of histamine, moth retina is almost devoid of acetylcholine, γ-aminobutyrate, and 5-hydroxytryptamine (Maxwell & Hildebrand 1981). Labeled histamine is metabolized to imidazoleacetic acid and N-acetylhistamine by neural tissue of the locust (Elias & Evans 1983) and cockroach (Huggins & Woodruff 1968). Evidence is lacking that histamine is a transmitter in arthropods, but its high endogenous levels and the high capacity for synthesis and metabolism make it a candidate for transmitter function.

Histamine has also been postulated to function as a neurotransmitter in the somatogastric nervous system of the spiny lobster, *Panulirus interruptus* (Claiborne & Silverston 1984). Histamine has a nonuniform distribution among the nerves and ganglia examined, and it was not associated with connective tissue. The distribution correlates with the known axonal pathways and terminals of the inferior ventricular nerve. Extracellular stimulation of this nerve produced an inhibitory response in identified neurons of the somatogastric ganglion, a response mediated by elevated Cl^- conductance. The inhibitory response and the elevated Cl^- conductance were mimicked by application of histamine to the somatogastric ganglion, and repeated application of histamine reduced the response, as though the histamine receptor were desensitized. Very high concentrations of four H_1 antagonists and two H_2 antagonists failed to block the response (Claiborne & Silverston 1984).

On several neurons of the circumesophageal ganglia of the marine mollusk, *Onchidium verruculatum*, histamine produced both excitatory and inhibitory responses. The excitatory response was blocked by pyrilamine (0.2 mM), and the inhibitory by metiamide (0.2 mM) (Gotow et al 1980a,b). The excitatory response was associated with enhanced Na^+ conductance; the inhibitory response showed little change in membrane conductance and clear hyperpolarization in the absence of K^+. Some evidence suggests that the latter response is mediated by cyclic nucleotides (Gotow & Hashimura 1982). In the absence of known histamine-specific circuits, follower cells, or synapses, the physiological relevance of these findings and those of others on various species is not clear (Kerkut et al 1968, Carpenter & Gaubatz 1975, Takecuchi et al 1976, Ku & Takeuchi 1983).

CONCLUDING REMARKS

The diffuse pattern of histaminergic projections in mammalian brain and the turnover of histamine imply that histamine is released in all parts of the mammalian brain. In some parts of the mammalian brain, histamine may directly control some physiological functions. In other parts of the nervous system, the neurotransmitter function of histamine may be to serve as a modulator (as proposed for 5-hydroxytryptamine and norepinephrine), as suggested in mammalian brain by the potentiation by histamine of some electrophysiological responses to other transmitters without alone producing a response, and in *Aplysia*, where a histaminergic neuron is contributory to, but not obligatory for, the eating response. This modulator function may become especially prominent in perturbed states such as disease, perhaps caricatured by the sensitivity of osmosensitive cells in the hypothalamus to histamine after, but not before, injection of a hypertonic solution. Other actions of histamine and of other biogenic amines may be similarly state-dependent.

ACKNOWLEDGMENTS

A grant (MH-31805) from the National Institute of Mental Health supported the work by J. P. G. cited here. G. D. P. is a postdoctoral fellow supported by a training grant (5-T32-DA-07135) from the National Institute on Drug Abuse. We thank Drs. K. Weiss and I. Kupfermann for discussions on invertebrates, Dr. J. Goldfarb for discussions on mammalian electrophysiology, and Drs. M. C. Gershengorn and L. E. Hokin for reviewing the section on phosphatidylinositol turnover. We are grateful to Ms. Bernice Y. Martin for excellent word processing.

Literature Cited

Abel, J. J., Kubota, S. 1919. On the presence of histamine (β-iminazolylethylamine) in the hypophysis cerebri and other tissues of the body and its occurrence among the hydrolytic decomposition products of proteins. *J. Pharmacol. Exp. Ther.* 13:243–300

Abou, Y. Z., Adam, H. M., Stephen, W. R. G. 1973. Concentration of histamine in different parts of the brain and hypophysis of rabbit: Effect of treatment with histidine, certain other amino acids and histamine. *Br. J. Pharmacol.* 48:577–89

Allison, J. H., Blisner, M. E., Holland, W. H., Hipps, P. P., Sherman, W. R. 1976. Increased brain myo-inositol 1-phosphate in lithium-treated rats. *Biochem. Biophys. Res. Commun.* 71:664–70

Almeida, A. P., Beaven, M. A. 1981. Phylogeny of histamine in vertebrate brain. *Brain Res.* 208:244–50

Anderson, E. G., Haas, H. L., Hosli, L. 1973. Comparison of effects of noradrenaline and histamine with cyclic AMP on brain stem neurones. *Brain Res.* 49:471–75

Angus, J. A., Black, J. W. 1980. Pharmacological assay of cardiac H_2-receptor blockade by amitriptyline and lysergic acid diethylamide. *Circ. Res.* (Suppl. 1) 46:64–69

Arrang, J.-M., Garbarg, M., Schwartz, J.-C. 1983. Autoinhibition of brain histamine release mediated by a novel class (H_3) of histamine receptors. *Nature* 302:832–37

Armstrong, W. E., Sladek, C. D. 1984. Excitation by histamine of supraoptic neurons *in vitro. Soc. Neurosci. Abstr.* 10:91

Arunlakshana, O., Schild, H. O. 1959. Some quantitative uses of drug antagonists. *Br. J. Pharmacol.* 14:48–59

Ascher, P., Chesnoy-Marchais, D. 1982. Interactions between three slow potassium responses controlled by three distinct receptors in *Aplysia* neurones. *J. Physiol.* 324:67–92

Atack, C., Carlsson, A. 1972. *In vitro* release of endogenous histamine, together with noradrenaline and 5-hydroxytryptamine, from slices of mouse cerebral hemispheres. *J. Pharm. Pharmacol.* 24:990–92

Bailey, C. H., Chen, M. C., Weiss, K. R., Kupfermann, I. 1982. Ultrastructure of a histaminergic synapse in *Aplysia. Brain Res.* 238:205–10

Barbin, G., Garbarg, M., Schwartz, J.-C., Storm-Mathisen, J. 1976. Histamine synthesizing afferents to the hippocampal region. *J. Neurochem.* 26:259–63

Barker, L. A. 1981. Histamine H_1- and muscarinic receptor antagonist activity of cimetidine and tiotidine in the guinea pig isolated ileum. *Agents Actions* 11:699–705

Barker, L. A., Ebersole, J. 1982. Histamine H_2-receptors on guinea-pig ileum myenteric plexus neurones mediate the release of contractile agents. *J. Pharmacol. Exp. Ther.* 221:69–75

Barnett, A., Malick, J. B., Taber, R. I. 1971. Effects of antihistamines on isolation-induced fighting in mice. *Psychopharmacologica* 19:359–65

Barth, H., Lorenz, W. 1978. Structural requirements of imidazole compounds to be inhibitors or activators of histamine methyltransferase: Investigation of histamine analogues and H_2-receptor antagonists. *Agents Actions* 8:359–65

Bartholeyns, J., Bouclier, M. 1984. Involvement of histamine in growth of mouse and rat tumors: Antitumoral properties of monofluoromethylhistidine, an enzyme-activated irreversible inhibitor of histidine decarboxylase. *Canc. Res.* 44:639–45

Beaven, M. A., Roderick, N. B. 1980. Impromidine, a potent inhibitor of histamine methyltransferase and diamine oxidase. *Biochem. Pharmacol.* 29:2897–2900

Ben-Ari, Y., LaSalle, G. L., Barbin, G., Schwartz, J.-C., Garbarg, M. 1977. Histamine synthesizing afferents within the amygdaloid complex and bed nucleus of the stria terminalis of the rat. *Brain Res.* 138:285–94

Berridge, M. J. 1984. Inositol triphosphate and diacylglycerol as second messengers. *Biochem. J.* 220:345–60

Berridge, M. J., Irvine, R. F. 1984. Inositol triphosphate, a novel second messenger in cellular signal transduction. *Nature* 312:315–21

Berridge, M. J., Downes, C. P., Hanley, M. R. 1982. Lithium amplifies agonist dependent phosphatidylinositol responses in brain and salivary glands. *Biochem. J.* 206:587–95

Bertaccini, G., Coruzzi, G. 1981. Evidence for and against heterogeneity in the histamine H_2-receptor population. *Pharmacology* 23:1–13

Bhargava, K. P., Palit, G., Dixit, K. S. 1982. Behavioral effects of intracerebroventricular injection of histamine in dogs. In *Advances in Histamine Research, Advances in the Biosciences*, Vol. 33, ed. B. Uvnas, K. Tasaka, pp. 115–26. Oxford: Pergamon

Biggs, M. J., Johnson, E. S. 1980. Electrically-evoked release of [^{3}H]-histamine from the guinea-pig hypothalamus. *Br. J. Pharmacol.* 70:555–60

Bischoff, S., Korf, J. 1978. Different localization of histidine decarboxylase and histamine-N-methyltransferase in the rat brain. *Brain Res.* 141:375–79

Black, J. W., Duncan, W. A. M., Durant, C. J., Ganellin, C. R., Parsons, E. M. 1972. Definition and antagonism of histamine H$_2$-receptors. *Nature* 236: 385–90

Boissier, J. R., Guernet, M., Tillement, J. P., Blanco, I., Blanco, M. 1970. Variations des taux cérébraux d'histamine provoquées par la diphenhydramine et la L-histidine chez le rat. *Life Sci.* 9(Part 2): 249–56

Bolton, T. B., Clark, J. P., Kitamura, K., Lang, R. J. 1981. Evidence that histamine and carbachol may open the same ion channels in longitudinal smooth muscle of guinea-pig ileum. *J. Physiol.* 320: 363–79

Bonkowski, L., Dryden, W. F. 1976. The effects of putative neurotransmitters on the resting membrane potential of dissociated brain neurones in culture. *Brain Res.* 107: 69–84

Bonkowski, L., Dryden, W. F. 1977. Effects of iontophoretically applied neurotransmitters on mouse brain neurones in culture. *Neuropharmacology* 16: 89–97

Bouclier, M., Jung, M. J., Gerhart, F. 1983. Effect of prolonged inhibition of histidine decarboxylase on tissue histamine concentrations. *Experientia* 39: 1303–5

Bradley, P. B. 1968. Synaptic transmission in the central nervous system and its relevance for drug action. *Int. Rev. Neurobiol.* 11: 1–56

Bradley, P. B., Jones, H., Roberts, F. 1984. Histamine potentiates the effects of excitatory amino acids on quiescent neurones in the rat medulla. *Br. J. Pharmacol.* 82: 195P

Brand, J. 1968. The pharmacologic basis for the control of motion sickness by drugs. *Pharmacol. Physicians* 2: 1–5

Brimblecombe, R. W., Duncan, W. A. M., Durant, G. J., Emmett, J. C., Ganellin, C. R., et al. 1975. Cimetidine, a non-thiourea H$_2$-receptor antagonist. *J. Int. Med. Res.* 3: 86–92

Brimble, M. J., Wallis, D. I. 1973. Histamine H$_1$ and H$_2$-receptors at a ganglionic synapse. *Nature* 246: 156–58

Bristow, D. R., Hare, J. R., Hearn, J. R., Martin, L. E. 1981. Radioligand binding studies using [^{3}H] cimetidine and [^{3}H] ranitidine. *Br. J. Pharmacol.* 72: 547P–48P

Brown, P. A., Vernikos, J. 1980. Antihistamine effect on synaptosomal uptake of serotonin, norepinephrine and dopamine. *Eur. J. Pharmacol.* 65: 89–92

Brown, D. D., Tomchick, R., Axelrod, J. 1959. The distribution and properties of a histamine-methylating enzyme. *J. Biol. Chem.* 234: 2948–50

Brown, E., Kendall, D. A., Nahorski, S. R. 1984. Inositol phospholipid hydrolysis in rat cerebral cortical slices: I. Receptor characterization. *J. Neurochem.* 42: 1379–87

Brownstein, M. J., Palkovits, M., Tappaz, M. L., Saavedra, J., Kizer, J. S. 1976. Effect of surgical isolation of the hypothalamus on its neurotransmitter content. *Brain Res.* 117: 287–95

Brownstein, M. J., Saavedra, J. M., Axelrod, J., Zeman, G. H., Carpenter, D. O. 1974. Coexistence of several putative neurotransmitters in single identified neurons of *Aplysia*. *Proc. Natl. Acad. Sci. USA* 71: 4662–65

Carette, B. 1978. Responses of preoptic-septal neurons to iontophoretically applied histamine. *Brain Res.* 145: 391–95

Carlini, E. A., Green, J. P. 1963. The subcellular distribution of histamine, slow-reacting substance and 5-hydroxytryptamine in the brain of the rat. *Br. J. Pharmacol.* 20: 264–77

Carlsson, A., Lindquist, M. 1969. Central and peripheral monoaminergic membrane-pump blockade by some addictive analgesics and antihistamines. *J. Pharm. Pharmacol.* 21: 460–64

Carpenter, D. O., Gaubatz, G. L. 1975. H$_1$ and H$_2$ histamine receptors on *Aplysia* neurones. *Nature* 254: 343–44

Carruthers, S. G., Shoeman, D. W., Hignite, C. E., Azarnoff, D. L. 1978. Correlation between plasma diphenhydramine level and sedative and antihistamine effects. *Clin. Pharmacol. Ther.* 23: 375–82

Carswell, H., Nahorski, S. R. 1982. Distribution and characteristics of histamine H$_1$-receptors in guinea-pig airways identified by [^{3}H] mepyramine. *Eur. J. Pharmacol.* 81: 301–7

Chang, R. S. L., Tran, V. T., Snyder, S. H. 1979a. Characteristics of histamine H$_1$-receptors in peripheral tissues labeled with [^{3}H]mepyramine. *J. Pharmacol. Exp. Ther.* 209: 437–42

Chang, R. S. L., Tran, V. T., Snyder, S. H. 1979b. Heterogeneity of histamine H$_1$-receptors: Species variations in [^{3}H]mepyramine binding of brain membranes. *J. Neurochem.* 32: 1653–63

Chang, R. S. L., Tran, V. T., Snyder, S. H. 1980. Neurotransmitter receptor localizations: Brain lesion induced alterations in benzodiazepine, GABA, β-adrenergic, and histamine H$_1$-receptor binding. *Brain Res.* 190: 95–110

Chasin, M., Mamrak, F., Samaneigo, S. G. 1974. Preparation and properties of a cell-free, hormonally responsive adenylate cyclase from guinea pig brain. *J. Neurochem.* 22: 1031–38

Chesnoy-Marchais, D., Ascher, P. 1983. Effects of various cations on the slow K$^+$ conductance increases induced by carbachol, histamine and dopamine in *Aplysia* neurones. *Brain Res.* 259: 57–67

Chiel, H. J., Weiss, K. R., Kupfermann, I. 1983. An identified histaminergic neuron acts pre- and post-synaptically to inhibit outputs of identified buccal-cerebral interneurons. *Soc. Neurosci. Abstr.* 9:913

Chronister, R. B., Palmer, G. C., DeFrance, J. F., Sikes, R. W., Hubbard, J. I. 1982. Histamine: Correlative studies in nucleus accumbens. *J. Neurobiol.* 13:23–37

Chung-Hwang, E., Khurana, H., Fisher, H. 1976. The effect of dietary histidine level on the carnosine concentration of rat olfactory bulbs. *J. Neurochem.* 26:1087–91

Claiborne, B. J., Selverston, A. I. 1984. Histamine as a neurotransmitter in the stomatogastric nervous system of the spiny lobster. *J. Neurosci.* 4:708–21

Clark, W. G., Clark, Y. L. 1980. Changes in body temperature after administration of acetylcholine, histamine, morphine, prostaglanins and related agents. *Neurosci. Biobehav. Rev.* 4:175–240

Cockroft, S. 1981. Does phosphatidylinositol breakdown control the Ca^{2+}-gating mechanism? *Trends Pharmacol. Sci.* 2:340–42

Constanti, A., Nistri, A. 1976. Antagonism by some antihistamines of the amino acid-evoked responses recorded from the lobster muscle fibre and the frog spinal cord. *Br. J. Pharmacol.* 58:583–92

Coyle, J. T., Snyder, S. H. 1969. Antiparkinsonian drugs: Inhibition of dopamine uptake in the corpus striatum as a possible mechanism of action. *Science* 166:899–901

Cummins, C. J., Lust, W. D., Passonneau, J. V. 1983. Regulation of glycogenolysis in transformed astrocytes *in vitro*. *J. Neurochem.* 40:137–44

Daly, J. W. 1977. *Cyclic Nucleotides in the Nervous System*. New York: Plenum

Daly, J. W., McNeal, E., Partington, C., Neuwirth, M., Creveling, C. R. 1980. Accumulations of cyclic AMP in adenine-labeled cell-free preparations from guinea pig cerebral cortex: Role of α-adrenergic and H_1-histaminergic receptors. *J. Neurochem.* 35:326–37

Dashputra, P., Sharma, M., Jagtap, M., Khapre, M., Rajapurkar, M. 1966. Modification of metrazol-induced convulsions in rats by antihistamines. *Arch. Int. Pharmacodyn. Ther.* 160:106–12

Daum, P. R., Hill, S. J., Young, J. M. 1982. Histamine H_1-agonist potentiation of adenosine-stimulated cyclic AMP accumulation in slices of guinea-pig cerebral cortex: Comparison of response and binding parameters. *Br. J. Pharmacol.* 77:347–57

Daum, P. R., Downes, C. P., Young, J. M. 1984. Histamine stimulation of inositol 1-phosphate accumulation in lithium-treated slices from regions of guinea pig brain. *J. Neurochem.* 43:25–32

Davis, R. J., Ganong, B. R., Bell, R. M., Czech, M. P. 1985. sn-1,2-Dioctanoylglycerol: A cell-permeable diaclyglycerol that mimics phorbal diester action on the epidermal growth factor receptor and mitogenesis. *J. Biol. Chem.* 260:1562–66

De Riemer, S. A., Strong, J. A., Albert, K. A., Greengard, P., Kaczmarek, L. K. 1985. Enhancement of calcium current in *Aplysia* neurones by phorbol ester and protein kinase. *Nature* 313:313–16

Dirks, R., Steinbusch, H. W. M., Bol, J. G. J. M., Mulder, A. H. 1984. Distribution of histamine-, in relation to dopamine- and noradrenline-immunoreactive cell bodies in the central nervous system of the rat. *Soc. Neurosci. Abstr.* 10:63

Dismukes, K., Kuhar, M. J., Snyder, S. H. 1974. Brain histamine alterations after hypothalamic isolation. *Brain Res.* 78:144–51

Dismukes, R. K., Ghosh, P., Creveling, C. R., Daly, J. W. 1975. Altered responsiveness of adenosine 3′,5′-monophosphate-generating systems in rat cortical slices after lesions of the medial forebrain bundle. *Exp. Neurol.* 49:725–35

Dismukes, K., Rogers, M., Daly, J. W. 1976. Cyclic adenosine 3′,5′-monophosphate formation in guinea-pig brain slices: effect of H_1- and H_2-histaminergic agonists. *J. Neurochem.* 26:785–90

Donoso, A. O., Alvarez, E. O. 1984. Brain histamine as neuroendocrine transmitter. *Trends Pharmacol. Sci.* 5:98–100

Downes, C. P. 1982. Receptor-stimulated inositol phospholipid metabolism in the central nervous system. *Cell Calcium* 3:413–28

Dropp, J. J. 1972. Mast cells in the central nervous system of several rodents. *Anat. Rec.* 174:227–38

Dropp, J. J. 1976. Mast cells in mammalian brain. I. Distribution. *Acta Anat.* 94:1–21

Duggan, D. E., Hooke, K. F., Maycock, A. L. 1984. Inhibition of histamine synthesis *in vitro* and *in vivo* by S-α-fluoromethylhistidine. *Biochem. Pharmacol.* 33:4003–9

Durant, G. J., Ganellin, C. R., Parson, M. E. 1975. Chemical differentiation of histamine H_1- and H_2-receptor agonists. *J. Med. Chem.* 18:905–9

Dzielak, D. J., Thureson-Klein, A., Klein, R. L. 1983. Local modulation of neurotransmitter release in bovine splenic vein. *Blood Vessels* 20:122–34

Edvinsson, L., MacKenzie, E. T. 1977. Amine mechanisms in the central circulation. *Pharmacol. Rev.* 28:275–348

El-Ackad, T. M., Brody, M. J. 1975a. Fluorescence histochemical localization of non-mast cell histamine. In *Neuropsychopharmacology*, ed. J. R. Boissier, H.

Hippius, P. Pichot, pp. 551–59. New York: American Elsevier

El-Ackad, T. M., Brody, M. J. 1975b. Evidence for non-mast cell histamine in the vascular wall. *Blood Vessels* 12:181–91

Elias, M. S., Evans, P. D. 1983. Histamine in the insect nervous system: Distribution, synthesis and metabolism. *J. Neurochem.* 41:562–68

Engberg, I., Flatman, J. A., Kadzielawa, K. 1976. Lack of specificity of motoneurone responses to microiontophoretically applied phenolic amines. *Acta Physiol. Scand.* 96:137–39

Epstein, C. M. 1984. Ranitidine and confusion. *Lancet* i:1071

Faingold, C. L., Berry, C. A. 1972a. A comparison of the EEG effects of the potent antihistaminic (DL-chlorpheniramine) with a less potent isomer (L-chlorpheniramine). *Arch. Int. Pharmacodyn.* 199:213–18

Faingold, C. L., Berry, C. A. 1972b. Effects of antihistaminic agents upon the electrographic activity of the cat brain: A power spectral density study. *Neuropharmacology* 11:491–98

Ferrendelli, J. A., Kinscherf, D. A., Chang, M.-M. 1975. Comparison of the effects of biogenic amines on cyclic GMP and cyclic AMP levels in mouse cerebellum *in vitro*. *Brain Res.* 84:63–73

Fram, D. H., Green, J. P. 1965. The presence and measurement of methylhistamine in urine. *J. Biol. Chem.* 240:2036–42

Fram, D. H., Green, J. P. 1968. Methylhistamine excretion during treatment with a monoamine oxidase inhibitor. *Clin. Pharmacol. Ther.* 9:355–57

Friedel, R. O., Schanberg, S. M. 1975. Effects of histamine on phospholipid metabolism of rat brain in vivo. *J. Neurochem.* 24:819–20

Fukui, H., Watanabe, T., Wada, H. 1980. Immunochemical cross reactivity of the antibody elicited against L-histidine decarboxylase purified from the whole bodies of fetal rats with the enzyme from rat brain. *Biochem. Biophys. Res. Commun.* 93:333–39

Furano, A. V., Green, J. P. 1964a. The uptake of biogenic amines by mast cells of the rat. *J. Physiol.* 170:263–71

Furano, A. V., Green, J. P. 1964b. The compartmentation and elimination of ^{14}C-histamine by neoplastic mast cells in culture. *Biochem. Biophys. Acta* 86:596–603

Gajtkowski, G. A., Norris, D. B., Rising, T. J., Wood, T. P. 1983. Specific binding of ^3H-tiotidine to histamine H_2 receptors in guinea pig cerebral cortex. *Nature* 304:65–67

Garbarg, M., Krishnamoorthy, M. S., Feger, J., Schwartz, J.-C. 1973. Effects of mesencephalic and hypothalamic lesions on histamine levels in rat brain. *Brain Res.* 50:361–67

Garbarg, M., Barbin, G., Feger, J., Schwartz, J.-C. 1974. Histaminergic pathway in rat brain evidenced by lesions of the medial forebrain bundle. *Science* 186:833–35

Garbarg, M., Barbin, G., Bischoff, S., Pollard, H., Schwartz, J.-C. 1976. Dual localization of histamine in an ascending neuronal pathway and in non-neuronal cells evidenced by lesions in the lateral hypothalamic area. *Brain Res.* 106:333–48

Garbarg, M., Barbin, G., Llorens, C., Palacios, J. M., Pollard, H., et al. 1980a. Recent developments in brain histamine research: Pathways and receptors. In *Transmitters, Receptors and Drug Action*, ed. W. B. Essman, pp. 179–202. New York: Spectrum

Garbarg, M., Barbin, G., Rodergras, E., Schwartz, J.-C. 1980b. Inhibition of histamine synthesis in brain by α-fluoromethylhistidine, a new irreversible inhibitor: *In vitro* and *in vivo* studies. *J. Neurochem.* 35:1045–52

Geller, H. M. 1976. Effects of some putative neurotransmitters on unit activity of tuberal hypothalamic neurons *in vitro*. *Brain Res.* 108:423–30

Geller, H. M. 1979. Are histamine H_2-receptor depressions of neuronal activity in tissue cultures from rat hypothalamus mediated through cyclic adenosine monophosphate? *Neurosci. Lett.* 14:49–53

Geller, H. M. 1981. Histamine actions on activity of cultured hypothalamic neurons: Evidence for mediation by H_1- and H_2-histamine receptors. *Dev. Brain Res.* 1:89–101

Geller, H. M., Hoffer, B. J. 1977. Effect of calcium removal on monoamine-elicited depressions of cultured tuberal neurons. *J. Neurobiol.* 8:43–55

Geller, H. M., Springfield, S. A., Tiberio, A. R. 1984. Electrophysiological actions of histamine. *Can. J. Physiol. Pharmacol.* 62:715–19

Goldberg, D. J., Schwartz, J. H. 1980. Fast axonal transport of foreign transmitters in an identified serotonergic neurone of *Aplysia californica*. *J. Physiol.* 307:259–72

Goldschmidt, R. C., Hough, L. B., Glick, S. D., Padawer, J. 1984a. Mast cells in rat thalamus: Nuclear localization, sex difference and left-right asymmetry. *Brain Res.* 323:209–17

Goldschmidt, R. C., Khandelwal, J. K., Hough, L. B. 1984b. Presence and characterization of *tele*-methylhistamine in mast cells. *Agents Actions* 14:174–78

Goldschmidt, R. C., Hough, L. B., Glick, S. D. 1985. Rat brain mast cells: Contribution to brain histamine levels. *J. Neurochem.* 44:1943–47

Gotoh, H., Schwartz, J. H. 1982. Specificity of axonal transport in C2, a histaminergic neuron of *Aplysia californica. Brain Res.* 242:87–98

Gotow, T., Hashimura, S. 1982. Modulation of the histamine-induced inhibitory response in an identified Onchidium neuron by cyclic nucleotides. *Brain Res.* 239:634–38

Gotow, T., Kirkpatrick, C. T., Tomita, T. 1980a. Excitatory and inhibitory effects of histamine on molluscan neurons. *Brain Res.* 196:151–67

Gotow, T., Kirkpatrick, C. T., Tomita, T. 1980b. An analysis of histamine-induced inhibitory response in molluscan neurons. *Brain Res.* 196:169–82

Green, J. P. 1964. Histamine and the nervous system. *Fed. Proc.* 23:1095–1102

Green, J. P. 1966a. Uptake and binding of histamine. *Fed. Proc.* 26:211–18

Green, J. P. 1966b. Synthesis, uptake and binding of histamine and 5-hydroxytryptamine in mast cells. In *Mechanisms of Release of Biogenic Amines*, ed. U. S. von Euler, S. Rosell, B. Uvnas, pp. 125–45. London: Pergamon

Green, J. P. 1970. Histamine *Handb. Neurochem.* 4:221–50

Green, J. P. 1983. Histamine receptors in brain. *Handb. Psychopharmacol.* 17:385–420

Green, J. P., Day, M. 1963. Biosynthetic pathways in mastocytoma mast cells in culture and *in vivo. Ann. NY Acad. Sci.* 103:334–50

Green, J. P., Hough, L. B. 1980. Histamine receptors. In *Cellular Receptors of Hormones and Neurotransmitters*, ed. D. Schulster, A. Levitzki, pp. 287–305. New York: Wiley

Green, J. P., Khandelwal, J. K. 1985. Histamine turnover in regions of rat brain. In *Frontiers in Histamine Research, Advances in the Biosciences*, Vol. 51, ed. C. R. Ganellin, J.-C. Schwartz, pp. 185–95. Oxford: Pergamon

Green, J. P., Maayani, S. 1977. Tricyclic antidepressant drugs block histamine H_2 receptor in brain. *Nature* 269:163–65

Green, J. P., Fram, D. H., Kase, N. 1964. Methylhistamine and histamine in the urine of women during the elaboration of estrogen. *Nature* 204:1165–68

Green, J. P., Johnson, C. L., Weinstein, H., Maayani, S. 1977. Antagonism of histamine-activated adenylate cyclase in brain by d-lysergic acid diethylamide. *Proc. Natl. Acad. Sci. USA* 74:5697–5701

Green, J. P., Johnson, C. L., Weinstein, H. 1978. Histamine as a neurotransmitter. In *Psychopharmacology: A Generation of Progress*, ed. M. A. Lipton, A. DiMascio, K. F. Killam, pp. 319–32. New York: Raven

Green, J. P., Johnson, C. L., Weinstein, H. 1979. Histamine activation of adenylate cyclase in brain: An H_2-receptor and its blockade by LSD. In *Histamine Receptors*, ed. T. O. Yellin, pp. 185–210. New York: Spectrum

Green, J. P., Maayani, S., Weinstein, H., Hough, L. B. 1980. Histamine and psychotropic drugs. *Psychopharmacol. Bull.* 16:36–38

Gross, P. M. 1981. Histamine H_1- and H_2-receptors are differentially and spatially distributed in cerebral vessels. *J. Cereb. Blood Flow Metabol.* 1:441–46

Gross, P. M. 1982. Cerebral histamine: Indications for neuronal and vascular regulation. *J. Cereb. Blood Flow Metab.* 2:3–23

Gruol, D. L., Weinreich, D. 1979a. Two pharmacologically distinct histamine receptors mediating membrane hyperpolarization on identified neurons of *Aplysia californica. Brain Res.* 162:281–301

Gruol, D. L., Weinreich, D. 1979b. Cooperative interactions of histamine and competitive antagonism by cimetidine at neuronal histamine receptors in the marine mollusc, *Aplysia californica. Neuropharmacology* 18:415–21

Gustafsson, A., Plym Forshell, G. 1964. Distribution of histamine-N-methyltransferase in nature. *Acta Chem. Scand.* 18:2098–2102

Haas, H. L. 1974. Histamine: Action on single hypothalamic neurones. *Brain Res.* 76:363–66

Haas, H. L. 1981a. Analysis of histamine action by intra- and extra-cellular recording in hippocampal slices of the rat. *Agents Actions* 11:125–28

Haas, H. L. 1981b. Histamine hyperpolarizes hippocampal neurones *in vitro. Neurosci. Lett.* 22:75–78

Haas, H. L. 1984. Histamines potentiates neuronal excitation by blocking a calcium-dependent potassium conductance. *Agents Actions* 14:534–37

Haas, H. L. 1985. Histamine. In *Neurotransmitter Actions in the Vertebrate Nervous System*, ed. M. Rogawski, J. Barker, pp. 315–31. New York: Plenum

Haas, H. L., Bucher, U. M. 1975. Histamine H_2-receptors on single central neurones. *Nature* 255:634–35

Haas, H. L., Geller, H. M. 1982. Electrophysiology of histaminergic transmission in the brain. In *Advances in Histamine Research, Advances in the Biosciences*, Vol.

33, ed. B. Uvnas, K. Tasaka, pp. 81–91. New York: Pergamon

Haas, H. L., Konnerth, A. 1983. Histamine and noradrenaline decrease calcium-activated potassium conductance in hippocampal pyramidal cells. *Nature* 302: 432–34

Haas, H. L., Wolf, P. 1977. Central actions of histamine: Microelectrophoretic studies. *Brain Res.* 122: 269–79

Haas, H. L., Anderson, E. G., Hosli, L. 1973. Histamine and metabolites: Their effects and interactions with convulsants on brain stem neurones. *Brain Res.* 51: 269–78

Haas, H. L., Wolf, P., Nussbaumer, J.-C. 1975. Histamine: Action on supraoptic and other hypothalamic neurones of the cat. *Brain Res.* 88: 166–70

Haas, H. L., Wolf, P., Palacios, J. M., Garbarg, M., Barbin, G., et al. 1978. Hypersensitivity to histamine in the guinea-pig brain: Microiontophoretic and biochemical studies. *Brain Res.* 156: 275–91

Haas, H. L., Jefferys, J. G. R., Slater, N. T., Carpenter, D. O. 1984. Modulation of low calcium induced field bursts in the hippocampus by monoamines and cholinomimetics. *Pflugers Arch.* 400: 28–33

Hadfield, A. J., Robinson, N. R., Hill, S. J. 1983. The nature of the binding of [³H] mepyramine to homogenates of guinea-pig cerebral cortex at different [³H] ligand concentrations. *Biochem. Pharmacol.* 32: 2449–51

Hallcher, L. M., Sherman, W. R. 1980. The effects of lithium and other agents on the activity of *myo*-inositol-1-phosphatase from bovine brain. *J. Biol. Chem.* 255: 10896–10901

Hawthorne, J. N., Azila, N. 1982. Phosphatidylinositol and calcium gating: Some difficulties. In *Phospholipids in the Nervous System*, ed. L. A. Horrocks, G. B. Ansell, G. Porcellati, 1: 265–70. New York: Raven

Head, R. J., Hjelle, J. T., Jarrott, B., Berkowitz, B., Cardinale, G., et al. 1980. Isolated brain microvessels: Preparation, morphology, histamine and catecholamine contents. *Blood Vessels* 17: 173–86

Hegstrand, L. R., Kanof, P. D., Greengard, P. 1976. Histamine-sensitive adenylate cyclase in mammalian brain. *Nature* 260: 163–65

Heinrich, M. A. 1953. The effect of the antihistaminic drugs on the central nervous system in rats and mice. *Arch. Int. Pharmacodyn.* 92: 444–63

Hill, S. J., Young, J. M. 1980. Histamine H₁-receptors in the brain of the guinea pig and the rat: Differences in ligand binding properties and regional distribution. *Br. J. Pharmacol.* 68: 687–96

Hill, S. J., Daum, P., Young, J. M. 1981. Affinities of histamine H₁-antagonists in guinea-pig brain: Similarity of values determined from [³H] mepyramine binding and from inhibition of a functional response. *J. Neurochem.* 37: 1357–60

Hokin, L. E. 1985. Receptors and phosphoinositide-generated second messengers. *Ann. Rev. Biochem.* 54: 205–35

Hokin, M. R., Hokin, L. E. 1953. Enzyme secretion and the incorporation of P³² into phospholipids of pancreas slices. *J. Biol. Chem.* 203: 967–77

Hosli, L., Haas, H. L. 1971. Effects of histamine, histidine and imidazole acetic acid on neurones of the medulla oblongata of the cat. *Experientia* 27: 1311–12

Hough, L. B., Barker, L. A. 1981. Histamine H₂-receptor antagonism by propantheline and derivatives. *J. Pharmacol. Exp. Ther.* 219: 453–58

Hough, L. B., Domino, E. F. 1979a. Tele-methylhistamine distribution in rat brain. *J. Neurochem.* 32: 1865–66

Hough, L. B., Domino, E. F. 1979b. Tele-methylhistamine oxidation by type B monoamine oxidase. *J. Pharmacol. Exp. Ther.* 208: 422–28

Hough, L. B., Green, J. P. 1980. Possible functions of brain histamine. *Psychopharmacol. Bull.* 16: 42–44

Hough, L. B., Green, J. P. 1981. Histamine-activated adenylate cyclase in brain homogenates of several species. *Brain Res.* 219: 363–70

Hough, L. B., Green, J. P. 1984. Histamine and its receptors in the nervous system. *Handb. Neurochem.* 6: 145–211. 2nd ed.

Hough, L. B., Weinstein, H., Green, J. P. 1980. One agonist and two receptors mediating the same effect: Histamine receptors linked to adenylate cyclase in the brain. *Adv. Biochem. Psychopharmacol.* 21: 183–92

Hough, L. B., Khandelwal, J. K., Morrishow, A. M., Green, J. P. 1981. An improved GCMS method to measure tele-methylhistamine. *J. Pharmacol. Methods* 5: 143–48

Hough, L. B., Khandelwal, J. K., Green, J. P. 1982. Ontogeny and subcellular distribution of rat brain tele-methylhistamine. *J. Neurochem.* 38: 1593–99

Hough, L. B., Khandelwal, J. K., Green, J. P. 1984a. Histamine turnover in regions of rat brain. *Brain Res.* 291: 103–9

Hough, L. B., Khandelwal, J. K., Goldschmidt, R. C., Diomande, M., Glick, S. D. 1984b. Normal levels of histamine and tele-methylhistamine in mast cell-deficient mouse brain. *Brain Res.* 292: 133–38

Hough, L. B., Glick, S. D., Su, K. 1985a. A role for histamine and histamine H₂-

receptors in non-opiate footshock-induced analgesia. *Life Sci.* 36:859–66

Hough, L. B., Goldschmidt, R. C., Glick, S. D., Padawer, J. 1985b. Mast cells in rat brain: Characterization, localization, and histamine content. In *Frontiers in Histamine Research, Advances in the Biosciences*, Vol. 51, ed. C. R. Ganellin, J.-C. Schwartz, pp. 131–40. London: Pergamon

Howland, R. D., Spector, S. 1972. Disposition of histamine in mammalian blood vessels. *J. Pharmacol. Exp. Ther.* 182:239–45

Huggins, A. K., Woodruff, G. N. 1968. Histamine metabolism in invertebrates. *Comp. Biochem. Physiol.* 26:1107–11

Hughes, J. D., Reed, W. D., Serjeant, C. S. 1983. Mental confusion associated with ranitidine. *Med. J. Australia* ii:12–13

Ibrahim, M. Z. M. 1970. The immediate and delayed effects of compound 48/80 on the mast cells and parenchyma of rabbit brain. *Brain Res.* 17:348–50

Ibrahim, M. Z. M., Waziri, R., Kamath, S. 1979. The mast cells of the mammalian central nervous system. IV. Culture of neurolipomastcytoid cells from rabbit and rat leptomeninges. *Cell Tissue Res.* 204:217–32

Irvine, R. F., Dawson, R. M. C., Freinkel, N. 1982. Stimulated phosphatidylinositol turnover—a brief appraisal. *Contemp. Metab.* 2:301–42

Jarrott, B., Hjelle, J., Spector, S. 1979. Association of histamine with cerebral microvessels in regions of bovine brain. *Brain Res.* 168:323–30

Johnson, C. L. 1982a. Histamine receptors and cyclic nucleotides. In *Pharmacology of Histamine Receptors*, ed. C. R. Ganellin, M. E. Parsons, pp. 146–216. Bristol: Wright

Johnson, C. L. 1982b. Biochemical mechanisms of action of histamine. In *The Chemical Regulation of Biological Mechanisms*, ed. A. M. Creighton, S. Turner, pp. 16–26. London: Royal Soc. Chem.

Jones, H., Bradley, P. B., Roberts, F. 1983. Excitatory effects of microiontophoretically applied histamine in the rat medulla may be mediated via histamine H_2-receptors. *Br. J. Pharmacol.* 79:282P

Jones, L. M., Cockcroft, S., Michell, R. H. 1979. Stimulation of phosphatidylinositol turnover in various tissues by cholinergic and adrenergic agonists, by histamine and by caerulein. *Biochem. J.* 182:669–76

Kahlson, G., Rosengren, E. 1971. Biogenesis and physiology of histamine. *Monogr. Physiol. Soc.* 21:235–76. Baltimore: Williams & Wilkins

Kanof, P. D., Greengard, P. 1978. Brain histamine receptors as targets for anti-depressant drugs. *Nature* 272:329–33

Karnushina, I. L., Palacios, J. M., Barbin, G., Dux, E., Joo, F., et al. 1979. Histamine-related enzymes and histamine receptors in isolated brain capillaries. *Agents Actions* 9:89–90

Karnushina, I. L., Palacios, J. M., Barbin, G., Dux, E., Joo, F., et al. 1980. Studies on a capillary-rich fraction isolated from brain: Histaminic components and characterization of the histamine receptors linked to adenylate cyclase. *J. Neurochem.* 34:1201–8

Kataoka, K., DeRobertis, E. 1967. Histamine in isolated small nerve endings and synaptic vesicles of rat brain cortex. *J. Pharmacol. Exp. Ther.* 156:114–25

Kehoe, J., Marder, E. 1976. Identification and effects of neural transmitters in invertebrates. *Ann. Rev. Pharmacol.* 16:245–68

Kenakin, T. P. 1984. The classification of drugs and drug receptors in isolated tissues. *Pharmacol. Rev.* 36:165–222

Kendall, D. A., Nahorski, S. R. 1984. Inositiol phospholipid hydrolysis in rat cerebral cortical slices. II. Calcium requirement. *J. Neurochem.* 42:1388–94

Kerkut, G. A., Walker, R. J., Woodruff, G. N. 1968. The effects of histamine and other naturally occurring imidazoles on neurones of *Helix aspersa*. *Br. J. Pharmacol. Chemother.* 32:241–52

Kety, S. S. 1978. Strategies of basic research. In *Psychopharmacology: A Generation of Progress*, ed. M. A. Lipton, A. DiMascio, K. F. Killam, pp. 7–11. New York: Raven

Khandelwal, J. K., Hough, L. B., Pazhenchevsky, B., Morrishow, A. M., Green, J. P. 1982. Presence and measurement of methylimidazoleacetic acids in brain and body fluids. *J. Biol. Chem.* 257:12815–19

Khandelwal, J. K., Hough, L. B., Green, J. P. 1984. Regional distribution of the histamine metabolite, *tele*-methylimidazoleacetic acid, in rat brain: Effects of pargyline and probenecid. *J. Neurochem.* 42:519–22

Kirsten, E. B., Sharma, J. N. 1976. Microiontophoresis of acetylcholine, histamine and their antagonists on neurones in the medial and lateral vestibular nuclei of the cat. *Neuropharmacology* 15:743–53

Kolesnick, R. N., Gershengorn, M. C. 1985. Arachidonic acid inhibits thyrotropin-releasing hormone-induced elevation of cytoplasmic free calcium in GH_3 pituitary cells. *J. Biol. Chem.* 260:707–13

Kollonitsch, J., Patchett, A. A., Marburg, S., Maycock, A. L., Perkins, L. M., et al. 1978. Selective inhibitors of biosynthesis of aminergic neurotransmitters. *Nature* 274:906–8

Krnjevic, K., Phillis, J. W. 1963. Actions of

certain amines on cerebral cortical neurones. *Br. J. Pharmacol.* 20:471–90

Ku, B. S., Takeuchi, H. 1983. Identification of three further giant neurons, r-APN, INN and FAN, in the caudal part on the dorsal surface of the suboesophageal ganglia of *Achatina fulica* Férussac. *Comp. Biochem. Physiol.* 76c:99–106

Kuhar, M. J., Taylor, K. M., Snyder, S. H. 1971. The subcellular localization of histamine and histamine methyltransferase in rat brain. *J. Neurochem.* 18:1515–27

Kuhn, R. 1970. The imipramine story. In *Discoveries in Biological Psychiatry*, ed. F. J. Ayd, B. Blackwell, pp. 205–17. Philadelphia: Lippincott

Kuo, J.-F., Lee, T.-P., Reyes, P. L., Walton, K. G., Donnelly, T. E., et al. 1972. Cyclic nucleotide-dependent protein kinases. X. An assay method for the measurement of guanosine 3′,5′-monophosphate in various biological materials and a study of agents regulating its levels in heart and brain. *J. Biol. Chem.* 247:16–22

Kwiatkowski, H. 1943. Histamine in nervous tissue. *J. Physiol.* 102:32–41

Laduron, P. M. 1984. Criteria for receptor sites in bindings studies. *Biochem. Pharmacol.* 33:833–39

Laduron, P. M., Janssen, P. F. M., Gommeren, W., Leysen, J. E. 1982. *In vitro* and *in vivo* binding characteristics of a new long-acting histamine H_1 antagonist, astemizole. *Mol. Pharmacol.* 21:294–300

Leake, L. D., Walker, R. J. 1980. *Invertebrate Neuropharmacology*, pp. 214–17, 344. Glasgow: Blackie

Leibowitz, S. F. 1979. Histamine: Modification of behavioral and physiological components of body fluid homeostasis. In *Histamine Receptors*, ed. T. O. Yellin, pp. 219–53. New York: Spectrum

Leibowitz, S. F. 1980. Neurochemical systems of the hypothalamus—Control of feeding and drinking behavior and water-electrolyte excretion. In *Handbook of the Hypothalamus*, ed. P. J. Morgane, J. Panksepp, 3A:299–437. New York: Dekker

Leslie, R., Osborne, N. N., Patel, S., Peard, A. 1984. An immunohistochemical study to localise histaminergic neurones in invertebrate and vertebrate nervous systems. *Br. J. Pharmacol.* 82:250P

Lidbrink, P., Jonsson, G., Fuxe, K. 1971. The effect of imipramine-like drugs and antihistamine drugs on uptake mechanisms in the central noradrenaline and 5-hydroxytryptamine neurons. *Neuropharmacology* 10:521–36

Lindl, T. 1979. Cyclic AMP and its relation to ganglionic transmission. A combined biochemical and electrophysiological study of the rat superior cervical ganglion *in vitro*.

Neuropharmacology 18:227–35

Liou, S. Y., Shibata, S., Yamakawa, K., Ueki, S. 1983. Inhibitory and excitatory effects of histamine on suprachiasmatic neurons in rat hypothalamic slice preparation. *Neurosci. Lett.* 41:109–13

Lomax, P., Green, M. D. 1981. Histaminergic neurons in the hypothalamic thermoregulatory pathways. *Fed. Proc.* 40:2741–45

Maayani, S., Hough, L. B., Weinstein, H., Green, J. P. 1982. Response of the histamine H_2-receptor in brain to antidepressant drugs. *Adv. Biochem. Psychopharmacol.* 31:133–47

Madison, D. V., Nicoll, R. A. 1982. Noradrenaline blocks accommodation of pyramidal cell discharge in the hippocampus. *Nature* 299:636–38

Maeyama, K., Watanabe, T., Taguchi, Y., Yamatodani, A., Wada, H. 1982. Effect of α-fluoromethylhistidine, a suicide inhibitor of histidine decarboxylase, on histamine levels in mouse tissues. *Biochem. Pharmacol.* 31:2367–70

Maeyama, K., Watanabe, T., Yamatodani, A., Taguchi, Y., Kambe, H., et al. 1983. Effect of α-fluoromethylhistidine on the histamine content of the brain of W/W^v mice devoid of mast cells: Turnover of brain histamine. *J. Neurochem.* 41:128–34

Mani, R. B., Spellum, J. S., Frank, J. H., Laureno, R. 1984. H_2-receptor blockers and mental confusion. *Lancet* ii:98

Martin, S. M., Spencer, A. N. 1983. Neurotransmitters in coelenterates. *Comp. Biochem. Physiol.* 74C:1–14

Martres, M. P., Baudry, M., Schwartz, J.-C. 1975. Histamine synthesis in the developing rat brain: Evidence for a multiple compartmentation. *Brain Res.* 83:265–75

Maxwell, G. D., Hildebrand, J. G. 1981. Anatomical and neurochemical consequences of deafferentation in the development of the visual system of the moth *Manduca sexta*. *J. Comp. Neurol.* 195:667–80

Maxwell, G. D., Tait, J. F., Hildebrand, J. G. 1978. Regional synthesis of neurotransmitter candidates in the CNS of the moth *Manduca sexta*. *Comp. Biochem. Physiol.* 61c:109–19

Mazurkiewicz-Kwilecki, I. M. 1984. Possible role of histamine in brain function: Neurochemical, physiological, and pharmacological indications. *Can. J. Physiol. Pharmacol.* 62:709–14

McCall, R. B., Aghajanian, G. K. 1979. Serotonergic facilitation of facial motoneuron excitation. *Brain Res.* 169:11–27

McCaman, R. E., McKenna, D. G. 1978. Monosynaptic connections between histamine-containing neurons and their various follower cells. *Brain Res.* 141:165–71

McCaman, R. E., Weinreich, D. 1982. On the nature of histamine-mediated slow hyperpolarizing synaptic potentials in identified molluscan neurones. *J. Physiol.* 328:485–506

McCaman, R. E., Weinreich, D. 1985. Histaminergic synaptic transmission in the cerebral ganglion of *Aplysia. J. Neurophysiol.* 53:1016–37

McCulloch, M. W., Medgett, I. C., Rand, M. J. 1979. Effects of the histamine H_2-receptor blocking drugs burimamide and cimetidine on noradrenergic transmission in the isolated aorta of the rabbit and atria of the guinea pig. *Br. J. Pharmacol.* 67:535–43

Meibach, R. C., Maayani, S., Green, J. P. 1980. Characterization and radioautography of [^{3}H]LSD binding by rat brain slices *in vitro*: The effect of 5-hydroxytryptamine. *Eur. J. Pharmacol.* 67:371–82

Michaelson, I. A., Coffman, P. Z. 1967. The subcellular localization of histamine in guinea pig brain—a re-evaluation. *Biochem. Pharmacol.* 16:2085–90

Michell, R. 1983. Ca^{2+} and protein kinase C: Two synergistic cellular signals. *Trends Biochem. Sci.* 8:263–65

Monnier, M., Sauer, R., Hatt, A. M. 1970. The activating effect of histamine on the central nervous system. *Int. Rev. Neurobiol.* 12:265–305

Mueller, A. L., Hoffer, B. J., Dunwiddie, T. V. 1981. Noradrenergic responses in rat hippocampus: Evidence for mediation by α and β receptors in the *in vitro* slice. *Brain Res.* 214:113–26

Nemeth, P. R., Ort, C. A., Wood, J. D. 1984. Intracellular study of effects of histamine on electrical behaviour of myenteric neurones in guinea-pig small intestine. *J. Physiol.* 355:411–25

Nishibori, M., Oishi, R., Saeki, K. 1984. Histamine turnover in the brain of different mammalian species: Implications for neuronal histamine half-life. *J. Neurochem.* 43:1544–49

O'Brien, R. D. ed. 1979. *The Receptors—A Comprehensive Treatise*, Vol. 1. New York: Plenum

Ohga, Y., Daly, J. W. 1977. The accumulation of cyclic AMP and cyclic GMP in guinea pig brain slices. Effect of calcium ions, norepinephrine, and adenosine. *Biochem. Biophys. Acta* 498:46–60

Oishi, R., Nishibori, M., Saeki, K. 1984. Regional differences in the turnover of neuronal histamine in the rat brain. *Life Sci.* 34:691–99

Olianas, M., Oliver, A. P., Neff, N. H. 1982. Biochemical and electrophysiological studies on the mechanism of action of typical and atypical antidepressants on the H_2-histamine receptor complex. *Adv. Biochem. Psychopharmacol.* 31:149–56

Olianas, M., Oliver, A. P., Neff, N. H. 1984. Correlation between histamine-induced neuronal excitability and activation of adenylate cyclase in the guinea pig hippocampus. *Neuropharmacology* 23:1071–74

Olsson, Y. 1965. Storage of monoamines in mast cells of normal and sectioned peripheral nerve. *Z. Zellforsch.* 68:255–65

Olsson, Y. 1968. Mast cells in the nervous system. *Int. Rev. Cytol.* 24:27–70

Ono, J. K., McCaman, R. E. 1979. Measurement of endogenous transmitter levels after intracellular recording. *Brain Res.* 165:156–60

Ono, J. K., McCaman, R. E. 1980. Identification of additional histaminergic neurons in *Aplysia*: Improvement of single cell isolation techniques for *in tandem* physiological and chemical studies. *Neuroscience* 5:835–40

Oron, Y., Dascal, N., Nadler, E., Lupu, N. 1985. Inositol 1,4,5-triphosphate mimics muscarinic response in *Xenopos* oocytes. *Nature* 313:141–43

Orr, E. L. 1984. Dural mast cells: Source of contaminating histamine in analyses of mouse brain histamine levels. *J. Neurochem.* 43:1497–99

Orr, E. L., Pace, K. R. 1984. The significance of mast cells as a source of histamine in the mouse brain. *J. Neurochem.* 42:727–32

Osborne, N. N. 1977. Do snail neurones contain more than one neurotransmitter? *Nature* 270:622–23

Osborne, N. N., Wolter, K.-D., Neuhoff, V. 1979. *In vitro* experiments on the accumulation and release of ^{14}C-histamine by snail (*Helix pomatia*) nervous tissue. *Biochem. Pharmacol.* 28:2799–2805

Padawer, J. ed. 1963. Mast cells and basophils. *Ann. NY Acad. Sci.* 103:1–492

Pakkari, I., Karppanen, H. 1982. Analeptic effect of centrally administered histamine H_2-receptor agonist dimaprit but not impromidine in urethane-anaesthetized rats. *Neuropharmacology* 21:171–78

Palacios, J. M., Mengod, G., Picatoste, F., Grau, M., Blanco, I. 1976. Properties of rat brain histidine decarboxylase. *J. Neurochem.* 27:1455–60

Palacios, J. M., Garbarg, M., Barbin, G., Schwartz, J.-C. 1978. Pharmacological characterization of histamine receptors mediating the stimulation of cyclic AMP accumulation in slices from guinea-pig hippocampus. *Mol. Pharmacol.* 14:971–82

Palacios, J. M., Young, W. S., Kuhar, M. J. 1979. Autoradiographic localization of H_1-histamine receptors in brain using [^{3}H]-mepyramine: Preliminary studies. *Eur. J. Pharmacol.* 58:295–304

Palacios, J. M., Wamsely, J. K., Kuhar, M. J. 1981a. The distribution of histamine H_1-receptors in the rat brain: An autoradiographic study. *Neuroscience* 6:15–37

Palacios, J. M., Wamsley, J. K., Kuhar, M. J. 1981b. GABA, benzodiazepine, and histamine-H_1 receptors in the guinea pig cerebellum: Effects of kainic acid injections studied by autoradiographic methods. *Brain Res.* 214:155–62

Panula, P., Yang, H.-Y. T., Costa, E. 1984. Histamine-containing neurons in the rat hypothalamus. *Proc. Natl. Acad. Sci. USA* 81:2572–76

Parsons, M. E. 1977. The antagonism of histamine H_2-receptors *in vitro* and *in vivo* with particular reference to the actions of cimetidine. In *Cimetidine: Proceedings of the Second International Symposium on Histamine H_2-Receptor Antagonists*, ed. W. L. Burland, M. Alison Simkins, pp. 13–20. Amsterdam: Excerpta Medica

Peck, A. W., Fowle, A. S. E., Bye, C. 1975. A comparison of triprolidine and clemastine on histamine antagonism and performance tests in man: Implications for the mechanism of drug-induced drowsiness. *Eur. J. Clin. Pharmacol.* 8:455–63

Pellmar, T. C. 1984. Histamine and norepinephrine decrease calcium current in hippocampal pyramidal cells. *Soc. Neurosci. Abstr.* 10:203

Persinger, M. A. 1977. Mast cells in the brain: Possibilities for physiological psychology. *Physiol. Psychol.* 5:166–76

Persinger, M. A. 1979. Brain mast cell numbers in the albino rat: Sources of variability. *Behav. Neural Biol.* 25:380–86

Phillis, J. W., Kostopoulos, G. K., Odutola, A. 1975. On the specificity of histamine H_2-receptor antagonists in the rat cerebral cortex. *Can. J. Physiol. Pharmacol.* 53:1205–9

Phillis, J. W., Tebecis, A. K., York, D. H. 1968a. Histamine and some antihistamines: Their actions on cerebral cortical neurones. *Br. J. Pharmacol.* 33:426–40

Phillis, J. W., Tebecis, A. K., York, D. H. 1968b. Depression of spinal motoneurones by noradrenaline, 5-hydroxytryptamine and histamine. *Eur. J. Pharmacol.* 4:471–75

Picastoste, F., Blanco, I., Palacios, J. M. 1977. The presence of two cellular pools of rat brain histamine. *J. Neurochem.* 29:735–38

Pollard, H. S., Bischoff, S., Schwartz, J.-C. 1974. Turnover of histamine in rat brain and its decrease under barbiturate anaesthesia. *J. Pharmacol. Exp. Ther.* 190:88–99

Pollard, H. S., Bischoff, S., Llorens-Cortes, C., Schwartz, J.-C. 1976. Histidine decarboxylase and histamine in discrete nuclei

of rat hypothalamus and the evidence for mast-cells in the median eminence. *Brain Res.* 118:509–13

Pollard, H., Llorens-Cortes, C., Barbin, G., Garbarg, M., Schwartz, J.-C. 1978. Histamine and histidine decarboxylase in brain stem nuclei: Distribution and decrease after lesions. *Brain Res.* 157:178–81

Pollard, H., Pachot, I., Legrain, P., Buttin, G., Schwartz, J.-C. 1985. Development of a monoclonal antibody against L-histidine decarboxylase as a selective tool for the localization of histamine-synthesizing cells. In *Frontiers in Histamine Research, Advances in the Biosciences*, Vol. 51, ed. C. R. Ganellin, J.-C. Schwartz, pp. 103–17. London: Pergamon

Psychoyos, S. 1978. H_1- and H_2-histamine receptors linked to adenylate cyclase in cell-free preparations of guinea pig cerebral cortex. *Life Sci.* 23:2155–62

Quach, T. T., Duchemin, A. M., Rose, C., Schwartz, J.-C. 1980. ^{3}H-Glycogen hydrolysis elicited by histamine in mouse brain slices: selective involvement of H_1 receptors. *Mol. Pharmacol.* 17:301–8

Rand, M. J., Story, D. F., Wong-Dusting, H. 1982. Effects of impromidine, a specific H_2-receptor agonist and 2-(2-pyridyl)-ethylamine, an H_1-receptor agonist, on stimulation-induced release of $[^3H]$-noradrenaline in guinea-pig isolated atria. *Br. J. Pharmacol.* 76:305–11

Reite, O. B. 1972. Comparative physiology of histamine. *Physiol. Rev.* 52:778–819

Renaud, L. P. 1976. Histamine microiontophoresis on identified hypothalamic neurons: 3 patterns of response in the ventromedial nucleus of the rat. *Brain Res.* 115:339–44

Rising, T. J., Norris, D. B., Warrander, S. E., Wood, T. P. 1980. High-affinity $[^3H]$cimetidine binding in guinea pig tissues. *Life Sci.* 27:199–206

Roberts, F., Calcutt, C. R. 1983. Histamine and the hypothalamus. *Neuroscience* 9:721–39

Robinson, J. D., Green, J. P. 1962. Sulfomucopolysaccharides in brain. *Yale J. Biol. Med.* 35:248–57

Robinson, J. D., Anderson, J. H., Green, J. P. 1965. The uptake of 5-hydroxytryptamine and histamine by particulate fractions of brain. *J. Pharmacol. Exp. Ther.* 147:236–43

Robinson-White, A., Beaven, M. A. 1982. Presence of histamine and histamine-metabolizing enzyme in rat and guinea-pig microvascular endothelial cells. *J. Pharmacol. Exp. Ther.* 223:440–46

Rumore, M. M., Schlichting, D. A. 1985. Analgesic effects of antihistamines. *Life Sci.* 36:403–16

Ryan, M. J., Brody, M. J. 1972. Neurogenic and vascular stores of histamine in the dog. *J. Pharmacol. Exp. Ther.* 181:83–91

Sastry, B. S. R., Phillis, J. W. 1976a. Depression of rat cerebral cortical neurones by H_1 and H_2 histamine receptor agonists. *Eur. J. Pharmacol.* 38:269–73

Sastry, B. S. R., Phillis, J. W. 1976b. Evidence for an ascending inhibitory histaminergic pathway to the cerebral cortex. *Can. J. Physiol. Pharmacol.* 54:782–86

Satayavivad, J., Kirsten, E. B. 1977. Iontophoretic studies of histamine and histamine antagonists in the feline vestibular nuclei. *Eur. J. Pharmacol.* 41:17–26

Schentag, J. J., Cerra, F. B., Calleri, G., DeGlopper, E., Rose, J. G., et al. 1979. Pharmacokinetic and clinical studies in patients with cimetidine-associated mental confusion. *Lancet* i:177–81

Schild, H. O. 1947. pA, a new scale for the measurement of drug antagonism. *Br. J. Pharmacol.* 2:189–206

Schwabe, U., Ohga, Y., Daly, J. W. 1978. The role of calcium in the regulation of cyclic nucleotide levels in brain slices of rat and guinea-pig. *Naunyn-Schmiedebergs Arch. Pharmacol.* 302:141–51

Schwartz, J.-C. 1975. Histamine as a neurotransmitter in brain. *Life Sci.* 17:503–18

Schwartz, J.-C. 1979. Histamine receptors in brain. *Life Sci.* 25:895–912

Schwartz, J.-C., Lampart, C., Rose, C. 1970. Properties and regional distribution of histidine decarboxylase in rat brain. *J. Neurochem.* 17:1527–34

Schwartz, J.-C., Pollard, H., Bischoff, S., Rehault, M. C., Verdiere-Sahuque, M. 1971. Catabolism of ^{3}H-histamine in the rat brain after intracisternal administration. *Eur. J. Pharmacol.* 16:326–35

Schwartz, J.-C., Lampart, C., Rose, C. 1972. Histamine formation in rat brain in vivo: Effects of histidine loads. *J. Neurochem.* 19:801–10

Schwartz, J.-C., Baudry, M., Chast, F., Pollard, H., Bischoff, S., et al. 1973. Histamine in the brain: Importance of transmethylation processes and their regulation. In *Metabolic Regulation and Functional Activity in the Central Nervous System*, ed. E. Genazzani, H. Herken, pp. 172–84. Berlin: Springer-Verlag

Schwartz, J.-C., Barbin, G., Bischoff, S., Garbarg, M., Pollard, H., et al. 1975. Histamine in an ascending pathway in the brain: Localization, turnover and effects of psychopharmacological agents. See El-Ackad & Brody 1975a, pp. 575–83

Schwartz, J.-C., Barbin, G., Garbarg, M., Pollard, H., Rose, C., et al. 1976a. Neurochemical evidence for histamine acting as a transmitter in mammalian brain. *Adv. Biochem. Psychopharmacol.* 15:111–26

Schwartz, J.-C., Baudry, M., Bischoff, S., Martres, M.-P., Pollard, H., et al. 1976b. Pharmacological studies of histamine as a central neurotransmitter. In *Drugs and Central Synaptic Transmission*, ed. P. B. Bradley, B. N. Dhawan, pp. 371–82. Baltimore: Univ. Park Press

Schwartz, J.-C., Barbin, G., Baudry, M., Garbarg, M., Martres, M.-P., et al. 1979a. Metabolism and functions of histamine in the brain. *Curr. Dev. Psychopharmacol.* 5:173–261

Schwartz, J.-C., Barbin, G., Garbarg, M., Llorens, C., Palacios, J.-M., et al. 1979b. Histaminergic systems in brain. *Adv. Pharmacol. Ther.* 2:171–80

Schwartz, J.-C., Pollard, H., Quach, T. T. 1980. Histamine as a neurotransmitter in mammalian brain: Neurochemical evidence. *J. Neurochem.* 35:26–33

Schwartz, J.-C., Barbin, G., Duchemin, A.-M., Garbarg, M., Pollard, H., et al. 1981. Functional role of histamine in the brain. In *Neuropharmacology of Central Nervous System and Behavioral Disorders*, ed. G. C. Palmer, pp. 539–70. New York: Academic

Schwartz, J.-C., Barbin, G., Duchemin, A. M., Garbarg, M., Llorens, C., et al. 1982. Histamine receptors in the brain and their possible functions. In *Pharmacology of Histamine Receptors*, ed. C. R. Ganellin, M. E. Parsons, pp. 351–91. Bristol: Wright

Segal, M. 1980. Histamine produces a Ca^{2+}-sensitive depolarization of hippocampal pyramidal cells in vitro. *Neurosci. Lett.* 19:67–71

Segal, M. 1981. Histamine modulates reactivity of hippocampal CA3 neurons to afferent stimulation in vitro. *Brain Res.* 213:443–48

Segal, M. 1982. Norepinephrine modulates reactivity of hippocampal cells to chemical stimulation in vitro. *Exp. Neurol.* 77:86–93

Seppala, T., Nuotto, E., Korttila, K. 1981. Single and repeated dose comparison of three antihistamines and phenylpropanolamine: Psychomotor performance and subjective appraisals of sleep. *Br. J. Clin. Pharmacol.* 12:179–88

Siggins, G. R., Hoffer, B. J., Bloom, F. E. 1971. Studies of norepinephrine-containing afferents to Purkinje cells of rat cerebellum. III. Evidence for mediation of norepinephrine effects by cyclic 3',5'-adenosine monophosphate. *Brain Res.* 25:535–53

Silverstone, P. H. 1984. Ranitidine and confusion. *Lancet* i:1071

Smith, I. R., Cleverley, M. T., Ganellin, C. R., Metters, K. M. 1980. Binding of [^{3}H]cimetidine to rat brain tissue. *Agents Actions* 10:422–26

252 PRELL & GREEN

Snider, R. M., McKinney, M., Forray, C., Richelson, E. 1984. Neurotransmitter receptors mediate cyclic GMP formation by involvement of arachidonic acid and lipoxygenase. *Proc. Natl. Acad. Sci. USA* 81:3905–9

Snyder, S. H., Bennett, J. P. 1976. Neurotransmitter receptors in the brain: Biochemical identification. *Ann. Rev. Physiol.* 38:153–75

Snyder, S. H., Taylor, K. M. 1972. Histamine in the brain: A neurotransmitter? In *Perspectives in Neuropharmacology—A Tribute to Julius Axelrod*, ed. S. H. Snyder, pp. 43–73. New York: Oxford Univ. Press

Snyder, S. H., Baldessarini, R. J., Axelrod, J. 1966. A sensitive and specific enzymatic isotopic assay for tissue histamine. *J. Pharmacol. Exp. Ther.* 153:544–49

Snyder, S. H., Brown, B., Kuhar, M. J. 1974. The subsynaptosomal localization of histamine, histidine decarboxylase and histamine methyltransferase in rat hypothalamus. *J. Neurochem.* 23:37–45

Spreeg, K. V., Wang, S., Avant, G. R., Parker, R., Schenker, S. 1981. In vitro antagonism of benzodiazepine binding to cerebral receptors by H_1 and H_2 antihistamines. *J. Lab. Clin. Med.* 97:112–22

Sperk, G., Hortnagl, H., Reither, H., Hornykiewicz, O. 1981a. Evidence for neuronal localization of histamine-N-methyltransferase in rat brain. *J. Neurochem.* 37:525–26

Sperk, G., Hortnagl, H., Reither, H., Hornykiewicz, O. 1981b. Changes in histamine in the rat striatum following local injection of kainic acid. *Neuroscience* 6:2669–75

Stein, C., Weinreich, D. 1982. An in vitro characterization of γ-glutamylhistamine synthetase: A novel enzyme catalyzing histamine metabolism in the central nervous system of the marine mollusk, *Aplysia californica*. *J. Neurochem.* 38:204–14

Stein, C., Weinreich, D. 1983. Metabolism of histamine in the CNS of *Aplysia californica*: Cellular distribution of γ-glutamylhistamine synthetase. *Comp. Biochem. Physiol.* 74C:79–83

Steinbusch, H. W. M., Mulder, A. H. 1985. Localization and projections of histamine-immunoreactive neurons in the central nervous system of the rat. In *Frontiers in Histamine Research, Advances in the Biosciences*, Vol. 51, ed. C. R. Ganellin, J.-C. Schwartz, pp. 119–30. London: Pergamon

Study, R. E., Greengard, P. 1978. Regulation by histamine of cyclic nucleotide levels in sympathetic ganglia. *J. Pharmacol. Exp. Ther.* 207:767–78

Sturrock, R. R. 1980. A morphological lifespan study of neurolipomastocytes in various regions of the mouse forebrain.

Neuropathol. Appl. Neurobiol. 6:211–19

Subramanian, N., Mulder, A. H. 1976. Potassium-induced release of tritiated histamine from rat brain tissue slices. *Eur. J. Pharmacol.* 35:203–6

Subramanian, N., Whitmore, W. L., Seidler, F. J., Slotkin, T. A. 1980. Histamine stimulates brain phospholipid turnover through a direct, H-1 receptor mediated mechanism. *Life Sci.* 27:1315–19

Subramanian, N., Whitemore, W. L., Seidler, F. J., Slotkin, T. A. 1981. Ontogeny of histaminergic neurotransmission in the rat brain: Concomitant development of neuronal histamine, H-1 receptors, and H-1 receptor-mediated stimulation of phospholipid turnover. *J. Neurochem.* 36:1137–41

Swazey, J. P. 1974. *Chlorpromazine in Psychiatry: A Study of Therapeutic Innovation.* Cambridge: MIT Press

Sweatman, P., Jell, R. M. 1977. Dopamine and histamine sensitivity of rostral hypothalamic neurones in the cat: Possible involvement in thermoregulation. *Brain Res.* 127:173–78

Tagami, H., Sunami, A., Akagi, M., Tasaka, K. 1984. Effects of histamine on the hippocampal neurons in guinea-pigs. *Agents Actions* 14:538–42

Takai, Y., Kikkawa, U., Kaibuchi, K., Nishizuka, Y. 1984. Membrane phospholipid metabolism and signal transduction for protein phosphorylation. *Adv. Cyclic Nucleotide Protein Phosphoryl. Res.* 18:119–58

Takeda, N., Inagaki, S., Shiosaka, S., Taguchi, Y., Oertel, W. H., et al. 1984a. Immunohistochemical evidence for the coexistence of histidine decarboxylase-like and glutamate dehydrogenase-like immunoreactivities in nerve cells of the magnocellular nucleus of the posterior hypothalamus of rats. *Proc. Natl. Acad. Sci. USA* 81:7647–50

Takeda, N., Inagaki, S., Taguchi, Y., Tohyama, M., Watanabe, T., et al. 1984b. Origins of histamine-containing fibers in the cerebral cortex of rats studied by immunohistochemistry with histidine decarboxylase as a marker and transection. *Brain Res.* 323:55–63

Takeuchi, H., Yokoi, I., Mori, A., Horisaka, K. 1976. Effets de l'histamine et de ses derives sur l'excitabilité d'un neurone geant identifiable d'*Achatina fulca* Férussac. Un récepteur histaminergique different d'H_1 et d'H_2. *CR Acad. Sci. Paris* 170:1118–26

Taylor, J. E., Richelson, E. 1979. Desensitization of histamine H_1 receptor-mediated cyclic GMP formation in mouse neuroblastoma cells. *Mol. Pharmacol.* 15:462–71

Taylor, J. E., Richelson, E. 1980. High-affinity binding of tricyclic antidepressants to histamine H_1-receptors: Fact and Artifact. *Eur. J. Pharmacol.* 67:41–46

Taylor, J. E., Richelson, E. 1982. High-affinity binding of [^{3}H] doxepin to histamine H_1-receptors in rat brain: Possible identification of a subclass of histamine H_1-receptors. *Eur. J. Pharmacol.* 78:279–85

Taylor, K. M. 1975. Brain histamine. *Handb. Psychopharmacol.* 3:327–79

Taylor, K. M., Snyder, S. H. 1972. Dynamics of the regulation of histamine levels in mouse brain. *J. Neurochem.* 19:341–54

Taylor, K. M., Snyder, S. H. 1973. The release of histamine from tissue slices of rat hypothalamus. *J. Neurochem.* 21:1215–23

Tran, V. T., Snyder, S. H. 1981. Histidine decarboxylase: Purification from fetal rat liver, immunologic properties, and histochemical localization in brain and stomach. *J. Biol. Chem.* 256:680–86

Tran, V. T., Lebovitz, R., Toll, L., Snyder, S. H. 1981. [^{3}H]Doxepin interactions with histamine H_1-receptors and other sites in guinea pig and rat brain homogenates. *Eur. J. Pharmacol.* 70:501–9

Tuomisto, L., Eriksson, L., Fyhrquist, F. 1984. Plasma vasopressin levels after i.c.v. infusion of histamine agonists in the conscious goat. *Agents Actions* 14:558–60

Turner, J. D., Cottrell, G. A. 1977. Properties of an identified histamine-containing neurone. *Nature* 267:447–48

Turner, J. D., Powell, B., Cottrell, G. A. 1980. Morphology and ultrastructure of an identified histamine-containing neuron in the central nervous system of the pond snail, *Lymnaea stagnalis* L. *J. Neurocytol.* 9:1–14

Tuttle, R. S. 1967. Physiological release of histamine-^{14}C in the pyramidal cat. *Am. J. Physiol.* 213:620–24

Uchida, M. 1980. Histamine-induced decrease of membrane-bound calcium ions in the membrane fraction of rabbit taenia coli. *Eur. J. Pharmacol.* 64:357–60

van den Brink, F. G., Lein, E. J. 1978. Competitive and non-competitive antagonism. *Handb. Exp. Pharmacol.* 18(2):333–67

Verdiere, M., Rose, C., Schwartz, J.-C. 1975. Synthesis and release of histamine studied on slices from rat hypothalamus. *Eur. J. Pharmacol.* 34:157–68

Vyas, S., Verma, S. 1981. The direct excitatory effects of cimetidine on the smooth muscle of guinea pig ileum. *Agents Actions* 11:193–95

Wada, H., Watanabe, T., Yamatodani, A., Maeyama, K., Itoi, N., et al. 1985. Physiological functions of histamine in the brain.

In *Frontiers in Histamine Research, Advances in the Biosciences*, Vol. 51, ed. C. R. Ganellin, J.-C. Schwartz, pp. 225–35. London: Pergamon

Wahlestedt, C., Skagerberg, G., Hakanson, R., Sundler, F., Wada, H., et al. 1985. Spinal projection of hypothalamic histidine decarboxylase-immunoreactive neurones. *Agents Actions* 16:231–33

Watanabe, T., Taguchi, Y., Hayashi, H., Wada, H., Tanaka, J., Shiosaka, S., et al. 1983. Evidence for the presence of a histaminergic neuron system in the rat brain: An immunohistochemical analysis. *Neurosci. Lett.* 39:249–54

Watanabe, T., Taguchi, Y., Shiosaka, S., Tanaka, J., Kubota, H., et al. 1984. Distribution of the histaminergic neuron system in the central nervous system of rats: A fluorescent immunohistochemical analysis with histidine decarboxylase as a marker. *Brain Res.* 295:13–25

Wauquier, A., van den Broeck, W. A. E., Awouters, F., Janssen, P. A. J. 1981. A comparison between astemizole and other antihistamines on sleep-wakefulness cycles in dogs. *Neuropharmacology* 20:853–59

Weinreich, D. 1977. Synaptic responses mediated by identified histamine-containing neurones. *Nature* 267:854–56

Weinreich, D. 1978. Histamine-containing neurons in *Aplysia*. In *Biochemistry of Characterized Neurons*, ed. N. Osborne, pp. 153–75. New York: Plenum

Weinreich, D. 1979. γ-Glutamylhistamine: A major product of histamine metabolism in ganglia of the marine mollusk, *Aplysia californica*. *J. Neurochem.* 32:363–69

Weinreich, D. 1985. Neurobiology of a histaminergic neuron in the CNS of the mollusk *Aplysia californica*. In *Frontiers in Histamine Research, Advances in the Biosciences*, Vol. 51, ed. C. R. Ganellin, J.-C. Schwartz, pp. 205–14. London: Pergamon

Weinreich, D., Rubin, L. 1981. Irreversible inhibitors of histidine decarboxylase in *Aplysia* ganglia: A tool for the identification of histaminergic synapses. *Comp. Biochem. Physiol.* 69C:383–85

Weinreich, D., Weinreich, C. A. 1977. Endogenous levels of histidine in histamine containing neurons and other identified nerve cells of *Aplysia californica*. *Comp. Biochem. Physiol.* 56C:1–4

Weinreich, D., Yu, Y.-T. 1977. The characterization of histidine decarboxylase and its distribution in nerves, ganglia and in single neuronal cell bodies from the CNS of *Aplysia californica*. *J. Neurochem.* 28:361–69

Weinreich, D., Weiner, C., McCaman, R. 1975. Endogenous levels of histamine in single neurons isolated from CNS of

Aplysia californica. Brain Res. 84:341–45

Weiss, K. R., Cohen, J. L., Kupfermann, I. 1978a. Modulatory control of buccal musculature by a serotonergic neuron (metacerebral cell) in *Aplysia. J. Neurophysiol.* 41:181–203

Weiss, K. R., Shapiro, E., Koester, J., Kupfermann, I. 1978b. A histaminergic synaptic potential produced by a voltage-dependent apparent decrease of conductance in the metacerebral cell of *Aplysia. Soc. Neurosci. Abstr.* 4:657

Weiss, K. R., Chiel, H. J., Kupfermann, I. 1983. The histaminergic neuron C2 of *Aplysia* is a mechanoafferent cell whose output can be gated synaptically. *Soc. Neurosci. Abstr.* 9:913

Wilcox, B. J., Seybold, V. S. 1982. Localization of neuronal histamine in rat brain. *Neurosci. Lett.* 29:105–10

Yamatodani, A., Maeyama, K., Watanabe, T., Wada, H., Kitamura, Y. 1982. Tissue distribution of histamine in a mutant mouse deficient in mast cells: Clear evidence for the presence of non-mast-cell histamine. *Biochem. Pharmacol.* 31:305–9

Young, A. B., Pert, C. D., Brown, D. G., Taylor, K. M., Snyder, S. H. 1971. Nuclear localization of histamine in neonatal rat brain. *Science* 173:247–49

Ann. Rev. Neurosci. 1986. 9 : 255–76

GENETICS AND MOLECULAR BIOLOGY OF IONIC CHANNELS IN *DROSOPHILA*

Mark A. Tanouye, C. A. Kamb, and Linda E. Iverson

Division of Biology, California Institute of Technology, Pasadena, California 91125

Lawrence Salkoff

Department of Anatomy and Neurobiology, Washington University School of Medicine, St. Louis, Missouri 63110

INTRODUCTION

In this review we examine mutations that alter electrical excitability in the nervous system of the fruitfly, *Drosophila melanogaster*, and discuss how these mutations may be used to approach the molecular basis of ionic channel function.

Electrical activity in the nervous system depends on the transient and selective movement of ions across nerve cell membranes. It is best illustrated by the action potential (Hodgkin & Huxley 1952), where the axonal membrane becomes selectively permeable to Na^+ to generate the rising phase. The falling phase is due to loss of Na^+ permeability (inactivation) and increased permeability to K^+. The key to action potential genesis lies in the nature of ionic permeability changes. Membrane macromolecules that act as ionic channels underlie these changes. In the low permeability condition, the channels are closed. Following an activation event, channels open into a conformation that allows ions to flow through (see Hille 1984, for a recent review).

Many of the most important questions in excitable membrane neurobiology now focus on the molecular nature of ionic channels. One would like to know the molecular features responsible for different ionic

255

0147–006X/86/0301–0255$02.00

selectivities and how channels open and close. Also, because excitability varies among different types of neurons (see Bullock & Horridge 1965, Serrano 1982, for example), they must have different distributions and/or densities of channels. One would like to approach the problem of how a given cell controls channel distributions. To address these issues, it is imperative to isolate channel proteins and the genes that encode them.

Molecular analyses of channels such as the acetylcholine receptor and the Na^+ channel have been facilitated by specific ligands such as α-bungarotoxin (Vandlen et al 1979, Conti-Tronconi & Raftery 1982) and tetrodotoxin (Agnew et al 1978, Barchi 1982). These toxins bind the channels with high affinity, allowing biochemical purification. Purification of channel molecules has led to the cloning of ACh receptor (Noda et al 1983) and Na^+ channel (Noda et al 1984) genes.

Because most ionic channels are not abundant molecules and high affinity ligands are not generally available, a majority have not been characterized biochemically. For example, although K^+-selective channels have been well characterized electrophysiologically (reviewed in Adams et al 1980, Hagiwara 1983, Latorre & Miller 1983), their structure is unknown. Pharmacological agents such as tetraethylammonium ion (Armstrong & Hille 1972) and the aminopyridines (Yeh et al 1976) have been useful in studying K^+ channel function but lack the affinity and specificity required for channel purification.

An alternative approach for studying the molecular basis of ionic channels is to make use of genetic mutations. A major advantage of this method is that prior biochemical purification of the channel is not required. Instead, mutations are used as a type of functional assay for the molecule in question. The complete rationale for the approach may be outlined as follows:

1. Mutations affecting a particular ionic channel are identified using a combination of behavioral, electrophysiological, and pharmacological methods.
2. Genetic methods are used to identify the gene encoding the channel and to provide markers or genetic tags to facilitate cloning of the gene.
3. The gene is cloned using recombinant DNA methods. The nucleotide sequence is determined from the cloned DNA. The amino acid sequence of the channel protein may be deduced from the nucleotide sequence.
4. Amino acid sequence data alone may be sufficient to allow the eventual biochemical purification of the channel molecule. For example, synthetic peptides might be synthesized and used as ionic channel specific antigens to generate antibodies that bind channel proteins. Channels may then be purified by immunoaffinity chromatography.

This type of approach is being used to study ionic channels in *Drosophila*. A number of excitability mutants have been characterized electrophysiologically and genetically. Genetic analyses have provided a basis for molecular cloning, and several genes are currently being cloned. This paper reviews these genes.

IONIC CHANNELS IN *DROSOPHILA* NERVE AND MUSCLE

Membrane excitability in normal *Drosophila* is similar to that of other arthropods (see, for example, Julian et al 1962, Shrager 1974, Pichon & Boistel 1976, Connor 1975, Quinta-Ferreira et al 1982, Cull-Candy 1967, Crawford & McBurney 1976, Hagiwara & Byerly 1981, Washio 1972, Patlak 1976, Ashcroft & Stanfield 1982). Propagated axonal action potentials in *Drosophila* are regenerative Na^+ spikes that are blocked by tetrodotoxin (Jan et al 1977, Wu & Ganetzky 1980, Tanouye et al 1981, Tanouye & Ferrus 1985). K^+-selective channels are important for action potential repolarization (Jan et al 1977, Tanouye et al 1981, Ganetzky & Wu 1983). Acetylcholine is a major central nervous system neurotransmitter (Hall & Greenspan 1979, Wu et al 1983b,c). Although other central nervous system transmitters undoubtedly exist, they have not been characterized. Glutamate is the transmitter at the neuromuscular junction, and its release is Ca^{2+}-dependent (Jan & Jan 1967a,b, Ikeda 1980). Inhibitory neuromuscular transmission has not been described. The major voltage-dependent inward current of *Drosophila* muscle is carried by Ca^{2+} (Ikeda 1980, Salkoff & Wyman 1983a,b, Wu & Haugland 1985). Outward currents in muscle are via K^+-selective channels (Suzuki & Kano 1977, Salkoff & Wyman 1983a,b, Wu & Haugland 1985).

EXCITABILITY MUTANTS IN *DROSOPHILA*

Mutants in several *Drosophila* genes have been implicated in abnormal ionic channel function (for details, see below). All of the mutants were originally isolated as behavioral mutants. Three general classes of mutants have been described.

Leg-shaking Mutants

Normal *Drosophila* are immobile under ether anesthesia. In contrast, several behavioral mutants twitch their abdomens and shake all appendages when etherized. Of these behaviors, leg-shaking is the easiest to score and serves as the basis of mutant isolation (Kaplan & Trout 1969, Trout & Kaplan 1973). Mutants of two loci, *Shaker* and *ether-a-go-go*, appear to have abnormalities in K^+-selective ionic channels (see below). Other leg-shaking loci, *Hyperkinetic* and *Shudderer* (Kaplan & Trout 1969, Williamson 1982), have not been extensively characterized.

Temperature-sensitive Paralytics

A number of behavioral mutants are temperature-dependent conditional mutants (Suzuki et al 1971, Grigliatti et al 1973, Siddiqi & Benzer 1976, Wu et al 1978). At a permissive temperature (usually room temperature) the animals behave normally. At a restrictive temperature (usually high temperature) the animals are paralyzed. The original impetus for examining temperature-sensitive mutations was that they might encode thermolabile channels that were conductive at permissive temperatures and non-conductive at restrictive temperatures. Mutants of three loci, *no action potential*, *paralyzed*, and *seizure*, may have Na^+ channel defects (see below). Other paralytics include: *shibire*, *comatose*, and *out-cold*.

Bang-sensitive Mutants

The mutants, *bang-sensitive*, *bang-senseless*, *technical knockout*, *knockdown*, and *easily-shocked* (Ganetzky & Wu 1982b), are paralyzed following a sudden jolt or vibration of the culture vial (usually applied by a vortex mixer). A specific ionic channel defect for these mutants has not been demonstrated; however, double mutant genetic analyses suggest alterations in membrane excitability (see below).

SHAKER

The best characterized *Drosophila* excitability mutants are those of the *Shaker* (*Sh*) locus. The main interest in *Sh* comes from electrophysiological evidence demonstrating that *Sh* mutants encode abnormal K^+ channels. Classical genetic analysis suggests that *Sh* is a complex genetic locus and provides the basis for cloning the *Sh* gene(s). Molecular experiments indicate that part of the *Sh* complex has been cloned.

Behavior of Sh Mutants

Sh mutants are behavioral mutants that shake their legs vigorously under ether anesthesia (Catsch 1944, Kaplan & Trout 1969, Trout & Kaplan 1973). At least 25 *Sh* alleles have been isolated, virtually all on the basis of abnormal leg-shaking behavior. In addition to leg-shaking, other behavioral defects such as wing-scissoring, abdominal spasms, and antennal twitching are observed in etherized animals (Ganetzky & Wu 1982b, Tanouye & Ferrus 1985). Fully awake mutant flies twitch and shudder on occasion. The leg-shaking defect does not appear to be caused exclusively by synaptic interactions within the central nervous system. If one cuts off the legs of a *Sh* fly, the severed legs continue to shake, suggesting that defects in peripheral nerve and muscle are sufficient to account for the leg-shaking phenotype.

Genetic and Cytogenetic Localization of Sh

Sh mutations occur in a small region (Figure 1) that includes several bands within division 16F on the salivary X chromosome (Tanouye et al 1981). Fine structure genetic mapping suggests that several loci in the 16F region may affect nerve excitability (see below). These loci are collectively called the *Shaker* gene complex. The term "*Sh* locus" generally refers to the site occupied by the Sh^{KS133} mutation. Most *Sh* mutations are dominant for the leg-shaking phenotype, although some weak alleles (e.g. Sh^{rK0120}, Sh^{E62}) may be partially dominant or recessive (Jan et al 1977, Tanouye et al 1981, Jan et al 1983, Tanouye & Ferrus 1985). Because they are leg-shaking

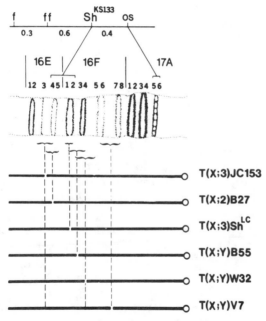

Figure 1 Genetic and cytogenetic localization of *Sh*. Sh^{KS133} maps by recombination between the genes *ff* and *os* on the X chromosome. A drawing of the salivary chromosome for this region is shown. Six chromosomal rearrangements used to localize *Sh* cytogenetically are indicated by *open circles* representing the centromere (proximal end). The cytogenetic locations of their breakpoints are shown, with *brackets* indicating the degree of uncertainty. Note that the translocated chromosomes also contain other breakpoints not associated with *Sh*, which have been omitted for clarity. Adapted from Tanouye et al (1981).

dominants, *Sh* mutations cannot be easily organized into genetic complementation groups. Thus, *Sh* mutant alleles refer to any leg-shaking mutation that maps genetically to the region occupied by the *Sh* complex. Fine structure genetic mapping on a limited number of alleles shows that many, but not all, of the mutations map to the *Sh* locus (see below).

The genetic location of Sh^{KS133} (map position 1–57.6) is determined by recombination analyses that place it 0.9 map units to the right of *forked. Sh* is located 0.6 map units to the right of the marker *fluff* (ff) and 0.4 map units left of *out-stretched-small eyes* (*os*) (Tanouye et al 1981). The physical location of Sh^{KS133} is determined by cytogenetic methods using translocations that have chromosome breaks at different points in the 16E and 16F regions of the salivary X chromosome. Sh^{KS133} maps to the right of 16E2-4, which is a breakpoint of translocation $T(X;3)JC153$. It maps to the left of 16F1-4, which is the breakpoint of translocation $T(X;Y)B55$.

Evidence that Sh Encodes K^+ Channels

Several studies, using different electrophysiological preparations, have suggested that *Sh* mutations encode abnormal K^+ channels or channel subunits. The evidence is of two general types: 1. Pharmacological agents are used to examine *Sh* defects in neurons. K^+ channel blockers mimic *Sh* defects in normal flies. Agents that block other channels eliminate other possibilities for the *Sh* defect. 2. *Sh* muscle may be examined by voltage clamp. The individual ionic currents are separated, and a specific K^+ conductance is found to be altered by *Sh* mutations.

Jan et al (1977) first described physiological effects of *Sh* mutations at the larval neuromuscular junction. The Sh^{KS133} mutant allele causes abnormally large and asynchronous transmitter release under low Ca^{2+} conditions (Figure 2). The abnormal release is due to prolonged Ca^{2+} conductance at the mutant nerve terminal. The time course of the conductance increase is measured by iontophoretically applying Ca^{2+} pulses to the nerve terminal at various times following nerve stimulation. In normal larvae, the increase in Ca^{2+} conductance following nerve stimulation lasts 1–2 ms. In Sh^{KS133} it lasts 60 ms.

Three possible explanations for the *Sh* defect were considered by Jan et al (1977). They were (*a*) a prolonged presynaptic depolarization due to abnormal Na^+ conductance, (*b*) defective K^+ channels that fail to repolarize the nerve terminal properly, and (*c*) abnormal Ca^{2+} channels that remain open after terminal repolarization. The possibility of a Na^+ channel defect was eliminated by using the Na^+ channel blocker tetrodotoxin. Since, action potentials had been eliminated, transmitter release was evoked by direct depolarization of the terminals. Under these conditions, the Ca^{2+} conductance increase was still prolonged in the mutant compared to normal, suggesting that regenerative Na⁺ channels are not necessary for expression of the *Sh* defect. The possibility of abnormal Ca^{2+} channels was tested by applying repolarizing pulses to mutant nerve terminals electrotonically and showing that Ca^{2+} conductance can be turned off. These results suggest that the mutant presynaptic terminal repolarizes slowly following an action potential and that defective Na⁺ and Ca^{2+} channels are not the primary cause.

Evidence that the repolarization defect is due to an abnormal K⁺ conductance comes from experiments using the K⁺ channel blocker, 4-aminopyridine (Jan et al 1977, Ganetzky & Wu 1983). When the drug is applied to normal larvae, it mimics the Sh^{KS133} neuromuscular transmission defect. In contrast, the drug has little or no effect on mutant larvae (Ganetzky & Wu 1983).

Direct evidence for a repolarization defect in mutant nerve fibers comes from intracellular recordings of *Sh* action potentials (Tanouye et al 1981, Tanouye & Ferrus 1985b). Recordings made from the cervical giant fiber

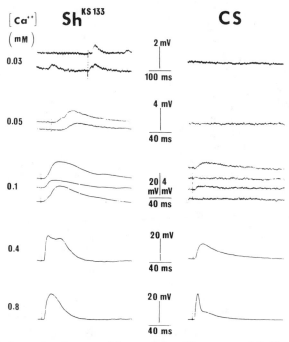

Figure 2 Excitatory junction potentials recorded at different external Ca^{2+} concentrations for normal (CS) and mutant (Sh^{KS133}) larvae. The abnormally large and broad EJPs seen in the mutant can also be produced in normal larvae by treating the preparation with 4-AP. Reproduced with permission from Jan et al (1977).

axons of *Drosophila* adults show that Sh^{KS133} action potential repolariz-ation is delayed (Figure 3). Whereas action potential durations in normal animals are about 0.5 ms, durations in Sh^{KS133} flies are about 5 ms. Ca^{2+} channel blockers such as Co^{2+} and Mn^{2+} fail to remove the mutant repolarization delay, so the defect is not due to abnormally large inward Ca^{2+} currents acting to keep the nerve depolarized. As was the case for neuromuscular transmission, Sh^{KS133} action potential defects are mimicked by applying 4-aminopyridine to normal giant fibers. The drug has little effect when applied to Sh^{KS133} mutant giant fibers.

The most complete analyses of *Sh* mutants are voltage clamp experi-ments on *Drosophila* pupal and adult flight muscle by Salkoff & Wyman (1981a,b, 1983a,b) and on larval muscle by Wu and co-workers (1983a, 1985). Normal, developing pupal flight muscle contains only a single voltage-activated ionic current, a fast transient K^+ current (Salkoff 1983a,b, Salkoff & Wyman 1981a,b, 1983a,b, Jan et al 1983). The current is similar to A-type K^+ current, first characterized in molluscan neurons (Connor & Stevens 1971, Neher 1971). A-type K^+ current activates rapidly and inactivates during a maintained voltage step pulse. In Sh^{KS133} mutant pupal muscle, the current is completely absent (Figure 4). The defect is seen only in A-type K^+ current. Normal mature adult flight muscle contains, in addition to A-type K^+ current, a number of other ionic currents. They are a voltage-activated inward Ca^{2+} current, a voltage-activated delayed recti-fier K^+ current, a Ca^{2+}-activated transient K^+ current, and a glutamate-activated synaptic current. Of these different ionic currents, only the A-type K^+ current is affected by *Sh* mutations. The Sh^{KS133} A-type K^+ current defect may be mimicked by the application of 3- or 4-aminopyridine to pupal or adult flight muscle.

Similar results are obtained from voltage clamp preparations of larval muscle (Wu et al 1983a, Wu & Ganetzky 1984, Wu & Haugland 1985). Normal larval muscle fibers contain four ionic currents similar to those

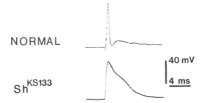

NORMAL

Sh^{KS133}

40 mV

4 ms

Figure 3   Typical action potentials recorded from normal and Sh^{KS133} mutant flies. An intracellular microelectrode records the action potential from the cervical giant fiber of adult *Drosophila*, evoked by a 0.1 ms stimulus to the brain. Sh^{KS133} shows a delay in repolarization. The mutant phenotype is mimicked in normal animals by treatment with 4-AP. From Tanouye et al (1981).

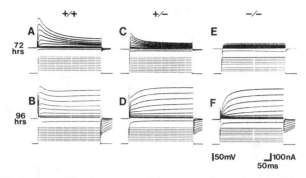

Figure 4 Comparison of flight muscle membrane properties during pupal development in normal ($+/+$), heterozygous $+/Sh$ ($+/-$), and homozygous mutant Sh/Sh ($-/-$) pupae. The alleles, Sh^{KS133}, Sh^{102}, and one stock of Sh^{rK0120}, as shown here, all have a similar phenotype. (*A*) In normal pupae, A-type K^+ current is mature at 72 hr of pupal development. (*B*) Delayed rectifier K^+ current is mature and is seen added to A-type K^+ current at 96 hr of development. (*C, D*) Membrane current response of heterozygous pupae. At 72 and 96 hr, only half the A-type K^+ current is present compared to normal. Delayed rectifier K^+ current is unaffected by the mutant gene. At 96 hr, the full current response of the delayed rectifier K^+ current is present in all animals. (*E, F*) Membrane current response of homozygous mutant pupae. A-type K^+ current is not present at either 72 or 96 hr. In *F* at 96 hr, the full current response of the delayed rectifier K^+ current is present. In *A–F*, the leak current elicited by a hyperpolarizing pulse to -180 mV is shown (below *baseline trace*) but the voltage record is not shown. The mutant phenotype is mimicked in normal animals by treatment with 4-AP. Adapted from Salkoff & Wyman (1981a).

described for adult flight muscle: a glutamate-activated synaptic current, a Ca^{2+} current, an A-type K^+ current, and a delayed rectifier K^+ current. A Ca^{2+}-activated K^+ current has not been described. In Sh^{KS133} larval muscle, A -type K^+ current is eliminated with no alterations of the other ionic currents.

Other Sh Alleles

If a gene encodes a single function, like a particular K^+ conductance, then: (*a*) all mutant alleles of the gene should show a common general defect, and (*b*) the specific defects may not be identical, since independently-derived alleles are probably mutated at different places in the gene(s). Experiments on different alleles indicate a range of mutant defects.

All of the several different *Sh* mutants tested appear to have the same general defect. In voltage clamp experiments, mutants all have defective A-type K^+ currents (see, for example, Salkoff & Wyman 1981a, Salkoff 1983b, Wu & Haugland 1985). Other ionic conductances are not affected by any *Sh* mutation. Giant fiber recordings show that *Sh* defects in different alleles are always limited to action potential repolarization (Tanouye et al 1981,

Tanouye & Ferrus 1985). Studies on larval neuromuscular transmission show that all *Sh* mutations tested cause abnormal transmitter release (Jan et al 1977, 1983, Ganetzky & Wu 1982b, 1983).

Although *Sh* mutants have the same general defect, there are allelic differences. Tanouye & Ferrus (1985) described these differences in action potential repolarization. Five alleles had delayed repolarization and increased action potential durations. Going from most to least extreme, these alleles were: $Sh^{102} \geq Sh^{KS133} > Sh^M > Sh^{E62} > Sh^{rKO120}$. Compared to normal action potentials, the durations in the extreme mutants, Sh^{102} and Sh^{KS133}, were longer by an order of magnitude or more. In Sh^{rKO120} the duration was increased by about a factor of two. Behavioral differences among the different mutants were difficult to quantify, but were in general agreement with Sh^{102} being the most extreme leg-shaker and Sh^{rKO120} being the least extreme. A sixth allele, Sh^5, differed from the others. Rather than having delayed repolarization, this mutant showed incomplete repolarization and multiple spikes.

Sh allelic differences have also been shown by voltage clamp. Results are in general agreement with action potential recordings and show that Sh^{102}, Sh^{KS133}, and Sh^M are the most extreme alleles. Each of these mutants completely lacks A-type K^+ currents (Salkoff & Wyman 1981a, Salkoff 1983b, Wu & Haugland 1985). Sh^{E62} is less extreme; currents are present but reduced in animals carrying this allele (Jan et al 1983). Voltage clamp results on Sh^{rKO120} are not completely consistent. Wu & Haugland (1985) found that A-type K^+ currents in Sh^{rKO120} mutants are qualitatively similar to those of normal flies except for a slight change in voltage dependence of inactivation. Other analyses (Salkoff & Wyman 1981a, Salkoff 1983b) showed that Sh^{rKO120} mutants completely lack A-type K^+ currents. Some of these results appear to be due to different Sh^{rKO120} stocks (Wu & Haugland 1985, L. Salkoff, unpublished results; see also below).

The results of voltage clamp experiments on Sh^5 mutants are also not completely consistent. Salkoff & Wyman (1981a, Salkoff 1983b) showed that in Sh^5 pupal flight muscle, A-type K^+ currents have normal activation and peak conductance. However, inactivation kinetics and recovery from inactivation are altered (see below). Wu & Haugland (1985) reported that in Sh^5 larval muscle, A-type K^+ currents are reduced in amplitude and have alterations in both activation and inactivation voltages (see below).

Evidence that Sh Mutants Produce an Altered Gene Product

Sh defects could be due to underproduction of a molecule; e.g. a component of a channel. Alternatively, the defect(s) could be due to abnormal channel molecules produced by the mutated gene. Two types of evidence suggest that at least some *Sh* mutations encode abnormal molecules. These are: (*a*) voltage clamp experiments on Sh^5 mutants, and (*b*) gene dosage analyses.

Voltage clamp preparations of Sh^5 muscle provide electrophysiological evidence of an altered gene product because defects seen are most easily explained as alterations in intrinsic channel properties. In Sh^5 pupal muscle, the peak conductance for A-type K^+ currents is similar to normal (Salkoff & Wyman 1981a, Salkoff 1983b). However, the kinetics of the currents are altered in the mutant, so that channels both inactivate more rapidly and recover from inactivation faster than in normal flies. In Sh^5 larval muscle, A-type K^+ currents have alterations in activation and inactivation voltages (Wu & Haugland 1985). These results are difficult to explain solely on the basis of differences in the number of channels in the membrane. They argue instead for alterations in the Sh^5 gene product.

Gene dosage analysis provides genetic evidence that some *Sh* alleles encode channel structural defects. The experiments are based on the observation in *Drosophila* that the amount of gene product is generally proportional to the number of gene copies (see for example, O'Brien & MacIntyre 1978). If the *Sh* mutation is due to the underproduction of a channel, then normal behavior and physiology should be restored in an animal that has a normal complement of Sh^+ genes in addition to the mutant gene. If the *Sh* mutation encodes an altered gene product, the defect should still be present, albeit reduced. The addition of normal Sh^+ genes by using a small duplication of the region (for example, the proximal portion of $T(X;4)B^S$ or $Dp(X;3)JC153$, see below) to Sh^{KS133}, Sh^5, or Sh^{rKO120} reduces the severity of leg-shaking under ether anesthesia, but does not restore normal behavior (Tanouye et al 1981). Voltage clamping pupal muscle shows that a normal complement of Sh^+ genes restores only 50% of the A-type K^+ current to Sh^{KS133} mutants (Salkoff 1983b). These results suggest that at least some *Sh* alleles encode an abnormal gene product. They suggest additionally that normal and defective *Sh* gene products must compete with one another. This competition may be for a limited number of channel sites in the membrane or as subunits in a multimeric channel protein.

The most direct electrophysiological evidence for abnormal ionic channels in *Sh* mutants should ultimately come from patch clamp recordings of single ionic channels (Hamill et al 1981). With patch clamp techniques, recordings can be obtained from the channel of interest without contamination from other channels. This should allow unequivocal identification of the channel classes affected by *Sh* mutations. Also, within certain technical constraints, recordings may be obtained from mutant channels that have a decreased current conductance. Thus, these methods can be used to distinguish mutant *Sh* alleles that have decreased conductance from those that have altered opening or closing kinetics. Patch clamp recordings have not been reported for *Sh* mutants; however, recent reports have described normal *Drosophila* single K^+ channel currents (Sun 1984, Wu et al 1983c).

Genetic Organization of the Sh Region

The genetic organization of the *Sh* region has been well-characterized (Figure 5; Tanouye et al 1981, Ferrus & Tanouye 1981, A. Ferrus and M. A. Tanouye, unpublished results). The region may contain a cluster of functionally related genetic loci. Figure 5 shows a representation of the *Sh* region. It is flanked on the left by the 16E2-4 breakpoint of the insertional translocation $T(X;3)JC153$. This breakpoint causes neither behavioral nor physiological abnormalities and thus probably lies just outside of the *Sh* complex.

To the right of the $T(X;3)JC153$ breakpoint are three complementation groups containing recessive lethal mutations (Ferrus & Tanouye 1981). In Figure 5 the complementation groups are denoted l_1, l_2, and l_3. To date, three l_1 alleles, one l_2 allele, and six l_3 alleles have been isolated. The time of lethality varies among the different mutants, but in general it occurs relatively late: in the first or second larval instar.

The lethals are viable as *lethal*/+ heterozygotes. The mutations are not leg-shaking dominants and, thus, differ from the leg-shaking mutants described above. Two types of evidence, however, suggest that lethality is due to improper nervous system function.

1. Mosaic analyses show that lethality foci for the mutations are located in

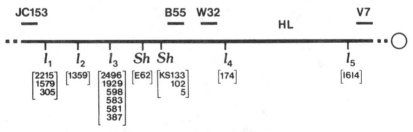

Figure 5   Genetic organization of the *Sh* region. *Horizontal bars* show the relative locations of translocation breakpoints for $T(X;3)JC153$, $T(X;Y)B55$, $T(X;Y)W32$, $T(X;Y)V7$. A haplolethal region (HL) lies between the breakpoints of W32 and V7. The *vertical bars* show the locations of lethal complementation groups l_1 to l_5 and sites occupied by *Sh* mutations. Letters and numbers in brackets designate different alleles. Note that additional complementation groups may lie between l_4 and l_5. Adapted from Tanouye et al (1981), Ferrus & Tanouye (1981), and A. Ferrus and M. A. Tanouye (unpublished observations).

the ventral area of the gynandromorph fate map, a region from which the nervous system and muscles originate.

2. Intracellular microelectrode recordings from *lethal*/+ heterozygotes show abnormal giant fiber action potentials.

Sh leg-shaking mutations are located to the right of the l_3 complementation group. Two alleles, Sh^{E62} and Sh^{KS133}, may be separated by recombination and occupy positions shown in Figure 5. Recombination experiments have failed to separate Sh^{102} and Sh^5 from Sh^{KS133}; they must be located within about 0.001 map units of each other.

The breakpoints of two X–Y translocations, $T(X;Y)B55$ and $T(X;Y)W32$, cause *Sh* behavioral and electrophysiological abnormalities (Tanouye et al 1981, Salkoff 1983b). These breakpoints map genetically to the right of Sh^{KS133}, as shown in Figure 5 (see Tanouye et al 1981). Recombination distances cannot be determined for the breakpoints, since they interfere with recombination. Therefore, it is not possible to determine the map distances between Sh^{KS133} and the breakpoints. Two additional chromosomal breakpoints, $T(X;3)Sh^{LC}$ and $T(X;2)B27$, also cause *Sh* abnormalities. These chromosomes are not easily manipulated genetically, but cytogenetic analysis indicates that their breakpoints are close to that of $T(X;Y)B55$.

Located to the right of the $T(X;Y)W32$ breakpoint are two recessive lethal complementation groups, l_4 and l_5, and the $T(X;Y)V7$ translocation breakpoint. The l_4 and l_5 complementation groups each contain one allele. These lethals have not been studied extensively. The $T(X;Y)V7$ breakpoint causes weak leg-shaking and may have slightly altered giant fiber action potentials (Tanouye et al 1981). Voltage clamp analysis indicates that this breakpoint does not appear to alter A-type K^+ currents (Salkoff 1983b). Sh^{rK0120} has not been mapped by recombination but also appears to map to the right of the $T(X;Y)W32$ breakpoint (Tanouye et al 1981).

Small deficiencies and duplications of the *Sh* region may be constructed using X–Y translocations (Tanouye et al 1981, Salkoff, 1983b). A hemizygous deficiency that completely deletes the material between the $T(X;Y)B55$ and $T(X;Y)W32$ breakpoints is viable and eliminates A-type K^+ currents (Salkoff 1983b). These deficiency animals shake their legs vigorously and have abnormal giant fiber action potentials (Tanouye et al 1981). The region between the $T(X;Y)W32$ and $T(X;Y)V7$ breakpoints, when deleted, results in haplolethality, i.e. a heterozygous deficiency is lethal, and animals die in the embryonic stage. Haplolethals are relatively rare in *Drosophila*; only two others are known (Lindsley et al 1972, Spencer et al 1982).

Several of the above observations suggest that the *Sh* region might be a gene complex containing a number of functionally related genetic loci (Judd 1976). *Sh* mutant phenotypes are associated with several mutations and chromosome breakpoints that are located at distinct sites within the region. These include sites occupied by Sh^{E62}, Sh^{KS133}, and Sh^{rK0120} alleles, and the $T(X;Y)B55$, $T(X;Y)W32$, and $T(X;Y)V7$ translocation breakpoints. At present, however, it is dificult to determine whether mutations at these different sites correspond to abnormalities in different gene products (indicating a cluster of genes) or represent alterations in the expression of a single gene product. Lethal mutations at l_1, l_2, and l_3 also appear to cause nervous system–related effects. These effects may involve membrane excitability, but their relationships to *Sh* are not yet clear.

Molecular Biology of the Sh Region

Genetic and cytogenetic analysis (see above) shows that the *Sh* region, located between the $T(X;3)JC153$ and $T(X;Y)V7$ translocation breakpoints, contains approximately six average-sized salivary chromosome bands and about 0.2 recombination map units. The entire region is of manageable size for molecular cloning experiments, since it should contain only a few hundred kilobases (kb) of DNA.

The strategy for cloning *Sh* is similar to that used for other genetically well-characterized *Drosophila* genes whose gene products have not yet been identified biochemically (e.g. the *white locus*, Levis et al 1982; the *Bithorax* gene complex, Bender et al 1983):

1. The first requirement is a cloned segment of DNA mapping in or near *Sh* that may be used as a hybridization probe to initiate chromosomal walking experiments.
2. The cloned region is extended by chromosomal walking.
3. The sites occupied by mutations or chromosome breakpoints are identified by gene transfer, in situ hybridization, and/or Southern blot analysis in order to align the genetic map of the region with a molecular map.
4. Messenger RNA transcription sites corresponding to mutations may be identified by Northern blot analysis.

The cDNA Clone Adm 135 H4

Wolfner (1980) isolated a number of cDNA clones from *Drosophila* third instar larval messenger RNA. In situ hybridization to salivary gland chromosomes shows that one of the clones, adm 135 H4, hybridizes to a single site at 16F. Additional in situ hybridization analysis on chromosomes broken in and around *Sh* (Ferrus et al 1982) shows that this clone hybridizes to a site located between the $T(X;3)JC153$ and $T(X;Y)B55$ translocation breakpoints.

Gene transfer experiments demonstrate that the gene corresponding to adm 135 H4 is identical to the gene identified by the l_2 lethal mutation (Kamb et al 1984). The cDNA clone was used to isolate a genomic DNA clone containing the entire adm 135 H4 transcription unit. The genomic DNA was inserted into a vector (Goldberg et al 1983) that is suitable for P-element–mediated transformation in *Drosophila*. Flies injected with this construct were used to show that the transferred adm 135 H4 gene rescues animals carrying a lethal l_2 mutation.

Three groups have initiated a chromosomal walk from the site of adm 135 H4. A *Drosophila* genomic DNA library was screened with the cDNA probe, adm 135 H4, to isolate a family of overlapping clones from the region. Restriction fragments from the extreme ends of the cloned DNA segments were used to extend the cloned region further. About 70 kb of DNA in the *Sh* region has thus far been cloned and analyzed (Kamb et al 1984, Jan et al 1984, C. V. Cabrera and A. C. Ferrus, personal communication). The site encoding Sh^{KS133} has not yet been identified.

Sh *Mutations Induced by Transposable Elements*

P-elements are special DNA sequences in *Drosophila* that get moved ("transposed"), infrequently and at random, from one location to another in the genome (Engels 1981, Rubin et al 1982). The frequency of transposition may be experimentally manipulated, usually by crossing appropriate strains. One important use of P-elements, mentioned above, is in *Drosophila* gene transfer. P-elements are also useful as a means of tagging genes of interest in order to clone them. When a P-element moves into a particular gene, that gene's coding sequence is disrupted and the gene product cannot be produced. For example, transposition of a P-element into the *Sh* locus of a normal fly may produce the *Shaker* phenotype. Because the P-element DNA sequence is known, the result, then, is to label the *Sh* locus by inserting a bit of DNA that can be recognized. Attempts have been made to clone *Sh* in this fashion by isolating *Sh* mutations due to the transposition of a P-element into the *Sh* locus. Using P-element sequences as hybridization probes, it should be possible to identify DNA clones corresponding to *Sh* coding regions. Several *Sh* mutations induced by P-element transposition have been reported (Jan et al 1983), but DNA clones of the *Sh* region have not yet been isolated using this technique.

NAP^{ts}

The mutant *nap*^{ts} (*no action potential*) is a temperature-sensitive paralytic (Wu et al 1978). Recombination tests show that it maps to the second chromosome (map position 2–56.2). It is located cytologically in the 41B to

42A region of the salivary second chromosome. Only a single recessive allele is known. Mutant animals (both larvae and adults) are rapidly paralyzed at temperatures greater than 37.5°C. The paralysis reverses instantly when animals are returned to room temperature.

Several lines of evidence suggest that nap^{ts} mutants have a Na^+ channel abnormality. Electrophysiological studies show that mutant nerve action potentials fail at restrictive temperatures (Wu et al 1978). Even at room temperature, however, nap^{ts} nerve excitability is not completely normal (Wu & Ganetzky 1980). There is an increased action potential refractory period and an increased sensitivity to blockage by the Na^+ channel blocker, tetrodotoxin (TTX). The mutant refractory period defect may be mimicked in normal animals by applying low doses of TTX.

Experiments on *Drosophila* neurons cultured in vitro also suggest that nap^{ts} mutants have a Na^+ channel defect (Wu et al 1983b, Suzuki & Wu 1984). Veratradine is a drug that causes persistent Na^+ influx through voltage-sensitive Na^+ channels (Catterall 1981). The drug is toxic to cultured *Drosophila* neurons; however, nap^{ts} neurons are more resistant to the toxic effects than are normal neurons. Normal neurons may be made more resistant by the application of TTX.

Normal and nap^{ts} Na^+ channels have also been examined in ligand-binding studies using radioactively-labeled TTX or STX (saxitoxin) (Hall et al 1982, Kauvar 1982, Jackson et al 1984). These studies show that the maximum specific binding of ligand is reduced by about 40% in nap^{ts} mutants. There appears to be little effect on the dissociation constant. Thus, it is now fairly clear that the nap^{ts} mutation affects voltage-sensitive Na^+ channels. It is less clear whether it encodes abnormal channel proteins (i.e. a structural gene) or alters Na^+ channel density (i.e. a regulatory gene) (Ganetzky 1984, Jackson et al 1984).

DOUBLE MUTANT ANALYSES; OTHER EXCITABILITY MUTANTS

Our understanding of the defects associated with *Sh* and nap^{ts} mutations is reasonably good; that is, nap^{ts} appears to affect Na^+ channels and *Sh* affects A-type K^+ channels. Our understanding is less complete, however, for a host of other *Drosophila* behavioral mutants. One way of sorting through the different mutants is by double mutant analyses. The rationale is that two mutations that alter membrane excitability by affecting different functional components might be expected to interact. An animal carrying both of the mutations may manifest this interaction as either a suppressed mutant phenotype or an enhanced phenotype.

This prediction has been tested using *Sh*; nap^{ts} double mutants

(Ganetzky & Wu 1982a,b). The leg-shaking defects and abnormal transmission at the larval neuromuscular junction generally associated with *Sh* mutations are completely suppressed by *nap*[ts] in double mutant combinations. This suppression is unconditional; it occurs at temperatures permissive for the *nap*[ts] mutation (i.e. room temperature). (Note, however, that *nap*[ts] excitability is not completely normal at room temperature; see above). At restrictive temperatures, *nap*[ts] continues to cause paralysis in the double mutants. Suppression of *Sh* mutant phenotypes appears to be due to the longer action potential refractory periods associated with *nap*[ts] (Ganetzky & Wu 1982a). The suggestion is that *Sh* leg-shaking and transmission defects are caused largely by multiple firing of the respective motor axons.

Double mutant combinations using different mutations have implicated several other genes in excitable membrane function. The best studied are *ether-a-go-go* (*eag*) and *paralytic* (*para*[ts]).

Mutations of the *eag* locus cause behavioral leg shaking. The locus lies at X-chromosome map position 45.9. At least five alleles have been reported, including several induced by transposable P-elements (Kaplan & Trout 1969, Ganetzky & Wu 1983, 1984, Wu & Ganetzky 1984).

Sh and *eag* interact synergistically in double mutant combinations. This interaction was discovered in a very elegant series of experiments by Ganetzky & Wu (1983) that were designed to examine variability in *Sh*[rK0120] mutant phenotypes. Earlier analyses of larval neuromuscular transmission had shown that *Sh*[rK0120] caused the most extreme *Sh* abnormalities (Jan et al 1977, Ganetzky & Wu 1982a). Experiments in other preparations, however, suggested a more modest *Sh*[rK0120] effect (see for example, Tanouye et al 1981, Tanouye & Ferrus 1985, Wu & Haugland 1985). Ganetzky & Wu (1983) found that extreme defects were due to a synergistic second site mutation carried in some *Sh*[rK0120] stocks.

The second site mutation mapped to the *eag* locus. *Sh*[rK0120] now appears to be consistently the weakest of the *Sh* alleles (Tanouye & Ferrus 1985b, Wu & Haugland 1985; see, however, Salkoff 1983b). The synergistic interaction is not allele-dependent; all *Sh* and *eag* alleles probably interact (Ganetzky & Wu 1983). Recent voltage clamp experiments suggest that *eag* mutations also affect K^+ currents (Wu et al 1983a, Wu & Ganetzky 1984). Both A-type K^+ currents and delayed rectifier K^+ currents are altered. The *eag* locus has been cloned using P-element–induced mutations (Ganetzky & Wu 1984; B. Ganetzky, personal communication).

Mutants of the *para*[ts] locus are temperature-sensitive paralytics. The locus lies at X-chromosome position 53.9; the cytological location is in the 14C7 to 14F1 region of the salivary X chromosome (Ganetzky 1984). Six mutant alleles have been reported (Suzuki et al 1971, Siddiqi & Benzer 1976,

Ganetzky 1984). Behavioral and electrophysiological analyses show that the *para*[ts1] allele has properties similar to those of *nap*[ts]. The restrictive temperature is 29°C for adults and 37°C for larvae. Paralysis and recovery are both rapid (within seconds). Action potentials fail at restrictive temperatures in both larvae and adults (Siddiqi & Benzer 1976, Wu & Ganetzky 1980, Benshalom & Dagan 1981).

Double mutant combinations between *para*[ts] and *nap*[ts] show an unconditional lethal interaction (Ganetzky 1984). In *para*[ts1]; *nap*[ts] combinations no progeny are seen. There is some variation among different *para*[ts] alleles, but in each case viability is considerably reduced in the double mutants. The results on *para*[ts] suggest that it may also correspond to a Na$^+$ channel defect. Thus far, experiments using TTX and STX have not shown a binding defect associated with *para*[ts] mutants (Kauvar 1982, Jackson et al 1984).

Double mutant analyses suggest that other genes may be involved in excitable membrane function. These include the leg-shaking locus, *Hyperkinetic*, and the bang-sensitive loci, *bang-sensitive, technical knockout, easily shocked, knockdown*, and *bang senseless* (Ganetzky & Wu 1982b). Each of the mutants is suppressed in double mutant combinations with *nap*[ts], thus suggesting that their defects may be due to abnormal multiple firing.

Two additional genetic loci may encode ionic channel functions. Mutants of the *seizure* locus are temperature-sensitive paralytics. These mutants have abnormal TTX and STX binding properties, suggesting a possible Na$^+$ channel defect (Jackson et al 1984). Voltage clamp analysis of the mutant *slowpoke* shows that it may be defective in Ca^{2+}-dependent K$^+$ channels (Elkins et al 1984).

CONCLUSION

Our understanding of ionic channel genes in *Drosophila* has progressed remarkably in recent years through work done by a small handful of investigators. Continued work on these genes will probably provide important insight into the functioning of the nervous system. Mutants provide one way of approaching the molecular biology of different ionic channels without prior biochemical purification of channel proteins. In some cases, for example the different classes of K$^+$ channels, homology at the DNA or protein level may provide a means of identifying genes that encode all the different classes. In combination with electrophysiological methods, such as voltage clamp and patch clamp, molecular studies of mutants should allow detailed channel structure/function analyses. Other genetic methods, such as double mutant analyses, should help to begin unraveling the complexity of excitable membrane, in much the same way as was done for various biochemical pathways, for ribosome assembly, and for phage morphogenesis (see for example, Hartman & Roth 1973).

ACKNOWLEDGMENTS

We thank M. Ramaswami and A. Ferrus for valuable comments and C.-F. Wu for providing a copy of a manuscript in press. The preparation of this review and some of the work presented in it was supported by the Pfeiffer Research Foundation and USPHS grant NS21327-01 to M. T. L. S. was supported by NSF grant BNS8311024. C. K. was supported by NRSA training grant GM07616. L. I. was supported by a Muscular Dystrophy Association Fellowship. M. T. is a McKnight Foundation Scholar and a Sloan Foundation Fellow.

Literature Cited

Adams, D., Smith, S., Thompson, S. 1980. Ionic currents in molluscan soma. *Ann. Rev. Neurosci.* 3:141–67

Agnew, W. S., Levinson, S. R., Brabson, J. S., Raftery, M. A. 1978. Purification of the tetrodotoxin-binding component associated with the voltage-sensitive sodium channel from *Electrophorus electricus* electroplax membranes. *Proc. Natl. Acad. Sci. USA* 75:2606–10

Armstrong, C. M., Hille, B. 1972. The inner quaternary ammonium ion receptor in potassium channels of the node of Ranvier. *J. Gen. Physiol.* 59:388–400

Ashcroft, F. M., Stanfield, P. R. 1982. Calcium and potassium currents in muscle fibers of an insect (*Carausius morosus*). *J. Physiol.* 323:93–115

Barchi, R. L. 1982. Biochemical studies of the excitable membrane sodium channel. *Int. Rev. Neurobiol.* 263:69–101

Bender, W., Akam, M., Kareh, F., Beachy, P. A., Peifer, M., et al. 1983. Molecular genetics of the *Bithorax* complex in *Drosophila melanogaster*. *Science* 221:23–29

Benshalom, G., Dagan, D. 1981. Electrophysiological analysis of the temperature-sensitive paralytic *Drosophila* mutant, *para*[ts]. *J. Comp. Physiol.* 144:409–17

Bullock, T. H., Horridge, G. A. 1965. *Structure and Function in the Nervous System of Invertebrates*. San Francisco: Freeman. 1161 pp.

Catsch, A. 1944. Eine erbliche Storung des Bewegungsmechanismus bei *Drosophila melanogaster*. *Z. Induk. Abstamm. Vererbungsl.* 82:64–66

Catterall, W. A. 1981. Studies of voltage-sensitive sodium channels in cultured cells using ion flux and ligand-binding methods. In *Excitable Cells in Tissue Culture*, ed. P. G. Nelson, M. Lieberman, pp. 279–314. New York: Plenum

Connor, J. A. 1975. Neural repetitive firing: A comparative study of membrane properties of crustacean walking leg axons. *J. Neurophysiol.* 38:922–32

Connor, J. A., Stevens, C. F. 1971. Voltage clamp studies of a transient outward current in gastropod neural somata. *J. Physiol.* 213:21–30

Conti-Tronconi, B. M., Raftery, M. A. 1982. The nicotinic cholinergic receptor: Correlation of molecular structure with functional properties. *Ann. Rev. Biochem.* 51:491–530

Crawford, A. C., McBurney, R. N. 1976. On the elementary conductance event produced by L-glutamate and quanta of the natural transmitter at the neuromuscular junction of *Maia squinado*. *J. Physiol.* 258:205–25

Cull-Candy, S. G. 1976. Two types of extrajunctional L-glutamate receptors in locust muscle. *J. Physiol.* 255:449–64

Elkins, T. T., Ganetzky, B., Wu, C.-F. 1984. A *Drosophila* mutation with a defect in Ca^{2+}-dependent outward current. *Soc. Neurosci. Abstr.* 10:1090

Engels, W. R. 1981. Hybrid dygenesis in *Drosophila* and the stochastic loss hypothesis. *Cold Spring Harbor Symp. Quant. Biol.* 45:561–65

Ferrus, A. C., Cabrera, C. V., Tanouye, M. A. 1982. Cloning the *Shaker* gene. *Calif. Inst. Tech. Ann. Rep.* 160

Ferrus, A., Tanouye, M. A. 1981. Fine genetic analysis of the *Shaker* locus. *Calif. Inst. Techn. Ann. Rep.* 247

Ganetzky, B. 1984. Genetic studies of membrane excitability in *Drosophila*: Lethal interaction between two temperature-sensitive paralytic mutations. *Genetics* 108:897–911

Ganetzky, B., Wu, C.-F. 1982a. *Drosophila* mutants with opposing effects on nerve excitability: Genetic and spatial interactions in repetitive firing. *J. Neurophysiol.* 47:501–14

Ganetzky, B., Wu, C.-F. 1982b. Indirect suppression involving behavioral mutants with altered nerve excitability in *Drosophila melanogaster*. *Genetics* 100:597–614

Ganetzky, B., Wu, C.-F. 1983. Neurogenetic analysis of potassium currents in *Drosophila*: Synergistic effects on neuromuscular transmission in double mutants. *J. Neurogenet.* 1:17–28

Ganetzky, B., Wu, C.-F. 1984. Mutations of a gene affecting potassium currents induced by transposable elements in *Drosophila*. *Soc. Neurosci. Abstr.* 10:1090

Goldberg, D. A., Posakony, J. W., Maniatis, T. 1983. Correct developmental expression of a cloned alcohol dehydrogenase gene transduced into the *Drosophila* germ line. *Cell* 34:59–73

Grigliatti, T. A., Hall, L., Rosenbluth, R., Suzuki, D. T. 1973. Temperature-sensitive mutations in *Drosophila melanogaster*. XIV. A selection of immobile adults. *Mol. Gen. Genet.* 120:107–14

Hagiwara, S. 1983. In *Membrane Potential-Dependent Ion Channels in Cell Membrane, Phylogenetic and Developmental Approaches*. New York: Raven. 118 pp.

Hagiwara, S., Byerly, L. 1981. Calcium channel. *Ann. Rev. Neurosci.* 4:69–125

Hall, J. C., Greenspan, R. J. 1979. Genetic analysis of *Drosophila* neurobiology. *Ann. Rev. Genet.* 13:127–95

Hall, L. M., Wilson, S. D., Gitschier, J., Martinez, N., Strickhartz, G. R. 1982. Identification of a *Drosophila melanogaster* mutant that affects the saxitoxin receptor of the voltage-sensitive sodium channel. *Ciba Found. Symp.* 88:207–20

Hamill, D., Marty, A., Neher, E., Sakmann, B., Sigworth, F. 1981. Improved patch-clamp techniques for high resolution current recording from cells and cell-free membrane patches. *Pfluegers Arch.* 391:85–100

Hartman, P. E., Roth, J. E. 1973. Mechanisms of suppression. *Adv. Genet.* 17:1–105

Hille, B. 1984. *Ionic Channels of Excitable Membranes*. Sunderland, Mass: Sinauer Assoc., Inc. 426 pp.

Hodgkin, A. L., Huxley, A. F. 1952. A quantitative description of membrane current and its application to conduction and excitation in nerve. *J. Physiol.* 117:500–44

Ikeda, K. 1980. Neuromuscular physiology. In *Genetics and Biology of Drosophila*, ed. M. Ashburner, T. R. F. Wright, pp. 369–405. New York: Academic

Jackson, F. R., Wilson, S. D., Strichartz, G. R., Hall, L. M. 1984. Two types of mutants affecting voltage-sensitive sodium channels in *Drosophila melanogaster*. *Nature* 308:189–91

Jan, L. Y., Barbel, S., Timpe, L., Laffer, C., Salkoff, L., et al. 1983. Mutating a gene for a potassium channel by hybrid dysgenesis: An approach to cloning of the *Shaker* locus in *Drosophila*. *Cold Spring Harbor Symp. Quant. Biol.* 48:233–45

Jan, L. Y., Jan, Y. N. 1976a. Properties of the larval neuromuscular junction in *Drosophila melanogaster*. *J. Physiol.* 262:189–214

Jan, L. Y., Jan, Y. N. 1976b. L-glutamate as an excitatory transmitter at the *Drosophila* larval neuromuscular junction. *J. Physiol.* 262:215–36

Jan, Y. N., Jan, L. Y., Dennis, M. J. 1977. Two mutations of synaptic transmission in *Drosophila*. *Proc. R. Soc. London Ser. B* 198:87–108

Jan, L. Y., Papazian, D. M., Jan, Y. N., O'Farrell, P. H. 1984. Cloning of potassium channel gene(s) in the *Shaker* locus of *Drosophila*. *Soc. Neurosci. Abstr.* 10:1089

Judd, B. 1976. Complex loci. In *The Genetics and Biology of Drosophila*, ed. M. Ashburner, E. Novitsky. 1b:767–99. London: Academic

Julian, F. J., Moore, J. W., Goldman, D. E. 1962. Membrane potentials of the lobster giant axon obtained by use of the sucrose-gap technique. *J. Gen. Physiol.* 45:1195–1216

Kamb, A., Iverson, L., Tanouye, M. 1984. Molecular analysis of the *Shaker* (*Sh*) gene complex in *Drosophila melanogaster*. *Soc. Neurosci. Abstr.* 10:1089

Kaplan, W. D., Trout, W. E. III. 1969. The behavior of four neurological mutants of *Drosophila*. *Genetics* 61:399–409

Kauvar, L. 1982. Reduced [^{3}H] tetrodotoxin binding in the *nap*[ts] paralytic mutant of *Drosophila*. *Mol. Gen. Genet.* 187:172–73

Latorre, R., Miller, C. 1983. Conduction and selectivity in potassium channels. *J. Memb. Biol.* 71:11–30

Lindsley, D., Sandler, L., Baker, B., Carpenter, A., Dennell, R., et al. 1972. Segmental aneuploidy and the genetic gross structure of the *Drosophila* genome. *Genetics* 71:157–84

Neher, E. 1971. Two fast transient current components during voltage clamp on snail neurons. *J. Gen. Physiol.* 58:36–53

Noda, M., Shimizu, S., Tanabe, T., Takai, T., Kayano, T., et al. 1984. Primary structure of *Electrophorus electricus* sodium channel deduced from cDNA sequence. *Nature* 312:121–27

Noda, M., Takahashi, H., Tanabe, T., Toyosato, M., Kikyotani, S., et al. 1983. Structural homology of *Torpedo californica* acetylcholine receptor subunits. *Nature* 302:528–32

O'Brien, S. J., MacIntyre, R. J. 1978. Genetics and biochemistry of enzymes and specific proteins of *Drosophila*. In *The Genetics and Biology of Drosophila*, ed. M. Ashburner, T. Wright, 2a:395–551. London: Academic

Patlak, J. B. 1976. The ionic basis for the action potential in the flight muscle of the fly, *Sarcophaga bullata*. *J. Comp. Physiol.* 107:1–11

Pichon, Y., Boistel, J. 1967. Current-voltage relations in the isolated giant axon of the cockroach under voltage-clamp conditions. *J. Exp. Biol.* 47:343–55

Quinta-Ferreira, M. E., Rojas, E., Arispe, N. 1982. Potassium currents in the giant axon of the crab *Carcinus maenas*. *J. Memb. Biol.* 66:171–81

Rubin, G. M., Kidwell, M. G., Bingham, P. M. 1982. The molecular basis of hybrid dysgenesis: The nature of induced mutations. *Cell* 29:987–94

Salkoff, L. 1983a. *Drosophila* mutations reveal two components of fast outward current. *Nature* 302:249–51

Salkoff, L. 1983b. Genetic and voltage clamp analysis of a *Drosophila* potassium channel. *Cold Spring Harbor Symp. Quant. Biol.* 48:221–31

Salkoff, L., Wyman, R. J. 1981a. Genetic modification of potassium channels in *Drosophila Shaker* mutants. *Nature* 293:228–30

Salkoff, L., Wyman, R. J. 1981b. Outward currents in developing *Drosophila* flight muscle. *Science* 212:461–63

Salkoff, L., Wyman, R. J. 1983a. Ion currents in *Drosophila* flight muscles. *J. Physiol.* 337:687–709

Salkoff, L., Wyman, R. J. 1983b. Ion channels in *Drosophila* muscle. *Trends Neurosci.* 6:128–33

Serrano, E. E. 1982. *Variability in molluscan neuron soma currents*. PhD thesis. Stanford Univ., Stanford, Calif.

Shrager, P. 1974. Ionic conductance changes in voltage clamped crayfish axons at low pH. *J. Gen. Physiol.* 64:666–90

Siddiqi, Q., Benzer, S. 1976. Neurophysiological defects in temperature-sensitive mutants of *Drosophila melanogaster*. *Proc. Natl. Acad. Sci. USA* 73:3253–57

Spencer, F. A., Hoffmann, F. M., Gelbart,

W. M. 1982. Decapentaplegic: A gene complex affecting morphogenesis in *Drosophila melanogaster*. *Cell* 28:451–61

Sun, Y.-A., Wu, C.-F. 1984. Voltage-dependent single-channel currents in dissociated CNS neurons of *Drorophila*. *Soc. Neurosci. Abstr.* 10:1090

Suzuki, D. T., Grigliatti, T., Williamson, R. 1971. Temperature-sensitive mutants in *Drosophila melanogaster*: A mutation (*para*) causing reversible adult paralysis. *Proc. Natl. Acad. Sci. USA* 68:890–93

Suzuki, N., Kano, M. 1977. Development of action potential in larval muscle fibers in *Drosophila melanogaster*. *J. Cell Physiol.* 93:383–88

Suzuki, N., Wu, C.-F. 1984. Altered sensitivity to sodium channel-specific neurotoxins in cultured neurons from temperature-sensitive paralytic mutants of *Drosophila*. *J. Neurogenet.* 1:225–38

Tanouye, M. A., Ferrus, A. 1985. Action potentials in normal *Shaker* mutant *Drosophila*. *J. Neurogenet.* In press

Tanouye, M. A., Ferrus, A., Fujita, S. C. 1981. Abnormal action potentials associated with the *Shaker* complex locus of *Drosophila*. *Proc. Natl. Acad. Sci. USA* 78:6548–52

Trout, W. E. III, Kaplan, W. D. 1973. Genetic manipulation of motor output in *Shaker* mutants of *Drosophila*. *J. Neurobiol.* 4:495–512

Vandlen, R. L., Wu, W., Eisenach, J. C., Raftery, M. A. 1979. Studies of the composition of purified *Torpedo californica* acetylcholine receptor and of its subunits. *Biochemistry* 18:1845–54

Washio, H. 1972. The ionic requirements for the initiation of action potentials in insect muscle fibers. *J. Gen. Physiol.* 59:121–34

Williamson, R. L. 1982. Lithium stops hereditary shuddering in *Drosophila melanogaster*. *Psychopharmacology* 76:265–68

Wolfner, M. 1980. Ecdysone-responsive genes of the salivary gland of *Drosophila melanogaster*. PhD thesis. Stanford Univ., Stanford, Calif.

Wu, C.-F., Ganetzky, B. 1980. Genetic alteration of nerve membrane excitability in temperature-sensitive paralytic mutants of *Drosophila melanogaster*. *Nature* 286:814–16

Wu, C.-F., Ganetzky, B. 1984. Properties of potassium channels altered by mutations of two genes in *Drosophila*. *Biophys. J.* 45:77–78

Wu, C.-F., Ganetzky, B., Haugland, F., Liu, A.-X. 1983a. Potassium currents in *Drosophila*: Different components affected by mutations of two genes. *Science* 220:1076–78

Wu, C.-F., Ganetzky, B., Jan, L. Y., Jan, Y.

N., Benzer, S. 1978. A *Drosophila* mutant with a temperature-sensitive block in nerve conduction. *Proc. Natl. Acad. Sci. USA* 75:4047–51

Wu, C.-F., Haugland, F. 1985. Voltage clamp analysis of membrane currents in larval muscle fibers of *Drosophila*: Alteration of potassium currents in *Shaker* mutants. *J. Neurosci.* In press

Wu, C.-F., Suzuki, N., Poo, M.-m. 1983b. Dissociated neurons from normal and mutant *Drosophila* larvae central nervous system in cell culture. *J. Neurosci.* 3:1888–89

Wu, C.-F., Young, S. H., Tanouye, M. A. 1983c. Single-channel recording of α-bungarotoxin resistant acetylcholine channels in dissociated CNS neurons of *Drosophila. Soc. Neurosci. Abstr.* 9:507

Yeh, J. Z., Oxford, G. S., Wu, C. H., Narahashi, T. 1976. Dynamics of aminopyridine block of potassium channels in squid axon membrane. *J. Gen. Physiol.* 68:519–35

Ann. Rev. Neurosci. 1986. 9:277–304

HYBRIDIZATION APPROACHES TO THE STUDY OF NEUROPEPTIDES*

Joan P. Schwartz and Erminio Costa

Laboratory of Preclinical Pharmacology, National Institute of Mental Health, Saint Elizabeths Hospital, Washington DC 20032

INTRODUCTION

The discovery that the brain contains the opioid peptides methionine[5] (met)-enkephalin and leucine[5] (leu)-enkephalin by Hughes et al (1975) has triggered an exponential increase in the number of new brain peptides postulated to function as neurotransmitters and/or neuromodulators. In this review we highlight how various technologies of molecular biology are being applied to the most critical areas of neuropeptide research today. First, the cloning and sequencing of genes and/or mRNAs for neuropeptide precursors has provided evidence for a variety of mechanisms responsible for the generation of structural variation in regulatory neuropeptides, ranging from diversity in the evolution of gene families, to differential RNA splicing mechanisms, to the existence of polyproteins associated with differential processing. Second, brain-specific cloning approaches are leading to the discovery of new, functionally significant neuropeptides. Third, the DNA probes generated are being used to study those changes in the dynamics of neuropeptide biosynthesis that are important to assess the participation of neuropeptides in brain function. We discuss both neuropeptides and neuroendocrine peptides, because reasons for a distinction between the two classes of peptides are gradually disappearing and because neuroendocrine tissues, particularly the pituitary, are being used as model systems for the study of physiological regulation of neuropeptides in

* The US Government has the right to retain a nonexclusive, royalty-free license in and to any copyright covering this paper.

general. Furthermore, we limit this review to studies on vertebrate neuropeptides, despite fascinating work on invertebrate peptides and on other proteins, such as receptors and growth factors, which are also important in synaptic function.[1]

The kinds of cloning technologies and hybridization analyses currently used for neuroscience research are those developed by molecular biologists over the past years: detailed descriptions are available in several sources (Maniatis et al 1982, Perbal 1984). Two basic types of cloning can be carried out, to produce libraries that contain either copies of all the mRNAs of a given cell (a *cDNA library*) or all the genes of an organism (a *genomic library*). A cDNA is prepared by copying mRNA with reverse transcriptase to produce a complementary (c)DNA copy. In most tissues, mRNA is distinguished by a poly(A)$^+$ tail at the 3' end: however, brain contains in addition a population of poly(A)$^-$ mRNAs, as is discussed below. A cDNA or a genomic probe, isolated from a library, consists of the DNA sequence coding for a specific protein and can be used for hybridization analyses. A *Southern blot* is a hybridization analysis of DNA fragments, size-separated on a gel, in order to determine, for example, the number of genes that exist for a specific protein. A *Northern or RNA blot* is a hybridization analysis of mRNA, size-separated on a gel, to determine the number and/or quantity of messages expressed in a tissue. *Solution hybridization* is another quantitative analysis for the number of copies of DNA or RNA.

To establish the biosynthesis of a neuropeptide in a given neuron, *in situ hybridization* of the specific radioactive cDNA to mRNA in tissue sections has been proposed (Brahic & Haase 1978, Hudson et al 1981). More recently a method using biotinylated cDNA probes and streptavidin-horseradish peroxidase or -fluorescence for detection has been developed (Langer et al 1981). In addition to specificity, in situ hybridization provides great sensitivity (down to the level of a few copies of a specific mRNA per cell) as well as the possibility for quantification. In situ hybridization can be used to complement immunohistochemistry of neuropeptides, in order to obtain absolute identification. The combination of the two procedures has already proven useful for the study of which cells are involved in synthesis of a specific protein and how they are regulated developmentally and hormonally (Gee & Roberts 1983, Griffin et al 1983). Gee et al (1983) used in situ hybridization with a cDNA probe for proopiomelanocortin (POMC) to demonstrate the presence of cell bodies in the periarcuate nucleus and nucleus tractus solitarius of the hypothalamus. The mRNA was colocalized with ACTH staining. Somatostatin mRNA-containing cell bodies have

---

[1] The literature search for references cited was concluded in December 1984.

been detected in human hippocampus and cortex (W. S. T. Griffin, personal communication), and the distribution was similar to that determined by immunohistochemistry (Johansson et al 1984). The technique should also be useful for demonstrating co-synthesis of two neuropeptides within the same neuron, particularly with a new double-label method (Haase et al 1985). It should provide resolution sufficient, for example, to determine whether specific neurons synthesize α-preprotachykinin (substance P), β-preprotachykinin (substance P-substance K), or both mRNAs (Nawa et al 1984).

In vitro nuclear transcription run-off (McKnight & Palmiter 1979) is a hybridization analysis that allows one to determine the rate of transcription of a specific gene. Isolated nuclei complete transcription in the presence of radioactive nucleotides, and the specific RNA (now radioactive) is hybridized to a cDNA probe to separate it from all the others. *DNA-mediated gene transfer or transfection* of mammalian cells provides an assay for DNA sequences in a cloned gene that function as regulatory signals, for example, for initiation of transcription or for regulation of transcription by hormones, transmitters, or other agents. Work of Moore et al (1983) has demonstrated that this methodology also has the potential to allow one to study DNA sequences that act as signals for sorting proteins into specific cellular compartments or as signals for processing of polyproteins. They prepared an SV40-pBR322 recombinant vector containing a cDNA for human proinsulin and transfected it into either AtT-20, a mouse anterior pituitary cell line that synthesizes and processes POMC and secretes ACTH, or into the fibroblast L-cell line. Both cell types synthesized proinsulin from the transfected cDNA but only the AtT-20 cells processed it to insulin. The L cells secreted essentially all of the proinsulin synthesized, as proinsulin, by a pathway the authors defined as "constitutive." The AtT-20 cells processed proinsulin to insulin and contained a stored form of the insulin that could be released by the secretagogue 8-Br-cyclic AMP, via a regulated pathway of secretion. Moore et al (1983) therefore suggested that the gene must contain a sorting signal recognized by the secretory AtT-20 cells that directed the proinsulin into granules, as well as processing information that allowed the POMC-processing enzymes to act on proinsulin. Mutation and deletion studies should allow the definition of DNA sequences containing such information. Since some of the diversity inherent in neuropeptide expression derives from tissue-specific differential processing of the precursor, as is seen, for example, with POMC (O'Donohue & Dorsa 1982) and proenkephalin (Liston et al 1984), evidence for differential processing signals in precursor genes would be very useful. All of these methods have been applied to the study of neuropeptides, and we discuss specific examples in the remainder of the review.

MECHANISMS THAT GENERATE NEUROPEPTIDE DIVERSITY

The complexity of functions that our brain carries out—from motor control, to sensory perception, to learning, memory, and behavior—requires great complexity in terms of methods for storage and retrieval of information. The functional basis of brain information processing is dependent on the signaling capabilities of neurons. Synaptic transmission, the process of communication between neurons, utilizes multiple chemical signals in the exchange of information. Although we can describe with a certain precision how synaptic communication relates to neuronal excitability, we do not understand how generation of action potentials relates to memory, learning, and behavioral output. Currently, variability in synaptic communication is considered to play a role in producing the phenotypic diversity of brain function. This can be generated at many levels, in the diversity of cell types, as well as in the diversity of transmitters, their modulation by cotransmitters, and receptor phenotypes. The participation of multiple chemical signals in synaptic communication, the graded release of these specific signals at a synapse, and the modulation of specific receptor responsiveness further contribute to the variety of functional output that makes each brain unique.

The application of molecular biology technologies to neuroscience has expanded our understanding of the ways in which neuropeptides can be generated and their generation can be modulated. Utilization of the known amino acid sequence of a peptide to clone and sequence both the mRNA and the gene for its precursor has produced an impressive amount of new information. A list of cloned peptide precursors and their tissue distributions is presented in Table 1. These studies have yielded information on gene families and predicted evolutionary relationships between genes, have shown that regulation can occur at the level of RNA processing to produce different peptides from the same gene, have corroborated the findings of polyproteins containing multiple copies of one or several peptides, and have led to the predictions of new peptides. The alternative approach has been to look for brain-specific messages and proteins. Some of this work has centered on the presence of poly(A)$^-$ mRNAs, which are essentially specific to brain and are primarily expressed after birth as the brain is maturing. In addition, examination of cloned brain-specific mRNAs has led to the prediction of new neuroproteins and putative neuromodulatory peptides, for which sequences and immunohistochemical localizations are available but whose functional profiles remain to be established.

Genomic Changes

Numerous peptide precursors have been cloned to date (Table 1) and the sequences of their genes and/or mRNAs determined. Computer analyses and comparisons of neuropeptide precursor gene sequences have revealed that genetic mechanisms have been involved in the generation of neuro-peptide diversity. This can be inferred from the existence of families of genes, as well as the evolutionary relatedness of genes, and from the evidence for gene conversion. There are three families of opioid peptides in vertebrate tissues and each family is characterized by a specific precursor. The sequences of these precursors have been determined. The three opioid peptide precursor genes represent the best examples of a set of genes that appear to have evolved by a series of internal duplications as well as from a common ancestral gene. Proopiomelanocortin (POMC) was the first to be cloned and sequenced, at both the mRNA (Nakanishi et al 1979, Uhler & Herbert 1983) and gene levels (Cochet et al 1982, Nakanishi et al 1980, Notake et al 1983b, Whitfeld et al 1982). This precursor contains within its sequence the opioid peptide, β-endorphin, but also adrenocortic-otropin (ACTH), β-lipotropin, and α-, β-, and γ-melanocyte stimulating hormone (MSH). The 31 amino acid β-endorphin peptide has the pentapeptide met[5]-enkephalin sequence at its amino-terminal end. The sequence of POMC is 80–90% conserved in mammals but is quite different in chum salmon, in which only the sequences for α- and β-MSH are relatively homologous, and only the met[5]-enkephalin portion of β-endorphin is conserved (Soma et al 1984). The complete sequence of proenkephalin (proenkephalin A) cDNAs (Comb et al 1982, Gubler et al 1982, Howells et al 1984, Noda et al 1982a, Yoshikawa et al 1984) and genes (Comb et al 1983, Legon et al 1982, Noda et al 1982b) appeared soon thereafter and demonstrated that the four biologically active enkephalin peptides, met[5]-enkephalin, leu[5]-enkephalin, met[5]-enkephalin-arg[6]-phe[7] and met[5]-enkephalin-arg[6]-gly[7]-leu[8] are encoded in one precursor termed proenkephalin (PE). The recent cloning of the PE gene from *Xenopus laevis* (Martens & Herbert 1984) has provided the interesting result that this precursor contains no leu[5]-enkephalin but contains seven copies of met[5]-enkephalin, of which two are the carboxy-terminal extended forms. This gene structure thus predicts that met[5]-enkephalin was the original opioid peptide whose sequence was duplicated to produce the PE gene, and that the change to leu-enkephalin must have occurred since the time at which *Xenopus* split off evolutionarily from the main branch of vertebrates (~ 350

Myr) (Martens & Herbert 1984). The last opioid peptide precursor to be cloned, prodynorphin (proenkephalin B) (Horikawa et al 1983, Kakidani et al 1982), contains three known peptides—dynorphin (1–17 and 1–8), α- and β-neo-endorphin—all of which contain an amino-terminal leu[5]-enkephalin sequence. In addition, the sequence of the mRNA and gene predicts the

Table 1 Cloned neuropeptides—mRNA and peptide distribution

Precursor	Brain and/or pituitary		Other tissues
	mRNA	Peptides	
Luteinizing hormone releasing hormone (LHRH)		+ (1)[a]	Placenta (1)
Corticotropin releasing factor (CRF)	+	+ (2)	
Arginine vasopressin (AVP)	+	+ (3)	
Oxytocin (OX)	+	+ (4)	Corpus luteum (21)
Prolactin (PRL)	+	+ (5)	Placenta (22)
Growth hormone (GH)	+	+ (6)	
Lutropin (LH)	+	+ (7, 8)	
Proopiomelanocortin (POMC)	+	+ (9, 10)	Gut (12) Thyroid (23) Testis (24, 25)
Proenkephalin (PE)	+	+ (11)	Adrenal medulla (26–28) Gut (12)
Prodynorphin	+	+ (12)	Adrenal medulla (12) Gut (12)
Neuropeptide Y	+	+ (13)	Adrenal medulla (13)
Somatostatin		+	Pancreas (29)
Cholecystokinin	+	+ (14, 15)	Gut (14, 15) Thyroid (14, 15)
Substance P– Substance K	+	+ (16, 17)	Thyroid (17) Gut (17)
Calcitonin-calcitonin gene related peptide (CGRP)	+	+ (18)	Thyroid (18, 30)
Growth hormone releasing factor (GHRF)	+	+ (19)	Pancreas (31, 32)
Vasoactive intestinal peptide (VIP)	+	+ (20)	

[a] **References:** 1. Seeburg & Adelman (1984). 2. Furutami et al (1983). 3. Land et al (1982). 4. Land et al (1983). 5. Schuler et al (1983). 6. Seeburg et al (1977). 7. Chin et al (1983). 8. Jameson et al (1984). 9. Civelli et al (1982). 10. Mocchetti et al (1984c). 11. Tang et al (1983). 12. Jingami et al (1984). 13. Minth et al (1984). 14. Deschenes et al (1984). 15. Gubler et al (1984). 16. Nawa et al (1983). 17. Nawa et al (1984b). 18. Amara et al (1984). 19. Mayo et al (1983). 20. Itoh et al (1983). 21. Ivell & Richter (1984). 22. Taii et al (1984). 23. Steenbergh et al (1984). 24. Chen et al (1984). 25. Pintar et al (1984). 26. Comb et al (1982). 27. Gubler et al (1982). 28. Noda et al (1982a). 29. Shen et al (1982). 30. Rosenfeld et al (1981). 31. Gubler et al (1983). 32. Mayo et al (1983).

possible existence of a fourth dynorphin peptide, leumorphin (29 amino acids), at the carboxy-terminal of the precursor.

Cloning and sequence analysis of the genes for the three opioid precursors had the immediate result of clarifying the interrelationships among them; it became clear, for example, that β-endorphin was not the precursor for met^5-enkephalin nor dynorphin for leu^5-enkephalin. In addition, the concept of a polyprotein, containing several active peptides and/or hormones, was substantiated, as were the concepts that the sequences of the biologically active peptides were set off by pairs of basic amino acids recognizable by processing enzymes and that a glycine next to the carboxyl-terminal pair of basic amino acids was a signal for the formation of an amidated peptide. But an additional observation has been the striking similarity between the structures of the three genes. In all three, there are an exon containing the 5'-untranslated region, a second exon encoding the signal peptide, and a third that encodes almost all the coding region of the precursor. All three precursors have five to six cysteines in essentially identical locations at the amino-terminus, and the biologically active peptides are primarily in the carboxy-terminal half of the protein. The precursors are very similar in size (235–269 amino acids). Further sequence homology is seen within each gene, suggesting that a common sequence (i.e. for met^5-enkephalin or MSH) was duplicated several times during the course of evolution to produce the genes that have now been sequenced. These results thus raise the possibilities that the PE and prodynorphin genes evolved from a common ancestral gene through multiple duplications and alterations or that all three genes evolved by a common mechanism. It would therefore be extremely interesting to know whether comparable genes or sequences exist in the invertebrates and also whether prodynorphin represents a branch point in evolution since *Xenopus* split off.

Gene Families

The genes for growth hormone (GH), prolactin (PRL), and chorionic somatomammotropin (CS) show a high degree of sequence homology (Barta et al 1981, Niall et al 1971, Selby et al 1984) and represent a gene family that is probably evolving by concerted mechanisms (Selby et al 1984) and contains gene duplications (in human, two GH genes and three CS genes). Each gene is expressed in only one type of cell, however, thus raising the possibility that gene duplication and evolution could result in the production of new peptide products expressed in specific cells of the nervous system. Another example is the bombesin-gastrin-releasing peptide (GRP) family. Bombesin was originally isolated in amphibia but bombesin-like peptides have been found in the mammalian nervous system

(Moody & Pert 1979) and gastrointestinal tract (Dockray et al 1979). GRP, isolated from stomach and intestine, is very similar in structure to bombesin. The recent cloning of the GRP precursor (Spindel et al 1984) should provide a probe useful for cloning the bombesin-like precursor from mammalian brain and thus identifying another neuropeptide precursor and/or family.

Gene Conversion

Gene conversion is actually a mechanism for the conservation of a specific DNA sequence between genes, presumably because the sequence encodes a region essential for function. Cloning and sequencing of the oxytocin-neurophysin I and arginine vasopressin (AVP)-neurophysin II precursor genes (Land et al 1982, 1983) revealed the presence of three exons, each encoding a separate functional domain, for both genes. The second exon of each gene, encoding amino acids 10–74 of the neurophysins, is identical. These results suggest that the two genes arose by gene duplication and that a more recent gene conversion event has resulted in their second exons' being identical (Ruppert et al 1984). The overall result has been the generation of two distinct neuropeptides, oxytocin and vasopressin, which share almost identical "carrier" proteins, the neurophysins, but are synthesized by different cells within the hypothalamus and have different biological functions.

Differential Splicing of Nuclear Precursor RNAs

The existence of split genes, composed of exons and introns, which are transcribed as a complete entity, following which the hnRNA is then processed by splicing to remove the intron sequences, has opened the possibility of regulation of gene expression at the level of splicing, with the generation of alternative mRNAs and proteins. Two examples in mammalian cells suggest that differential splicing offers a powerful method for generating diversity of neuropeptide expression. The first example was discovered when a calcitonin cDNA probe was used to study rat medullary thyroid carcinomas as they switched from being high to low producers of calcitonin. Associated with this switch was the production of a new mRNA (Rosenfeld et al 1981). Further work demonstrated that the calcitonin gene contained the coding sequence for both mRNAs and that the coding sequence for both RNAs was also present in the nuclear transcripts. However, in vitro translation of the mRNAs produced two different proteins (Rosenfeld et al 1982). The gene consists of four coding exons, two of which are common; a third contains the coding sequence for calcitonin and the fourth the coding sequence for the new peptide, named calcitonin gene-related peptide (CGRP). The two precursor proteins share the same amino-terminal 76 amino acids but differ in their carboxyl-terminal ends: thus, procalcitonin (136 amino acids) generates the 32 amino acid calcitonin while pro-CGRP (128 amino acids) generates the 37 amino acid CGRP. These findings have been confirmed for the human gene (Nelkin et al 1984).

Not only are two mRNAs that encode unique peptides produced, but the expression of these mRNAs is tissue-specific. Whereas calcitonin mRNA is found only in the thyroid and in human lung carcinomas (Nelkin et al 1984), CGRP mRNA is widely distributed in the rat nervous system, with the highest concentrations in the trigeminal nucleus, the hypothalamus, and the midbrain (Rosenfeld et al 1983). The peptide itself has been localized by immunohistochemistry to a number of brain nuclei, as well as to spinal cord, sensory ganglia, and fibers in many peripheral tissues (Gibson et al 1984, Rosenfeld et al 1983), thus indicating a possible role of CGRP in ingestive behavior and in sensory relay of pain and thermal information (Rosenfeld et al 1983). This tissue-specific expression of these mRNAs (no calcitonin mRNA has been found in brain) appears to be the result of a selection of polyadenylation site once the complete nuclear transcript is synthesized, which is then associated with a differential splicing pattern (Amara et al 1984). Understanding the molecular basis for choice of polyadenylation sites and differential processing should provide a great deal of information on the role of splicing choices as a mechanism for the generation of cell- or neuron-specific diversity in gene expression.

The second example of differential splicing is one more strictly defined as exon inclusion/exclusion. Cloning and sequencing of the substance P precursor mRNA led to the discovery of two mRNAs, which differed by 54 nucleotides: The unique sequence in the longer mRNA coded for a decapeptide, set off by pairs of basic amino acids, which showed homology to the amphibian peptide, kassinin, and was named substance K (Nawa et al 1983). The two precursors were called α- and β-preprotachykinin. The gene was shown to consist of seven exons, with substance P encoded by exon 3 and substance K encoded by exon 6 (Nawa et al 1984b). The two mRNAs are generated by differential splicing, in which exon 6 is either included (β-preprotachykinin) or excluded (α-preprotachykinin), and this choice also appears to be tissue-specific. Within the brain, the ratio of α- to β-preprotachykinin mRNA varies over a three-fold range from region to region, while in other tissues β-preprotachykinin predominates by seven- to nine-fold (intestine) to more than 30-fold (thyroid) (Nawa et al 1984b). It will be particularly interesting to see whether individual neurons express only one of the two mRNAs, and what the molecular basis of such a cell-specific splicing choice is.

The peptide substance K has been isolated from porcine spinal cord, as a gut-contracting peptide, and named neurokinin α (Kimura et al 1983). Furthermore, chemically synthesized substance K has been shown to exhibit a variety of biological activities, different in range and potency from those of substance P (Hunter & Maggio 1984, Nawa et al 1984a).

Polyproteins: Differential Distribution and Processing

High molecular weight peptide precursors containing multiple copies of one or several biologically active peptides have been termed polyproteins. The peptides are released by enzymatic processing, frequently carried out by trypsin-like enzymes acting at pairs of dibasic amino acids. Cloning and sequencing of mRNAs and/or genes for specific peptides has provided support for the existence of polyproteins and has clarified the confusion related to the opioid peptides by providing the structures of the three precursors (see above). The sequences of cloned polyproteins have led in several cases to the predictions of new biologically active peptides, most of which have subsequently been identified in tissue extracts. In addition to the precursors for calcitonin-CGRP and for substance P–substance K, discussed in the section on differential splicing, the existence of new peptides was predicted from sequences determined by cDNA cloning for three other precursors: γ-MSH in POMC (Nakanishi et al 1979); leumorphin in prodynorphin (Kakidani et al 1982); and PHM-27 in provasoactive intestinal peptide (VIP) (Itoh et al 1983). The sequences for procorticotropin-releasing factor (CRF) (Furutani et al 1983), proluteinizing hormone releasing hormone (LHRH) (Seeburg & Adelman 1984) and progrowth hormone releasing factor (GHRF) (Mayo et al 1983) all contain potential additional peptides, none of which has yet been isolated. Proglucagon represents one of several precursors known to be synthesized in non-neuronal tissues but for which some evidence exists of CNS distribution also (Tager et al 1980). The cloned sequence of proglucagon (Bell et al 1983, Heinrich et al 1984, Lopez et al 1983) predicts two additional glucagon-like peptides and contains the sequence for an amino-terminally extended form of glucagon called glicentin, any of which may finally be identified in brain. No hybridization studies with the proglucagon cDNA in brain have yet been published.

Immunological or immunohistochemical evidence has suggested the presence in brain of a number of "non-neuronal" peptides in addition to glucagon. One of the powerful uses of cDNA hybridization has been to demonstrate the presence of mRNAs for these peptides' precursors in the central nervous system, thus demonstrating that such peptides are synthesized locally and are not artifacts due to blood-borne or pituitary contamination. Thus the identification of POMC mRNA not only in the hypothalamus but other brain regions as well (Civelli et al 1982, Mocchetti et al 1984c), along with the presence of POMC-derived peptides, conclusively demonstrates local synthesis of POMC within the brain.

HYBRIDIZATION: A TOOL FOR THE IDENTIFICATION OF BRAIN-SPECIFIC PROTEINS

Several laboratories have taken a more direct approach to study those genes which are only expressed in the brain, by analyzing the complexity of the brain RNA population relative to other tissues, or by screening either cDNA or genomic libraries for the presence of brain-specific sequences. Hybridization analysis has shown that brain expresses more genetic information than any other tissue (Bantle & Hahn 1976, Chikaraishi et al 1978, Grouse et al 1978, Kaplan et al 1978). Furthermore, about 50% of brain mRNAs are not polyadenylated (Chikaraishi 1979, Van Ness et al 1979), and many mRNAs begin to be expressed only after birth (Chaudhari & Hahn 1983). The original estimates of the complexity of polysomal mRNA in brain predicted 150,000 distinct molecules, of 1500 bases average size, but more recent work suggests that brain mRNAs tend to be larger with at least 30,000 different polyadenylated mRNAs expressed in brain (Milner & Sutcliffe 1983). More than 50% of these are estimated to be brain-specific.

The existence of a class of non-polyadenylated mRNAs, equal in number to the poly(A)$^+$-mRNAs, doubles the potential for diversity in terms of gene expression. The evidence that the poly(A)$^-$ molecules do function as mRNAs, their presence on ribosomes and capability to be translated in vitro (Chikaraishi 1979, Van Ness et al 1979), has been further substantiated by Brilliant et al (1984). They have screened a genomic library for brain-specific messages and identified two clones, out of seven, whose transcripts are poly(A)$^-$. One of these clones is brain-specific, whereas the other is expressed in brain, liver, and kidney. The proteins for which the poly(A)$^-$ mRNAs code, and their functions, remain to be determined.

Milner & Sutcliffe (1983) have generated a cDNA library from rat brain poly(A)$^+$ mRNA and screened 191 random clones for expression in brain versus liver or kidney. Their analysis and calculations suggest that 56% of the messages are brain-specific, whereas the remainder are expressed in two or all three tissues. Two of the brain-specific clones have been sequenced, open-reading frames determined, and antibodies generated against specific peptides whose sequences were determined from the DNA sequences (Sutcliffe et al 1983). One clone, called p1A75, codes for a 28 kD protein, whose mRNA was found throughout the brain as well as in PC12 pheochromocytoma cells. The protein was localized by immunohistochemistry to large neurons in all regions of the brain. The staining was granular and confined to the cytoplasm, leading the authors to suggest that the protein "could be involved in the synthesis or directional export of proteins destined for dendrites or could be a component of cytoplasmic organelles such as mitochondria" (Sutcliffe et al 1983). A second brain-specific clone, p1B236, coded for a 318 amino acid protein, with several pairs of dibasic amino acids in the sequence, therefore suggesting the possibility that it was a polyprotein precursor for one or more neuropeptides. Its mRNA was most abundant in midbrain and hindbrain, with lower amounts in cortex and olfactory regions. Antisera raised against three of the potential peptides produced staining in fibers throughout the brain but most abundantly in pons, cerebellum, hypothalamus, and parts of the neocortex and hippocampus. A fourth antiserum, raised against a fourth potential peptide, showed no staining in any brain area. Colchicine treatment of the rats revealed cell bodies stained by the antisera in the same areas where mRNA had been detected (Sutcliffe et al 1983).

Whether any of these authors have identified a brain-specific protein whose processing will generate brain-specific, biologically active peptides remains to be established. The nonexpression of a mRNA in liver or kidney does not guarantee that it is brain-specific. Almost all of the neuropeptides whose precursors have been cloned (Table 1) are found in tissues other than brain, none of which include liver or kidney. The use of liver or kidney for comparison thus excludes a number of genes that have no specific role in brain function but it does not necessarily define brain-specific genes. From a functional standpoint a more meaningful approach appears to be that of searching for brain region–specific clones by subtractive hybridization techniques (Davis et al 1984, Sargent & Dawid 1983): for example, that population of mRNAs left in a striatal sample after removal of all messages expressed in the cortex or even in the remainder of the brain.

We are left finally with the question: Have any brain-specific neuropeptides been identified, or does such a class exist? The candidates for brain-specific peptides are CRF, AVP, CGRP, and the peptides from clone p1B236; for neuroendocrine pituitary-specific peptides, GH; and for brain-adrenal medulla specific peptides, neuropeptide Y. Exhaustive studies of tissue-specific mRNA expression have not been carried out for any of them. In fact, one of the more striking findings to emerge from research on peptides to date has been the relative lack of tissue-specific localization, although in some instances the presence of a peptide in a nonneuronal tissue may be the result of its localization in nerve fibers innervating that tissue. Hybridization studies of mRNA expression in a given tissue, and ultimately in situ hybridization analysis of mRNA expression in a specific cell type, will provide the final answer as to whether there exist neuropeptides that are brain-specific. Furthermore, the recent proposal that brain-specific genes contain an identifier DNA sequence (ID sequence) in one of their introns that makes them brain-specific (Milner et al 1984, Sutcliffe et al 1982) does not appear to be holding up, since virtually the same DNA sequence has been identified in genes for prolactin (PRL) (Schuler et al 1983), GH (Barta et al 1981), rat UI, the Harvey sarcoma virus-ras gene, and a tubulin pseudogene (Milner et al 1984). As a neurobiologist, one believes that there must be brain-specific proteins and peptides—perhaps the correct strategy for identifying them has not yet been found.

HYBRIDIZATION TO ASSESS THE DYNAMIC STATE OF NEUROPEPTIDE STORES

Since every cell is totipotent for the genome, specific regulatory mechanisms must govern which tissue or cell expresses any given gene. In addition to the "off" or "on" state, many genes are expressed in a regulated state, in which the rate of transcription of the gene can be modulated by the appropriate hormone, by nerve impulse traffic, or by specific transsynaptic modulation. The availability of cDNA probes for various neuropeptides has led to many studies on regulation of the mRNA content and of the transcription rate for specific neuropeptide genes. In this section we discuss current research trends without attempting to provide comprehensive details of all such experiments. Much of the work has been carried out in the pituitary, but recent studies have looked at both the brain and adrenal medulla.

A problem of great significance in neuroscience is the development of appropriate methodology to assess the functional participation of a given neuronal pathway in a behavioral or pharmacological modification of brain function. Thus, one needs to estimate changes in the dynamic state of the neurotransmitter that is located in this pathway. For neuropeptides, measurement of changes in peptide content are not sufficient because they are not readily interpretable in dynamic terms. For example, peptide content could increase either because of a decrease in utilization, resulting in an accumulation of the peptide, or because of an increased rate of formation, compensatory to an increased rate of utilization. Since the peptides are synthesized as parts of polyproteins, an increased rate of formation of peptide could occur at the level of gene transcription, mRNA translation, or precursor processing. In order to estimate in vivo changes in the dynamic state of neuropeptides, one needs parallel measurements of mRNA as well as of the precursor and the peptide itself. To determine whether a hormone or transmitter is modulating the expression of the gene itself, one must estimate transcription rates. Detection of changes in mRNA content, including the content of nuclear precursors, by hybridization analysis of either gel blots (Northern blot) or dot blots, is suggestive of a change in the rate of transcription but could also occur as a result of other mechanisms, such as mRNA stabilization or changes in mRNA degradation. At present in vitro nuclear transcription run-off (McKnight & Palmiter 1979) is the only experimental method available to study the regulation of the transcription rate of a specific gene. However, for the purpose of estimating changes in the dynamic state of a neuropeptide, what is important is to demonstrate an increase of mRNA and to document that the increase is of functional significance by showing an increased content of the translation products, including that of the biologically active final product. Thus, hybridization analysis of mRNA with cDNA probes, together with radioimmunoassays of the precursors and neuropeptide, can determine the direction of changes in the dynamic state of a peptide due to behavioral or physiological stimuli, but in order to determine the mechanism whereby the changes became operative, one has to use a more direct approach to study regulation of gene expression, such as nuclear transcription run-off.

Pituitary

A large body of work has established that both the tissue content and the release of the pituitary neuroendocrine peptides are regulated by their releasing factors, as well as by peptidergic and nonpeptidergic transmitters and steroids reaching the gland through the portal circulation. DNA hybridization techniques are now being applied to determine which of these effects involve changes in the rate of gene expression.

Regulation of gene expression in anterior pituitary cells by peptides and steroids has been demonstrated for prolactin (PRL), growth hormone (GH), and POMC. Much of this work has utilized the GH cell lines (derived from a rat anterior pituitary tumor), which will ultimately allow an analysis of the molecular events occurring in the cell between activation of the receptor and a change of transcription rate in the nucleus. For prolactin (PRL), the content of nuclear precursor RNA was stimulated by thyrotropin-releasing hormone (TRH) (Biswas et al 1982) and by epidermal growth factor (EGF) (Murdoch et al 1982). EGF was shown to stimulate the rate of transcription as well (Murdoch et al 1982). Increases of the levels of cytoplasmic mRNA and of prolactin followed. These results demonstrate that EGF regulates prolactin gene expression at the level of transcription and suggest that TRH does likewise. The powerful approach of DNA transfection, basically an assay for transcription of a cloned gene, is being applied in the case of EGF to determine the regulatory sequence in the PRL gene responsible for EGF effects (Supowit et al 1984). A DNA vector was constructed with the SV40 plasmid vector, $pSV2_{neo}$, which contains the selectable marker for G418 resistance, into which was inserted a fusion gene containing the 5'-flanking region of the PRL gene and the protein-coding sequences of the GH gene. This vector was inserted into the human cell line, A431, which contains EGF receptors, by DNA-mediated gene transfer. Neomycin-resistant cells produced an authentic GH mRNA: EGF increased both the content of this mRNA and the rate of transcription of the GH gene with kinetics similar to those seen for EGF regulation of the PRL gene in GH4 cells. If the 5' GH DNA sequence was substituted for the 5' PRL DNA sequence, EGF regulation was lost. This methodology will ultimately allow an exact determination of the DNA sequence responsible for conferring EGF sensitivity on the PRL gene. Transcriptional regulation of the GH gene by a pancreatic peptide GH-releasing factor and by thyroid hormone has been demonstrated in primary cultures of anterior pituitary cells (Barinaga et al 1983, Spindler et al 1982). Constant in vivo infusion of corticotropin-releasing factor (CRF) increased POMC mRNA and β-endorphin in the anterior lobe of the pituitary but decreased POMC mRNA in the intermediate lobe: both effects took several days to be detected, and no transcriptional analyses were carried out (Höllt & Haarmann 1984).

Glucocorticoids exert transcriptional regulation on at least two of the anterior pituitary neuroendocrine peptides, GH and POMC. Dexamethasone treatment of GH cells increased GH mRNA and nuclear precursor RNA content (Dobner et al 1981, Spindler et al 1982) as well as stimulating the rate of transcription (Evans et al 1982). Furthermore, treatment of adrenalectomized rats with dexamethasone increased GH transcription in pituitary nuclei (Evans et al 1982). Steroids exert their actions by binding to nuclear receptors, which then translocate to and bind to nuclear acceptor sites, probably in the 5'-flanking region of the gene in question. Several laboratories have begun DNA-cell transfection experiments using different vectors in order to analyze the glucocorticoid regulatory DNA sequence in the GH gene and have demonstrated steroid regulation of GH mRNA and peptide content (Doehmer et al 1982, Miller et al 1984, Robins et al 1982). Since purified steroid receptors are becoming available, such experiments should ultimately lead to an understanding of the molecular basis for the interaction between a specific DNA sequence and a steroid-receptor protein complex.

A series of in vivo studies demonstrated that adrenalectomy resulted in an increase of POMC gene transcription in the anterior pituitary within an hour (Birnberg et al 1983). The mRNA content began to increase 24 hr later and continued to increase for at least 18 days (Civelli et al 1983, Herbert et al 1981, Schachter et al 1982). Administration of dexamethasone to control rats resulted in a 75–80% decrease of POMC mRNA whereas it blocked the effects of adrenalectomy (Birnberg et al 1983, Schachter et al 1982). One advantage of studies using pituitary is that plasma levels of the peptides can be assayed as an index of secretion, while tissue content provides a measure of synthesis and processing. Thus in these studies secretion of ACTH and synthesis of POMC were correlated with changes in POMC mRNA. Adrenalectomy caused rapid release of ACTH, with a concomitant fall in tissue content, which was not restored until a day after POMC mRNA content increased; in contrast, dexamethasone decreased ACTH release, resulting in an initial increase in the tissue content, which then declined as the mRNA decreased (Birnberg et al 1983, Herbert et al 1981, Schachter et al 1982). These results thus demonstrate that the plasma steroid content modulates the transcription rate of the POMC gene. Initial transfection experiments and in vitro transcription studies using HeLa cell extracts (Manley et al 1980) have not yet detected a glucocorticoid regulatory sequence in the POMC gene (Mishina et al 1982, Notake et al 1983a), although deletion of a segment from 53 to 59 bases upstream of the CAP site resulted in enhanced transcription. It is possible that this sequence forms part of the inhibitory steroid site, which may be tonically activated otherwise by steroids in the culture medium.

Although the POMC gene is also expressed in the intermediate lobe of the pituitary, it is not under steroid regulation (Herbert et al 1981, Schachter et al 1982), presumably because there are no glucocorticoid receptors in these cells. However, intermediate lobe POMC is known to be under the inhibitory control of dopamine, released from the median eminence into the portal circulation of the pituitary. Hybridization analysis of POMC mRNA in rats treated with haloperidol (a dopamine receptor antagonist) for three days showed that the POMC mRNA content increased three–five-fold (Chen et al 1983). The effect was dose- and time-dependent and was correlated with increases in ^{35}S-POMC synthesis and β-endorphin tissue content. Treatment of rats with ergocryptine (a dopamine receptor agonist) for 3 days led to a decrease of POMC mRNA. The results not only confirmed those of an earlier study in which mRNA was measured by in vitro translation (Höllt et al 1982) but showed the greatly enhanced sensitivity of hybridization methods: Chen et al (1983) could detect an increase in POMC mRNA within 6 hr, whereas Höllt et al (1982) saw no change until 7 days (maximal increase 100–150%).

Adrenal Medulla

The discovery that the adrenal medulla contains enkephalin peptides (Costa et al 1979, Schultzberg et al 1978) has led to several studies on the regulation of proenkephalin synthesis and peptide content in that tissue. The enkephalins are found in the same granules as the catecholamines and can be released simultaneously by either splanchnic nerve stimulation (Govoni et al 1981, Lewis et al 1979) or by nicotinic receptor activation (Livett et al 1981, Stine et al 1980, Viveros et al 1980). Chromaffin cells thus appear to be a good model system in which to study the regulation of expression of peptides coexisting with classical transmitters (Costa et al 1979, Kataoka et al 1984, Schultzberg et al 1978). Since cAMP appears to regulate tyrosine hydroxylase synthesis (Guidotti & Costa 1973, Kumakura et al 1979), cDNA hybridization analyses of PE mRNA and its regulation by cAMP in cultured chromaffin cells were undertaken. Treatment of the cells with 8-Br-cyclic AMP resulted in a time- and dose-dependent increase of PE mRNA within 12 hr, with an increase in the cellular content of both high- and low-molecular weight enkephalin peptides and an increase in released enkephalins apparent by two days (Quach et al 1984, Schwartz et al 1984). Similar results were obtained by Eiden et al (1984a,b) using forskolin, cholera toxin, or nicotine to elevate intracellular cyclic AMP. The effect of cyclic AMP is due to an increase in PE gene transcription (J. P. Schwartz, unpublished results). Comparable changes were seen in tyrosine hydroxylase mRNA (J. R. Naranjo and J. P. Schwartz, unpublished results) by using a rat tyrosine hydroxylase probe (Lewis et al 1983). Denervation of the spanchnic nerve caused an increase of PE mRNA (Kilpatrick et al 1984), whereas destruction of catecholamine stores by reserpine either in vivo (Naranjo et al 1984) or in cultured cells (Eiden et al 1984a, Naranjo et al 1984) resulted in a decrease of PE mRNA. In both instances, the content of enkephalin peptides increased, thus clearly demonstrating that measurement of the peptide content alone is not a suitable index for estimating the dynamic state of the peptide content.

Brain

The neuropeptide content of neurons can be changed by many CNS-acting drugs or behavioral paradigms. Hybridization analyses of mRNAs for specific neuropeptide precursors in various brain regions have just begun to give indications of the potentials and limitations of hybridization techniques in assessing the dynamic state of a peptide. Changes in the content of neuropeptide precursor mRNAs have been correlated with changes in precursor and peptide content on a regional basis. However, the technique of in situ hybridization will be necessary to provide similar information on a cellular basis. In this section, we discuss the use of hybridization analyses to analyze the regional distribution of neuropeptide precursor mRNAs and therefore the cell bodies of the peptidergic neurons. We also discuss pharmacological studies that have shown the great specificity in terms of which peptide is affected in which brain area by which drug. These studies have begun to define interactions between classical neurotransmitters and neuropeptides and have shown transsynaptic regulation of neuropeptide-containing neurons at the level of transcription of the precursor gene. Finally we discuss a series of experiments that demonstrate clearly how necessary it is to measure the specific mRNA together with both the precursor and the peptide in order to analyze changes in the dynamic state of a peptide.

The brain distribution of the mRNAs for the three opioid peptide precursors has been determined. POMC mRNA was detected in the hypothalamus, in confirmation of in situ hybridization and immunohisto-chemical studies showing the hypothalamic location of POMC-containing cell bodies (Gee et al 1983). In addition, however, POMC mRNA was detected by RNA gel blot (Northern) hybridization in the amygdala and cortex (Civelli et al 1982) as well as the midbrain, brain stem and cerebellum (Mocchetti et al 1984c). Although the hypothalamic content of POMC mRNA was 20-fold lower than that of the anterior pituitary, the content in the other brain regions was 10–20-fold lower than that of the hypothalamus.

PE mRNA was found in every brain region examined (striatum, hypothalamus, midbrain, brain stem, cortex, cerebellum, hippocampus), as well as in spinal cord and the neurointermediate pituitary, and the relative content of mRNA correlated well with that of the enkephalin peptides (Mocchetti et al 1984c, Tang et al 1983). Enkephalinergic neurons are thought to be primarily small interneurons, and these results provide corroboration that the cell bodies and terminals are closely associated within a given brain region. Prodynorphin mRNA has been identified in porcine hypothalamus and spinal cord (Jingami et al 1984), although the distribution of dynorphin peptides in the brain includes the hypothalamus, as well as striatum, hippocampus, neocortex, medulla-pons, cerebellum, and posterior pituitary (Goldstein & Ghazarossian 1980). Although the results of Jingami et al (1984) suggest that one set of cell bodies in the hypothalamus projects to many other brain regions, the results obtained for POMC suggest that small groups of prodynorphin neurons may be located in other brain areas. The regional distribution of mRNA has been determined for only two other neuropeptides for which cDNAs are available, CGRP (Rosenfeld et al 1983) and substance P–substance K (Nawa et al 1984b): the results were discussed above in the section on differential splicing.

A neuropharmacological approach was taken to study regulation of the synthesis of opioid peptides in various brain areas. A cDNA probe for human pheochromocytoma PE (Comb et al 1982), used to analyze PE mRNA in rat brain, detected the presence of an $\sim 1400b$ mRNA in all brain regions, which appeared to be the same as that found in the adrenal, providing the first evidence that the brain and adrenal enkephalin peptides were derived from the same PE precursor (Tang et al 1983). This has recently been confirmed with the cloning and sequencing of rat brain PE mRNA (Howells et al 1984, Yoshikawa et al 1984) and of the rat PE gene (Rosen et al 1984). Earlier work had shown that repeated injections of antipsychotics elicited a selective increase of enkephalins in the striatum (Hong et al 1978). Indirect evidence had suggested that the increase was associated with an increase in biosynthesis (Hong et al 1979). Chronic treatment of rats with the dopamine blocker, haloperidol, led to a four-fold increase of PE mRNA content in the striatum, with no change occurring in the hypothalamus (Tang et al 1983). The effect in the striatum was confirmed by Sabol et al (1983), using in vitro translation which had a lower sensitivity than hybridization analysis. These studies have now been extended to show that there is also no change in the PE mRNA content of brain stem or cortex (Mocchetti et al 1985). Neither an acute nor a one-day exposure to haloperidol had any effect on striatal PE mRNA or enkephalin peptide content but by four days both were increased: the increase became maximal by two to three weeks. In addition, the haloperidol effect is dose-dependent (I. Mocchetti, unpublished results). The increase of PE mRNA content is paralleled by an increase in high molecular weight enkephalin precursors (separated on a Bio-Gel P-2 column and analyzed by radio-immunoassay after trypsin-carboxypeptidase B treatment to reveal cryptic enkephalin peptides) and in the low molecular weight biologically active peptides, met^5-enkephalin, met^5-enkephalin-arg^6-phe^7 and met^5-enkephalin-arg^6-gly^7-leu^8. One can therefore infer that the dynamic state of the enkephalin peptides is changed, because of an increase in the rates of formation and utilization. Results from in vitro nuclear transcription run-off experiments show that haloperidol increases the rate of PE gene transcription (J. P. Schwartz and I. Mocchetti, unpublished results). We interpret these studies to indicate that dopamine exerts a tonic inhibition, either directly or transsynaptically, on the activity of the striatal enkephalinergic neuron: relief of this inhibition by blockade of dopaminergic function with haloperidol turns on neuronal activity and the synthesis of PE. Further pharmacological support comes from studies with reserpine and 6-hydroxydopamine (I. Mocchetti, in preparation). Although reserpine given systemically depletes catecholamines in all areas of the brain, PE mRNA increased only in the striatum, suggesting that a catecholamine-enkephalin transsynaptic connection is operative only in this brain region.

In confirmation of these findings, injection of 6-hydroxydopamine unilaterally into the substantia nigra led to an increase of PE mRNA only on the ipsilateral side of the striatum. The specificity extends to POMC: haloperidol had no effect on POMC mRNA content in any brain region but increased it in the neurointermediate lobe of the pituitary (Mocchetti et al 1984c), in confirmation of earlier results (Chen et al 1983, Höllt et al 1982). Thus, just as dopamine released from the median eminence regulates POMC synthesis in the pituitary, dopamine secreted from nigro-striatal neurons transsynaptically regulates PE synthesis in the striatum.

Two additional studies investigated changes of opioid peptide precursor synthesis elicited pharmacologically. Sabol et al (1983) reported that chronic lithium treatment increased striatal PE mRNA 25–29%, measured by in vitro translation. No other brain regions were studied. Mocchetti et al (1984b) reported that in rats made tolerant to and dependent on morphine, there was a 50% decrease of hypothalamic POMC mRNA content, with no change in β-endorphin content. No changes were detected in other brain regions. An acute dose of morphine was ineffective. Neither PE mRNA nor enkephalin peptide content was altered in any brain region.

Burbach et al (1984) have demonstrated that osmotic stress of rats, induced by drinking 2% saline for 14 days, resulted in an increase of vasopressin-neurophysin (VP) mRNA in specific hypothalamic nuclei. The VP mRNA content of the supraoptic nucleus increased five-fold, and that of the paraventricular nucleus two-fold. These results are in good agreement with a nine-fold increase in plasma vasopressin content, since these are the hypothalamic nuclei that project to the posterior pituitary. The content of VP mRNA in the suprachiasmatic nucleus did not change, suggesting that the VP neurons of this nucleus project elsewhere and are differently regulated.

These findings point out the potential value of cDNA hybridization techniques for the detection of changes in the dynamic equilibrium of neuropeptides. One can ask questions about transsynaptic modulation of peptide utilization and probe mechanisms of transsynaptic regulation associated with changes in brain function. Such information is beginning to emerge concerning interactions between serotonergic and opioid peptide-containing neurons. Treatment of rats with any one of three drugs that reduce serotonin content by different biochemical mechanisms— fenfluramine, 5,7-dihydroxytryptamine, or p-chlorophenylalanine— increased the enkephalin peptide content of striatum and hypothalamus and the β-endorphin content of hypothalamus. However, no change was detected in either PE mRNA or POMC mRNA content in these brain regions (Mocchetti et al 1984a,c, 1985). In addition, fenfluramine treatment had no effect on the content of HMW enkephalin precursors in either striatum or hypothalamus. These results demonstrate that serotonergic neurons transsynaptically activate the utilization of enkephalins and β-endorphin: reduction of serotonergic transmission by three different mechanisms caused a decrease in the rate of opioid peptide utilization and a resultant accumulation of the active form of the peptides.

These findings have important implications in understanding the functional role of neuropeptides in the CNS and evaluating intrinsic mechanisms of transsynaptic regulation of gene expression. They demonstrate that changes in peptide content, as measured by radioimmunoassay, can occur as a result of decreased utilization without a change in the synthesis rate or content of the precursor. But they have also led to the proposal that one can begin to estimate the dynamic state of a neuropeptide by combining cDNA hybridization techniques to measure the content of precursor mRNA with radioimmunoassay of column-separated high molecular weight precursors and low molecular weight peptides. With the addition of the in vitro transcription run-off assay in isolated nuclei, one can then determine whether a drug or transmitter is affecting a neuropeptide at the level of transcription, translation, processing, or utilization.

SUMMARY

During the course of evolution, species have increased in complexity, and their nervous systems have evolved correspondingly with an increase in the diversity of their capabilities to respond. Part of that diversity has resulted from an increase in cell types and numbers and their interconnections. In addition, much of it comes from the panoply of neurotransmitters available, of which the neuropeptides represent a major portion. The application of the techniques of molecular biology to the nervous system has led to an appreciation of some of the genetic means by which such diversity can be generated.

The cloning and sequencing of peptide precursor genes has shown the existence of gene families, genes with duplications of internal sequences, and genes evolutionarily related to one another, suggesting that one response to the increasing complexity of the organism has been a genetic diversification of the precursor population for peptides. As the precursor genes evolved and thereby provided increasing numbers of peptides, the receptor genes may have evolved simultaneously to provide diversification in the responses to these peptides (for example, the opioid peptide precursors) (Comb et al 1983). The precursor sequences obtained have led not only to the predictions of new peptides but also to the discovery of alternative methods of generating diversity from a single gene. At one extreme, the gene is translated into a polyprotein containing several peptides, which are produced in and released from the same cell. At the other extreme, the nuclear transcript of the gene is differentially spliced such that one peptide is expressed in one tissue and another in a different tissue (Calcitonin-CGRP), or one peptide may be expressed with or without a second peptide in different cells (substance P–substance K). The net result is either one neuron producing a multiplicity of responses to several co-released peptides derived from a polyprotein (POMC or PE) or a tissue- or cell-specificity in terms of which peptide is produced and released.

Numerous applications have been made utilizing the cDNA probes generated from the cloning of neuropeptide precursors. Hybridization analyses, including in vitro transcription run-off, have demonstrated that the transcription of neuropeptide genes is regulated by transsynaptic activation of transmitter receptors located in the neuronal membrane, or by hormones, or by as yet unveiled mechanisms. Hybridization techniques have allowed assessment of the dynamic state of neuropeptides functioning as neuromodulators. In turn, such an assessment has created an appropriate background against which to analyze the participation of a given neuronal system storing a putative peptide neurotransmitter in the mode of action of a drug or in the response of an animal to the environment. Detection by in situ hybridization of a mRNA for a specific peptide precursor, along with the peptide, confirms synthesis of that peptide in a particular brain nucleus or area. As a result of such studies, "non-neuronal" peptides have been shown to be synthesized in the brain, while synthesis of "neuronal" peptides has been demonstrated in non-neuronal tissues. The technique of in situ hybridization will ultimately provide evidence for synthesis on a cell-by-cell basis and should greatly expand our ability to identify co-transmitter location.

The technology of DNA mediated-cell transfection will ultimately allow us to correlate the structure of a gene with its function and expression. With the various techniques of site-specific mutagenesis available, it should be possible to define which parts of a protein are necessary for a specific function, how changes in its structure affect the processing pattern or post-translational processing events such as glycosylation or phosphorylation, how changes in regulatory sequences of the gene affect its expression or stimulation by hormones or transmitters, and how tissue-specific expression of a gene is determined. As these answers come in, cDNA hybridization techniques should become applicable to the study of human diseases involving neuropeptides and of possible methods for preventing or reversing the diseases.

Literature Cited

Amara, S. G., Evans, R. M., Rosenfeld, M. G. 1984. Calcitonin/calcitonin gene-related peptide transcription unit: Tissue-specific expression involves selective use of alternative polyadenylation sites. *Mol. Cell Biol.* 4:2151–60

Bantle, J. A., Hahn, W. E. 1976. Complexity and characterization of polyadenylated RNA in the mouse brain. *Cell* 8:139–50

Barinaga, M., Yamonoto, G., Rivier, C., Vale, W., Evans, R., Rosenfeld, M. G. 1983. Transcriptional regulation of growth hormone expression by growth hormone-releasing factor. *Nature* 306:86–87

Barta, A., Richards, R. I., Baxter, J. D., Shine, J. 1981. Primary structure and evolution of rat growth hormone gene. *Proc. Natl. Acad. Sci. USA* 78:4867–71

Bell, G. I., Santene, R. F., Mullenbach, G. T. 1983. Hamster preproglucagon contains the sequence of glucagon and two related peptides. *Nature* 302:716–18

Birnberg, N. C., Lissitzky, J.-C., Hinman, M., Herbert, E. 1983. Glucocorticoids regulate proopiomelanocortin gene expression *in vivo* at the levels of transcription and secretion. *Proc. Natl. Acad. Sci. USA* 80:6982–86

Biswas, D. K., Hanes, S. D., Brennessel, B. A. 1982. Mechanism of induction of prolactin synthesis in GH cells. *Proc. Natl. Acad. Sci. USA* 79:66–70

Brahic, M., Haase, A. T. 1978. Detection of viral sequences of low reiteration frequency by in situ hybridization. *Proc. Natl. Acad. Sci. USA* 75:6125–29

Brilliant, M. H., Sueoka, N., Chikaraishi, D. M. 1984. Cloning of DNA corresponding to rare transcripts of rat brain: Evidence of transcriptional and posttranslational control and of the existence of nonpolyadenylated transcripts. *Mol. Cell. Biol.* 4:2187–97

Burbach, J. P. H., De Hoop, M. J., Schmale, H., Richter, D., De Kloet, E. R., Ten Haaf, J., De Wied, D. 1984. Differential responses to osmotic stress of vasopressin-neurophysin mRNA in hypothalamic nuclei. *Neuroendocrinology* 39:582–84

Chaudhari, N., Hahn, W. E. 1983. Genetic expression in the developing brain. *Science* 220:924–28

Chen, C.-L. C., Dionne, F. T., Roberts, J. L. 1983. Regulation of the proopiomelanocortin mRNA levels in rat pituitary by dopaminergic compounds. *Proc. Natl. Acad. Sci. USA* 80:2211–15

Chen, C.-L. C., Mather, J. P., Morris, P. L., Bardin, C. W. 1984. Expression of pro-opiomelanocortin-like gene in the testis and epididymis. *Proc. Natl. Acad. Sci. USA* 81:5672–75

Chikaraishi, D. M. 1979. Complexity of cytoplasmic polyadenylated and nonpolyadenylated rat brain ribonucleic acids. *Biochemistry* 18:3249–56

Chikaraishi, D. M., Deeb, S. S., Sueoka, N. 1978. Sequence complexity of nuclear RNAs in adult rat tissues. *Cell* 13:111–20

Chin, W. W., Godine, J. E., Klein, D. R., Chang, A. S., Tan, L. K., Habener, J. F. 1983. Nucleotide sequence of the cDNA encoding the precursor of the β subunit of rat lutropin. *Proc. Natl. Acad. Sci. USA* 80:4649–53

Civelli, O., Birnberg, N., Comb, M., Douglass, J., Lissitzky, J. C., Uhler, M., Herbert, E. 1983. Regulation of opioid gene expression. *Peptides* 4:651–56

Civelli, O., Birnberg, N., Herbert, E. 1982. Detection and quantitation of pro-opiomelanocortin mRNA in pituitary and brain tissues from different species. *J. Biol. Chem.* 257:6783–87

Cochet, M., Chang, A. C. Y., Cohen, S. N. 1982. Characterization of the structural gene and putative 5′-regulatory sequences for human proopiomelanocortin. *Nature* 297:335–39

Comb, M., Seeburg, P. H., Adelman, J., Eiden, L., Herbert, E. 1982. Primary structure of the human Met- and Leu-enkephalin precursor and its mRNA. *Nature* 295:663–66

Comb, M., Rosen, H., Seeburg, P., Adelman, J., Herbert, E. 1983. Primary structure of the human proenkephalin gene. *DNA* 2:213–29

Costa, E., Di Giulio, A. M., Fratta, W., Hong, J., Yang, H.-Y. T. 1979. Interactions of enkephalinergic and catecholaminergic neurons in CNS and in periphery. In *Catecholamines: Basic and Clinical Frontiers*, ed. E. Usdin, I. J. Kopin, J. Barchas, pp. 1020–25. New York: Pergamon

Davis, M. M., Cohen, D. I., Nielsen, E. A., Steinmetz, M., Paul, W. E., Hood, L. 1984. Cell-type-specific cDNA probes and the murine I region: The localization and orientation of A_α^d. *Proc. Natl. Acad. Sci. USA* 81:2194–98

Deschenes, R. J., Lorenz, L. J., Haren, R. S., Roos, B. A., Collier, K. J., Dixon, J. E. 1984. Cloning and sequence analysis of a cDNA encoding rat preprocholecystokinin. *Proc. Natl. Acad. Sci. USA* 81:726–30

Dobner, P. R., Kawasaki, E. S., Yu, L.-Y., Bancroft, F. C. 1981. Thyroid or glucocorticoid hormone induces pregrowth-hormone mRNA and its probable nuclear precursor in rat pituitary cells. *Proc. Natl. Acad. Sci. USA* 78:2230–34

Dockray, G. J., Vaillant, C., Walsh, J. H. 1979. The neuronal origin of bombesin-like immunoreactivity in the rat gastrointestinal tract. *Neuroscience* 4:1561–68

Doehmer, J., Barinaga, M., Vale, W., Rosenfeld, M. G., Verma, I. M., Evans, R. M. 1982. Introduction of rat growth hormone gene into mouse fibroblasts via a retroviral DNA vector: Expression and regulation. *Proc. Natl. Acad. Sci. USA* 79: 2268–72

Eiden, L. E., Giraud, P., Affolter, H.-U., Herbert, E., Hotchkiss, A. J. 1984a. Alternative modes of enkephalin biosynthesis regulation by reserpine and cyclic AMP in cultured chromaffin cells. *Prod. Natl. Acad. Sci. USA* 81:3949–53

Eiden, L. E., Giraud, P., Dave, J. R., Hotchkiss, A. J., Affolter, H.-U. 1984b. Nicotinic receptor stimulation activates enkephalin release and biosynthesis in adrenal chromaffin cells. *Nature* 312:661–63

Evans, R. M., Birnberg, N. C., Rosenfeld, M. G. 1982. Glucocorticoid and thyroid hormones transcriptionally regulate growth hormone gene expression. *Proc. Natl. Acad. Sci. USA* 79:7659–63

Furutani, Y., Morimoto, Y., Shibahara, S., Noda, M., Takahashi, H., et al. 1983. Cloning and sequence analysis of cDNA for ovine corticotropin-releasing factor precursor. *Nature* 301:537–40

Gee, C. E., Chen, C.-L. C., Roberts, J. L., Thompson, R., Watson, S. J. 1983. Identification of proopiomelanocortin neurones in rat hypothalamus by *in situ* cDNA-mRNA hybridization. *Nature* 306:374–76

Gee, C. E., Roberts, J. L. 1983. *In situ* hybridization histochemistry: A technique for the study of gene expression in single cells. *DNA* 2:157–63

Gibson, S. J., Polak, J. M., Bloom, S. R., Sabate, I. M., Muldeny, P. M., et al. 1984. Calcitonin gene-related peptide immunoreactivity in the spinal cord of man and of eight other species. *J. Neurosci.* 4:3101–11

Goldstein, A., Ghazarossian, V. E. 1980. Immunoreactive dynorphin in pituitary and brain. *Proc. Natl. Acad. Sci. USA* 77:6207–10

Govoni, S., Hanbauer, I., Hexum, T. D., Yang, H.-Y. T., Kelly, G. D., Costa, E. 1981. In vivo characterization of the mechanisms that secrete enkephalin-like peptide stores in dog adrenal medulla. *Neuropharmacology* 20:639–45

Griffin, W. S. T., Alejos, M., Nilaver, G., Morrison, M. R. 1983. Brain protein and messenger RNA identification in the same cell. *Brain Res. Bull.* 10:597–601

Grouse, L. D., Schrier, B. K., Bennett, E. L.,

Rosenzweig, M. R., Nelson, P. G. 1978. Sequence diversity studies of rat brain RNA: Effects of environmental complexity on rat brain RNA diversity. *J. Neurochem.* 30:191–203

Gubler, U., Chua, A. O., Hoffman, B. J., Collier, K. J., Eng, J. 1984. Cloned cDNA to cholecystokinin mRNA predicts an identical preprocholecystokinin in pig brain and gut. *Proc. Natl. Acad. Sci. USA* 81:4307–10

Gubler, U., Monahan, J. J., Lomedico, P. T., Bhatt, R. S., Collier, K. J., et al. 1983. Cloning and sequence analysis of cDNA for the precursor of human growth hormone-releasing factor, somatocrinin. *Proc. Natl. Acad. Sci. USA* 80:4311–14

Gubler, U., Seeburg, P., Hoffman, B. J., Gage, L. P., Udenfriend, S. 1982. Molecular cloning establishes proenkephalin as precursor of enkephalin-containing peptides. *Nature* 295:206–8

Guidotti, A., Costa, E. 1973. Involvement of adenosine-3',5'-monophosphate in the activation of tyrosine hydroxylase elicited by drugs. *Science* 197:902–4

Haase, A. T., Walker, D., Stowring, L., Ventura, P., Geballe, A., et al. 1985. Detection of two viral genomes in single cells by double-label hybridization *in situ* and color microradioautography. *Science* 227:189–92

Heinrich, G., Gros, P., Lund, P. K., Bentley, R. C., Habener, J. F. 1984. Pre-proglucagon messenger RNA: Nucleotide and encoded amino acid sequences of the rat pancreatic complementary DNA. *Endocrinology* 115:2176–81

Herbert, E., Birnberg, N., Lissitzky, J.-C., Civelli, O., Uhler, M. 1981. Pro-opiomelanocortin: A model for the regulation of expression of neuropeptides in pituitary and brain. *Neurosci. Newslett.* 12:16–27

Höllt, V., Haarmann, I. 1984. Corticotropin-releasing factor differentially regulates proopiomelanocortin messenger ribonucleic acid levels in anterior as compared to intermediate pituitary lobes of rats. *Biochem. Biophys. Res. Commun.* 124:407–15

Höllt, V., Haarmann, I., Seizinger, B. R., Herz, A. 1982. Chronic heloperidol treatment increases the level of *in vitro* translatable messenger RNA coding for the β-endorphin/adrenocorticotropin precursor proopiomelanocortin in the pars intermedia of the rat pituitary. *Endocrinology* 110:1885–91

Hong, J. S., Yang, H.-Y. T., Fratta, W., Costa, E. 1978. Rat striatal met-enkephalin content after chronic treatment with cataleptogenic and non-cataleptogenic antischizophrenic drugs. *J. Pharmacol. Exp. Ther.* 205:141–47

Hong, J. S., Yang, H.-Y. T., Gillin, J. C., Di Giulio, A. M., Fratta, W., Costa, E. 1979. Chronic treatment with haloperidol accelerates the biosynthesis of enkephalins in rat striatum. *Brain Res.* 160:192–95

Horikawa, S., Takai, T., Toyosato, M., Takahashi, H., Noda, M., et al. 1983. Isolation and structural organization of the human preproenkephalin B gene. *Nature* 306:611–14

Howells, R. D., Kilpatrick, D. L., Bhatt, R., Monahan, J. J., Poonian, M., Udenfriend, S. 1984. Molecular cloning and sequence determination of rat preproenkephalin cDNA: Sensitive probe for studying transcriptional changes in rat tissues. *Proc. Natl. Acad. Sci. USA* 81:7651–55

Hudson, P., Penschow, J., Shine, J., Ryan, G., Niall, H., Coghlan, J. 1981. Hybridization histochemistry: Use of recombinant DNA as a "homing probe" for tissue localization of specific mRNA populations. *Endocrinology* 108:353–56

Hughes, J., Smith, T. W., Kosterlitz, H. W., Fothergill, L. A., Morgan, B. A., Morris, H. R. 1975. Identification of two related pentapeptides from the brain with potent opiate agonist activity. *Nature* 258:577–79

Hunter, J. C., Maggio, J. E. 1984. Pharmacological characterization of a novel tachykinin isolated from mammalian spinal cord. *Eur. J. Pharmacol.* 97:159–60

Itoh, N., Obata, K., Yanaihara, N., Okamoto, H. 1983. Human preprovasoactive intestinal peptide contains a novel PHI-27-like peptide, PHM-27. *Nature* 304:547–49

Ivell, R., Richter, D. 1984. The gene for the hypothalamic peptide hormone oxytocin is highly expressed in the bovine corpus luteum: Biosynthesis, structure and sequence analysis. *EMBO J.* 3:2351–54

Jameson, L., Chin, W. W., Hollenberg, A. N., Chang, A. S., Habener, J. F. 1984. The gene encoding the β-subunit of rat luteinizing hormone. *J. Biol. Chem.* 259:15474–80

Jingami, H., Nakanishi, S., Imura, H., Numa, S. 1984. Tissue distribution of messenger RNAs coding for opioid peptide precursors and related RNA. *Eur. J. Biochem.* 142:441–47

Johansson, O., Hökfelt, T., Elde, R. P. 1984. Immunohistochemical distribution of somatostatin-like immunoreactivity in the central nervous system of the adult rat. *Neuroscience* 13:265–339

Kakidani, H., Furutani, Y., Takahashi, H., Noda, M., Morimoto, Y., et al. 1982. Cloning and sequence analysis of cDNA for porcine β-neo-endorphin/dynorphin precursor. *Nature* 298:245–49

Kaplan, B. B., Schachter, B. S., Osterburg, H. H., de Vellis, J. S., Finch, C. E. 1978. Sequence complexity of polyadenylated RNA obtained from rat brain regions and cultured rat cells of neural origin. *Biochemistry* 17:5516–24

Kataoka, Y., Gutman, Y., Guidotti, A., Panula, P., Wroblewski, J., Cosenza-Murphy, D., Wu, J. Y., Costa, E. 1984. Intrinsic GABAergic system of adrenal chromaffin cells. *Proc. Natl. Acad. Sci. USA* 81:3218–22

Kilpatrick, D. L., Howells, R. D., Fleminger, G., Udenfriend, S. 1984. Denervation of rat adrenal glands markedly increases preproenkephalin mRNA. *Proc. Natl. Acad. Sci. USA* 81:7221–23

Kimura, S., Okada, M., Sugita, Y., Kanazawa, I., Munekata, E. 1983. Novel neuropeptides, neurokinin α and β, isolated from porcine spinal cord. *Proc. Jpn. Acad. Ser. B* 59:101–4

Kumakura, K., Guidotti, A., Costa, E. 1979. Primary cultures of chromaffin cells: Molecular mechanisms for the induction of tyrosine hydroxylase mediated by 8-Br-cyclic AMP. *Mol. Pharmacol.* 16:865–76

Land, H., Grez, M., Ruppert, S., Schmale, H., Rehbein, M., Dichter, D., Schütz, G. 1983. Deduced amino acid sequence from the bovine oxytocin-neurophysin I precursor cDNA. *Nature* 302:342–44

Land, H., Schütz, G., Schmale, H., Richter, D. 1982. Nucleotide sequence of cloned cDNA encoding bovine arginine vasopressin-neurophysin II precursor. *Nature* 295:299–303

Langer, P. R., Waldrop, A. A., Ward, D. C. 1981. Enzymatic synthesis of biotin-labeled polynucleotides: Novel nucleic acid affinity probes. *Proc. Natl. Acad. Sci. USA* 78:6633–37

Legon, S., Glover, D. M., Hughes, J., Lowry, P. J., Rigby, P. W. J., Watson, C. J. 1982. The structure and expression of the preproenkephalin gene. *Nucl. Acids Res.* 10:7905–18

Lewis, E. J., Tank, A. W., Weiner, N., Chikaraishi, D. M. 1983. Regulation of tyrosine hydroxylase mRNA by glucocorticoid and cyclic AMP in a rat pheochromocytoma cell line: Isolation of a cDNA clone for tyrosine hydroxylase mRNA. *J. Biol. Chem.* 258:14632–37

Lewis, R. V., Stern, A. S., Rossier, J., Stein, S., Udenfriend, S. 1979. Putative enkephalin precursors in bovine adrenal medulla. *Biochem. Biophys. Res. Commun.* 89:822–29

Liston, D., Patey, G., Rossier, J., Verbanck, P., Vanderhaeghen, J.-J. 1984. Processing of proenkephalin is tissue-specific. *Science* 225:734–37

Livett, B. G., Dean, D. M., Whelan, L. G., Udenfriend, S., Rossier, J. 1981. Co-release of enkephalin and catecholamines from cultured adrenal chromaffin cells. *Nature* 289:317–19

Lopez, L. C., Frazier, M. L., Su, C.-J., Kumar, A., Saunders, G. F. 1983. Mammalian pancreatic preproglucagon contains three glucagon-related peptides. *Proc. Natl. Acad. Sci. USA* 80:5485–89

Maniatis, T., Fritsch, E. F., Sambrook, J. 1982. *Molecular Cloning—A Laboratory Manual*. Cold Spring Harbor, NY: Cold Spring Harbor Lab.

Manley, J. L., Fire, A., Cano, A., Sharp, P. A., Gefter, M. L. 1980. DNA-dependent transcription of adenovirus genes in a soluble whole-cell extract. *Proc. Natl. Acad. Sci. USA* 77:3855–59

Martens, G. J. M., Herbert, E. 1984. Polymorphism and absence of Leu-enkephalin sequences in proenkephalin genes in *Xenopus laevis*. *Nature* 310:251–54

Mayo, K. E., Vale, W., Rivier, J., Rosenfeld, M. G., Evans, R. M. 1983. Expression-cloning and sequence of a cDNA encoding human growth hormone-releasing factor. *Nature* 306:86–88

McKnight, G. S., Palmiter, R. D. 1979. Transcriptional regulation of the ovalbumin and conalbumin genes by steroid hormones in chick oviduct. *J. Biol. Chem.* 254:9050–58

Miller, A. D., Ong, E. S., Rosenfeld, M. G., Verma, I. M., Evans, R. M. 1984. Infectious and selectable retrovirus containing an inducible rat growth hormone minigene. *Science* 225:993–98

Milner, R. J., Bloom, F. E., Lai, C., Lerner, R. A., Sutcliffe, J. G. 1984. Brain-specific genes have identifier sequences in their introns. *Proc. Natl. Acad. Sci. USA* 81:713–17

Milner, R. J., Sutcliffe, J. G. 1983. Gene expression in rat brain. *Nucl. Acids Res.* 11:5497–5520

Minth, C. D., Bloom, S. R., Polak, J. M., Dixon, J. E. 1984. Cloning, characterization and DNA sequence of a human cDNA encoding neuropeptide tyrosine. *Proc. Natl. Acad. Sci. USA* 81:4577–81

Mishina, M., Kurosaki, T., Yamamoto, T., Natake, M., Masu, M., Numa, S. 1982. DNA sequences required for transcription *in vivo* of the human corticotropin-β-lipotropin precursor gene. *EMBO J.* 1:1533–38

Mocchetti, I., Giorgi, O., Schwartz, J. P., Costa, E. 1984a. A reduction of the tone of 5-hydroxytryptamine neurons decreases utilization rates of striatal and hypo-

thalamic enkephalins. *Eur. J. Pharmacol.* 106:427–30

Mocchetti, I., Giorgi, O., Schwartz, J. P., Fratta, W., Costa, E. 1984b. Morphine pellets lower the hypothalamic content of proopiomelanocortin mRNA but not that of proenkephalin mRNA. *Soc. Neurosci. Abstr.* 10:1111

Mocchetti, I., Schwartz, J. P., Costa, E. 1984c. Studies of brain neuropeptide dynamics utilizing cDNA probes: Pharmacological implications. In *Physiological and Pharmacological Control of Nervous System Development*, ed. F. Caciagli, pp. 77–80. The Netherlands: Elsevier

Mocchetti, I., Schwartz, J. P., Costa, E. 1985. Use of mRNA hybridization and radioimmunoassay to study mechanisms of drug-induced accumulation of enkephalins in rat brain structures. *Mol. Pharmacol.* 28:86–91

Moody, T. W., Pert, C. B. 1979. Bombesin-like peptides in rat brain: Quantitation and biochemical characterization. *Biochem. Biophys. Res. Commun.* 90:7–14

Moore, H.-P. H., Walker, M. D., Lee, F., Kelly, R. B. 1983. Expressing a human proinsulin cDNA in a mouse ACTH-secreting cell. Intracellular storage, proteolytic processing, and secretion on stimulation. *Cell* 35:531–38

Murdoch, G. H., Potter, E., Nicolaisen, A. K., Evans, R. M., Rosenfeld, M. G. 1982. Epidermal growth factor rapidly stimulates prolactin gene transcription. *Nature* 300:192–94

Nakanishi, S., Inone, A., Kita, T., Nakamura, M., Chang, A. C. Y., Cohen, S. N., Numa, S. 1979. Nucleotide sequence of cloned cDNA for bovine corticotropin-β-lipotropin precursor. *Nature* 278:423–27

Nakanishi, S., Teranishi, Y., Noda, M., Notake, M., Watanabe, Y., Kakidani, H., Jingami, H., Numa, S. 1980. The protein-coding sequence of the bovine ACTH-β-LPH precursor gene is split near the signal peptide region. *Nature* 287:752–55

Naranjo, J. R., Mocchetti, I., Kageyama, H., Guidotti, A., Schwartz, J. P., Costa, E. 1984. Action of reserpine on met[5]-enkephalin utilization and proenkephalin mRNA content in adrenal medulla and striatum. *Soc. Neurosci. Abstr.* 10:284

Nawa, H., Doteuchi, M., Igano, K., Inouye, K., Nakanishi, S. 1984a. Substance K: A novel mammalian tachykinin that differs from substance P in its pharmacological profile. *Life Sci.* 34:1153–60

Nawa, H., Hirose, T., Takashima, H., Inayama, S., Nakanishi, S. 1983. Nucleotide sequences of cloned cDNAs for two types of bovine brain substance P precursor. *Nature* 306:32–36

Nawa, H., Kotani, H., Nakanishi, S. 1984b. Tissue-specific generation of two prepro-tachykinin mRNAs from one gene by alternative RNA splicing. *Nature* 312: 729–34

Nelkin, B. D., Rosenfeld, K. I., de Bustros, A., Leong, S. S., Roos, B. A., Baylin, S. B. 1984. Structure and expression of a gene encoding human calcitonin and calcitonin gene-related peptide. *Biochem. Biophys. Res. Commun.* 123: 648–55

Niall, H. D., Hogan, M. L., Sauer, R., Rosenblum, I. Y., Greenwood, F. C. 1971. Sequences of pituitary and placental lactogenic and growth hormones: Evolution from a primordial peptide by gene reduplication. *Proc. Natl. Acad. Sci. USA* 68: 866–69

Noda, M., Furutani, Y., Takahashi, H., Toyosato, M., Hirose, T., Inayama, S., Nakanishi, S., Numa, S. 1982a. Cloning and sequence analysis of cDNA for bovine adrenal preproenkephalin. *Nature* 295: 202–6

Noda, M., Teranishi, Y., Takahashi, H., Toyosato, M., Notake, M., Nakanishi, S., Numa, S. 1982b. Isolation and structural organization of the human prepro-enkephalin gene. *Nature* 297: 431–34

Notake, M., Kurosaki, T., Yamamoto, T., Handa, H., Mishina, M., Numa, S. 1983a. Sequence requirement for transcription *in vitro* of the human corticotropin-β-lipotropin precursor gene. *Eur. J. Biochem.* 133: 599–605

Notake, M., Tobimatsu, T., Watanabe, Y., Takahashi, H., Mishina, M., Numa, S. 1983b. Isolation and characterization of the mouse corticotropin-β-lipotropin precursor gene and a related pseudogene. *FEBS Lett.* 156: 67–71

O'Donohue, T. L., Dorsa, D. M. 1982. The opiomelanotropinergic neuronal and endocrine systems. *Peptides* 3: 353–95

Perbal, B. 1984. *A Practical Guide to Molecular Cloning.* New York: Wiley

Pintar, J. E., Schachter, B. S., Herman, A. B., Durgerian, S., Krieger, D. T. 1984. Characterization and localization of proopiomelanocortin messenger RNA in the adult rat testis. *Science* 225: 632–34

Quach, T. T., Tang, F., Kageyama, H., Mocchetti, I., Guidotti, A., Meek, J. L., Costa, E., Schwartz, J. P. 1984. Enkephalin synthesis in adrenal medulla. Modulation of proenkephalin mRNA content of cultured chromaffin cells by 8-bromo-adenosine-3',5'-monophosphate. *Mol. Pharmacol.* 26: 255–60

Robins, D. M., Paek, I., Seeburg, P. H., Axel, R. 1982. Regulated expression of human growth hormone genes in mouse cells. *Cell* 29: 623–31

Rosen, H., Douglass, J., Herbert, E. 1984. Isolation and characterization of the rat proenkephalin gene. *J. Biol. Chem.* 259: 14309–13

Rosenfeld, M. G., Amara, S. G., Roos, B. A., Ong, E. S., Evans, R. M. 1981. Altered expression of the calcitonin gene associated with RNA polymorphism. *Nature* 290: 63–65

Rosenfeld, M. G., Lin, C. R., Amara, S. G., Stolarsky, L., Roos, B. A., Ong, E. S., Evans, R. M. 1982. Calcitonin mRNA polymorphism: Peptide switching associated with alternative RNA splicing events. *Proc. Natl. Acad. Sci. USA* 79: 1717–21

Rosenfeld, M. G., Mermod, J.-J., Amara, S. G., Swanson, L. W., Sawchenko, P. E., Rivier, J., Vale, W. W., Evans, R. M. 1983. Production of a novel neuropeptide encoded by the calcitonin gene via tissue-specific RNA processing. *Nature* 304: 129–35

Ruppert, S., Scherer, G., Schutz, G. 1984. Recent gene conversion involving bovine vasopressin and oxytocin precursor genes suggested by nucleotide sequence. *Nature* 308: 554–57

Sabol, S. L., Yoshikawa, K., Hong, J.-S. 1983. Regulation of methionine-enkephalin precursor messenger RNA in rat striatum by haloperidol and lithium. *Biochem. Biophys. Res. Commun.* 113: 391–99

Sargent, T. D., Dawid, I. B. 1983. Differential gene expression in the gastrula of *Xenopus laevis. Science* 222: 135–39

Schachter, B. S., Johnson, L. K., Baxter, J. D., Roberts, J. L. 1982. Differential regulation by glucocorticoids of proopiomelanocortin mRNA levels in the anterior and intermediate lobes of the rat pituitary. *Endocrinology* 110: 1442–44

Schuler, L. A., Weber, J. L., Gorski, J. 1983. Polymorphism near the rat prolactin gene caused by insertion of an Alu-like element. *Nature* 305: 159–60

Schultzberg, M., Lundberg, J. M., Hökfelt, T., Terenius, L., Brandt, J., Elde, R. P., Goldstein, M. 1978. Enkephalin-like immunoreactivity in gland cells and nerve terminals of the adrenal medulla. *Neuroscience* 3: 1169–86

Schwartz, J. P., Quach, T. T., Tang, F., Kageyama, H., Guidotti, A., Costa, E. 1984. Cyclic AMP-mediated modulation of gene expression in adrenal chromaffin cell cultures. *Adv. Cycl. Nucl. Res.* 17: 529–34

Seeburg, P. H., Adelman, J. P. 1984. Characterization of cDNA for precursor of human luteinizing hormone releasing hormone. *Nature* 311: 666–68

Seeburg, P. H., Shine, J., Martial, J. A.,

304   SCHWARTZ & COSTA

Baxter, J. D., Goodman, H. M. 1977. Nucleotide sequence and amplification in bacteria of the structural gene for rat growth hormone. *Nature* 270:486–94

Selby, M. J., Barta, A., Baxter, J. D., Bell, G. I., Eberhardt, N. L. 1984. Analysis of a major human chorionic somatomammotropin gene: Evidence for 2 functional promoter elements. *J. Biol. Chem.* 259:13131–38

Shen, L.-P., Pictet, R. L., Rutter, W. J. 1982. Human somatostatin I: Sequence of the cDNA. *Proc. Natl. Acad. Sci. USA* 79: 4575–79

Soma, G.-I., Kitahara, N., Nishizawa, T., Nanami, H., Kotake, C., Okazaki, H., Andoh, T. 1984. Nucleotide sequence of a cloned cDNA for proopiomelanocortin precursor of chum salmon, *Onchorynchus keta. Nucl. Acids Res.* 12:8029–41

Spindel, E. R., Chin, W. W., Price, J., Rees, L. H., Besser, G. M., Habener, J. F. 1984. Cloning and characterization of cDNAs encoding human gastrin-releasing peptide. *Proc. Natl. Acad. Sci. USA* 81:5699–5703

Spindler, S. R., Mellon, S. H., Baxter, J. D. 1982. Growth hormone gene transcription is regulated by thyroid and glucocorticoid hormones in cultured rat pituitary tumor cells. *J. Biol. Chem.* 257:11627–32

Steenbergh, P. H., Höppener, J. W. M., Zandberg, J., Roos, B. A., Jansz, H. S., Lips, C. J. M. 1984. Expression of the proopiomelanocortin gene in human medullary thyroid carcinoma. *J. Clin. Endocrinol. Metab.* 58:904–8

Stine, S. M., Yang, H.-Y. T., Costa, E. 1980. Release of enkephalin-like immunoreactive material from isolated bovine chromaffin cells. *Neuropharmacology* 19: 683–85

Supowit, S. C., Potter, E., Evans, R. E., Rosenfeld, M. G. 1984. Polypeptide hormone regulation of gene transcription: Specific 5′ genomic sequences are required for epidermal growth factor and phorbol ester regulation of prolactin gene expression. *Proc. Natl. Acad. Sci. USA* 81:2975–79

Sutcliffe, J. G., Milner, R. J., Bloom, F. E., Lerner, R. A. 1982. Common 82-nucleotide sequence unique to brain RNA. *Proc. Natl. Acad. Sci. USA* 79:4942–46

Sutcliffe, J. G., Milner, R. J., Shinnick, T. M., Bloom, F. E. 1983. Identifying the protein products of brain-specific genes with antibodies to chemically synthesized peptides. *Cell* 33:671–82

Tager, H., Hohenboken, M., Markese, J., Dinerstein, R. J. 1980. Identification and localization of glucagon-related peptides in rat brain. *Proc. Natl. Acad. Sci. USA* 77:6229–33

Taii, S., Thara, Y., Mori, T. 1984. Identification of the mRNA coding for prolactin in the human decidua. *Biochem. Biophys. Res. Commun.* 124:530–37

Tang, F., Costa, E., Schwartz, J. P. 1983. Increase of proenkephalin mRNA and enkephalin content of rat striatum after daily injection of haloperidol for 2 to 3 weeks. *Proc. Natl. Acad. Sci. USA* 80: 3841–44

Uhler, M., Herbert, E. 1983. Complete amino acid sequence of mouse pro-opiomelanocortin derived from the nucleotide sequence of pro-opiomelanocortin cDNA. *J. Biol. Chem.* 258:257–61

Van Ness, J., Maxwell, I. H., Hahn, W. E. 1979. Complex population of nonpolyadenylated messenger RNA in mouse brain. *Cell* 18:1341–49

Viveros, O. H., Diliberto, E. J., Hazum, E., Chang, K.-J. 1980. Enkephalins as possible adrenomedullary hormones: Storage, secretion and regulation of synthesis. *Adv. Biochem. Psychopharmacol.* 22:191–204

Whitfeld, P. L., Seeburg, P. H., Shine, J. 1982. The human pro-opiomelanocortin gene: Organization, sequence, and interspersion with repetitive DNA. *DNA* 1:133–43

Yoshikawa, K., Williams, C., Sabol, S. L. 1984. Rat brain preproenkephalin mRNA. cDNA cloning, primary structure, and distribution in the central nervous system. *J. Biol. Chem.* 259:14301–8

Ann. Rev. Neurosci. 1986. 9 : 305–28

LINKAGE BETWEEN AXONAL ENSHEATHMENT AND BASAL LAMINA PRODUCTION BY SCHWANN CELLS

Richard P. Bunge, Mary Bartlett Bunge, and Charles F. Eldridge

Department of Anatomy and Neurobiology, Washington University School of Medicine, St. Louis, Missouri 63110

INTRODUCTION

Theodor Schwann (1839) identified the cell that now bears his name in his efforts to establish the cellular nature of all body parts. He considered the demonstration that the fatty sheaths of peripheral nerve were related to cells particularly compelling support for his cell theory. The question whether the fatty myelin sheath derived from the cell of Schwann during development was debated during the ensuing century and finally was settled when the resolution of the electron microscope allowed demonstration that the myelin sheath was formed from Schwann cell plasma membrane.

The resolution of the electron microscope was also required to demonstrate the precise nature of ensheathment of unmyelinated nerve fibers (fibers of Remak). Ultrastructural studies established that these nerve fibers are provided with an investment of Schwann cell cytoplasm rather than a thin layer of myelin, as was often stated earlier. This cytoplasmic ensheathment has now been studied in considerable detail, although its functional contribution to the enclosed axon remains enigmatic.

We know now that, with rare exceptions, Schwann cells encircle all nerve fibers in peripheral nerve. One Schwann cell either encloses one axon, with a myelin sheath and associated cytoplasm, or a number of unmyelinated axons, each one harbored within a furrow of cytoplasm. These Schwann cell–neurite units are further enveloped by a sleeve of basal lamina, and this

305

0147–006X/86/0301–0305$02.00

triumvirate is suspended in endoneurial extracellular matrix, which consists largely of collagenous fibrils. Techniques are now available to allow the development of peripheral nervous tissue in culture (reviewed by Bunge et al 1983). When a portion of a sensory ganglion or dissociated cells derived from the ganglion are placed in culture, neuronal processes extend from the neuronal somata, and Schwann cells proliferate and ensheathe the neurites with either myelin or cytoplasm. Extracellular matrix, including collagen fibrils and the basal lamina that covers Schwann cells, is also formed in the culture dish and, moreover, appears in the absence of fibroblasts (Figure 1) (Bunge et al 1980).

Through using these and additional specialized tissue culture techniques workers have recently recognized that a major part of Schwann cell function relates to extracellular matrix production. Whereas this was perhaps not unexpected, considering the amount of extracellular matrix within the peripheral nerve trunk, it was surprising to realize that the Schwann cell is unable to function normally in its role in myelination and ensheathment unless it is able to produce and relate to extracellular matrix components. The purpose of this chapter is to review briefly how the role of the Schwann cell in extracellular matrix production was discovered and how the linkage between Schwann cell (SC)[1] function and extracellular matrix (ECM) production came to be recognized. A more general review of SC biology appears elsewhere (Bunge & Bunge 1981).

SECRETORY PRODUCTS OF SCHWANN CELLS

The following observations derive from primary culture preparations that allow separation and recombination of SCs and neurons (free of fibro-blasts); methods for obtaining these types of cultures and the characteristics of these culture preparations are reviewed elsewhere (Bunge et al 1983).

When cultures containing neurons and SCs are given (for 18 hr) a medium containing radiolabeled leucine, methionine, proline, or gluco-samine, more than 25 labeled polypeptides (ranging in size from M_r 15,000 to > 250,000) may be released into the medium as assessed by fluorography of sodium dodecyl sulfate polyacrylamide gel electrophoresis (SDS-PAGE) preparations (Carey & Bunge 1981). The gel pattern is different from that obtained by SDS-PAGE analysis of the cell layer polypeptides (Carey & Bunge 1981), thus indicating that they are not simply products of cell death. The radiolabeled substances released into the medium are primarily secreted by the SCs and not the neurons because, if neuronal somata are

---

[1] Abbreviations used: ECM, extracellular matrix, SC, Schwann cell; SDS-PAGE, sodium dodecyl sulfate polyacrylamide gel electrophoresis; TCA, trichloroacetic acid.

removed immediately before labeling, the trichloroacetic acid- (TCA) precipitable counts in labeled polypeptides are not substantially diminished and the gel pattern of radiolabeled polypeptides is not significantly altered (Carey & Bunge 1981). Also, one of the proteins that we now know to be type IV procollagen is not found by electrophoresis of the medium of metabolically labeled neuron-only cultures (Carey et al 1983). In addition, neurons cultured in the absence of SCs do not stain positively for the ECM components we know to be produced by SCs (Carey et al 1983, Cornbrooks et al 1983a).

As we detail below, the identification of some of the bands observed by SDS-PAGE and also by immunostaining (as in Figures 2 and 3) has led to the conclusion that neuron-related SCs secrete types I, III, IV, and V collagens, laminin, entactin, and heparan sulfate proteoglycan. Antibodies against entactin and heparan sulfate proteoglycan, provided by Drs. Chung (Carlin et al 1981) and Hassell et al (1980), respectively, stain the basal lamina of neuron-related SCs (C. Eldridge, unpublished observations; Figure 3). In an analysis of proteoglycan synthesis, Mehta et al (1985) found that SCs synthesize two heparan sulfate–containing proteoglycans; the larger one is a component of the SC basal lamina, whereas the smaller one appears to be associated with the SC plasmalemma. In these neuron-SC preparations, a small amount of secreted fibronectin can be detected in the culture medium; affinity-purified antihuman plasma fibronectin antibody (McDonald et al 1981) specifically immunoprecipitates a M_r 220,000 (reduced) polypeptide (Cornbrooks et al 1983a). Since one of the major secretory products of fibroblasts is fibronectin, we cannot be certain that the fibronectin in the medium does not derive from a small number of fibroblasts in some neuron-SC cultures. Immunostaining of the cell layer with fibronectin antibody is not seen, however, under the culture conditions utilized (Cornbrooks et al 1983a). The synthesis of laminin and type IV collagen (McGarvey et al 1984) and the lack of antifibronectin antibody staining (Brockes et al 1979, Baron-Van Evercooren et al 1982) have also been observed for cultured SCs derived from sciatic nerve and prepared by techniques that remove them from neuronal contact.

Collagen Secretion

That SCs in the presence of neurons secrete collagen has been determined by a number of techniques. When metabolically radiolabeled culture medium is treated with purified bacterial collagenase before analysis by SDS-PAGE, at least three high-molecular weight proteins are greatly diminished or are no longer present (Bunge et al 1983, Carey et al 1983). A prominent collagenase-sensitive protein (M_r 190,000) comigrates with the major collagenous protein released from cultured rat parietal endoderm

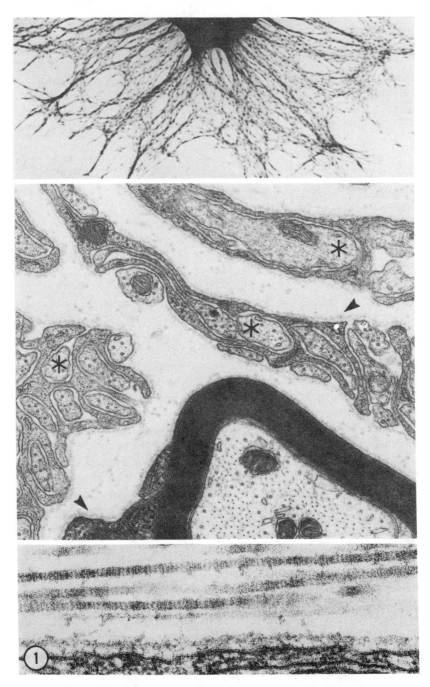

cells, identified as type IV procollagen (Carey et al 1983). This protein in the SC radiolabeled medium is also specifically immunoprecipitated by affinity-purified antibody to mouse type IV procollagen (Carey et al 1983). Also, ^{3}H-proline-labeled polypeptides in the medium of SC-neuron cultures may be subjected to limited pepsin digestion, which hydrolyzes nearly all polypeptides except triple-helical portions of native collagen molecules (Bruckner & Prockop 1981). The resulting pepsin-resistant polypeptides manifest mobilities in SDS-PAGE that are characteristic of types I, III, and V as well as IV collagens (Bunge et al 1983, Carey et al 1983). Using antibodies against type IV (Figure 2) (Carey et al 1983) and V (C. Eldridge, unpublished observations; Figure 3) collagens from Drs. Timpl and Madri (Roll et al 1980), respectively, the SC basal lamina is stained by the immunofluorescence technique. When a radiolabeled neuron/SC culture is treated with trypsin to remove the basal lamina (M. Bunge et al 1982), type IV collagen is removed from the cell layer, as examined by SDS-PAGE (Carey et al 1983). When intact neuron-SC cultures are treated with purified bacterial collagenase just before fixation and examined in the electron microscope, cross-striated extracellular fibrils are no longer seen and SC basal lamina appears disrupted (Bunge et al 1980).

Laminin Secretion

Antibodies have also been used to determine whether SCs secrete laminin into the culture medium. Affinity-purified rabbit antimouse laminin antibody (Timpl et al 1979) specifically immunoprecipitates from metabolically-labeled medium radioactive polypeptides with $M_r >$ 250,000, 200,000, and 150,000. The M_r 150,000 protein may be identical to laminin C (Cooper et al 1981) or entactin (Carlin et al 1981).

SCs produce laminin as detected by immunostaining using the Timpl antibody (Cornbrooks et al 1983a). When the SCs are grown in a culture situation in which they form basal lamina, the staining pattern takes the

---

Figure 1 Morphological characteristics of neuron-SC cultures. The *top figure* illustrates, in a whole mount, an area of outgrowth emanating from the dense sensory ganglion explant where the neuronal somata are clustered. The outgrowth consists of nerve fibers ensheathed by SCs; the nuclei but not the cytoplasm of the SCs are visible. (OSO$_4$ fixation, Sudan black staining; 7 days in vitro ×75. Courtesy of Dr. Patrick Wood.) The *middle figure* is an electron micrograph of the outgrowth, showing ensheathment of neurites (*) by cytoplasm or myelin. Basal lamina (▶) is present on the SC surface, and transversely-sectioned collagen fibrils occur in the intercellular space. (Ten weeks in vitro; ×40,000. By David Copio.) The *bottom panel* illustrates, in an electron micrograph at higher magnification, the SC basal lamina and cross striated collagenous fibrils. Lengths of fibrils are visible here because the fascicle was longitudinally sectioned, in contrast to the area in the middle figure, which was transversely sectioned. (Twelve weeks in vitro; ×132,000.)

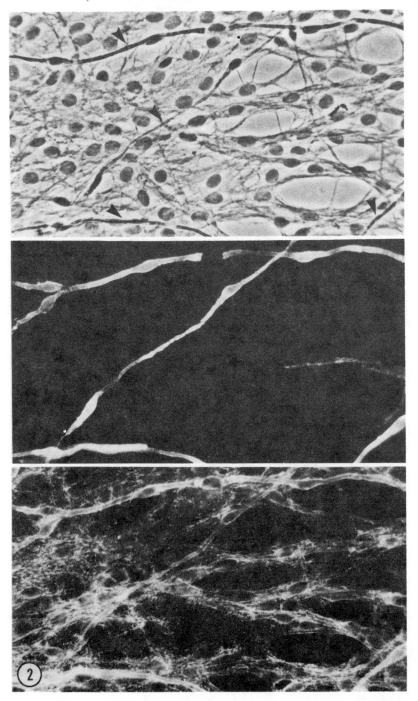

2

form of linear arrays that outline SC-neuron units; in the case of the larger and singly-occurring myelinated axons, the fluorescent antibody clearly outlines the internode. This configuration is consistent with the localization of laminin to the basal lamina, as would be expected (Courtoy et al 1982, Laurie et al 1982). However, when SCs do not have basal lamina, because they are not cultured with neurons or they are cultured with neurons in serum-free medium, the staining is no longer linear but is particulate instead (Cornbrooks et al 1983a). This is in agreement with work by others studying SCs grown in isolation, bereft of neurons (McGarvey et al 1984). Antilaminin antibody does not stain neuronal somata or their processes (Cornbrooks et al 1983a).

The release of laminin by SCs in every culture system studied thus far differs from that of type IV procollagen (see below); the release of these two components is apparently regulated differently. Also, the finding that laminin is present on the SC surface at all stages of SC differentiation differs from observations on other SC antigens. The cell surface antigens, galactocerebroside (Mirsky et al 1980) and C4 (Cornbrooks & Bunge 1982), and the extracellular (basal lamina) antigen, B3 (Cornbrooks et al 1981; C. Cornbrooks, personal communication), are not expressed by SCs lacking contact with neurons. Two of these antigens, galactocerebroside (Eldridge et al 1984) and C4 (C. Cornbrooks, personal communication), are expressed by SCs growing in contact with neurons but in a serum-free medium. The monoclonal antibody, B3, stains SCs only when the basal lamina is present (C. Cornbrooks, personal communication) and thus does not stain SCs in contact with neurons in a serum-free medium. B3 (Figure 3) and anti-BM1 (Hassell et al 1980) both recognize a basal lamina–associated heparan sulfate proteoglycan produced by SCs (C. Eldridge, unpublished observations).

Regulation of Secretion

When SCs are grown in a serum-containing medium but without neurons for two weeks, the pattern of electrophoresed radiolabeled medium

---

Figure 2   Staining of a neuron-SC culture outgrowth region for galactocerebroside and for the NCl domain of type IV collagen. The outgrowth area is shown by phase microscopy in the *top figure*; a few internodes of myelin (►), many phase dense SC nuclei, and numerous unmyelinated axons ensheathed by SC cytoplasm are present. The surfaces of myelinating SCs only are stained by anti-galactocerebroside monoclonal antibody, as shown in the *middle panel*. The exterior of these myelinated fibers as well as of nonmyelinating SCs is stained by collagen IV antiserum, in the position of the basal lamina (*bottom figure*). The primary antibodies were added to the culture simultaneously; the secondary antibodies, fluorescein-goat antirabbit and rhodamine goat antimouse, recognized the α-type IV and α-galactocereroside antibodies, respectively. (Cultures grown in complete medium for two weeks after having been raised in serum-free N2 medium; × 470. Anti-galactocerebroside antibody provided by Dr. B. Ranscht; Ranscht et al 1982.)

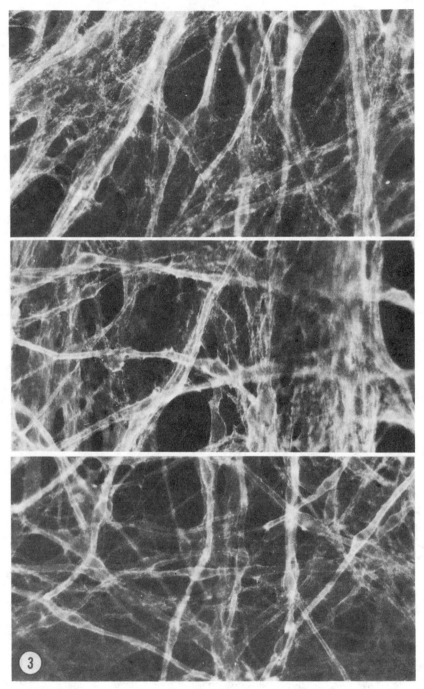

polypeptides after SDS-PAGE is different from that observed after radiolabeling a culture containing both SCs and neurons (or containing SCs divested of neurons immediately before the radiolabeling). Whereas most bands (such as type I collagen and laminin) appear unchanged, two new bands appear and two bands are diminished in the lanes containing SC-only culture medium (Carey et al 1983). One of the polypeptides reduced in amount has been identified as type IV procollagen. Thus, axonal contact may regulate matrix assembly by selectively modulating procollagen IV accumulation. Because type IV procollagen is an important structural component of basal laminae (Kleinman et al 1982), this regulatory mechanism may be responsible for the effect of the nerve cell on the formation of SC basal lamina.

The conclusion regarding axonal contact influencing type IV procollagen release and thus basal lamina formation is supported by electron microscopic studies of SC and SC-neuron cultures. Removal of neurons from month-old SC-neuron cultures leads to the disappearance of basal lamina; the reintroduction of neurons into SC-only cultures leads to the reappearance of lamina (M. Bunge et al 1982). If basal laminae are removed by means of trypsin from cultures of neurons and SCs or cultures of SCs (depleted of neurons just before trypsinization), laminae reappear only in the neuron-SC cultures (M. Bunge et al 1982). Other laboratories have likewise found that SCs cultured without neurons do not acquire basal laminae (Armati-Gulson 1980, Dubois-Dalcq et al 1981, Fields & Raine 1982, McGarvey et al 1984). If neurons are removed from older neuron-SC cultures, however, SC basal lamina may persist in the absence of neurons, even though formation of a new lamina does not occur (M. Bunge et al 1982). This conclusion, that maintenance but not formation of the SC basal lamina occurs without axonal contact, is in agreement with work done in animals rather than culture (Billings-Gagliardi et al 1974, Weinberg & Spencer 1978, Payer 1979).

Schwann Cell Basal Lamina in Vivo

As pointed out above, the basal lamina of cultured neuron-related SCs contains types IV and V collagens, laminin, entactin, and heparan sulfate proteoglycan, as detected by immunofluorescence techniques. This is also true for SC basal laminae in the animal (Shellswell et al 1979, Sanes 1982,

---

Figure 3 Staining of neuron-SC culture outgrowth regions for type V collagen (*top figure*), B3 (*middle figure*), and entactin (*bottom figure*). The basal lamina-coated external surfaces are stained. The "railroad track" staining pattern bulges along the myelinated axon, where the SC nucleus is positioned. The monoclonal antibody, B3, recognizes a heparan sulfate proteoglycan of basement membrane. (Cultures grown in complete medium for two weeks after having been raised in serum-free N2 medium; × 470.)

Cornbrooks et al 1983a, Palm & Furcht 1983, Sanes & Chiu 1983). Though it is not yet understood how much fibronectin contributes to the basal lamina, fibronectin surrounds myelinated axons in vivo (Sanes 1982); the staining of basal lamina is not as distinct as when antilaminin or anticollagen IV antibody is utilized (Cornbrooks et al 1983a, Palm & Furcht 1983) probably in part because, unlike laminin and type IV collagen, fibronectin is dispersed throughout the endoneurium. Type IV collagen, laminin, entactin, fibronectin and heparan sulfate proteoglycan are major components of basal laminae found throughout the body (Timpl & Martin 1982), including both the synaptic and extrasynaptic regions of muscle fiber basal laminae (Sanes & Chiu 1983). Recently it has been found that basal laminae of different cell types may have different compositions. The SC basal lamina does not contain a number of components present in the synaptic regions of muscle cell basal laminae (Sanes & Chiu 1983). In addition, SC basement membranes are stained by four monoclonal antibodies that also bind to all areas of muscle fiber basal lamina but not to blood vessels or perineurium within muscle or to kidney glomerular/ tubular, lens capsule, or Reichert's basement membranes (Sanes & Chiu 1983).

LINKAGE OF SCHWANN CELL FUNCTION AND EXTRACELLULAR MATRIX ORGANIZATION

There are a number of in vitro situations in which SCs (in culture with neurons) do not complete differentiation to express full function as expected. Because the most obvious function of SCs is to ensheathe and myelinate neurites, the development of myelin in the culture serves as an assay for normal SC function at the light microscope level. When SCs are stranded on guy-roping neurite fascicles and consequently do not contact the substratum, they huddle together in clusters and do not extend along, ensheathe, or myelinate neurites. ECM appears abnormal electron microscopically. When secretion is perturbed by growing SCs and neurons in a defined (serum-free) medium or a medium containing the proline analogue, cis-hydroxyproline, SC development is again arrested and, in addition, ECM generation is deficient. We now discuss these three situations.

Failure of Schwann Cell Function Related to Faulty Alliance with Extracellular Matrix

A particularly dramatic example of SC functional failure is observed in cultures in which conditions for SC functional expression are suitable (including contact with axons), except that in certain areas SCs do not have access to the substratum of reconstituted type I collagen provided in the culture dish (Figure 4). This condition may develop in culture because during early development of the neuritic outgrowth the nerve fiber growth cone is more adhesive than the neurite shaft (Bray et al 1979). Thus, neuritic shafts may become suspended in the culture medium. In these guy-roping regions SCs become aggregated into grape-like clusters on the surfaces of axonal fascicles and fail to ensheathe and myelinate the adjacent axons; more distally, where these same axons are in contact with substratum, they are normally ensheathed and myelinated (Bunge & Bunge 1978). The localized SC dysfunction can be rapidly corrected by forcing the suspended elements onto the collagen coated substratum. Within hours the aggregated SCs separate from one another and extend along the length of the axon; ensheathment, myelination, and normal basal lamina organization follow within several days (Bunge & Bunge 1978).

We have interpreted these observations as indicating that during an early stage of development the SC requires contact not only with the axon surface but also with a second surface that provides contact for the organization of basal lamina materials. In this way the cell develops its "sidedness," with an adaxonal region contacting the axon surface and an abaxonal surface contacting basal lamina. The cell thus becomes polarized, a necessary step for functional expression by many epithelial cells, as we discuss below. Once this polarization is accomplished, the dual contact may no longer be needed; when fascicles become secondarily detached following ensheathment and myelination, they do not develop manifest abnormalities of ensheathment.

If the SC has the capability of producing basal lamina components (as documented above), why cannot these constituents provide for the organization of the abaxonal surface of the SC without provision of an additional surface for contact? One must assume that the basal lamina can only be organized between the cell surface and an adjacent surface. That this surface need not be coated with exogenous collagen is indicated by the observation that SCs can myelinate axons in cultures utilizing tissue culture plastic rather than a collagen-coated surface (C. Eldridge, unpublished observations). Perhaps in the culture dish the second surface allows adequate concentrations of basal lamina components to accumulate locally, thus leading to precise organization of these components. Whereas immunostaining indicates the presence of the usual SC basal lamina constituents in guy-roping regions (C. Eldridge, unpublished observations), electron microscopically the ECM appears abnormal; the basal lamina is patchy on some areas of the SC and redundant on others, and external to the lamina, often associated with redundant lamina material, there are abnormal-appearing particles in lieu of the usual recognizable collagen fibrils (Bunge & Bunge 1978, Figure 3; Bunge et al 1983).

Failure of Schwann Cell Function Related to Perturbed Secretion

One would then expect deficiencies in the production of ECM components manufactured by the SC to result in abnormalities of axonal ensheathment. This is very dramatically seen when SCs are allowed to interact with axons in cultures given a serum-free medium that does not support ECM production (Moya et al 1980). The serum-free defined medium designated N2 by Bottenstein & Sato (1979) is adequate to permit neuronal growth and SC proliferation. With time, however, the progression to axon ensheathment and myelination does not occur, and after many weeks in culture SCs show only partial contact with axons and develop no basal lamina covering. The lack of a basal lamina is related to the paucity of secreted triple helical collagen molecules; pepsin-resistant collagenous polypeptides are not detectable in N2 medium (Bunge et al 1983, Eldridge et al 1984).

When this medium is supplemented with components known to promote myelination in culture, e.g. serum and embryo extract, rapid progression to ensheathment, myelination, and basal lamina production ensue. Pepsin-resistant (triple-helical) collagenous polypeptides are secreted into the medium (Bunge et al 1983, Eldridge et al 1984). A variety of supplements have been tried in order to simplify the medium composition needed for SC function. The bovine serum protein, fetuin, when added with ascorbate, promotes both myelination (Carey 1983) and basal lamina production (Eldridge et al 1984). Fetuin alone does not improve SC function in N2 medium; ascorbate alone leads to a partial expression of function (see below).

Recent experiments have clarified the roles of undefined supplements in serum-free medium. It has been intermittently observed that the addition of embryo extract along with serum is not always required for myelination; an occasional lot of serum (in these experiments, human placental serum) is alone adequate to promote myelination. C. Eldridge (unpublished observations) finds that, when such a batch of serum is dialyzed, the ability to

---

Figure 4 These light micrographs show regions of myelin formation juxtaposed to regions of SC ensheathment failure. Fascicles of axons course from a region containing neuronal somata above the area illustrated. In the center of the *top figure*, SCs are aggregated on fascicles and do not extend along the axons; SCs in these small clusters do not have simultaneous access to both the axonal and substratal surfaces. In adjacent regions numerous segments of myelin are present. The *bottom figure* is from the region designated by an asterisk in the upper photograph. This culture is essentially stable as shown, but SC functional deficits can be corrected by a collagen gel overlay. (Neuron-SC culture grown in complete medium for two weeks after having been raised in serum-free N2 medium; myelin began to appear one week after the shift to complete medium.)

foster not only myelination but also basal lamina assembly in culture is lost. The ability of dialyzed serum to promote these expressions of SC function is fully restored by the addition of 50 μg/ml ascorbate. Thus, by the use of dialyzed serum it is possible to regulate SC myelination and basal lamina production by the withdrawal or addition of ascorbate to culture medium. A component in serum is needed in addition to the vitamin because ascorbate added to serum-free medium leads to only a partial expression of SC function.

The addition of ascorbate to serum-free defined medium leads to the accumulation of pepsin-resistant collagenous polypeptides in the medium of neuron-SC cultures (Bunge et al 1983, Eldridge et al 1984), and basal lamina and collagen fibrils are visible electron microscopically (Bunge et al 1983). This may be explained by the known role of ascorbate in collagen synthesis. The proline (and lysine) residues in procollagen chains are converted to hydroxyproline (and hydroxylysine) by the action of hydroxy-lases associated with ferrous ions at their active sites. This hydroxylation must occur in the presence of a reducing agent (ascorbate) to retain the iron atoms in their active state. If this hydroxylation is incomplete, normal triple helical organization of the procollagen molecule cannot be attained. Normal triple helical structure is needed for normal secretion of collagen-ous polypeptides (Prockop et al 1976).

The proline analogue, *cis*-hydroxyproline, has been used to perturb collagen secretion (Uitto & Prockop 1974, Wicha et al 1980). This analogue occupies some of the positions normally filled by proline in procollagen chains. The presence of *cis*-hydroxyproline precludes full hydroxylation to *trans*-hydroxyproline during procollagen synthesis and also directly inter-feres with triple helix formation (Inouye et al 1976). Consequently, these collagenous chains are not secreted normally. Eldridge et al (1983) have found that, in this SC-neuron system, *cis*-hydroxyproline (200 μg/ml) reduces the accumulation of secreted collagenous proteins by 57% but also reduces the accumulation of noncollagenous proteins by 34%. When SC-neuron cultures capable of myelination are treated with *cis*-hydroxyproline, ECM deposition is reduced and ensheathment and myelination by SCs is deficient (Copio & Bunge 1980). These abnormalities occur when the analogue is added to the medium during development of the outgrowth. If the outgrowth forms in normal medium and is then given the analogue, the cultures do not appear affected by the *cis*-hydroxyproline.

Schwann Cell Ensheathment and Basal Lamina Deficiencies in Dystrophic Mice

The experiments described above were carried out in tissue culture. If SC-ECM alliance is critical to SC function, might we not expect a disease state to indicate this linkage? The recessive genetic disorder of mice termed "dystrophic" (Michelson et al 1955) reveals interesting parallels. The dystrophic mouse develops progressive muscle wasting early in life, particularly in the hind limbs (Banker 1967, 1968). Initially this was considered to derive from a primary lesion in muscle tissues. It was subsequently noted, however, that major portions of certain peripheral nerve root regions (both dorsal and ventral) contain aggregates of axons without any form of SC ensheathment or myelination; in these regions axons are in direct apposition to one another (Bradley & Jenkison 1973). When it was demonstrated that these regions are capable of initiating quite abnormal (including retrograde) impulses, it was recognized that the muscle abnormalities might be engendered by abnormal electrical activity in the peripheral nerves (Huizar et al 1975, Rasminsky 1978). Thus, attention has recently focused on the nerve lesion.

There are, in fact, small defects in basal lamina coverage of the SC surface in many parts of the dystrophic mouse peripheral nerve (Madrid et al 1975), as well as minor defects in axonal ensheathment and myelination (Okada et al 1976, Bradley et al 1977, Jaros & Bradley 1979). It is in the cervical and lumbosacral nerve roots where substantial numbers of axons are without ensheathment and are not separated by non-neuronal cell processes or ECM material. In these regions SCs are present but they remain perched on the perimeter of the bundled axons. These SCs show only occasional patches of basal lamina and are for some reason unable to insert themselves between the axons and to undertake the processes of ensheathment and myelination (references above). That these axons are capable of accepting myelination is indicated by the fact that they are myelinated in the more distal parts of the nerve trunk. The metabolic lesion responsible for these striking regions of ensheathment failure in the dystrophic mouse nerve root is unknown. The similarity in the conformations of SCs (and failure to organize basal lamina) in these regions in vivo and the SCs perched on bundles of axons suspended in culture medium (as detailed above) is very striking. It is interesting to note that, as discussed above, the SC deficit on suspended fascicles in culture can be corrected by forcing the SCs into contact with the underlying collagen substratum (Bunge & Bunge 1978); the abnormalities in the dystrophic nerve root can be considerably improved by crushing the root (Stirling 1975), thus inducing increased ECM formation (Bray et al 1983).

When peripheral nerve tissue from dystrophic mice is established in long-term tissue culture, modest deficiencies in basal lamina coverage of SCs are seen along with some abnormalities of myelin segments (Okada et al 1980). The basal lamina defects can be substantially corrected by culturing axons and SCs from dystrophic mice in the presence of fibroblasts from normal mice (R. Bunge et al 1982, Cornbrooks et al 1983b). Very recently, in adroit experiments utilizing chimeras of dystrophic and *Shiverer* (myelin basic protein deficient) mice to allow identification of the genotype of each myelin-related SC, Peterson & Bray (1984) showed that dystrophic SCs grown among tissues of *Shiverer* origin show neither basal lamina defects nor ensheathment abnormalities.

Other Examples of Linked Schwann Cell Ensheathment and Extracellular Matrix Deficits

Additional examples of an apparent linkage between SC dysfunction and ECM abnormalities have been found. First, SCs in the company of sympathetic neurons appear to have different requirements for differentiation than when these cells are grown with sensory neurons. Under certain culture conditions, SCs grown with sympathetic neurons in a complete (serum- and embryo extract-containing) medium and in contact with the substratum fail to ensheathe neurites normally and also lack continuous basal lamina (Roufa et al 1985). Whereas the conditions for normal SC function in this case are not yet known, the paired findings of SC functional deficit and ECM abnormality are again observed. In a second instance, Cochran & Black (1982) questioned whether the tumor cell line, PC12, when induced to extend neurites among normal SCs in coculture, would be ensheathed by these SCs. Although the neurites provided a mitogenic signal that stimulated SC proliferation (see also Ratner et al 1984), the SCs did not ensheathe or myelinate these neurites, and basal lamina was not deposited; once again the linkage between failure to ensheathe and lack of basal lamina production was observed. When SC-neuron cultures are shifted from a serum-free medium to a complete medium in the presence of 250 μg/ml castanospermine (Dr. A. Elbein) to prevent processing of glycoproteins (from high-mannose intermediates to complex oligosaccharides), ten days later it is observed electron microscopically that SCs remain in association with neurites but little progression of ensheathment and basal lamina formation have occurred. In cultures shifted to a medium lacking this substance, however, many segments of myelin have been formed, thus indicating normal progression of ensheathment (N. Ratner, S. Porter, R. Bunge, and L. Glaser, manuscript in preparation).

There are exceptional regions within the normal peripheral nervous system where ensheathment patterns by nonneuronal cells differ substantially from the usual pattern in the spinal nerve trunk (and ganglion). Striking examples are seen in the enteric nervous system of the gut wall (for references, see Gabella 1981, Jessen & Mirsky 1983). Whereas elsewhere in peripheral nerve each nerve cell and its process are separated from other neurons and axons by ensheathment by satellite cells or SCs, within the small ganglionic masses of the enteric plexuses the neuronal elements receive communal ensheathment by thin layers of externally applied non-neuronal cell cytoplasm. Internal to this sheath, the neuronal cell bodies and axons come into direct contact with one another. No visible ECM is formed within the ganglion, and the closely packed cells are reminiscent of the parenchyma of the CNS. The ensheathing cell has many characteristics of an astrocyte (rather than a SC), despite its neural crest origin. It provides a border for these small neural aggregates that resembles the glia limitans of CNS tissue, being overlain only externally by basal lamina. The ensheathing cell has been demonstrated to contain intermediate filaments characteristic of astrocytes. Thus, in this remote but important body region, nerve cells receive altered ensheathment. We would suggest that, although the peculiar glial cell in these regions is of the same lineage as SCs, its response to either unusual neuronal influences and/or much altered influences of ECM materials in the enteric tissues engender a unique course of differentiation. The observations cited above that SC behavior may be influenced by neuronal type, in concert with characteristics of the local ECM, would lend support to this view.

CELL POLARIZATION

The linkage between SC basal lamina production and ensheathment function suggests that polarization of the SC surface may be required for development of full SC maturation. Those surface components specialized for contact with ECM moieties (e.g. laminin receptors) may become concentrated on the surface facing away from the axon (where the basal lamina is normally found), and those surface components specialized for axon apposition may move to the adaxonal surface. With this polarization of surface domains, the adaxonal surface is then able to surround the axon (for unmyelinated axons) or to progress to elaborate spiral extension (for myelinated axons). Linkage between the development of polarization of the cell surface and functional expression is observed in many cells of the epithelial type. The SC, in its production of basal lamina constituents and its relationship to basal lamina, resembles epithelial cells (see also discussion in Bunge & Bunge 1983).

Epithelial cells typically dedicate part of the cell surface to contact with basal lamina; other regions of the surface are then devoted to alternate activities (for review see Sabatini et al 1983). Typically, the formation of microvilli and release of certain secretory products occur at the apical plasmalemmal domain. This polarization is often reinforced by regimentation in orderly cuboidal or columnar ranks, with the apical portions of the lateral cell borders joined by specialized junctions, particularly tight junctions. The apposed plasmalemmas of neighboring cells often inter-digitate along the basolateral borders. Internal architecture of the cell is concomitantly asymmetrical, with the nucleus positioned toward the basal surface and the centrosome and Golgi complex situated between the nucleus and the secreting apical surface. Many specialized functions of the cell depend upon this structural polarization. This polarization is triggered by the attachment of a cell to a surface or to another cell (or both, as in the case of the SC). Attachment to substratum and/or adjacent cells establishes domains of the cell surface devoted to specific function, an essential aspect of cell polarization. This has been particularly well demonstrated using the MDCK cell line.

When kidney-derived MDCK epithelial cell monolayers grow to confluency in culture on permeable supports, they become polarized and develop many of the properties found in natural epithelia, including unidirectional transport of water (Cereijido et al 1978). They are attached to the collagen substratum basally, linked to one another by tight junctions near the free surface, and they develop microvilli only on the free surface. When such polarized monolayers of epithelial cells are infected with enveloped RNA viruses, the distribution of budding virions reflects the polarized nature of the cells. A myxovirus (such as influenza) is assembled only on the apical domain, whereas rhabdoviruses (such as vesicular stomatitis virus) are budded only from the basolateral surface (Rodriguez-Boulan et al 1983). (This difference is also seen in natural epithelia; Sabatini et al 1983.)

Polarity is lost if the MDCK cell monolayer is dissociated and maintained in suspension; both types of virions form indiscriminately along the entire cell surface of single suspended cells. But, if single cells are allowed to reattach to a surface (without tight junction formation) or to contact another cell (in a cell cluster not in contact with substratum but with tight junction linkage), morphological and viral budding polarities are reestablished (Rodriguez-Boulan et al 1983). Thus, in this MDCK cell system, attachment to a substratum or to another cell (with ensuing tight junction formation) is sufficient to trigger the expression of plasmalemmal polarity as shown by asymmetric viral budding. Fibroblasts, in contrast to substratum-attached epithelial cells, do not manifest this viral budding polarity; both types of viruses form from all regions of the surface membrane (Rodriguez-Boulan et al 1983). Fibroblasts may contact other fibroblasts, but the junctions that link them, gap and attachment-type junctions, are not polarized to specific domains of the cell surface as they are in epithelial cells (Hay 1984).

The mammary gland epithelium in culture also demonstrates a depen-dence on substratum attachment for functional expression; in this instance the requirement also includes a specific substratum configuration. Cultures of mouse mammary gland epithelial cells express their characteristic morphology and ability to produce casein only if maintained on floating collagen gels, and not on attached collagen gels or plastic (Emerman & Pitelka 1977). Electron microscopy reveals basal lamina production below mammary epithelial cells grown on floating gels, but not under cells grown on plastic (Shannon & Pitelka 1981). In the floating gels, the basolateral surfaces of the cells have access to nutrients and hormones in the culture medium. The gels, however, must be allowed to contract after detachment to obtain full secretory activity; this contraction allows cell shape to change from a squamous to a cuboidal configuration. More recent studies (Lee et al 1984) have determined that the culture configuration influences secretion of all types of caseins; none are secreted by cells on plastic substrata, and only limited secretion is seen on attached collagen gels. Whey proteins are secreted poorly in culture under all substratal conditions tested, as is α-lactalbumin; thus, substratal relationships are not by themselves able to induce secretion of all milk proteins.

The importance of substratum configuration is also dramatically illustrated in studies of endothelial cells in culture. Clones of capillary endothelial cells have been derived from bovine adrenal cortex (Folkman & Haudenschild 1980). When grown on the surface of collagen gels these cells increase in number, and typical monolayers are formed. These monolayers undergo a dramatic reorganization when covered with a second collagen layer (Montesano et al 1983). The monolayer undergoes local retractions, which result, within two days, in networks of anastomosing cords of cells. Electron microscopy reveals that nearly all endothelial cells now delimit a narrow lumen, forming tubular structures resembling blood capillaries. Basal lamina-like material appears on the external aspect of these tubes, at the interface between the endothelial cell plasma membrane and the collagen matrix. Whereas some endothelial tube formation had previously been achieved in much longer term cultures, this is the first report of rapid conversion of endothelial cell monolayers into tubular elements. The similarity in timing and configuration to the SC as it converts from a non-ensheathing to an ensheathing cell when forced into simultaneous contact with both collagen and axon is striking.

Hay and co-workers have studied in detail the cellular and extracellular elements involved when epithelial cells contact the substratum (for review see Hay 1984). Their model has been the corneal epithelial cell. The chief aim of the embryonic avian corneal epithelium in vivo is to secrete collagen and proteoglycan in the absence of fibroblasts; thus the extent to which these proteins are produced may be assayed to assess differentiation of this epithelium in vitro. Basal lamina (type IV collagen-rich lens capsule) and types I and II collagen promote epithelial differentiation in vitro, and concomitantly promote the flattening of the epithelial basal surface from a blebbed configuration. Subsequently, it was found that fibrillar or solubilized collagen (I, II or IV), laminin, and fibronectin lead to basal flattening (Sugrue & Hay 1981, Hay 1984). Along with the withdrawal of blebs there is reorganization of the basal epithelial cytoskeleton into a dense cortical mat of actin.

Laminin and type I collagen bind to the basal surfaces of these corneal epithelial cells that are signalled to differentiate in culture. Whether these laminin (Sugrue 1984) or collagen (Sugrue & Hay 1983) receptors are cytoskeleton-connected is being investigated; it is possible that interaction with the cytoskeleton via these receptors is responsible for the stimulatory effect of laminin and collagen on ECM secretion by the corneal epithelial cells (Hay 1984). Critical elements in cell-to-substratum bonding would be those molecules which traverse the cell membrane and have sites of attachment for cytoskeletal elements internally and extracellular elements externally. Bernfield et al (1984) have postulated that a membrane-associated proteoglycan has this function in the mammary epithelial cell. Laminin receptors have also been postulated to have this characteristic transmembrane liganding function between cytoskeleton and laminin (Liotta et al 1984). It is of interest to note that a membrane-associated heparan sulfate proteoglycan has recently been described in Schwann cells (Mehta et al 1985). It seems reasonable to assume Schwann cells have a rich supply of laminin receptors, considering their involvement in the manufacture of laminin and the observation that laminin is associated with the SC surface under all functional states, as described in detail above.

SUMMARY

The availability of several methods for the preparation of SCs free of other cell types has allowed recent experimentation providing new insights into the capacity of SCs to synthesize, release, and organize extracellular matrix materials, particularly those of the basal lamina. When these SC populations are combined in tissue culture with pure populations of neurons capable of directing SC function (without fibroblasts), new aspects of interrelationships between these cell types have come to light. In this brief chapter we review the results from this experimental approach during the last decade, and suggest the implications these observations have for interpreting known differences in SC functional expression in various body regions as well as for understanding certain disease processes. Of particular note is the discovery of an apparently essential linkage between the function of the SCs in organizing and relating to basal lamina and their ability to ensheathe and myelinate axons. It now appears that SC functional expression requires an alliance not only with the nerve fiber but also with the ECM through the production and organization of a basal lamina.

ACKNOWLEDGMENTS

We gratefully acknowledge collaborators who shared in that part of the work which originated in the St. Louis laboratory. These include Drs. Carey, Cochran, Cornbrooks, Moya, Okada, and Wood and Mrs. Ann Williams. We are thankful to Drs. Chung, Hassell (via Dr. J. Sanes, in our Department of Physiology), Madri, Ranscht, and Timpl, who provided antibodies to specific extracellular (or cellular) components used in our studies.

We thank Susan Mantia for preparing the manuscript and Peggy Bates for help with the figures.

Work in the authors' laboratory is supported by NIH grant NS 09923. C.F.E. also received support from Neurobiology Training Grant NIH-NS 07071.

Literature Cited

Armati-Gulson, P. 1980. Schwann cells, basement lamina, and collagen in developing rat dorsal root ganglia in vitro. *Dev. Biol.* 77:213–17

Banker, B. Q. 1967. A phase and electron microscopic study of dystrophic muscle. I. The pathological changes in the two-week-old Bar Harbor 129 dystrophic mouse. *J. Neuropathol. Exp. Neurol.* 26:259–75

Banker, B. Q. 1968. A phase and electron microscopic study of dystrophic muscle. II. The pathological changes in the newborn Bar Harbor 129 mouse. *J. Neuropathol. Exp. Neurol.* 27:183–209

Baron-Van Evercooren, A., Kleinman, H. K., Seppä, H. E. J., Rentier, B., Dubois-Dalcq, M. 1982. Fibronectin promotes rat Schwann cell growth and motility. *J. Cell Biol.* 93:211–16

Bernfield, M., Banerjee, S. D., Koda, J. E., Rapraeger, A. C. 1984. Remodeling of the basement membrane as a mechanism of morphogenetic tissue interaction. In *The Role of Extracellular Matrix in Development*, ed. R. L. Trelstad, pp. 545–72. New York: Liss

Billings-Gagliardi, S., Webster, H. deF., O'Connell, M. F. 1974. *In vivo* and electron microscopic observations on Schwann cells in developing tadpole nerve fibers. *Am. J. Anat.* 141:375–92

Bottenstein, J. E., Sato, G. H. 1979. Growth of a rat neuroblastoma cell line in serum-free supplemented media. *Proc. Natl. Acad. Sci. USA* 76:514–17

Bradley, W. G., Jenkison, M. 1973. Abnormalities of peripheral nerve in murine muscular dystrophy. *J. Neurol. Sci.* 18:227–47

Bradley, W. G., Jaros, E., Jenkison, M. 1977. The nodes of Ranvier in the nerves of mice with muscular dystrophy. *J. Neuropathol. Exp. Neurol.* 36:797–806

Bray, D. 1979. Mechanical tension produced by nerve cells in tissue culture. *J. Cell Sci.* 37:391–410

Bray, G. M., David, S., Carlstedt, T., Aguayo, A. 1983. Effects of crush injury on the abnormalities in the spinal roots and peripheral nerves of dystrophic mice. *Muscle Nerve* 6:497–503

Brockes, J. P., Fields, K. L., Raff, M. C. 1979. Studies on cultured rat Schwann cells. I. Establishment of purified populations from cultures of peripheral nerve. *Brain Res.* 165:105–18

Bruckner, P., Prockop, D. J. 1981. Proteolytic enzymes as probes for the triple-helical conformation of procollagen. *Analyt. Biochem.* 110:360–68

Bunge, M. B., Williams, A. K., Wood, P. M. 1982. Neuron-Schwann cell interaction in basal lamina formation. *Dev. Biol.* 92:449–60

Bunge, M. B., Williams, A. K., Wood, P. M.,

326   BUNGE, BUNGE & ELDRIDGE

Uitto, J., Jeffrey, J. J. 1980. Comparison of nerve cell and nerve cell plus Schwann cell cultures, with particular emphasis on basal lamina and collagen formation. *J. Cell Biol.* 84:184–202

Bunge, M. B., Bunge, R. P., Carey, D. J., Cornbrooks, C. J., Eldridge, C. F., Williams, A. K., Wood, P. M. 1983. Axonal and nonaxonal influences on Schwann cell development. In *Developing and Regenerating Vertebrate Nervous Systems*, ed. P. W. Coates, R. R. Markwald, A. D. Kenny, pp. 71–105. New York: Liss

Bunge, R. P., Bunge, M. B. 1978. Evidence that contact with connective tissue matrix is required for normal interactions between Schwann cells and nerve fibers. *J. Cell Biol.* 78:943–50

Bunge, R. P., Bunge, M. B. 1981. Cues and constraints in Schwann cell development. In *Studies in Developmental Neurobiology*, ed. W. M. Cowan, pp. 322–53. New York: Oxford Univ. Press

Bunge, R. P., Bunge, M. B. 1983. Interrelationship between Schwann cell function and extracellular matrix production. *Trends Neurosci.* 6:499–505

Bunge, R. P., Bunge, M. B., Williams, A. K., Wartels, L. K. 1982. Does the dystrophic mouse nerve lesion result from an extracellular matrix abnormality? In *Disorders of the Motor Unit*, ed. D. L. Schotland, pp. 23–34. New York: Wiley

Carey, D. 1983. Differentiation of Schwann cells in tissue culture. *J. Cell Biol.* 97:369a

Carey, D. J., Bunge, R. P. 1981. Factors influencing the release of proteins in cultured Schwann cells. *J. Cell Biol.* 91:666–72

Carey, D. J., Eldridge, C. F., Cornbrooks, C. J., Timpl, R., Bunge, R. P. 1983. Biosynthesis of type IV collagen by cultured rat Schwann cells. *J. Cell Biol.* 97:473–79

Carlin, B., Jaffe, R., Bender, B., Chung, A. E. 1981. Entactin, a novel basal lamina-associated sulfated glycoprotein. *J. Biol. Chem.* 256:5209–14

Cereijido, M., Robbins, E. S., Dolan, W. J., Rotunno, C. A., Sabatini, D. D. 1978. Polarized monolayers formed by epithelial cells on a permeable and translucent support. *J. Cell Biol.* 77:853–80

Cochran, M., Black, M. M. 1982. Interactions between Schwann cells (SCs) and nerve growth factor (NGF)-responsive PC12 pheochromocytoma cells in culture. *Anat. Rec.* 202:33a

Cooper, A. R., Kurkinen, M., Taylor, A., Hogan, B. L. M. 1981. Studies on the biosynthesis of laminin by murine parietal endoderm cells. *Eur. J. Biochem.* 119:189–97

Copio, D. S., Bunge, M. B. 1980. Use of a proline analog to disrupt collagen synthesis prevents normal Schwann cell differentiation. *J. Cell Biol.* 87:114a

Cornbrooks, C. J., Bunge, R. P. 1982. A cell-surface specific monoclonal antibody to differentiating Schwann cells. *Trans. Am. Soc. Neurochem.* 13:171

Cornbrooks, C. J., Bunge, R. P., Gottlieb, D. I. 1981. Components of the peripheral nervous system defined by monoclonal antibodies. *Soc. Neurosci. Abstr.* 7:149

Cornbrooks, C. J., Carey, D. J., McDonald, J. A., Timpl, R., Bunge, R. P. 1983a. In vivo and in vitro observations on laminin production by Schwann cells. *Proc. Natl. Acad. Sci. USA* 80:3850–54

Cornbrooks, C. J., Mithen, F., Cochran, J. M., Bunge, R. P. 1983b. Factors affecting Schwann cell basal lamina formation in cultures of dorsal root ganglia from mice with muscular dystrophy. *Dev. Brain Res.* 6:57–67

Courtoy, P. J., Timpl, R., Farquhar, M. G. 1982. Comparative distribution of laminin, type IV collagen, and fibronectin in the rat glomerulus. *J. Histochem. Cytochem.* 30:874–86

Dubois-Dalcq, M., Rentier, B., Baron-Van Evercooren, A., Burge, B. W. 1981. Structure and behavior of rat primary and secondary Schwann cells in vitro. *Exp. Cell Res.* 131:283–97

Eldridge, C. F., Bunge, M. B., Bunge, R. P. 1983. Biochemical effects of cis-4-hydroxy-L-proline on Schwann cells differentiating in vitro. *J. Cell Biol.* 97:244a

Eldridge, C. F., Bunge, M. B., Bunge, R. P. 1984. The effects of ascorbic acid on Schwann cell basal lamina assembly and myelination. *J. Cell Biol.* 99:404a

Emerman, J. T., Pitelka, D. R. 1977. Maintenance and induction of morphological differentiation in dissociated mammary epithelium on floating collagen membranes. *In Vitro* 13:316–28

Fields, K. L., Raine, C. S. 1982. Ultrastructure and immunocytochemistry of rat Schwann cells and fibroblasts in vitro. *J. Neuroimmunol.* 2:155–66

Folkman, J., Haudenschild, C. 1980. Angiogenesis in vitro. *Nature* 288:551–56

Gabella, G. 1981. Ultrastructure of the nerve plexuses of the mammalian intestine: The enteric glial cells. *Neuroscience* 6:425–36

Hassell, J. R., Robey, P. G., Barrach, H. J., Wilczek, T., Rennard, S. I., Martin, G. R. 1980. Isolation of heparan-containing proteoglycans from basement membrane. *Proc. Natl. Acad. Sci. USA* 77:4494–98

Hay, E. D. 1984. Cell-matrix interaction in the embryo: Cell shape, cell surface, cell skeletons, and their role in differentiation. See Bernfield et al 1984, pp. 1–31

Huizar, P., Kuno, M., Miyata, Y. 1975. Electrophysiological properties of spinal motoneurones of normal and dystrophic mice. *J. Physiol.* 248:231–46

Inouye, K., Sakakibara, S., Prockop, D. J. 1976. Effects of the stereo-configuration of the hydroxyl group in 4-hydroxyproline on the triple helical structures formed by homogenous peptides resembling collagen. *Biochim. Biophys. Acta* 420:133–41

Jaros, E., Bradley, W. G. 1979. Atypical axon-Schwann cell relationships in the common peroneal nerve of the dystrophic mouse: An ultrastructural study. *Neuropath. Appl. Neurobiol.* 5:133–47

Jessen, K. R., Mirsky, R. 1983. Astrocyte-like glia in the peripheral nervous system: An immunohistochemical study of enteric glia. *J. Neurosci.* 3:2206–18

Kleinman, H. K., McGarvey, M. L., Liotta, L. A., Robey, P. G., Tryggvason, K., Martin, G. R. 1982. Isolation and characterization of type IV procollagen, laminin and heparan sulfate proteoglycan from the EHS sarcoma. *Biochemistry* 21:6188–93

Laurie, G. W., LeBlond, C. P., Martin, G. R. 1982. Localization of type IV collagen, laminin, heparan sulfate proteoglycan, and fibronectin to the basal lamina of basement membranes. *J. Cell Biol.* 95:340–44

Lee, E. Y.-H., Parry, G., Bissell, M. J. 1984. Modulation of secreted proteins of mouse mammary epithelial cells by the collagenous substrata. *J. Cell Biol.* 98:146–55

Liotta, L. A., Rao, C. N., Barsky, S. H. 1984. Tumor cell interaction with the extracellular matrix. See Bernfield et al 1984, pp. 357–71

Madrid, R. E., Jaros, E., Cullen, M. J., Bradley, W. G. 1975. Genetically determined defect of Schwann cell basement membrane in dystrophic mouse. *Nature* 257:319–21

McDonald, J. A., Broekelmann, T. J., Kelley, D. G., Villiger, B. 1981. Gelatin-binding domain-specific anti-human plasma fibronectin Fab' inhibits fibronectin-mediated gelatin binding but not cell spreading. *J. Biol. Chem.* 256:5583–87

McGarvey, M. L., Baron-Van Evercooren, A., Kleinman, H. K., Dubois-Dalcq, M. 1984. Synthesis and effects of basement membrane components in cultured rat Schwann cells. *Dev. Biol.* 105:18–28

Mehta, H., Orphe, C., Todd, M. S., Cornbrooks, C., Carey, D. J. 1985. Synthesis by Schwann cells of basal lamina and membrane associated heparan sulfate proteoglycans. *J. Cell Biol.* 101:660–66

Michelson, A. M., Russel, E. S., Harman, P. J. 1955. Dystrophia muscularis: A hereditary primary myopathy in the house mouse. *Proc. Natl. Acad. Sci. USA* 61:1079–84

Mirsky, R., Winter, J., Abney, E. R., Pruss, R. M., Gavrilovic, J., Raff, M. C. 1980. Myelin-specific proteins and glycolipids in rat Schwann cells and oligodendrocytes in culture. *J. Cell Biol.* 84:483–94

Montesano, R., Orci, L., Vassalli, P. 1983. In vitro rapid organization of endothelial cells into capillary-like networks is promoted by collagen matrices. *J. Cell Biol.* 97:1648–52

Moya, F., Bunge, M. B., Bunge, R. P. 1980. Schwann cells proliferate but fail to differentiate in defined medium. *Proc. Natl. Acad. Sci. USA* 77:6902–6

Okada, E., Bunge, R. P., Bunge, M. B. 1980. Abnormalities expressed in long-term cultures of dorsal root ganglia from the dystrophic mouse. *Brain Res.* 194:455–70

Okada, E., Mizuhira, V., Nakamura, H. 1976. Dysmyelination in the sciatic nerves of dystrophic mice. *J. Neurol. Sci.* 28:505–20

Palm, S. L., Furcht, L. T. 1983. Production of laminin and fibronectin by Schwannoma cells: Cell-protein interactions in vitro and protein localization in peripheral nerve in vivo. *J. Cell Biol.* 96:1218–26

Payer, A. F. 1979. An ultrastructural study of Schwann cell response to axonal degeneration. *J. Comp. Neurol.* 183:365–83

Peterson, A. C., Bray, G. M. 1984. Normal basal laminas are realized on dystrophic Schwann cells in dystrophic ↔ shiverer chimera nerves. *J. Cell Biol.* 99:1831–37

Prockop, D. J., Berg, R. A., Kivirikko, K. I., Uitto, J. 1976. Intracellular steps in the biosynthesis of collagen. In *Biochemistry of Collagen*, ed. G. N. Ramachandran, A. H. Reddi, pp. 163–273. New York: Plenum

Ranscht, B., Clapshaw, P. A., Price, J., Noble, M., Seifert, W. 1982. Development of oligodendrocytes and Schwann cells studied with a monoclonal antibody against galactocerebroside. *Proc. Natl. Acad. Sci. USA* 79:2709–13

Rasminsky, M. 1978. Ectopic generation of impulses and cross-talk in spinal nerve roots of 'dystrophic' mice. *Ann. Neurol.* 3:351–57

Ratner, N., Glaser, L., Bunge, R. P. 1984. PC12 cells as a source of neurite-derived cell surface mitogen, which stimulates Schwann cell division. *J. Cell Biol.* 98:1150–55

Rodriguez-Boulan, E., Paskiet, K. T., Sabatini, D. D. 1983. Assembly of enveloped viruses in Madin-Darby canine kidney cells: Polarized budding from single attached cells and from clusters of cells in suspension. *J. Cell Biol.* 96:866–74

Roll, F. J., Madri, J. A., Albert, J., Furthmayr, H. 1980. Codistribution of collagen types IV and AB_2 in basement membranes and mesangium of the kidney. An immunoferritin study of ultrathin frozen sections. *J. Cell Biol.* 85:597–616

Roufa, D., Bunge, M. B., Johnson, M. I., Cornbrooks, C. J. 1985. Variation in nonneuronal cell populations in the outgrowth from sympathetic ganglia of differing embryonic age. *J. Neurosci.* In press

Sabatini, D. D., Griepp, E. B., Rodriguez-Boulan, E. J., Dolan, W. J., Robbins, E. S., Papadopoulos, S., Ivanov, I. E., Rindler, M. J. 1983. Biogenesis of epithelial cell polarity. In *Modern Cell Biol.* Vol. 2, *Spatial Organization of Eukaryotic Cells*, ed. J. R. McIntosh, pp. 419–50. New York: Liss

Sanes, J. 1982. Laminin, fibronectin, and collagen in synaptic and extrasynaptic portions of muscle fiber basement membrane. *J. Cell Biol.* 93:442–51

Sanes, J. R., Chiu, A. Y. 1983. The basal lamina of the neuromuscular junction. *Cold Spring Harbor Symp. Quant. Biol.* 48:667–78

Schwann, T. 1839. *Mikroskopische Untersuchungen über die Uebereinstimmung in der Struktur und dem Wachstum der Tiere und Pflanzen.* Berlin: Sander

Shannon, J. M., Pitelka, D. R. 1981. The influence of cell shape on the induction of functional differentiation in mouse mammary cells in vitro. *In Vitro* 17:1016–28

Shellswell, G. B., Restall, D. J., Duance, V. C., Bailey, A. J. 1979. Identification and differential distribution of collagen types in the central and peripheral nervous systems. *FEBS Lett.* 106:305–8

Stirling, C. A. 1975. Experimentally induced myelination of amyelinated axons in dys-trophic mice. *Brain Res.* 87:130–35

Sugrue, S. P. 1984. The identification of a cell surface binding protein for laminin from embryonic corneal epithelia. *J. Cell Biol.* 99:165a

Sugrue, S. P., Hay, E. D. 1981. Response of basal epithelial cell surface and cytoskeleton to solubilized extracellular matrix molecules. *J. Cell Biol.* 91:45–54

Sugrue, S. P., Hay, E. D. 1983. Identification of extracellular matrix binding sites on the cell surface of isolated embryonic corneal epithelia. *J. Cell Biol.* 97:319a

Timpl, R., Martin, G. R. 1982. Components of basement membranes. In *Immunochemistry of the Extracellular Matrix*, ed. H. Furthmayr, 2:119–50. Boca Raton, Fla.: CRC Press

Timpl, R., Rohde, H., Gehron-Robey, P., Rennard, S. I., Foidart, J. M., Martin, G. R. 1979. Laminin—a glycoprotein from basement membranes. *J. Biol. Chem.* 254:9933–37

Uitto, J., Prockop, D. J. 1974. Incorporation of proline analogues into collagen polypeptides. Effects on the production of extracellular procollagen and on the stability of the triple-helical structure of the molecule. *Biochim. Biophys. Acta* 336:234–51

Weinberg, H. J., Spencer, P. S. 1978. The fate of Schwann cells isolated from axonal contact. *J. Neurocytol.* 7:555–69

Wicha, M. S., Liotta, L. A., Vonderhaar, B. K., Kidwell, W. R. 1980. Effect of inhibition of basement membrane collagen deposition on rat mammary gland development. *Dev. Biol.* 80:253–66

Ann. Rev. Neurosci. 1986. 9 : 329–55
Copyright © 1986 by Annual Reviews Inc.

VERTEBRATE OLFACTORY RECEPTION

Doron Lancet

Department of Membrane Research, The Weizmann Institute of Science, Rehovot, Israel

INTRODUCTION

Through his sense of smell man has long encountered the phenomenon of molecular recognition. As early as first century BC, the Roman poet and philosopher Lucretius wrote (based on Democritus' atomic theory), "Such substances as agreeably titilate the senses are composed of smooth round atoms; those that seem bitter and harsh are more tightly compacted of hooked particles . . ." (Lucretius 55 BC). The last century has brought the realization that similar recognition of atomic or molecular configurations constitutes the basis of all life processes. Probably because olfaction has been a subject of study and speculation long before the advent of modern biochemistry, numerous olfactory theories have been proposed (Davies 1971, Amoore 1982). However, in the eyes of a contemporary molecular biologist, a special theory accounting for the initial steps in olfaction may seem superfluous, as the sensory cells in the nose are not unique in their ability to receive chemical signals selectively. It would be most parsimonious to assume that odorant recognition involves membrane protein receptors and transduction components analogous to those which mediate the specific responses to hormones, growth factors, neurotransmitters, and antigens. This notion, which corresponds to the combined Stereochemical/Functional-group theories of olfaction (Amoore 1982, Beets 1971), is validated by many of the studies reviewed here.

Olfactory recognition is mediated by a large ensemble of sensory cells, each conveying a fraction of the information that signifies the nature of the odorant and its concentration. The sensory neurons receive chemical signals at their dendritic membrane and respond by firing action potentials down their axons. The olfactory response is therefore analogous to neurotransmitter reception, and the sensory organ is a simple but

329

interesting neural assembly. I elaborate such concepts here in an attempt to relate olfactory receptor mechanisms to those found elsewhere in the nervous system.

I discuss in this review only the function of vertebrate olfactory sensory cells, and focus on biophysical and biochemical aspects involved in the initial steps of odor reception. I only mention in passing the structures and events beyond the first synapse. Insect olfactory reception is reviewed elsewhere in this volume (Kaissling 1986), and analogies to bacterial chemotaxis are discussed separately (Kleene 1985).

Studies of olfaction have been carried out by specialists in many fields: physics, physical chemistry, biochemistry, cell biology, electrophysiology, neurochemistry, developmental neurobiology, genetics, and psychology. It is often difficult to relate results and concepts arising in one discipline to those prevalent in another. While I make no attempt to provide comprehensive coverage, I hope that this review will help to bridge some gaps, so as to facilitate future formulation of a unified view of olfactory mechanisms.

THE SENSORY ORGAN

General Structure

Vertebrate olfaction is mediated by the olfactory epithelium (or neuro-epithelium or mucosa), the structure of which has been described in detail (Moulton & Beidler 1967, Graziadei 1971, Menco 1983). It is 100–200 μm thick, and its area varies from a few square centimeters (in man or frog), to more than 100 cm^2 (in dog). It contains a characteristic yellow-brown pigment, which, despite earlier claims, does not appear to be related to sensory function (Moulton & Beidler 1967). In embryonal development it is generated from the ectodermal olfactory placode, separate from other regions of the nervous system (Graziadei 1971, Cuschieri & Bannister 1975). The olfactory epithelium has three major cell types, whose nuclei form fairly discrete layers (Figure 1). The supporting cells are glia-like in nature and have microvilli on their apical surface. These cells have a role in the secretion of mucus components, and they may also function as isolating current sinks, analogous to retinal Muller cells, in guiding the growing sensory neuronal processes and in phagocytosis of shedded dendritic fragments (Graziadei 1971, Rafols & Getchell 1983, Getchell et al 1984).

The sensory neurons[1] are bipolar nerve cells, with a nonbranching

[1] The use of the term "receptor" for cells, although widespread among sensory physiologists (e.g. Fuortes 1971), may be confusing when dealing in parallel with molecular receptors. "Receptor" is therefore reserved here to a ligand-binding molecular structure, as in most of the biochemical and neurobiological literature, and "sensory" is used to describe a stimulus-transducing cell.

dendrite that terminates at the epithelial surface with a slight enlargement (olfactory knob) bearing 5–20 long cilia. Tight junctions between olfactory dendrites and the supporting cells are impenetrable to charged molecules, but do not form a barrier to lipophylic, uncharged odorants. The axons (0.2 μm, nonmyelinated) have no collaterals, and they form bundles ensheathed in envelopes of Schwann cells. The bundles join to form the olfactory nerve (first cranial nerve) leading to the olfactory bulb. At its bulbar synaptic target, every axon terminates within one of roughly 1000 glomerular structures (80–150 μm in diameter), each containing synapses of

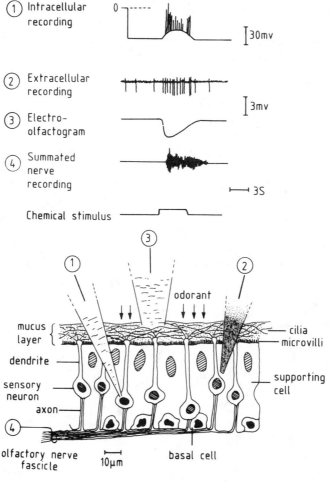

Figure 1 A schematic representation of structure and electrophysiology in olfactory epithelium.

10^4–10^5 sensory axons, with about 100 secondary neurons and mitral and tufted cells (Shepherd 1972). All of the 10^6–10^9 neurons in the epithelial sheet appear to be structurally similar when mature, and display no clear evidence for morphological classes.

The deepest layer of olfactory epithelium contains basal cells. Considerable evidence suggests that these are stem cells that divide and differentiate, to become functional sensory neurons. This is a rare case of a neuronal population undergoing such continuous renewal in adult vertebrates (Graziadei & Monti Graziadei 1979). When olfactory nerve is severed, or when the epithelial surface is irrigated with 0.1 M $ZnSO_4$, the sensory cells degenerate and disappear. Enhanced mitotic activity of the basal cell then gives rise to a newly formed population of sensory cells that send axons and form synaptic connections. A functional epithelium reforms after 30–60 days (Simmons & Getchell 1981).

Olfactory Cilia

The olfactory receptive apparatus probably resides in the extensions of the sensory dendrites, known as *olfactory cilia* (Reese 1965, Rhein & Cagan 1980, Menco 1980, 1983). These sensory organelles (similar to cilia from other tissues) are 0.25 μm in diameter, have a basal body and a tubulin-containing $9 \times 2+2$ microtubular core (axoneme), and are ensheathed in lipid bilayer. Olfactory cilia are distinct in their extended length (30–200 μm) and in tapering to a 0.15 μm diameter with loss of microtubuli along their distal length. Olfactory cilia in some species (including all mammals) are immotile, as they lack dinein, the ciliary energy-transducing ATPase (Lidow & Menco 1984). It is therefore unlikely that ciliary motility is important for sensory function. In being modified ciliary structures, olfactory cilia resemble other sensory organelles, such as rod and cone outer segments in the retina, or hair cell kinocilia in the inner ear (cf Vinnikov 1982, Fuortes 1971). The cilia are bathed in a layer (10–30 μm thick) of mucus, where odorants dissolve prior to their interaction with the ciliary membrane.

ELECTROPHYSIOLOGY

A Model Neuronal System

Stimulation with odorants leads to the generation of graded potentials across the sensory dendritic membrane, giving rise to action potential in the sensory axon (for review see Moulton & Tucker 1964, Moulton & Beidler 1967, Gesteland 1976, Holley & MacLeod 1977, Getchell & Getchell 1982, Getchell et al 1984). This sequence of events is analogous to that initiated by neurotransmitters at the postsynaptic dendritic membranes. Thus, the olfactory epithelium may serve as a model neuronal system (see Margolis 1975). In this respect it has several experimental advantages: (a) It is a natural nervous tissue "slice" that can be easily exposed, dissected, and manipulated without seriously affecting its functional integrity; (b) it contains only one type of neuron and mainly one type of glia-like cell; (c) its neurons receive no efferent synapses, and odorant binding appears to be their sole chemical input; (d) the dendritic membranes (olfactory cilia) are not obstructed by presynaptic elements, and can be detached and isolated; (e) the defined epithelial strata afford facile identification of extracellular electrophysiological recording sites; (f) the output nerve is easily accessible and contains only one type of nonmyelinated axon; (g) there is no problem of a search for stimulating ligands, as hundreds of compounds with low molecular weight (odorants) are excitatory, and each individual cell responds to many dozens. The main disadvantage of olfactory epithelial neurons is the small size of their somata; this makes intracellular recording relatively difficult.

Summated Recordings

Summated odor-evoked activity (Figure 1) can be studied (a) by recording of compound action potentials from olfactory nerve bundles or from the nerve layer of olfactory bulb (Adrian 1950, Gesteland 1976, Hornung & Mozell 1981, Kashiwayanagi & Kurihara 1984); and (b) by electroolfacto-gram (EOG) recordings from the ciliated epithelial surface (Ottoson 1971, Gesteland 1976, Holley & MacLeod 1977, Getchell & Getchell 1982, and Figure 1).

EOG responses involve transepithelial currents carried through the apical membrane by sodium ions and possibly potassium ions, and require extracellular calcium (Takagi et al 1969). It is widely accepted that the EOG represents chiefly a summation of generator potentials in the olfactory neurons, similar to those observed in other sensory cells (Fuortes 1971). EOG recordings constitute a highly useful experimental tool in physiological, physicochemical, and biochemical analyses of olfactory mechanisms (e.g. Menevse et al 1977, Goldberg et al 1979, Senf et al 1980, Simmons & Getchell 1981, Mackay-Sim et al 1982, Schafer et al 1984).

Cellular Recordings

Extracellular recordings from single olfactory epithelial cells (mainly with metal-filled micropipettes) have been instrumental in understanding the quantitative and qualitative properties of the sensory neurons (Gesteland et al 1965, Gesteland 1976, Holley & MacLeod 1977, Sicard & Holley 1984, and Figure 1). The following features are observed: (a) Most neurons increase their firing rate with increasing odorant concentration; (b) each sensory neurons responds (with different degrees) to many odorants, and each odorant stimulates a high proportion of the neurons tested; (c) in experiments involving many dozens of neurons, virtually no two cells are found to be identical.

Intracellular recordings from olfactory sensory neurons (Getchell 1977, Suzuki 1977, Trotier & MacLeod 1983, Masukawa et al 1985) demonstrate that the sensory neurons have a relatively high input resistance (~ 200 MOhm) and low resting potential (-45 ± 15 mV); that the neurons depolarize, decrease their input resistance, and increase their firing rate in response to most odorants; that these changes increase with odorant concentration; and that membrane conductances to sodium and possibly to potassium are involved. The modeled electrotonic properties of the sensory neuron suggest that very few dendritic ion channels suffice for the initiation of action potentials at the axon hillock (Hedlund et al 1984).

The neurotransmitter(s) at the first synapse in the olfactory pathway has not been unequivocally identified. The dipeptide, carnosine (β-alanyl-histidine), fulfills many of the required criteria (Margolis 1975a,b 1981, Rochel & Margolis 1982), but there is still a controversy with respect to its ability to affect the postsynaptic neurons (cf Frosch & Dichter 1984). Because of the functional diversity of the sensory neurons, an intriguing possibility is that different sensory neuron types utilize different neurotransmitters or neuropeptides at their central synapses. Several other types of neurotransmitter receptors (e.g. α- and β-adrenergic, muscarinic cholinergic and peripheral-type benzodiazepine receptors) have been found in olfactory epithelial tissue (Margolis 1981, Anholt et al 1984). Some may be related to perireceptor events such as mucosecretory activity (Getchell et al 1984). Acetylcholine and substance-P have been reported to be potent, odorant-like stimulatory ligands (Bouvet & Holley 1984).

QUANTITATIVE PARAMETERS

Threshold

The olfactory system is capable of responding at threshold to relatively small concentrations of airborne chemicals. This is often illustrated by the special case of insect pheromones, where a few molecules per animal may trigger a behavioral response (Kaissling 1986). Although vertebrates may also have specialized and sensitive machinery to detect behaviorally significant odorants, their olfactory system is mainly geared to deal with diverse chemical stimuli at considerably higher thresholds. The mean detection threshold in humans is 10^{-9} M in the air, and the range is 10^{-13} to 10^{-4} M (Patte et al 1975, Amoore 1982, Punter 1983). Similar or somewhat lower values have been measured in other vertebrate species (Moulton 1977, Slotnick & Schoonover 1984).

Amplification Mechanisms

Olfactory signal amplification, which is essential for the system's capability to detect odorants at low concentrations, occurs at several levels. The term, "amplification" is taken to mean anything that will increase the probability of response to weak stimuli.

Air-water partition   In terrestrial species, amplification is afforded by odorant passage from air to the aqueous mucus. Counter-intuitively, such partition is in favor of the water phase, even for rather hydrophobic odorants: the equilibrium water/air molar partition coefficient (K_p) for typical odorants is in the range of $10–10^3$ (Amoore & Buttery 1978, Hornung & Mozell 1981, Senf et al 1980).[2]

Area of receptive membrane   As in other sensory cells (e.g. rod cells in the retina, Vinnikov 1982, Stryer 1986) the receptive membrane area of olfactory sensory neurons is greatly augmented: The cilia increase the exposed membrane area of an olfactory knob (~ 5 μm^2) up to 100-fold (Reese 1965, Moulton & Beidler 1967, Menco 1980, 1983).

Density of receptor molecules   Olfactory cilia have a fairly large density ($800–2500/\mu m^2$) of freeze fracture intramembrane particles, suggested to correspond to olfactory receptor molecules (Menco 1980, Kerjaschki & Hörandner 1976, Masson et al 1977). This density is comparable to that found in other receptor-rich membranes, and could lead to enhanced signal detection capabilities (see *Affinity of Olfactory Receptors*).

Biochemical transduction mechanisms   Considerable evidence suggests that odorant molecules may activate a transductory enzyme cascade, so that each occupied receptor molecule would lead to the opening of many ion channels (see *Olfactory Transduction*).

Neuronal convergence   The high convergence ratio (100–1000:1) of sensory to secondary neurons (Shepherd 1972) makes it possible for very small deviations from spontaneous firing rates to be integrated and yield appreciable signals (van Drongelen 1978).

---

[2] Unless the binding site of olfactory receptor proteins is located within the lipid bilayer, the favorable lipid/water partition of most odorants would appear to be irrelevant to signal amplification.

Affinity of Olfactory Receptors

On the basis of intramembranous particle counts it has been calculated (Menco 1980) that the volume concentration of receptor molecules in the mucus may be in the range of 10^{-5} M, and that each sensory cell has about 10^6 receptor molecules on its ciliated dendrite. Even if only 1% of these receptors correspond to any particular odorant specificity (see *Receptor Specificity and Diversity*), the mucus concentration of an individual receptor type will still be 10^{-7} M. This is large compared to 10^{-9} M in a typical suspension (10^7/ml) of cells with the same receptor count.

With the suggested transductory and neuronal amplification mechanisms it is believed that a few occupied receptor molecules per sensory cell would suffice to elicit a response at threshold (Moulton 1977, Kaissling 1986). Under these conditions of low receptor occupancy, the relation between the equilibrium dissociation constant (K_d) and the total (free plus bound) mucus odorant concentration at threshold (H_t) can be derived from the law of mass action as follows:

$$H_t = X \cdot (1 + K_d/R)$$

where R is the free ($\cong$ total) receptor concentration, and X is the concentration of the receptor-odorant complex. This equation has two important consequences: 1. For $K_d \gg R$ ($K_d = 10^{-6}$ M and up) one gets $H_t = K_d \cdot Y$, where $Y = X/R$ is the fractional receptor saturation. Since $Y \ll 1$, then $H_t \ll K_d$ is obtained, i.e. the odorant molar concentration at threshold is much lower than the equilibrium dissociation constant. In the limiting case, where one occupied receptor per cell ($Y = 10^{-6}$) is sufficient to elicit a response (akin to single photon detection by a retinal rod, Stryer 1986), a moderate value of K_d (10^{-6} M) will give some of the lowest observed thresholds ($H_t = 10^{-12}$ in the mucus, 10^{-14}–10^{-13} in the air). For an average odorant (threshold mucus concentration = 10^{-8} M), an extremely low affinity value ($K_d = 10^{-2}$ M) might be sufficient for detection. This range of low expected affinities agrees with values obtained from suprathreshold measurements (Senf et al 1980) and is consistent with a broad receptor specificity (Lehmann 1978).

2. For $K_d \ll R$ one obtains $H_t = X = R \cdot Y$. Under these conditions all ligand is bound, and H_t is independent of K_d. In other words, any decrease of K_d below about 0.1 of the receptor concentration (10^{-8} M in our estimate) should not result in an improvement of olfactory sensitivity. The calculated lowest possible olfactory threshold for $R = 10^{-7}$ is $H_t = 10^{-13}$ in the mucus, or 10^{-15}–10^{-14} in air, comparable to the lowest reported thresholds (e.g. Moulton 1977).

Dynamic Range and Dose-response Curves

The maximal olfactory stimulus strength is physically limited by odorant vapor saturation (10^{-5} to 10^{-3} M).[3] Thus the dynamic range from threshold to physical saturation would be 5 orders of magnitude for a typical odorant, and 8–10 decades for odorants with lower thresholds. The latter is comparable to perceptual dynamic ranges in vision and audition (Loewenstein 1971). A single olfactory sensory neuron changes its firing frequency over a narrower concentration range of about 3 decades, consistent with a hyperbolic saturation curve of receptor binding (Moulton & Tucker 1964, Senf et al 1980), with a possible contribution of membrane conductance shunting (Lipetz 1971). The dynamic range would be augmented through integration of signals from sensory cells with different odorant affinities.

Suprathreshold log-log dose-response curves for olfactory responses obtained by electrophysiology (cf Ottoson 1971) or by psychophysical experiments (cf Patte et al 1975) have slopes in the range of 0.2–0.4. Such low slopes correspond to a Hill's slope < 1 (cf Levitzki 1978), and are consistent with receptor heterogeneity (see *Receptor Specificity and Diversity*).

Onset Kinetics

Several hundred milliseconds elapse between the onset of the odorous stimulus and the electrical response of the sensory cells (O'Connell & Mozell 1969, Ottoson 1971, Holley & MacLeod 1977, Getchell et al 1984). In electrophysiological experiments, odorants reach the epithelial surface within a few dozen milliseconds (Getchell et al 1980), and the formation of the odorant-receptor is most probably in the same time range or faster, depending on odorant concentration (Pecht & Lancet 1977). Thus, neither access time nor complex formation can readily account for the observed latencies. The slow onset kinetics has been attributed to the diffusion of odorants from the mucus surface, assuming that the receptors are located only at the proximal end of the sensory cilium (Getchell et al 1980). Other possible explanations are the slow enzymatic production of second messenger and its intraciliary diffusion, or the activation of more distal transductory enzymes (see *Olfactory Transductions*).

Adaptation

Adaptation in the olfactory system is manifested at two levels: (*a*) the slow (~ 1 min) suppression of perceived odor intensity during prolonged

---

[3] Water solubility is not an independent constraint, since a saturated aqueous solution is at equilibrium with the saturated vapors of a pure compound (Amoore & Buttery 1978).

exposure (Engen 1971), which is thought to arise from central neuronal processing; and (b) the fast (< 1 sec) change from phasic to tonic response during continuous stimulation, which happens at the sensory neuron level (Ottoson 1971, Getchell & Shepherd 1978, Van Boxtel & Köster 1978).[4] Fast adaptation may be similar to desensitization in other receptor systems (Cuatrecasas & Hollenberg 1976), a process usually mediated by receptor conformational transition or covalent modification. An alternative mechanism, namely receptor internalization (Cuatrecasas & Roth 1983), is less likely to occur in the cytoskeleton-rich cilium, whose diameter is roughly equal to or smaller than that of an endocytic vesicle.

Termination Kinetics

When the odorant is removed from the air over the epithelium after a long stimulation, the sensory response diminishes from tonic level to zero within a few hundred milliseconds (Ottoson 1971, Getchell & Shepherd 1978). Such relatively fast turnoff is essential for a sensory mechanism that responds to changes occurring between consecutive sniffs. The termination of olfactory signals must involve both the removal of odorant from the mucus layer and the dissociation of bound odorant from the receptors, since the tonic firing level already includes all fast adaptation processes (Ottoson 1971). The rate constant (k_{off}) for the dissociation of the odorant-receptor complex would be given by $k_{off} = K_d \cdot k_{on}$, where k_{on}, the association rate constant, is often diffusion controlled, and is maximally about 10^8 M^{-1} sec^{-1} (Pecht & Lancet 1977). For dissociation to be complete within ~ 1 sec ($k_{off} > 1$ sec^{-1}), K_d cannot be much smaller than 10^{-8} M (cf Mason & Morton 1984), in very good agreement with the minimal value reached by equilibrium considerations. This correspondence may be interpreted to mean that the high mucus receptor concentrations evolved to avoid exceeding a maximal receptor affinity, as dictated by the constraints of termination kinetics. A similar situation may hold for other fast-terminating receptors (e.g. the nicotinic acetylcholine receptor, Changeux et al 1984).

The mechanism of odorant removal is not clear. Although some odorants may be modified by the cytochrome P-450 oxydative enzyme system (Dahl et al 1982), it is unlikely that the olfactory epithelium would contain degradative enzymes specific for all odorants, and the products of such degradation would often be odorants as well. An alternative mechanism, repartition from mucus to air, is thermodynamically unfavorable ($K_p \gg 1$) and occurs with very low efficiency (Hornung & Mozell 1981). A third mechanism, and in my opinion the most likely, is odorant dilution from the

---

[4] Fast adaptation may be related to the documented decrease (within a fraction of a second) of odor perception upon cessation of breathing during a sniff (Moncrieff 1967).

mucus layer into the much larger volume of the underlying tissue, followed by clearance in the blood. Blood-epithelium passage is indicated by the reported responses to blood-borne odorants (Maruniak et al 1983).

QUALITATIVE PARAMETERS

Olfactory systems have exquisite recognition capabilities, such as distinguishing between optical isomers (Beets 1971), or, reported in mice, segregating odor signals produced by inbred strains of mice differing only at the major histocompatibility locus (Boyse et al 1982). A vast literature exists on the classification of odor quality and its relation to chemical structure (for review see Moncrieff 1967, Beets 1971, Amoore 1982). Olfactory quality is multidimensional (Engen 1971, Schiffman 1974, Sicard & Holley 1984), in contrast to vision or audition, where sensory quality can be related to a unidimensional physical scale (wavelength or frequency). Despite considerable effort, a widely accepted system for odor classification has not emerged (Amoore 1982), and tentatively defined odor quality groups often do not correspond to odorant chemical structure (Beets 1971), or to electrophysiologically-defined sensory neuron classifications (Gesteland 1976, Holley & MacLeod 1977, Sicard & Holley 1984).

Primary Odors and Specific Anosmia

Much effort has been invested in looking for primary odors that would serve the same role as primary hues in color vision (Amoore 1982). These studies have met with serious difficulties, and it has not been clearly demonstrated that the odor quality of one pure odorant can be reproduced at will by mixing several others, or that perceived odor-quality classes constitute primary odors (Engen 1971, Amoore 1982). A more recent line of studies is that of human specific anosmias, deficits in the individual's ability to detect an odorant or odorant group (Amoore 1971, 1982). Several dozen anosmia types have been discovered in humans, and some have been reported in mice (Price 1977, Wysocki et al 1977). There is good evidence for the genetic basis of specific anosmias (Amoore 1971), and the recently-described polymorphism in the ability to smell androstenone is a particularly clear example (Wysocki & Beauchamp 1984). Specific anosmias probably correspond to a modified or deficient gene for one type of olfactory receptor molecule, in analogy to color blindness. The study of specific anosmias has led to important clues with respect to the nature and diversity of the putative olfactory receptor proteins but has not been conclusive in resolving the primary odor question. This is because it remains unclear whether the molecular receptor types correspond in a simple way to psychophysically defined primary odors (Beets 1971).

Receptor Specificity and Diversity

The olfactory system of terrestrial vertebrates is capable of responding to practically all volatile organic compounds, and also to many nonvolatile ones, if appropriately delivered (Moulton & Tucker 1964, Getchell 1974). Different cells are activated by different odorants, but all cells respond in the same way (depolarization, action potentials). This situation can be most simply explained by the existence of a family of receptor molecules, having different odorant binding sites in a "variable region" of the molecule but sharing the ability to interact with common transduction components at a molecular "constant region" (Goldberg et al 1979, Boyse et al 1982, Chen & Lancet 1984, Lancet 1984). Such a design is entirely analogous to that of the immune system (cf Sigal & Klinman 1978, Hood 1982), which has similarly evolved to respond to extraneous (and sometimes "unpredicted") chemical configurations.

The size (M) of the olfactory receptor "repertoire" is not known. The specific anosmia data are consistent with there being at least a few dozen receptor protein types, but the number M could easily be 10–100 times larger, as even 10^4 receptor types would be very few compared to the antibody repertoire (10^7–10^8, Hood 1982). Combinatorial mechanisms at the DNA, RNA, and protein levels, similar to those found in immuno-globulins and other multigene families (Hood 1982), could allow the coding of the olfactory receptor family by a relatively small number of genes.

As in the case of immunoglobulin molecules in B lymphocytes (Sigal & Klinman 1978), olfactory receptor molecules could be "clonally excluded," i.e. each sensory cell would bear only one molecular receptor type. Then, in order to produce the documented single-unit response patterns, the receptor proteins should have rather broad specificity, not unlikely in view of their expected low affinity. Alternatively, each sensory neuron could carry a mixture of several receptor types of narrower specificity (cf Gesteland 1976).

Quality Coding and Information Processing

The epithelial sheet is a mosaic of neuronal elements with different odorant specificities (Moulton 1976). Not all regions of the epithelium are equally sensitive to a given odorant (Mackay-Sim et al 1982), thus suggesting nonhomogeneous distribution of neuron types. Such segregation is prob-ably related to the developmental sequence during which different neuronal clones are generated. Little evidence suggests that the actual location of each of these "specificity patches" in the epithelial sheet contributes significantly to olfactory information processing (cf Moulton 1976). In other words, it is unlikely that olfactory coding would be greatly affected by reshuffling the neurons in the epithelial sheet while keeping their central connections unchanged. This is in clear contrast to vision, where a cell's position in the retinal sheet is of utmost importance, and the above manipulation would be deleterious to pattern recognition.[5]

Thus, to a good approximation, the information conveyed in the olfactory nerve should be reducible to N numbers that correspond to the (time-averaged) firing frequencies in the axons of N different sensory neuron types ($N = M$ if clonal exclusion holds). This N-dimensional "activity vector," whose components represent N independent information channels, encodes both the quality and the intensity of the odorant (or odorant mixture). To a first approximation, the direction of the vector (i.e. the ratios of channel activities, or "across fiber" pattern), will signify odor quality, while the vector length (having to do with the sum of all channel activities) would be a measure of odor intensity (cf Hainer et al 1954, Pfaffman 1955, Beets 1971, Polak 1973). It is interesting that for saturable (nonlinear) sensory cells, both the direction of the vector and its length may change as the concentration of a given odorant is varied (Lancet et al 1984). An open question is how, through central processing, quality and quantity are then separately decoded.

The size of N is unknown. From the pure information theory point of view, a small number ($N = 3$–5) would suffice (Beets 1971, Persaud & Dodd 1982). Conservatively estimating that each information channel carries seven bits of information (128 gradations of firing frequency), five channels could produce $2^{35} = 3 \cdot 10^{10}$ different patterns, enough to encode practically all the known volatile chemicals together with many mixtures and intensity gradations. However, there seems to be a large amount of redundancy in the olfactory system, since the electrophysiological recordings suggest the existence of at least a few dozens sensory neuronal types (Gesteland 1976, Holley & MacLeod 1977, Sicard & Holley 1984).

Neuronal Network Specification

Two parameters may determine the projection target of an epithelial sensory neuron: its spatial coordinates and its odorant specificity (i.e. the receptor molecules it bears on its dendrite). Recent studies suggest that bulbar projection sites are partly specified by epithelial space coordinates, hence that bulbar neurons have anatomical receptive fields in the epithelium (Kauer 1981, Jastreboff et al 1984, Fujita et al 1985, Clancy et al 1985). Such receptive fields may reflect preferential projections from an

[5] Spatiotemporal patterns due to chromatographic distribution of odorants across the epithelium have been proposed to play a modulatory role in olfactory coding (Mozell 1971, Moulton 1976).

epithelial region rich in a given type of sensory cells to a functionally defined central target (Jastreboff et al 1984).

Specificity-determined contacts could come about through coexpression of axonal recognition molecule(s) (Goodman et al 1984) with the dendritic odorant receptor(s) prior to synapse formation. Alternatively, an "instructive" mechanism could be operative, whereby the sensory cells undergo random synaptogenesis and the target neuron specifies the odorant receptor(s) to be expressed. The latter mechanism is consistent with the finding that rat embryonal sensory neurons gain much of their odorant selectivity after their synapses form (Gesteland et al 1982).

Clonally excluded receptors afford the most economic way to specify a functional correlation between dendritic odorant specificity and axonal synaptic target. In the simplest case, a given member of the repertoire of olfactory receptor molecules will always coexpress with a given type of axonal recognition molecule. Alternative scenarios, where each sensory cell expresses several dendritic receptor protein types, would increase the complexity of the differentiative events that must lead to the formation of a functional network.

The glomeruli of olfactory bulb may be common projection sites for sensory axon dendrites with the same or similar odorant specificity. This is suggested by high-resolution 2-deoxyglucose mapping data, showing that glomeruli act as integral functional units (Lancet et al 1982). Thus, it is quite possible that each glomerulus constitutes a synaptic target for a different sensory neuron type or clone, whereby N (see previous section) would be equal to the number of glomeruli. The finding that growing sensory axons induce the formation of glomeruli in a central target (e.g. Graziadei et al 1979), while suggesting a degree of anterograde information flow, does not preclude the retrograde "instructive" model.

BIOCHEMISTRY OF OLFACTORY RECEPTION

Olfactory Receptor Proteins

The most important goal of all biochemical studies of olfaction to date has been the isolation of olfactory receptor proteins. Still, despite much effort, the proteins that bind odorants and generate the sensory signal have not been identified as distinct molecular species (Dodd & Persaud 1981, Price 1981, Lancet 1984). This, together with the possibility that odorants (mostly lipophylic) could generate membrane currents by directly interacting with the lipid bilayer, led some authors to suggest that olfactory receptor proteins may not exist (Davies 1971, Price 1984, Kashiwayanagi & Kurihara 1984). Yet, considerable evidence (as detailed in other sections) supports the involvement of proteins in olfactory reception: the stereospecificity of sensory responses; the abolishment of olfactory reactivity by protein-specific modifying reagents; the existence of odor-specific, genetically-determined sensory deficits; the demonstration of saturable odorant binding in some olfactory preparations; the large density of ciliary intramembranous particles; and the involvement of cyclic nucleotide processing enzymes, which in other cases have been found in conjunction with protein receptors.

The ciliary membrane is the most probable location of the chemoreceptive molecular apparatus, since mild removal of the cilia in vivo abolishes electrophysiological olfactory responses (Adamek et al 1984), and isolated cilia preparations have been shown to have amino acid odorant binding activity (Rhein & Cagan 1980) and high concentrations of transductory enzymes (Pace et al 1985).

Studies Related to Odorant Binding

GENERAL PROBLEMS   Olfactory receptor molecules pose several difficulties when studied by ligand-binding methods: (a) Odorant receptors probably have low equilibrium binding constants and high dissociation rate constants (see QUALITATIVE PARAMETERS), conditions known to be deleterious to most receptor-binding techniques (Cuatrecasas & Hollenberg 1976); (b) odorants are lipophylic, and their facile interaction with nonreceptor proteins and with membrane lipids would result in a high degree of nonspecific binding; (c) olfactory receptors are likely to be heterogeneous; a biochemical preparation from entire epithelium could contain numerous receptor types, each scantly represented (similar to "nonimmune" serum). Odorants will often bind to more than one receptor type (complex saturation curves), and most ligands tested will bind to at least one receptor (lack of "negative controls" and of defined pharmacological profiles).

ODORANT BINDING MEASUREMENTS   The most fruitful binding studies with radiolabeled odorants have been performed in fish, where many of the above-mentioned problems are alleviated: The odorants are few and nonlipophylic (mainly amino acids), and receptor heterogeneity is less pronounced. Several authors have demonstrated saturable binding of radiolabeled amino acid odorants to fish olfactory epithelial membrane preparations (Cagan & Zeiger 1978, Brown & Hara 1981, Fesenko et al 1983), and a correlation was found between ligand binding and electrophysiological potencies (Cagan & Zeiger 1978). The possible interference of amino acid uptake mechanisms in the binding measurements has been pointed out as a complicating factor of these studies (Brown & Hara 1981).

In frog and rat olfactory epithelial membranes, two classes of saturable binding sites for camphor ($K = 10^7$ and $K = 10^9$) were reported, and the higher affinity sites were suggested to correspond to olfactory receptors (Fesenko et al 1979). Camphor binding was best inhibited by compounds with similar odor quality in humans, and was absent in membranes from other tissues. Similar results were obtained for the binding of pyrazine derivatives to bovine olfactory epithelial extracts (Pelosi et al 1982). A weak point of some of the above studies is that the concentrations of odorant binding sites in crude preparations (50–500 pmole/mg protein) was high relative to the expected amount of a single receptor type.

CHEMICAL MODIFICATION   Protein modification by the thiol-blocking reagents N-ethylmaleimide (NEM) and mersalyl, and by enzymatic iodination was found to block the electroolfactogram response, thus suggesting the functional involvement of essential cysteine and tyrosine residues (Getchell & Gesteland 1972, Dodd & Persaud 1981, Shirley et al 1983a, Delaleu & Holley 1980). Blocking was partially inhibited by simultaneous application of an odorant, and in some cases odorant protection displayed a measure of specificity (Getchell & Gesteland 1972). Thiol-related inactivation was suggested to involve free sulfhydryls in the extracellular region of the receptor molecules (Shirley et al 1983a). A possible alternative (admittedly inconsistent with mersalyl inhibition if the latter is completely membrane-impenetrable) is that essential free SH groups are part of the intracellular transduction mechanism, and odorant protection is conformationally mediated, as in the case of adenylate cyclase coupled hormone receptors (Schramm & Selinger 1984).

Affinity and photoaffinity labeling by covalently reactive odorants has been utilized to inactivate selectively olfactory receptors with a particular specificity (Criswell et al 1980, Delaleu & Holley 1983, Schafer et al 1984, Mason et al 1984). In most cases, some protection by homologous nonreactive odorants could be demonstrated.

Other reagents that inhibit electrophysiological olfactory responses in vivo are the lectins, Concanavalin-A (Shirley et al 1983b) and wheat germ agglutinin (Chen et al 1985c). Ultraviolet irradiation is found to inhibit electroolfactogram responses reversibly (Kauer et al 1983). Both lectins and ultraviolet light affect the responses to many odorants, suggesting reactivity with "common denominator" components of the receptive apparatus.

SOLUBILIZATION AND PURIFICATION OF ODORANT-BINDING PROTEINS   An anisole-binding protein from olfactory epithelium from the dog has been affinity-purified, and rabbit antibodies against it specifically inhibited the electroolfactogram response to homologous as well as heterologous odorants, thus suggesting immunoreactivity with a "constant" receptor determinant (Goldberg et al 1979). The rat camphor-binding protein has been partially purified and shown to be membrane-associated and to have a molecular weight of 120 kD. Bovine and rat pyrazine-binding proteins have been purified (Pevsner et al 1985, Bignetti et al 1985) and found to be soluble proteins constituting a dimer of 19 kD polypeptides. Further research, together with the application of criteria for receptor identification (next section) should establish the possible role of the odorant-binding proteins described in olfactory reception.

Additional Criteria for Olfactory Receptor Identification

The above studies describe epithelial proteins that bind odorants and therefore constitute olfactory receptor candidates. However, it has been pointed out that the mere binding of one or a few odorous compounds is not a sufficient proof of receptor identification (Price 1981). In view of this difficulty, I propose here a set of additional criteria that stem from the notions that (a) olfactory receptor proteins of different specificities share "common denominator" molecular properties, and (b) that unless otherwise proven, these general properties can be assumed to be similar to those of other well-studied membrane receptors. It is suggested that fulfillment of most or all of these criteria should be taken as evidence for olfactory receptor identification, even in the absence of ligand-binding data or clear-cut pharmacological correlations. The criteria are as follows:

1. Tissue specificity.
2. Enrichment in the cilia (vs epithelium).
3. Glycosylation.
4. Transmembrane disposition (integral membrane protein).
5. Correct bilayer concentration (major component).
6. Diversity (sequence heterogeneity).
7. Specific recognition by function-modulating reagents (antibodies, lectins).
8. Interaction with transductory proteins.
9. Reconstitution of odorant modulation of enzymatic activities.

Such criteria should be applied to distinct polypeptide species that constitute candidate receptor proteins. Some of them have been proposed and utilized previously. These include criterion 1 (Goldberg et al 1979, Fesenko et al 1979, Rhein & Cagan 1980, Chen et al 1985b), criteria 2–5 (Chen & Lancet 1984, Chen et al 1984, 1985b), and criterion 7 (Goldberg et al 1979, Shirley et al 1983b, Chen et al 1985c). The latter can only be applied if the function-modulating reagent exclusively labels a given polypeptide. Criterion 4 (transmembrane disposition) is a property of practically all cell surface receptors (cf Cuatrecasas & Hollenberg 1976, Cuatrecasas & Roth

1983, Changeux et al 1984), and most probably applies to odorant receptors as well. While an extracellular soluble receptor protein (Pevsner et al 1985) is a possibility not yet ruled out, an intracellular soluble receptor (similar to that for steroid hormones) is much less likely, because many functional odorants are membrane impenetrable (Getchell 1974, Rhein & Cagan 1980, Brown & Hara 1981, Fesenko et al 1983). Criterion 5 may be applied using the data of Menco (1980, 1983), which suggest that the total olfactory receptor population should constitute roughly 1 μg protein per cm^2 of epithelium (based on 10^6 receptor molecules of 10^5 Dalton per neuron). This crudely corresponds to about 0.1% of the total epithelial protein, 1% of the epithelial membrane protein, and 10% of the ciliary membrane protein. Any given odorant specificity (estimated as 1% of the population) should show binding at roughly 1, 10, and 100 pmole/mg protein in crude extracts, total membranes, and ciliary membranes, respectively. Criterion 6 (heterogeneity) is a property of polypeptide products of multigene families (Hood 1982, Sigal & Klinman 1978), and could be probed by protein chemical and molecular genetic methods. The receptor polypeptide can be identified by its specific non-covalent interaction with its immediate neighbor(s) in the transduction chain, e.g. the G-protein (criterion 8). Since the receptor is the component that confers odorant sensitivity on ciliary transductory enzymes (adenylate cyclase, protein kinase), it should be possible to identify it within a mixture of solubilized proteins by reconstitution experiments as described by Schramm & Selinger 1984 (criterion 9). Odorant binding, a natural but controversial criterion for olfactory receptor identification, is discussed under *Studies Related to Odorant Binding*, above.

Receptor Candidates in Isolated Olfactory Cilia

Isolated olfactory cilia constitute a cell-free membrane preparation rich in receptive components, similar to isolated retinal rod outer segments (Stryer 1986). They can be detached and purified by a simple "calcium shock" procedure (Rhein & Cagan 1980, 1981, Chen & Lancet 1984). The first preparation of isolated olfactory cilia, obtained from fish olfactory organ (Rhein & Cagan 1980, 1981), was shown to contain binding activity toward amino acid odorants, but the ciliary polypeptides responsible for binding were not identified. A preparation of olfactory cilia from the frog, *Rana ridibunda* (Chen & Lancet 1984) was shown to contain several specific proteins (using nonsensory cilia as the control), four of which were glycosylated. Similar results were reported in another frog species, *Rana catesbeiana* (Anholt & Snyder 1985). In view of the effect of lectins on olfactory responses, it was proposed that one or more of them was a functional surface component. One ciliary glycoprotein, gp95 (of molecular weight 95 kD) also fulfilled the criteria (Table 1) of ciliary enrichment, transmembrane disposition, correct bilayer concentration, and specific reactivity with the function-modulating wheat germ agglutinin (Chen et al 1986b). Thus, gp95 constitutes a plausible candidate for an olfactory receptor protein.

Cell Biological Studies Relevant to Olfactory Reception

CELL CULTURE   Adult or embryonal olfactory epithelium can be maintained in short-term culture (Farbman 1977, Farbman & Margolis 1980), where they have been studied biochemically (Margolis & Tarnoff 1973, Chen et al 1984) and electrophysiologically (Masukawa et al 1985), and shown to be capable of neurite outgrowth (Gonzales & Farbman 1984). The sensory neurons can be isolated and identified (Hirsch & Margolis 1979, Kleene & Gesteland 1983). In isolation, they can be maintained in primary culture (Noble et al 1984) and studied electrophysiologically (Maue & Dionne 1984). Whether neurons are produced by cell division in these cultures is unclear (Farbman & Margolis 1980, Debagge et al 1982).

Primary cultures (Schubert et al 1985) and a cell line (Goldstein & Quinn 1981) of the stem sensory neuroblasts (basal cells) have been reported. The line cells may be induced to differentiate into cells resembling sensory neurons (Teeter & Goldstein 1984). The future development of cloned sensory cell lines expressing defined odorant receptor molecule(s) would be extremely useful in studying sensory transduction, receptor specificity, and clonal exclusion. Such cells could also serve as an unlimited source of material for biochemical studies, complementing the small amounts (10^6–10^8 cells/animal) afforded by dissected epithelia.

SPECIFIC MARKERS   Olfactory marker protein (OMP) a 19 kD cytoplasmic protein, has been established as a specific marker of the sensory neurons (Margolis 1975a,b). This protein is found to be expressed in a developing epithelium only after synapse formation, and thus constitutes a highly useful probe for neuronal differentiation (Farbman & Margolis 1980). The recent isolation of the mRNA species coding for this protein (Rogers et al 1985) may help to resolve its yet unknown function, and possibly relate it to receptor mechanisms.

Monoclonal antibodies and lectins have been used to label fluorescently some sensory neurons (Allen & Akeson 1985, Hempstead & Morgan 1985), nerve fascicles (Fujita et al 1985), surface epithelial structures (Hempstead & Morgan 1983), and specific ciliary proteins (Chen et al 1986a). Such reagents could be used to probe olfactory function (Lancet 1984).

Olfactory Transduction

Odorants are low-molecular-weight ligands, and are therefore most likely to exert their effect via an allosteric conformational transition (Dodd & Persaud 1981, Atema 1973) rather than through receptor cross linking (cf

Cuatrecasas & Roth 1983). One possible conformational mechanism is that of an odorant-gated ion channel (Vodyanoy & Murphy 1983), similar to the nicotinic acetylcholine receptor (Changeux et al 1984). This configuration involves little amplification (one bound ligand opens one ion channel), and may not be optimal for a sensory mechanism that requires high sensitivity. More plausibly, odorant binding would cause the activation of an enzymatic cascade, leading to changes in second messenger levels (cf Dodd & Persaud 1981, Figure 2). Such a transduction mechanism involves high amplification factors: more than 10^5 second messenger molecules may be affected per one activated receptor molecule (Stryer 1986).

SECOND MESSENGER   Considerable evidence has accumulated over the last decade suggesting that cyclic AMP is involved in olfactory transduction: Olfactory epithelium has a relatively high level of adenylate cyclase activity (Kurihara & Koyama 1972); its phosphodiesterase isozyme pattern changes when the neurons degenerate (Margolis 1975b); cAMP and its dibutyryl derivative (but not cGMP or its analogues) cause the appearance of electroolfactogram-like potentials; and phosphodiesterase inhibitors modulate the electrophysiological response to odorants (Minor & Sakina 1973, Menevse et al 1977). The last of Sutherland's criteria for cyclic nucleotide mediation (Robinson et al 1971) was fulfilled by the

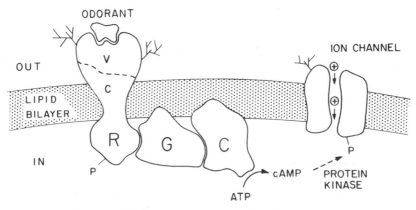

Figure 2   A most probable molecular model for olfactory reception and transduction, based on data and speculation as described in the text. The receptor molecule (R) is an integral membrane glycoprotein with a constant (c) and a variable (v) region. The odorant-occupied receptor activates a GTP-binding protein or G-protein (G), which modulates the activity of adenylate cyclase (C), an enzyme that produces the second messenger, cyclic AMP (cAMP). The latter molecule activates cAMP-dependent protein kinase to cause phosphorylation of other proteins. Phosphorylation (p) of ion channel polypeptides causes membrane potential changes, while that of the receptor polypeptide constitutes a feedback mechanism underlying adaptation.

recent demonstration that a cell-free olfactory cilia preparation has an extremely high activity of adenylate cyclase (>10 times that of brain membranes), which is specifically enhanced by odorants (Pace et al 1985). These findings establish the role of cyclic AMP as a second messenger in olfactory transduction. Odorant-induced transductory events may result also in the modulation of other membrane enzymes, such as the Na/K ATPase (Dreesen & Koch 1982).

Odorant stimulation of ciliary adenylate cyclase occurs only in the presence of GTP (Pace et al 1985), thus suggesting the involvement of a signal-coupling GTP-binding protein or G-protein (Gilman 1984, Schramm & Selinger 1984, Stryer 1986). The olfactory G-protein, a homologue of the hormone, neurotransmitter, and visual signal transducing proteins, can be directly identified in olfactory (but not in respiratory) cilia, via labeling with cholera toxin (Lancet & Pace 1984, Pace et al 1985).

PROTEIN PHOSPHORYLATION AND ION CHANNELS   The next step in olfactory transduction may involve cyclic AMP–dependent protein kinase and protein phosphorylation (cf Nestler et al 1984, Siegelbaum & Tsien 1983, and Figure 2). Olfactory cilia have such protein kinase activity, and contain two proteins of 56 kD and 24 kD whose phosphorylation is modulated by cyclic AMP (J. Heldman, Z. Chen, D. Lancet, unpublished). It is tempting to speculate that phosphorylation of an ion channel polypeptide (or a modulatory subunit thereof) is involved in ion gating, as in some other neurons (Siegelbaum & Tsien 1983). Alternatively, the cyclic nucleotide may directly gate the ion channel(s) (cf Stryer 1985).

The last components in the transductory chain are the membrane ion channel(s) that are affected by odorant binding. It has not been established whether these are located in the cilia, dendrite, soma, or a combination of these, but it has been argued that a small number of channels should suffice for signal generation (Hedlund et al 1984). Single channel currents in planar lipid bilayers reconstituted with rat olfactory epithelial membranes revealed a potassium channel, whose mean open time was very long (30–40 sec), and increased in the presence of odorants (Vodyanoy & Murphy 1983). Patch clamp recordings from whole isolated sensory neurons showed three types of potassium channels, but their relation to odorant response remains unknown (Maue & Dionne 1984). Direct application of the activated catalytic subunit of cyclic AMP–dependent protein kinase during membrane patch recordings could clarify the role of ion channel phosphorylation (cf Siegelbaum & Tsien 1983). The identification of the molecular components that directly underlie the conductance changes, and the elucidation of the role of various ions in olfactory transduction, require extensive future investigation.

CONCLUSION

The years to come will tell whether "we might fairly gauge the future of biological science, centuries ahead, by estimating the time it will take to reach a complete, comprehensive understanding of odor" (Thomas 1983). Clearly, olfactory mechanisms do not seem as intractable today as they did several years ago. Recent and future advances may owe their success to developments in other fields. For example, it is expected that molecular genetics will soon be recruited to help answer some of the pressing questions related to olfactory receptor specificity and function. In turn, olfactory neurons may become a useful system in which to study the role of gene diversity in neuronal network formation and information processing. The recently reported odorant-specific olfactory-deficient mutants in *Drosophila* (Rodrigues 1980) may be useful in this respect. In parallel, the study of transductory components in the dendritic membrane of olfactory neurons may help to understand molecular recognition and transmembrane signaling in neurons and other cells. I hope that this review will help further the interaction among disciplines necessary to achieve such goals.

Literature Cited

Adamek, G. D., Gesteland, R. C., Mair, R. G., Oakley, B. 1984. Transduction physiology of olfactory receptor cilia. *Brain Res.* 310: 87–97

Adrian, E. D. 1950. Sensory discrimination with some recent evidence from the olfactory organ. *Br. Med. Bull.* 6: 330–31

Allen, W. K., Akeson, R. 1985. Identification of a cell surface glycoprotein family of olfactory receptor neurons with a monoclonal antibody. *J. Neurosci.* 5: 284–96

Amoore, J. E. 1971. Olfactory genetics and anosmia. See Beidler 1971, pp. 245–56

Amoore, J. E. 1982. Odor theory and odor classification. See Theimer, pp. 27–76

Amoore, J. E., Buttery, R. G. 1978. Partition coefficients and comparative olfactometry. *Chem. Senses Flavour* 3: 57–71

Anholt, R. H., Murphy, K. M. M., Mack, G. E., Snyder, S. H. 1984. Peripheral type benzodiazepine receptors in the central nervous system: Localization to olfactory nerves. *J. Neurosci.* 4: 593–603

Anholt, R. H., Snyder, S. H. 1985. Biochemical characterization of isolated cilia from the olfactory epithelium of the bullfrog, *Rana catesbeiana. J. Neurosci.* In press

Atema, J. 1973. Microtubule theory of sensory transduction. *J. Theor. Biol.* 38: 181–90

Beets, J. T. 1971. Olfactory response and molecular structure. See Beidler 1971, pp. 257–321

Beidler, L. M., ed. 1971. Chemical senses: Olfaction. *Handb. Sensory Physiol.*, Vol. 4(1). 518 pp.

Bignetti, E., Cavaggioni, A., Pelosi, P., Persaud, K. C., Sorbi, R. T., Trindelli, R. 1985. *Eur. J. Biochem.* 149: 227–31

Boyse, E. A., Beauchamp, G. K., Yamazaki, K., Bard, J., Thomas, L. 1982. Chemosensory communication: A new aspect of the major histocompatibility complex and other genes in the mouse. *Oncodevel. Biol. Med.* 4: 101–16

Brown, S. B., Hara, T. J. 1981. Accumulation of chemostimulatory amino acids by a sedimentable fraction from olfactory rosettes of rainbow trout (*Salmo gairdueri*). *Biochim. Biophys. Acta* 675: 149–62

Bouvet, J.-F., Holley, A. 1984. Responses electriques olfactive de grenouille a l'application d'acetylcholine et de substance P. *CR Acad. Sci. Paris* 298(Ser. 3): 169–72

Cagan, R. H., Kare, M. K., eds. 1981. *Biochemistry of Taste and Olfaction*. New York/London: Academic. 539 pp.

Cagan, R. H., Zeiger, W. N. 1978. Biochemical studies of olfaction: Binding specificity of radioactively labeled stimuli to an isolated olfactory preparation from rainbow trout (*Salmo gairdueri*). *Proc. Natl. Acad. Sci. USA* 75:4679–83

Changeux, J.-P., Devillers-Thiery, A., Chemouilli, P. 1984. Acetylcholine receptor: An allosteric protein. *Science* 225: 1335–45

Chen, Z., Lancet, D. 1984. Membrane proteins unique to vertebrate olfactory cilia: Candidates for sensory receptor molecules. *Proc. Natl. Acad. Sci. USA* 81:1859–63

Chen, Z., Ophir, D., Lancet, D. 1984. A unique integral membrane glycoprotein of frog olfactory cilia: Biochemistry and immunofluorescence localization. *Soc. Neurosci. Abstr.* 10:861–61

Chen, Z., Greenberg, M., Pace, U., Lancet, D. 1985. A cell free assay for olfactory reactivity reveals properties of odorant receptor molecules. *Soc. Neurosci. Abstr.* 11: 970–70

Chen, Z., Ophir, D., Lancet, D. 1986a. Monoclonal antibodies to ciliary glycoproteins of frog olfactory neurons. *Brain Res.* In press

Chen, Z., Pace, U., Ronen, D., Lancet, D. 1986b. Polypeptide gp95: A unique glycoprotein of olfactory cilia with transmembrane receptor properties. *J. Biol. Chem.* In press

Criswell, D. W., McClure, F. L., Schafer, R., Brower, K. R. 1980. War gases as olfactory probes. *Science* 210:425–26

Cuatrecasas, P., Hollenberg, M. D. 1976. Membrane receptors and hormone action. *Adv. Protein Chem.* 30:241–51

Cuatrecasas, P., Roth, T., eds. 1983. *Receptor Mediated Endocytosis. Receptors and Recognition Series*, Vol. B15. London/New York: Chapman & Hall. 304 pp.

Cuschieri, A., Bannister, L. H. 1975. The development of the olfactory mucosa in the mouse: Light microscopy. *J. Anat.* 119:277–86

Dahl, R. A., Hadley, W. M., Hahn, F. F., Benson, J. M., McClellan, R. O. 1982. Cytochrome P-450-dependent monooxygenases in olfactory epithelium of dogs: Possible role in tumorigenicity. *Science* 216:57–59

Davies, J. T. 1971. Olfactory theories. See Beidler 1971, pp. 322–50

Debagge, P. L., Klein, N. J., O'Dell, D. S., Fraser, D. A., James, D. W. 1982. The culture of olfactory neurons. *J. Anat.* 135: 816–17

Delaleu, J. C., Holley, A. 1980. Modification of transduction mechanisms in the frog's olfactory mucosa using a thiol reagent as olfactory stimulus. *Chem. Senses Flavour* 3:205–18

Delaleu, J. C., Holley, A. 1983. Investigations

of the discriminative properties of the frog's olfactory mucosa using a photoactivable odorant. *Neurosci. Lett.* 37:251–56

Dodd, G., Persaud, K. 1981. Biochemical mechanisms in vertebrate primary olfactory neurons. See Cagan & Kare 1981, pp. 333–58

Dreesen, T. D., Koch, R. B. 1982. Odorous chemical perturbations of (Na$^+$+K$^+$)-dependent ATPase activities. *Biochem. J.* 203:69–75

Engen, T. 1971. Olfactory psychophysics. See Beidler 1971, pp. 216–44

Farbman, A. I. 1977. Differentiataon of olfactory receptor cells in organ culture. *Anat. Rec.* 189:187–99

Farbman, A. I., Margolis, F. L. 1980. Olfactory marker protein during ontogeny: Immunohistochemical localization. *Dev. Biol.* 74:205–15

Fesenko, E. E., Novoselov, V. I., Krapivinskaya, L. D. 1979. Molecular mechanisms of olfactory reception. IV. Some biochemical characteristics of the camphor receptor from rat olfactory epithelium. *Biochim. Biophys. Acta* 587:424–33

Fesenko, E. E., Novoselov, V. I., Krapivinskaya, L. D., Mjasoedov, N. F., Zolotarev, J. A. 1983. Molecular mechanisms of odor sensing. VI. Some biochemical characteristics of a possible receptor for amino acids from the olfactory epithelium of the skate *Dasyatis pastinaca* and *Carp cyprinus carpio*. *Biochim. Biophys. Acta* 759:250–56

Frosch, M. P., Dichter, M. A. 1984. Physiology and pharmacology of olfactory bulb neurons in dissociated cell culture. *Brain Res.* 290:321–32

Fujita, S. C., Mori, K., Imamura, K., Obata, K. 1985. Subclasses of olfactory receptor cells and their central projections demonstrated by monoclonal antibody. *Brain Res.* 326:192–96

Fuortes, M. G. F. 1971. Generation of responses in receptor. See Loewenstein 1971, pp. 243–68

Gesteland, R. C. 1976. Physiology of olfactory reception. In *Frog Neurobiology*, ed. R. Llinas, W. Precht, pp. 234–49. Berlin/Heidelberg: Springer-Verlag

Gesteland, R. C., Lettvin, J. Y., Pitts, W. H. 1965. Chemical transmission in the nose of the frog. *J. Physiol.* 181:525–59

Gesteland, R. C., Yancey, R. A., Farbman, A. I. 1982. Development of olfactory receptor neuron selectivity in the rat fetus. *Neuroscience* 7:3127–36

Getchell, M. L., Gesteland, R. C. 1972. The chemistry of olfactory reception: Stimulus-specific protection from sulphydryl reagent inhibition. *Proc. Natl. Acad. Sci. USA* 69:1494–98

Getchell, T. V. 1974. Unitary responses in frog olfactory epithelium to sterically related molecules at low concentrations. *J. Gen. Physiol.* 64:241–61

Getchell, T. V. 1977. Analysis of intracellular recordings from salamander olfactory epithelium. *Brain Res.* 123:275–86

Getchell, T. V., Getchell, M. L. 1982. Physiology of vertebrate olfactory chemoreception. See Theimer, pp. 1–25

Getchell, T. V., Heck, G. L., DeSimone, J. A., Price, S. 1980. The location of olfactory receptor sites. Inferences from latency measurements. *Biophys. J.* 29:397–472

Getchell, T. V., Margolis, F. L., Getchell, M. L. 1984. Perireceptor and receptor events in vertebrate olfaction. *Progr. Neurobiol.* 23:317–45

Getchell, T. V., Shepherd, G. M. 1978. Adaptive properties of olfactory receptors analysed with odour pulses of varying durations. *J. Physiol.* 282:541–60

Gilman, A. G. 1984. G proteins and dual control of adenylate cyclase. *Cell* 36:577–79

Goldberg, S. J., Turpin, J., Price, S. 1979. Anisole binding protein from olfactory epithelium: Evidence for a role in transduction. *Chem. Senses Flavour* 4:207–13

Goldstein, N. I., Quinn, M. R. 1981. A novel cell line isolated from the murine olfactory mucosa. *In Vitro* 17:593–97

Gonzales, F., Farbman, A. I. 1984. Developing olfactory receptor cells grow axons in tissue culture. *In Vitro* 20:268–68

Goodman, C. S., Bestiani, M. J., Doe, C. Q., du Lac, S., Helfand, S. L., Kuwada, J. Y., Thomas, J. B. 1984. Cell recognition during neuronal development. *Science* 225:1271–79

Graziadei, P. P. C. 1971. The olfactory mucosa of vertebrates. See Beidler 1971, pp. 27–58

Graziadei, P. P. C., Levine, R. R., Monti Graziadei, G. A. 1979. Plasticity of connections of the olfactory sensory neuron: Regeneration into the forebrain following bulbectomy in the neonatal mouse. *Neuroscience* 4:713–27

Graziadei, P. P. C., Monti Graziadei, G. A. 1979. Neurogenesis and neuron regeneration in the olfactory system of mammals. I. Morphological aspects of differentiation and structural organization of the olfactory sensory neurons. *J. Neurocytol.* 8:1–18

Hainer, R. M., Emslie, A. G., Jacobson, A. 1954. An information theory of olfaction. *Ann. NY Acad. Sci.* 58(2):158–74

Hedlund, B., Masukawa, L. M., Shepherd, G. M. 1984. The olfactory receptor cell: Electrophysiological properties of a small neuron. *Soc. Neurosci. Abstr.* 10:658

Hempstead, J. L., Morgan, J. I. 1983. Fluorescent lectins as cell-specific markers for rat olfactory epithelium. *Chem. Senses* 8:107–19

Hempstead, J. L., Morgan, J. I. 1985. A panel of monoclonal antibodies to rat olfactory epithelium. *J. Neurosci.* 5:438–49

Hirsch, J. D., Margolis, F. L. 1979. Cell suspensions from rat olfactory neuroepithelium: Biochemical and histochemical characterization. *Brain Res.* 161:277–91

Holley, A., MacLeod, P. 1977. Transduction et codage des informations olfactives chez les vertebres. *J. Physiol. (Paris)* 73:725–828

Hood, L. 1982. Antibody genes: Arrangements and rearrangements. In *Molecular Genetic Neuroscience*, ed. F. O. Schmitt, S. J. Bird, F. E. Bloom, pp. 75–85. New York: Raven

Hornung, D. E., Mozell, M. M. 1981. Accessibility of odorant molecules to the receptors. See Cagan & Kare 1981, pp. 33–45

Jastreboff, P. J., Pedersen, P. E., Greer, C. A., Stewart, W. B., Kauer, J. S., Benson, T. E., Shepherd, G. M. 1984. Specific olfactory receptor populations projecting to identified glomeruli in the olfactory bulb. *Proc. Natl. Acad. Sci. USA* 81:5250–54

Kaissling, K. E. 1986. Chemo-electrical transduction in insect olfactory receptors. *Ann. Rev. Neurosci.* 9:121–45

Kashiwayanagi, M., Kurihara, K. 1984. Neuroblastoma cells as model for olfactory cell: Mechanism of depolarization in response to various odorants. *Brain Res.* 293:251–58

Kauer, J. S. 1981. Olfactory receptor cell staining using horseradish peroxidase. *Anat. Rec.* 200:331–36

Kauer, J. S., Pretell, J. O., Hamilton, K. A. 1983. Ultraviolet radiation effects on olfactory receptor response. *Soc. Neurosci. Abstr.* 9:1025–25

Kerjaschki, D., Hörandner, H. 1976. The development of mouse olfactory vesicles and their cell contacts: A freeze-etching study. *J. Ultrastruct. Res.* 54:420–44

Kleene, S. J. 1985. Bacterial chemotaxis and vertebrate olfaction. *Experientia.* In press

Kleene, S. J., Gesteland, R. C. 1983. Dissociation of the frog olfactory epithelium. *J. Neurosci.* 9:173–83

Kurihara, K., Koyama, N. 1972. High activity of adenyl cyclase in olfactory and gustatory organs. *Biochem. Biophys. Res. Commun.* 48:30–34

Lancet, D. 1984. Molecular view of olfactory reception. *Trends Neurosci.* 7:35–36

Lancet, D., Greer, C. A., Kauer, J. S., Shepherd, G. M. 1982. Mapping of odor-related neuronal activity in the olfactory bulb by high-resolution 2-deoxyglucose autoradiography. *Proc. Natl. Acad. Sci. USA* 79:670–74

Lancet, D., Ben Simon, D., Chen Z. 1984. Computer modelling of neuronal activity in the olfactory pathway. *Chem. Senses* 8:255–56

Lancet, D., Pace, U. 1984. Proteins of olfactory cilia that may be involved in cyclic nucleotide-mediated sensory transduction. *Soc. Neurosci. Abstr.* 10:655–55

Lehmann, F. P. A. 1978. Stereoselective molecular recognition in biology. In *Receptors and Recognition* Vol. A5, ed. P. Cuatrecasas, M. F. Greaves, pp. 1–77. London: Chapman & Hall. 212 pp.

Levitzki, A. 1978. *Quantitative Aspects of Allosteric Mechanisms.* Mol. Biol. Biochem. Biophys. Ser. Vol. 28. Berlin/Heidelberg/New York: Springer-Verlag. 106 pp.

Lidow, M. S., Menco, B. Ph. M. 1984. Observations on axonemes and membranes of olfactory and respiratory cilia in frogs and rats using tannic acid-supplemented fixation and photographic rotation. *J. Ultrastruct. Res.* 86:18–30

Lipetz, L. E. 1971. The relation of physiological and psychological aspects of sensory intensity. See Loewenstein 1971, pp. 191–225

Loewenstein, W. R., ed. 1971. Principles of receptor physiology. *Handb. Sensory Physiol.*, Vol. 1. 600 pp.

Lucretius, T. C. 55 BC. *On the Nature of the Universe*, transl. R. E. Latham, 1951. Harmondworth/New York: Penguin. 264 pp. (From Latin)

Mackay-Sim, A., Shaman, P., Moulton, D. G. 1982. Topographic coding of olfactory quality: Odorant-specific patterns of epithelial responsivity in the Salamander. *J. Neurophysiol.* 48:584–95

Margolis, F. L. 1975a. Biochemical markers of the primary olfactory pathway: A model neuronal system. In *Adv. Neurochem.* 1:193–246

Margolis, F. L. 1975b. Biochemical studies of the primary olfactory pathway. In *Soc. Neurosci. Symp.* 3:167–88

Margolis, F. L. 1981. Neurotransmitter biochemistry of the mammalian olfactory bulb. See Cagan & Kare 1981, pp. 369–94

Margolis, F. L., Tarnoff, J. F. 1973. Site of biosynthesis of the mouse brain olfactory bulb protein. *J. Biol. Chem.* 248:451–55

Maruniak, J. A., Silver, W. L., Moulton, D. G. 1983. Olfactory receptors respond to blood-borne odorants. *Brain Res.* 265:312–16

Mason, J. R., Clark, L., Morton, T. H. 1984. Selective deficits in the sense of smell caused by chemical modification of the olfactory epithelium. *Science* 226:1092–94

Mason, J. R., Morton, T. H. 1984. Fast and loose covalent binding of ketones as a molecular mechanism in vertebrate olfactory receptors. *Tetrahedron* 40:483–92

Masson, C., Kouprach, I., Giachetti, I., MacLeod, P. 1977. Relation between intramembranous particle density of frog olfactory cilia and EOG response. In *Olfaction and Taste*, ed. J. Le Magnen, P. MacLeod, 6:195–95. London/Washington: Information Retrieval

Masukawa, L. M., Kauer, J. S., Shepherd, G. M. 1985. Electrophysiological properties of identified cells in the *in vitro* olfactory epithelium of the tiger salamander. *J. Neurosci.* 5:128–35

Maue, R. A., Dionne, V. E. 1984. Ion channel activity in isolated murine olfactory receptor neurons. *Soc. Neurosci. Abstr.* 10:655–55

Menco, B. Ph. M. 1980. Qualitative and quantitative freeze-fracture studies on olfactory and nasal respiratory epithelial surfaces of frog, ox, rat and dog. II. Cell apices, cilia and microvilli. *Cell Tissue Res.* 211:5–29

Menco, B. Ph. M. 1983. The structure of olfactory and nasal respiratory epithelial surfaces. In *Nasal Tumors in Animals and Man*, Vol. 1, ed. G. Reznik, S. F. Stinton, pp. 45–102. Boca Raton, Fla.: CRC Press

Menevse, A., Dodd, G., Poynder, T. M. 1977. Evidence for the specific involvement of cyclic AMP in the olfactory transduction mechanism. *Biochem. Biophys. Res. Commun.* 77:671–77

Minor, A. V., Sakina, N. L. 1973. Role of cyclic adenosine-3′,5′-monophosphate in olfactory reception. *Neurofysiologiya* 5:415–22

Moncrieff, R. W. 1967. *The Chemical Senses*. London: Leonard Hill. 730 pp.

Morgan, J., Hempstead, J. 1983. Culture and immunocytochemical characterization of the rat olfactory epithelium. *Soc. Neurosci. Abstr.* 9:464–64

Moulton, D. G. 1976. Spatial patterning of responses to odors in the peripheral olfactory system. *Physiol. Rev.* 56:578–93

Moulton, D. G. 1977. Minimum odorant concentrations detectable by the dog and their implications for olfactory receptor sensitivity. In *Chemical Signals in Vertebrates*, ed. D. Muller-Schwarz, M. M. Mozell, pp. 455–64. New York/London: Plenum. 610 pp.

Moulton, D. G., Beidler, L. M. 1967. Structure and function in the peripheral olfactory system. *Physiol. Rev.* 47:1–52

Moulton, D. G., Tucker, D. 1964. Electrophysiology of the olfactory system. *Ann. NY Acad. Sci.*, Vol. 116, Art. 2, *Recent advances in odor: theory measurement and control*, ed. H. E. Whipple, pp. 380–428

354 LANCET

Mozell, M. M. 1971. Spatial and temporal patterning. See Beidler 1971, pp. 95–131

Nestler, E. J., Walaas, S. I., Greengard, P. 1984. Neuronal phosphoproteins: Physiological and clinical implications. *Science* 225:1357–64

Noble, M., Mallaburn, P. S., Klein, N. 1984. The growth of olfactory neurons in short-term cultures of rat olfactory epithelium. *Neurosci. Lett.* 45:193–98

O'Connell, R. J., Mozell, M. M. 1969. Quantitative stimulation of frog olfactory receptors. *J. Neurophys.* 32:51–63

Ottoson, D. 1971. The Electro-olfactogram. See Beidler 1971, pp. 95–131

Pace, U., Hanski, E., Salomon, Y., Lancet, D. 1985. Odorant-sensitive adenylate cyclase may mediate olfactory reception. *Nature* 316:255–58

Patte, F., Etcheto, M., Laffort, P. 1975. Selected and standardized values of supra-threshold odor intensities for 110 substances. *Chem. Senses Flavour* 1:238–305

Pecht, I., Lancet, D. 1977. Kinetics of antibody-hapten interactions. In *Chemical Relaxation in Molecular Biology*. Mol. Biol. Biochem. Biophys. Ser., ed. I. Pecht, R. Rigler, 24:306–38. Berlin/Heidelberg/New York: Springer-Verlag. 418 pp.

Pelosi, P., Baldaccini, N. E., Pisanelli, A. M. 1982. Identification of a specific olfactory receptor for 2-isobutyl-3-methoxypyrazine. *Biochem. J.* 201:245–48

Persaud, K., Dodd, G. 1982. Analysis of discrimination mechanisms in the mammalian olfactory system using a model nose. *Nature* 299:352–55

Pevsner, J., Trifiletti, R. R., Strittmatter, S. M., Snyder, S. H. 1985. Isolation and characterization of an olfactory receptor protein for odorant pyrazines. *Proc. Natl. Acad. Sci. USA* 82:3050–54

Pfaffman, C. 1955. Gustatory nerve impulses in rat, cat and rabbit. *J. Neurophysiol.* 18:429–40

Polak, E. H. 1973. Multiple profile-multiple receptor site model for vertebrate olfaction. *J. Theor. Biol.* 40:469–84

Price, S. 1977. Specific anosmia to geraniol in mice. *Neurosci. Lett.* 4:49–50

Price, S. 1981. Receptor proteins in vertebrate olfaction. See Cagan & Kare 1981, pp. 69–84

Price, S. 1984. Mechanisms of stimulation of olfactory neurons: An essay. *Chem. Senses* 8:341–54

Punter, P. H. 1983. Measurement of human olfactory thresholds for several groups of structurally related compounds. *Chem. Senses* 7:215–35

Rafols, J. A., Getchell, T. V. 1983. Morphological relations between the receptor neurons, sustentacular cells and Schwann cells in the olfactory mucosa of the Salamander. *Anat. Rec.* 206:87–101

Reese, T. S. 1965. Olfactory cilia in the frog. *J. Cell Biol.* 25:209–30

Rhein, L. D., Cagan, R. H. 1980. Biochemical studies of olfaction: Isolation characterization, and odorant binding activity of cilia from rainbow trout olfactory rosettes. *Proc. Natl. Acad. Sci. USA* 77:4412–16

Rhein, L. D., Cagan, R. H. 1981. Role of cilia in olfactory recognition. See Cagan & Kare 1981, pp. 47–68

Robinson, G. A., Butcher, R. W., Sutherland, E. W. 1971. *Cyclic AMP*. New York/London: Academic. 531 pp.

Rochel, S., Margolis, F. L. 1982. Carnosine release from olfactory bulb synaptosomes is calcium-dependent and depolarization-stimulated. *J. Neurochem.* 38:1504–14

Rodrigues, V. 1980. Olfactory behavior of *Drosophila melanogaster*. In *Development and Neurobiology of Drosophila*, ed. O. Siddiqi, P. Babu, L. M. Hall, J. C. Hall, pp. 361–71. New York/London: Plenum

Rogers, K. E., Grillo, M., Sydor, W., Poonian, M., Margolis, F. L. 1985. Olfactory neuron-specific protein is translated from a large poly(A)$^+$ mRNA. *Proc. Natl. Acad. Sci. USA* 82:5218–22

Schafer, R., Fracek, S. P., Jr., Criswell, D. W., Brower, K. R. 1984. Protection of olfactory responses from inhibition by ethyl bromoacetate, diethylamine and other chemically active odorants by certain esters and other compounds. *Chem. Senses* 9:55–72

Schiffman, S. S. 1974. Physicochemical correlates of olfactory quality. *Science* 185:112–17

Schramm, M., Selinger, Z. 1984. Message transmission: Receptor controlled adenylate cyclase system. *Science* 225:1350–56

Schubert, D., Stallcup, W., LaCorbiere, M., Kidokoro, Y., Orgel, L. 1985. The ontogeny of electrically excitable cells in cultured olfactory epithelium. *Proc. Natl. Acad. Sci. USA*. In press

Shepherd, G. M. 1972. Synaptic organization of the mammalian olfactory bulb. *Physiol. Rev.* 52:864–917

Senf, W., Menco, B. Ph. M., Punter, P. H., Duyventeyn, P. 1980. Determination of odour affinities on the dose-response relationships of the frog's electro-olfactogram. *Experimentia* 36:213–15

Shirley, S., Polak, E., Dodd, G. 1983a. Chemical-modification studies on rat olfactory mucosa using a thiol-specific reagent and enzymatic iodination. *Eur. J. Biochem.* 132:485–94

Shirley, S., Polak, E., Dodd, G. 1983b. Selective inhibition of rat olfactory receptors by concanavalin A. *Biochem. Soc. Transact.* 11:780–81

Sicard, G., Holley, A. 1984. Receptor cell responses to odorants: Similarities and differences among odorants. *Brain Res.* 292:283–96

Siegelbaum, S. A., Tsien, R. W. 1983. Modulation of gated ion channels as a mode of transmitter action. *Trends Neurosci.* 6:307–13

Sigal, N. H., Klinman, N. R. 1978. The B-cell clonotype repertoire. *Adv. Immunol.* 26: 255–337

Simmons, P. A., Getchell, T. V. 1981. Neurogenesis in olfactory epithelium: Loss and recovery of transepithelial voltage transients following olfactory nerve section. *J. Neurophysiol.* 45:516–28

Slotnick, B. M., Schoonover, F. W. 1984. Olfactory thresholds in unilaterally bulbectomized rats. *Chem. Senses* 9:325–40

Stryer, L. 1986. Cyclic GMP cascade of vision. *Ann. Rev. Neurosci.* 9:87–119

Suzuki, N. 1977. Intracellular responses of lamprey olfactory receptors to current and chemical stimulation. In *Food Intake and Chemical Senses*, ed. Y. Katsuki, M. Sato, S. F. Takagi, Y. Oomura, pp. 13–22. Tokyo: Jpn. Sci. Soc. Press

Takagi, S. F., Kitamura, H., Imai, K., Takeuchi, H. 1969. Further studies on the roles of sodium and potassium in the generation of the electro-olfactogram, effects of mono-, di-, and trivalent cations. *J. Gen. Physiol.* 59:115–30

Teeter, J. H., Goldstein, N. I. 1984. Morpho-logical and electrophysiological differentiation of mouse olfactory cells in culture. *Assoc. Chemorecept. Sci. Abstr.* 6:127

Theimer, E. T., ed. 1982. *Fragrance Chemistry—the Science of the Sense of Smell.* New York/London: Academic. 635 pp.

Thomas, L. 1983. *Late Night Thoughts on Listening to Mahler's 9th Symphony.* New York: Viking

Trotier, D., MacLeod, P. 1983. Intracellular recordings from salamander olfactory receptor cells. *Brain Res.* 268:225–37

Van Boxtel, A., Köster, E. P. 1978. Adaptation the electroolfactogram in the frog. In *Chem. Senses Flavour* 3:39–44

van Drongelen, W. 1978. Unitary recordings of near threshold responses of receptor cells in the olfactory mucosa of the frog. *J. Physiol.* 277:423–35

Vinnikov, Y. A. 1982. *Evolution of Receptor Cells.* Mol. Biol. Biochem. Biophys. Ser., Vol. 34. Berlin/Heidelberg/New York: Springer-Verlag. 141 pp.

Vodyanoy, V., Murphy, R. B. 1983. Single channel fluctuations in bimolecular lipid membranes induced by rat olfactory epithelial homogenates. *Science* 220:717–19

Wysocki, C. J., Whitney, G., Tucker, D. 1977. Specific anosmia in the laboratory mouse. *Behav. Genet.* 7:171–88

Wysocki, Ch. J., Beauchamp, G. 1984. Ability to smell androstenone is genetically determined. *Proc. Natl. Acad. Sci. USA* 81: 4899–4902

Ann. Rev. Neurosci. 1986. 9 : 357–81

PARALLEL ORGANIZATION OF FUNCTIONALLY SEGREGATED CIRCUITS LINKING BASAL GANGLIA AND CORTEX*

Garrett E. Alexander

Department of Neurology, Johns Hopkins University School of Medicine, Baltimore, Maryland 21205

Mahlon R. DeLong

Departments of Neurology and Neuroscience, Johns Hopkins University School of Medicine, Baltimore, Maryland 21205

Peter L. Strick

Research Service, Veterans Administration Medical Center, Departments of Neurosurgery and Physiology, State University of New York–Upstate Medical Center, Syracuse, New York 13210

INTRODUCTION

Information about the basal ganglia has accumulated at a prodigious pace over the past decade, necessitating major revisions in our concepts of the structural and functional organization of these nuclei. From earlier data it had appeared that the basal ganglia served primarily to integrate diverse inputs from the entire cerebral cortex and to "funnel" these influences, via the ventrolateral thalamus, to the motor cortex (Allen & Tsukahara 1974, Evarts & Thach 1969, Kemp & Powell 1971). In particular, the basal

* The US Government has the right to retain a nonexclusive, royalty-free license in and to any copyright covering this paper.

357

ganglia were thought to provide a route whereby influences from the cortical association areas might be transmitted to the motor cortex and thereby participate in the initiation and control of movement.

Subsequent anatomical and physiological findings led to a revised view that stressed the apparent maintained segregation of influences from the sensorimotor and association cortices through the basal ganglia–thalamocortical pathways (DeLong & Georgopoulos 1981). On the basis of the then existing data, it was suggested that there might be two distinct loops through the basal ganglia : (a) a "motor" loop passing largely through the putamen, which received inputs from sensorimotor cortex and whose influences were ultimately transmitted to certain premotor areas, and (b) an "association" (or "complex") loop passing through the caudate nucleus, which received input from the association areas and whose influences were ultimately returned to portions of the prefrontal cortex (DeLong & Georgopoulos 1981, DeLong et al 1983b).

Recent anatomical and physiological findings have further substantiated the concept of segregated basal ganglia–thalamocortical pathways, and reinforced the general principle that basal ganglia influences are transmitted only to restricted portions of the frontal lobe (even though the striatum receives projections from nearly the entire neocortex). It has been shown, for example, that the thalamocortical portion of the "motor" circuit terminates largely within a restricted premotor region, the supplementary motor area (Schell & Strick 1984), while the corticostriate inputs to this circuit include projections not only from the supplementary motor area but from motor, arcuate premotor, and somatosensory cortex as well (Kunzle 1975, 1977, 1978). Using the "motor" circuit as a model, we have reexamined the available data on other portions of the basal ganglia–thalamocortical pathways and found that the evidence strongly suggests the existence of at least four additional circuits organized in parallel with the "motor" circuit. In the discussion that follows, we review some of the anatomic and physiologic features of the "motor circuit," as well as the data that support the existence of the other proposed parallel circuits, which we have designated the "oculomotor," the "dorsolateral prefrontal," the "lateral orbitofrontal," and the "anterior cingulate," respectively. Each of these five basal ganglia–thalamocortical circuits appears to be centered upon a separate part of the frontal lobe, as indicated in Figure 1.

This list of basal ganglia–thalamocortical circuits is not intended to be exhaustive. In fact, if the conclusions suggested in this review are valid, future investigations might be expected to disclose not only further details (or the need for revisions) of these five circuits, but perhaps also the existence of additional parallel circuits whose identification is currently precluded by a paucity of data.

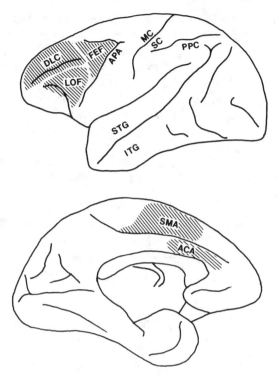

Figure 1   Frontal cortical targets of basal ganglia output. Schematic illustration of the five cortical areas that contribute to the "closed loop" portions of the basal ganglia–thalamocortical circuits discussed in this review.

Abbreviations are as follows: ACA: anterior cingulate area; APA: arcuate premotor area; DLC: dorsolateral prefrontal cortex; FEF: frontal eye fields; ITG: inferior temporal gyrus; LOF: lateral orbitofrontal cortex; MC: motor cortex; PPC: posterior parietal cortex; SC: somatosensory cortex; SMA: supplementary motor area; STG: superior temporal gyrus.

GENERALIZED BASAL GANGLIA–THALAMOCORTICAL CIRCUIT

Sufficient information has now accumulated to support the presentation of a generalized schema that emphasizes the basic features of the proposed basal ganglia–thalamocortical circuits. The elements of each circuit include discrete, essentially non-overlapping parts of the striatum, globus pallidus, substantia nigra, thalamus, and cortex. The basic design of each pathway is thought to be similar, as shown schematically in Figure 2. Each circuit receives multiple, partially overlapping corticostriate inputs, which are progressively integrated in their subsequent passage through pallidum and nigra to a restricted portion of the thalamus, and from there back to a single cortical area. Each circuit is, therefore, partially closed by the restricted thalamocortical projection returned to one of that circuit's multiple sources of corticostriate input. From the available evidence (discussed below), it would appear that within each of the proposed basal ganglia–thalamocortical circuits the integrative mechanisms that underly the "funneling" of multiple corticostriate inputs back to a single cortical area are progressively carried out at striatal, pallidal/nigral, and thalamic levels. In this limited sense, then, this organizational schema retains the concept of "funneling" that figured so prominently in earlier views. An important distinction, however, is that according to the present view such "funneling"

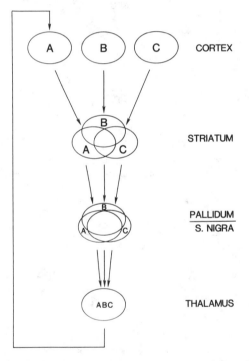

Figure 2 Generalized basal ganglia–thalamocortical circuit. "Skeleton" diagram of the proposed basal ganglia–thelamocortical circuits. Each circuit receives output from several functionally related cortical areas (A, B, C) that send partially overlapping projections to a restricted portion of the striatum. These striatal regions send further converging projections to the globus pallidus and substantia nigra, which in turn project to a specific region of the thalamus. Each thalamic region projects back to one of the cortical areas that feeds into the circuit, thereby completing the "closed loop" portion of the circuit.

occurs only *within* the segregated functional pathways. Moreover, it appears that each basal ganglia–thalamocortical circuit receives its multiple corticostriate inputs only from cortical areas that are functionally related (and usually interconnected). Thus, for example, the "motor" circuit receives inputs from at least four interconnected cortical areas that have been strongly implicated in the control of limb and orofacial movements, while the "oculomotor" circuit receives inputs from three interconnected areas implicated in the control of eye movements.

We emphasize at the outset that our use of the term "circuit" in this review is not intended to imply a rigidly self-enclosed pathway without substantial inputs and outputs to other structures. Rather, the concept we hope to convey is that, despite the influx of diverse influences to each of the basal ganglia–thalamocortical pathways, and the dispersal of influences from these pathways to other structures, there remains a central "closed loop" portion of each basal ganglia–thalamocortical pathway that receives input from and terminates within a single cortical area. It should also be stressed that the proposed linkages between different structures rest for the most part on comparisons of data from independent studies. Where published illustrations of these data have been limited, we have often had to rely on the assumption that different investigators have used the same anatomical terms to refer to the same areas. Also, because of the well-known species differences in the anatomical organization of these nuclei, we have relied primarily on studies in primates. Certain details of the proposed basal ganglia–thalamocortical circuits may eventually require revision. This is especially likely, for example, with respect to the pallido/nigro-thalamic connections, where the circuits are maximally compressed and detailed data are most lacking. Future studies, using double labeling and combined anterograde/retrograde techniques in the same animal, are likely to clarify many details that must now be considered provisional.

Because of the parallel nature of the basal ganglia–thalamocortical circuits and the apparent uniformity of synaptic organization at corresponding levels of these functionally segregated pathways (Nauta 1979, DeLong & Georgopoulos 1981), it would seem likely that similar neuronal operations are performed at comparable stages of each of the five proposed circuits. Thus, for example, the neural mechanisms mediating transmission of information through the pallidal portion of the "motor" circuit are likely to be comparable, if not identical, to those within the pallidal portion of the "dorsolateral prefrontal" circuit. If this assumption is correct, then detailed knowledge of the workings of one circuit may prove useful in attempts to clarify another. With this in mind, we shall examine certain physiological data obtained in two of the better understood circuits, the "motor" and oculomotor," in some detail.

"MOTOR" CIRCUIT

At the level of the striatum, the "motor" circuit is largely centered on the putamen, which receives substantial, somatotopically organized projections from the motor (Kunzle 1975) and somatosensory cortices (Kunzle 1977). While the corticostriate projections from the motor cortex overlap with those from the somatosensory cortex, the somatotopic features of both projections are in register. Both motor and somatosensory cortical "leg" areas project to a dorsolateral sector of the putamen, the "face" areas to a ventromedial sector, and the "arm" areas to a region in between (Kunzle 1975, 1977). Each of these terminal fields, like those of other corticostriate projections, is considerably elongated along the rostrocaudal axis. In double anterograde labeling studies, we have found that within the region of gross convergence of corticostriate projections from the "arm" areas of motor and somatosensory cortices, some of the patchy terminal fields from the two areas are in register, while others are not (J. Hedreen, M. R. DeLong, G. E. Alexander, and C. Kitt, unpublished data). Analogous observations have been made with respect to corticostriate projections to the caudate nucleus (Selemon & Goldman-Rakic 1985 (see below).

In addition to the motor and somatosensory projections, the putamen also receives topographically organized projections from area 5, from lateral area 6, including the arcuate premotor area, and from the supplementary motor area (Kunzle 1978, Selemon & Goldman-Rakic 1985, Jones et al 1977, Miyata & Sasaki 1984). While there is slight encroachment of each of these projections upon neighboring regions of the caudate nucleus, the terminal arborizations of each are confined principally to the putamen. It should be noted that published reports of corticostriate projections from the supplementary motor area (Kunzle 1978, Miyata & Sasaki 1984, Selemon & Goldman-Rakic 1985) appear to have been based principally upon injections of anterograde label into the rostrally located "face" representation (see Muakkassa & Strick 1979, Brinkman & Porter 1979), with resultant terminal labeling located mainly within the ventromedial putamen. These results, in combination with those involving injections of anterograde label into more caudal portions ("arm" representation) of the supplementary motor area (A. A. Martino and P. L. Strick, unpublished data), have confirmed the same pattern of somatotopic organization of corticostriate projections observed for motor and somatosensory cortex. It remains to be determined whether there are additional corticostriate inputs to the "motor" circuit from other functionally related regions, such as the precentral and ventral cingulate premotor areas (Muakkassa & Strick 1979), the supplementary somatosensory area, and certain parts of the superior and inferior parietal lobules.

The putamen sends topographically organized projections to the ventrolateral two-thirds of both the internal and the external segments of the globus pallidus (Cowan & Powell 1966, DeVito et al 1980, Johnson & Rosvold 1971, Nauta & Mehler 1966, Parent et al 1984a, Szabo 1962, 1967) and to caudolateral portions of the substantia nigra (Hedreen et al 1980, Nauta & Mehler 1966, Parent et al 1984b, Szabo 1962, 1967). The portion of the internal pallidal segment that receives putamen input projects in turn to the oral part of the ventrolateral nucleus of the thalamus (VLo) (DeVito & Anderson 1982, Kim et al 1976, Kuo & Carpenter 1973, Nauta & Mehler 1966). Recent studies have indicated that the VLo projects to the supplementary motor area (Schell & Strick 1984). Thus, as indicated in Figure 3, somatotopically organized corticostriate influences arising from the supplementary motor area, the arcuate premotor area, motor cortex, and somatosensory cortex are transmitted through the "motor" circuit and ultimately projected back upon a single cortical region, the supplementary motor area. This combination of "open-" and "closed-loop" features appears to be a general characteristic of all basal ganglia-thalamocortical circuits (Figures 2 and 3).

The contribution of the substantia nigra pars reticulata (SNr) to the "motor" circuit has not been fully resolved. Studies of neuronal activity in primates have indicated a prominent representation of orofacial structures in the lateral SNr (DeLong et al 1983b). On the basis of the topographic details revealed in studies reported thus far, it has been suggested (Schell & Strick 1984) that the "face" representation in the lateral SNr may project to the medial subdivision of the ventrolateral nucleus (VLm), but this has yet to be fully established.

A number of lines of evidence indicate that the supplementary motor area, the cortical terminus of the "motor" circuit, plays an important role in the programming and control of movement. The supplementary motor area sends projections not only to the motor cortex and the arcuate premotor area (Muakkassa & Strick 1979, Schell & Strick 1984) but also projects directly to the spinal cord (Biber et al 1978, Murray & Coulter 1981, Macpherson et al 1982, Palmer et al 1981). There is evidence for sparse projections from the supplementary motor area to extreme dorsolateral portions of the ventral horn in lower cervical segments of the spinal cord (G. R. Schell and P. L. Strick, unpublished data). This pattern of termination suggests the existence of direct projections to motor neurons innervating distal hand muscles, and raises the possibility of monosynaptic inputs to spinal motoneurons from the supplementary motor area, analogous to those arising in primary motor cortex. It has been shown that microstimulation of the supplementary motor area in the monkey produces movements of the limbs (Macpherson et al 1982). Neurons in the

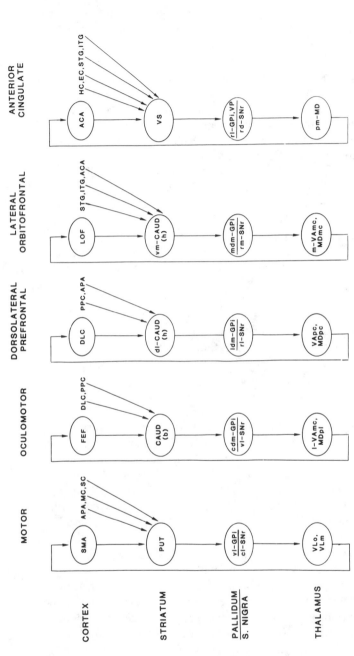

Figure 3 Proposed basal ganglia–thalamocortical circuits. Parallel organization of the five basal ganglia–thalamocortical circuits. Each circuit engages specific regions of the cerebral cortex, striatum, pallidum, substantia nigra, and thalamus.

Abbreviations are as follows: ACA: anterior cingulate area; APA: arcuate premotor area; CAUD: caudate, (b) body (h) head; DLC: dorsolateral prefrontal cortex; EC: entorhinal cortex; FEF: frontal eye fields; GPi: internal segment of globus pallidus; HC: hippocampal cortex; ITG: inferior temporal gyrus; LOF: lateral orbitofrontal cortex; MC: motor cortex; MDpl: medialis dorsalis pars paralamellaris; MDmc: medialis dorsalis pars magnocellularis; MDpc: medialis dorsalis pars parvocellularis; PPC: posterior parietal cortex; PUT: putamen; SC: somatosensory cortex; SMA: supplementary motor area; SNr: substantia nigra pars reticulata; STG: superior temporal gyrus; VAmc: ventralis anterior pars magnocellularis; Vapc: ventralis anterior pars parvocellularis; VLm: ventralis lateralis pars medialis; VLo: ventralis lateralis pars oralis; VP: ventral pallidum; VS: ventral striatum; cl-: caudolateral; cdm-: caudal dorsomedial; dl-: dorsolateral; l-: lateral; ldm-: lateral dorsomedial; m-: medial; mdm-: medial dorsomedial; pm: posteromedial; rd-: rostrodorsal; rl-: rostrolateral; rm-: rostromedial; vm-: ventromedial; vl-: ventrolateral.

supplementary motor area are somatotopically organized, and in single cell studies with behaving primates they have been shown to discharge in relation to limb movements (Brinkman & Porter 1979, Tanji & Kurata 1979) and during the preparation for such movements (Tanji et al 1980).

Two prominent features of the basal ganglia components of the "motor" circuit revealed by single cell recording studies in behaving primates are (a) the specificity of neuronal responses to active movements and passive manipulations of individual body parts, and (b) the maintained somatotopic organization of movement-related neurons throughout the circuit. The somatotopic organization in the putamen suggested by the topography of the corticostriate projections from motor and somatosensory cortex has been confirmed in studies of the sensorimotor response properties of neurons in awake, behaving primates (Alexander & DeLong 1985a, Crutcher & DeLong 1984, Liles 1979, 1985). Neurons related to active and/or passive movements of the lower extremity are found throughout a long rostrocaudal extent of the dorsolateral putamen; neurons related to orofacial movements are located ventromedially; and neurons related to movements of the upper extremity are located in an intermediate position. Recent studies in monkeys have revealed that movements of individual body parts can be evoked by microstimulation of the putamen (Alexander & DeLong 1985b). Moreover, the microexcitable zones within the putamen (which appear to correspond to clusters of functionally related putamen neurons) are somatotopically organized in precisely the same pattern as that revealed by the distributions of corticostriate terminals and the functional properties of putamen neurons (Alexander & DeLong 1985a). Neurons in both segments of the globus pallidus have also been found to exhibit discrete responses to active movements and passive manipulation of individual body parts (DeLong 1971, DeLong & Georgeopoulos 1979, DeLong et al 1985, Georgopoulos et al 1983) and to be somatotopically organized (DeLong et al 1985).

Neuronal activity in the putamen and globus pallidus has been shown to be related to specific aspects of limb movements, including direction, amplitude (or velocity), and load (Georgopoulos et al 1983, Crutcher & DeLong 1984b, Liles 1985). In the putamen, the directional specificity of movement-related neuronal discharge appears to be relatively independent of patterns of muscular activity (Crutcher & DeLong 1984b). These studies favor a role of the "motor" circuit in the control of movement direction and in the scaling of movement amplitude or velocity. It is noteworthy that in the putamen and globus pallidus the onset of neuronal discharge in relation to stimulus-triggered movements appears to follow that of the cortical motor areas (Crutcher & DeLong 1984b, Georgopoulos et al 1983, DeLong et al 1984). This finding suggests that the basal ganglia receive a "corollary discharge" or "efference copy" from the cerebral cortex. Although these findings do not support a role of the basal ganglia in the initiation of stimulus-triggered movements, it is suggested that these structures may influence the buildup of muscular activity in the first "agonist burst" and thereby participate in the control of movement speed and amplitude (Hallett & Khoshbin 1980, DeLong et al 1984). It is possible, however, that the "motor" circuit may also play a role in the programming and initiation of internally-generated movements (Evarts & Wise 1984, Neafsy et al 1978).

Recent studies have provided evidence of direct participation of the basal ganglia in the preparation for movement (Alexander 1984). These studies have revealed a population of neurons in the primate putamen that show instruction-dependent changes in discharge related to "motor set" that are similar to those found in motor cortex (Tanji & Evarts 1976), premotor cortex (Weinrich & Wise 1982), and the supplementary motor area (Tanji et al 1980), all of which project to the putamen. It is striking that, in general, set-related cells in the putamen do not show phasic changes during movement or exhibit any response to somatosensory stimulation, suggesting that they may not receive input from phasic movement-related or peripherally-driven neurons of the motor or somatosensory cortex. It remains to be determined whether such set-related influences are integrated at the level of the globus pallidus (and/or thalamus) with the movement- and peripherally-driven influences from the putamen.

Based on the fact that pallidal neurons have large, disk-shaped dendritic arborizations oriented in a plane orthogonal to incoming striatal fibers, it has recently been suggested that the globus pallidus might serve to integrate diverse influences from the striatum, with a resultant loss of specificity and degradation of information content in favor of a more global function (Percheron et al 1984). While theoretically possible on the basis of anatomy alone, this suggestion is not supported by the physiologic evidence indicating maintained specificity of neuronal functional properties within the globus pallidus. Although a single pallidal neuron appears to integrate the output of many putamen neurons, the physiologic findings suggest that at the neuronal level such integration within the "motor" circuit is carried out along lines of individual body parts, without loss of the somatotopic and functional specificity observed in the putamen (De Long et al 1985, Georgopoulos et al 1983).

The evidence for maintained somatotopy and neuronal specificity suggests that the "motor" circuit may be composed of multiple, parallel subcircuits or "channels" concerned with movements of individual body parts. Accordingly, in the generalized circuit depicted in Figure 2, it may be appropriate, depending on the level of analysis, to consider A, B, and C as representing information from somatotopically corresponding subregions of three separate, but functionally-related cortical areas (e.g. the "arm" representations of the primary motor, somatosensory, and supplementary motor areas). Moreover, within each broad somatotopic area (e.g. the "arm" area) of the putamen, neurons appear to be grouped into multiple functional clusters (with dimensions of 200–1000 μm) that represent a single body part (e.g. the wrist), and often a specific movement of that part (e.g. flexion) (Crutcher & DeLong 1984a, Liles 1979, 1985, Alexander & DeLong 1985a). These observations suggest that the "motor" circuit may be subdivided not only into three broad somatotopic channels representing "leg," "arm," and "face," but perhaps also into functionally-defined channels of an even finer grain, based upon an individual body part or even a specific movement of an individual body part. It remains to be determined whether the functional subunits (neuronal clusters) of the putamen project selectively to comparable subunits in the pallidum (and similarly for the pallidothalamic and thalamocortical projections). The answer to this question would help to clarify the "fine structure" and the nature of integration within the somatotopic channels of the "motor" circuit. Such information may prove helpful in analyzing the nature of information processing within the other basal ganglia–thalamocortical circuits.

The discontinuous distribution of putamen neuronal clusters and microexcitable zones suggests comparisons with certain anatomical and histochemical discontinuities recently identified in the neostriatum. Autoradiographic studies in the monkey have shown that projections to the putamen from the sensorimotor cortex (Jones et al 1977, Kunzle 1975) and from the centromedian nucleus of the thalamus (Kalil 1978) terminate in discontinuous patches and stripes. Moreover, histochemical, histofluorescence, and immunocytochemical studies have revealed similar discontinuities in the distributions of markers for the following putative neurotransmitters and neuromodulators: acetylcholine (Graybiel & Ragsdale 1978, Graybiel et al 1981), dopamine (Grabiel et al 1981), enkephalin (Graybiel et al 1981, Pickel et al 1980) and other opiates (Herkenham & Pert 1981), and substance P (Graybiel et al 1981). It has been suggested that the heterogeneous distribution of anatomical and biochemical markers may reflect an underlying cytoarchitectural organization of the neostriatum in terms of what Graybiel et al (1981) have called "striosomes" and what Goldman-Rakic (1982) has referred to as the "island" and "matrix" compartments. The dimensions of these compartments are remarkably similar to the dimensions of the clusters of functionally related putamen neurons and those of the striatal microexcitable zones (Alexander & DeLong 1985a,b, Crutcher & DeLong 1984a, Goldman-Rakic 1982, Liles 1979).

In both cats and monkeys, corticostriate fibers arising in the frontal cortex have been shown to terminate largely within the compartment that stains strongly for acetylcholinesterase (Ragsdale & Graybiel 1981). Furthermore, injections of horseradish peroxidase into pallidum and substantia nigra in cats have revealed preferential labeling of presumptive striatal projection neurons in acetylcholinesterase-rich zone (Graybiel et al 1979). Thus, these results suggest that the corticostriate terminal patches, the clusters of retrogradely labeled striatal projection neurons, and the zones of high acetylcholinesterase activity may occupy the same anatomic compartment. However, the degree to which corticostriate afferents may terminate selectively upon clusters of striatal projection neurons remains unknown. Nor is it known to what extent these anatomical findings, based primarily on studies of the caudate nucleus, are applicable to the putamen. Additional studies will be needed to clarify the relationships that may exist between the newly described physiological subunits and the various histochemical compartments of the primate neostriatum.

"OCULOMOTOR" CIRCUIT

The frontal eye fields (Brodmann's area 8) have been shown to project to a central portion of the body of the caudate nucleus (Kunzle & Akert 1977) that also receives projections from dorsolateral prefrontal cortex (areas 9 and 10) and posterior parietal cortex (area 7) (Yeterian & VanHoesen 1978, Selemon & Goldman-Rakic 1985). Each of these interconnected cortical areas has been implicated in oculomotor control mechanisms on the basis of single-cell recording studies in awake monkeys (Bizzi & Schiller 1970, Goldberg & Bushnell 1981, Lynch et al 1977), and the demonstration of direct projections to the superior colliculus (Fries 1984, Leichnetz et al 1981, Goldman & Nauta 1976, Kunzle et al 1976).

The body of the caudate projects to a caudal and dorsomedial sector of the internal pallidal segment and to the ventrolateral SNr (Parent et al 1984a, Szabo 1970). The latter projects in turn to parts of the magnocellular portion of the ventral anterior (VAmc) and the paralamellar portion of the mediodorsal (MDpl) thalamic nuclei (Carpenter et al 1976, Ilinsky et al 1985). Both of these thalamic areas project back to the frontal eye fields (Kievit & Kuypers 1977, Akert 1964, Barbas & Mesulam 1981), as indicated in Figure 3, thus partially closing the "oculomotor" circuit. There are indications that at least part of the nigrothalamic projection may represent branching collaterals from SNr neurons that also project to the superior colliculus (Anderson & Yoshida 1977, Parent et al 1984b, Beckstead et al 1981), suggesting that the nigrotectal pathway may provide an important direct output pathway for the "oculomotor" circuit.

Single cell studies in primates have revealed that frontal eye field neurons may discharge in relation to visual fixation, saccadic eye movements, or passive visual stimulation (Mohler et al 1973, Bizzi & Schiller 1970). The visual receptive field properties of some frontal eye field neurons have been shown to depend upon the animal's behavioral set, in that the cell's response to a visual stimulus within its receptive field may be enhanced when that stimulus serves as the target for a subsequent saccade (Goldberg & Bushnell 1981). Comparable studies of caudate neurons have yet to be reported. However, the ventrolateral SNr, which appears to receive the frontal eye field influences via projections from the body of the caudate, contains neurons that discharge selectively in relation to passive visual stimulation, fixation of gaze, and both visually-triggered and memory-contingent saccadic eye movements (Hikosaka & Wurtz 1983a–c). There is strong evidence that these neurons may participate in the control of saccadic eye movements via a GABAergic nigrotectal projection to the superior colliculus (Hikosaka & Wurtz 1983d, 1985a,b).

As noted above, the lateral SNr also receives projections from portions of the putamen involved in the "motor" circuit (Parent et al 1984a, Szabo 1967), but fibers from the putamen terminate more dorsally than do those from the body of the caudate (Szabo 1970). That these two fiber systems are merely closely juxtaposed rather than convergent is further suggested by the fact that neurons in the lateral SNr appear to discharge selectively either in relation to eye movements or to orofacial movements, but not both (DeLong et al 1983a,b, Hikosaka & Wurtz 1983a–c).

"DORSOLATERAL PREFRONTAL" CIRCUIT

It had been proposed previously that a single "complex" or "association" loop passed through the caudate nucleus and eventually influenced certain prefrontal "association" areas (DeLong & Georgopoulos 1981, DeLong et al 1983b). Subsequent anatomical findings have prompted a reappraisal of this scheme, and it now appears that there are at least two distinct basal ganglia–thalamocortical circuits that selectively influence separate prefrontal areas. Although there is considerable evidence indicating significant functional differentiation between these regions, the behavioral functions associated with each of the proposed "prefrontal" circuits have not yet been characterized to the same degree as have those of the "motor" and "oculomotor" circuits. Accordingly, we have chosen for the time being to give anatomical rather than functional designations to the two "prefrontal" circuits. We discuss each separately in this and the following sections.

The cortex within and around the principal sulcus and on the dorsal prefrontal convexity (Brodmann's areas 9, 10; Walker's area 46) provides the "closed loop" portion of the corticostriate input to the "dorsolateral prefrontal" circuit (see Figure 1). The projection from this cortical area terminates within the dorsolateral head of the caudate nucleus and throughout a continuous rostrocaudal expanse that extends to the tail of the caudate (Goldman & Nauta 1977, Selemon & Goldman-Rakic 1985, Yeterian & VanHoesen 1978). Grossly overlapping corticostriate projections to this same sector arise from the posterior parietal cortex (area 7) and the arcuate premotor area (Kunzle 1978, Miyata & Sasaki 1984, Selemon & Goldman-Rakic 1985).

Early anterograde transport studies were interpreted as indicating partial convergence of corticostriate projections from interconnected cortical association areas (Yeterian & VanHoesen 1978). Recently, it has been shown by means of double-label anterograde transport that although interconnected areas may project to the same general region in the neostriatum, the terminal arborizations of such projections frequently interdigitate rather than overlap (Selemon & Goldman-Rikac 1985). This finding suggests that the degree of maintained segregation within the basal ganglia–thalamocortical circuits, at least at the level of the striatum, may in fact be more pronounced than is evident from conventional single-label anatomical studies, which rely on comparisons of anterograde and/or retrograde labeling patterns across animals.

Rostral portions of the caudate nucleus project to the dorsomedial one-third of the globus pallidus and to rostral portions of the SNr (Szabo 1962, Johnson & Rosvold 1971, Cowan & Powell 1966, Parent et al 1984a). Within each of these projections there is a mediolateral gradient such that projections from the dorsolateral caudate (which is the recipient of projections from dorsolateral prefrontal cortex) are distributed to more lateral portions of the pallidum and nigra than are those from the ventromedial caudate. The dorsomedial one-third of the internal pallidal segment has been shown to project to the parvocellular portion of the ventral anterior thalamic nucleus (VApc) (Kuo & Carpenter 1973, Kim et al 1976). The VApc projects to regions on the convexity of the frontal lobe, including caudal prefrontal areas (Kievit & Kuypers 1977). The rostrolateral portions of the SNr have been shown to project to the MDpc (Ilinsky et al 1985), which in turn projects to dorsolateral prefrontal cortex in and around the principal sulcus (Akert 1964, Jacobson et al 1978, Pribram et al 1953). Thus, as indicated in Figure 3, the "dorsolateral prefrontal" circuit is partially closed by thalamocortical projections from both the VApc and MDpc.

The "dorsolateral prefrontal" circuit has not been functionally characterized to the same extent as have the "motor" or "oculomotor" circuits, but there are indications from lesioning and single cell recording studies that this system may participate in processes subserving spatial memory (Alexander et al 1980, Fuster & Alexander 1973, Fuster 1981, Goldman et al 1971, Isseroff et al 1982, Divac et al 1967).

"LATERAL ORBITOFRONTAL" CIRCUIT

Lateral orbitofrontal cortex (Brodmann's area 10, Walker's area 12) projects to a ventromedial sector of the caudate nucleus that extends from the head to the tail of this structure. This part of the caudate also receives input from the auditory and visual association areas of the superior and inferior temporal gyri, respectively (Selemon & Goldman-Rakic 1985, VanHoesen et al 1981, Yeterian & VanHoesen 1978).

Ventromedial portions of the caudate project to a dorsomedial sector of the internal pallidal segment that lies just medial to the sector innervated by the dorsolateral caudate, and to a rostromedial portion of the SNr (Johnson & Rosvold 1971, Szabo 1962). The latter projects in turn to medial parts of the VAmc and to the MDmc (Carpenter et al 1976, Ilinsky et al 1985). The "closed loop" portion of the "lateral orbitofrontal" circuit is thus completed by return projections from these two thalamic regions to the lateral orbitofrontal cortex (see Figure 1) (Ilinsky et al 1985), as indicated in Figure 3.

Like the other "prefrontal" basal ganglia loop, the "lateral orbitofrontal" circuit has yet to be fully characterized from a functional standpoint. It has been shown that bilateral lesions in primates restricted either to the lateral orbitofrontal area or to the portion of the caudate to which it projects appear to result in a perseverative interference with an animal's capacity to make appropriate switches in behavioral set (Divac et al 1967, Mishkin & Manning 1978). It remains to be seen whether lesions placed selectively at other points in this circuit will produce the same disruptions of behavior. With respect to possible future investigations of this pathway at the single cell level, it is likely that considerable ingenuity will be required to devise suitable behavioral paradigms for the functional characterization of the constituent neurons.

"ANTERIOR CINGULATE" CIRCUIT

Barely a decade has passed since the concept of the ventral striopallidal system was first proposed (Heimer & Wilson 1975). The concept was based upon striking parallels, in a number of species, between the connections and histochemical features of the neostriatum and those of the nucleus accumbens and the medium-celled portion of the olfactory tubercle

(Heimer & Wilson 1975, Heimer 1978, Heimer et al 1977, Nauta 1979). Because both subdivisions of the ventral striatum (accumbens and olfactory tubercle) receive extensive projections from so-called "limbic" structures, including the hippocampus, the amygdala, and entorhinal (area 28) and perirhinal cortices (area 35) (Heimer & Wilson 1975, Hemphill et al 1981, Kelley & Domesick 1982, Kelley et al 1982, Krayniak et al 1981, Nauta 1961), this portion of the striatum has been referred to as the "limbic" striatum (Nauta & Domesick 1984). Although the ventral striatum was once believed to receive its cortical input exclusively from nonisocortical areas, recent evidence indicates that there are also significant projections to this region from the anterior cingulate area (area 24) and widespread sources in the temporal lobe, including the temporal pole and the superior and inferior temporal gyri (Baleydier & Mauguiere 1980, Hemphill et al 1981, Powell & Leman 1976, Selemon & Goldman-Rakic 1985, VanHoesen et al 1976, 1981, Yeterian & VanHoesen 1978). There are indications that the accumbens may also receive projections from posterior portions of the medial orbitofrontal area (area 11 of Brodmann, Walker's area 13, area FF of Bonin & Bailey) (Nauta 1962, Yeterian & Van Hoesen 1978).

The ventral striatum has been shown in rats to project to the precommissural or ventral pallidum and to the substantia nigra (Heimer 1978, Nauta et al 1978). Recent anterograde transport studies have revealed that in the monkey the nucleus accumbens projects not only to the ventral pallidum and rostrodorsal substantia nigra, but also to a rostrolateral sector of the internal pallidal segment (J. Hedreen and M. R. DeLong, unpublished data), which projects in turn to a paramedian portion of the MDmc (Kuo & Carpenter 1973). The "anterior cingulate" circuit thus appears to be partially closed by the well-documented projections to the anterior cingulate area from posterior and medial portions of the mediodorsal nucleus (MD) (Baleydier & Mauguiere 1980, Jurgens 1983, Tobias 1975, Vogt et al 1979), as indicated in Figure 3.

The functional characteristics of this circuit cannot as yet be specified in any detail. The hippocampal and enterhinal inputs to this pathway are generally considered to be "limbic" structures, and by virtue of their connectivity and lamination (which is intermediate between that of isocortex and that of allocortex), several of the neocortical inputs (anterior cingulate, medial orbitofrontal, and temporal pole) have been designated as "paralimbic association" cortex (Pandya & Seltzer 1982). Considering the uncertainties surrounding the functions of the so-called "limbic" structures, and the paucity of behavioral and physiological data in the primate, it is difficult even to speculate on the possible functions that may be subserved by the "anterior cingulate" circuit. It is for this reason that we have employed a purely anatomical term to designate this proposed circuit.

SUBSIDIARY BASAL GANGLIA CIRCUITS

In addition to the five principal circuits outlined above, the individual nuclei of the basal ganglia also participate in several subsidiary circuits, which apparently serve, at least in part, to modify transmission through the basal ganglia–thalamocortical pathways. The nodal points of these subsidiary circuits include the subthalamic nucleus (Nauta & Cole 1978), the intralaminar nuclei of the thalamus (Kalil 1978, Parent et al 1983), the pedunculopontine nucleus (Parent et al 1983), and the dopaminergic nuclei of the mesencephalic tegmentum (Carpenter & Peter 1972, Parent et al 1983). It is beyond the scope of this review to consider these circuits in any detail. It should be noted, however, that the topographic features characteristic of the principal circuits are also found in these subsidiary circuits. Moreover, both the centromedian nucleus (Kunzle 1976) and the subthalamic nucleus (Monakow et al 1978) appear to be somatotopically organized by virtue of topographically organized projections from the motor cortex.

In addition to input from the cortex and thalamus, the striatum also receives projections from the nuclei of the dorsal raphe (DeVito et al 1980, Parent et al 1983), the locus coeruleus (Parent et al 1983), and the amygdala (Nauta 1961, Parent et al 1983, Russchen et al 1985). Presumably, these inputs also serve to modify the transfer of information through the various basal ganglia–thalamocortical circuits.

BEHAVIORAL AND CLINICAL IMPLICATIONS

We hope that the conceptual framework of segregated functional circuits that we have proposed may prove useful not only in attempts to clarify the normal functions of the basal ganglia, but also in efforts to understand the behavioral and motor disturbances that occur in disorders involving the basal ganglia, such as Parkinson's and Huntington's diseases. To the extent that this framework is an accurate representation of the structural and functional organization of the basal ganglia in normal individuals, it may prove useful in suggesting pathophysiologic mechanisms to account for some of the clinical manifestations of these disorders.

Several features of the proposed scheme of basal ganglia organization are of obvious relevance to the understanding of the symptoms of basal ganglia disorders. At the most basic level, the separation of the "motor" and "prefrontal" circuits provides a framework whereby relatively selective disturbances of "motor" or more "complex" behavior may occur from damage to different portions of the basal ganglia. As discussed above, there is considerable evidence for such dissociations in experimental studies in primates (DeLong & Georgopoulos 1981, Divac et al 1967). Furthermore, the existence of channels for the control of individual body parts in the "motor" circuit helps to clarify how involuntary movements or impairments of movements of a single body part (e.g. monochorea, focal dystonias, or focal dyskinesias) may result from restricted lesions or disturbances within these nuclei.

The finding that the supplementary motor area receives the output from the "motor" circuit provides a new perspective on the motor functions of the basal ganglia. Considerable evidence now exists for a role of the supplementary motor area in the programming, initiation, and execution of movement. For example, recent studies in humans indicate increased cerebral blood flow (and therefore increased metabolic activity) in the region of the supplementary motor area (but not in other regions) during the planning or rehearsal of complex movement sequences. Moreover, the supplementary motor area has been found to be active during the execution of complex but not simple finger movements (Roland et al 1980); this finding may reflect the relative requirements for programming in the two tasks. Additional evidence for a role of the supplementary motor area in the programming of movement comes from single cell studies in monkeys, which have revealed changes in neuronal activity during periods when the animal is preparing to make a movement (Tanji et al 1980). This activity appears to reflect the "motor set" of the animal. Thus, it would appear that the basal ganglia and the supplementary motor area form part of a system involved in the programming and execution of complex movements. It has been suggested that some of the motor abnormalities seen in Parkinson's disease might be understood in terms of the relations between the basal ganglia and the supplementary motor area (Schell & Strick 1984). In view of the supplementary motor area's apparent role in motor programming, it is of interest that patients with Parkinson's disease often show impairments in predictive tracking movements (Flowers 1978) and other complex aspects of movement that require motor programming (Marsden 1984). It is noteworthy that certain motor disturbances characteristic of Parkinson's disease have occasionally been observed in patients with lesions that involve the supplementary motor area. There are reports of global akinesia, maximal contralaterally, associated with lesions of the supplementary motor area in man (Damasio & VanHoesen 1980, LaPlane et al 1977). As one of the cardinal signs of Parkinson's disease, akinesia is generally attributed to the loss of dopaminergic input to the striatum that results from degeneration of the pars compacta of the substantia nigra. Thus, it is conceivable that the disruption of striatal dopaminergic transmission that occurs in patients with Parkinson's disease might result in akinesia as a consequence of disordered basal ganglia inputs to the supplementary motor area. This suggestion, however, is difficult to reconcile with the paucity of motor deficits observed in monkeys with lesions of the supplementary motor area (Brinkman 1984, Travis 1955), in whom the most consistent finding appears to be a moderate disruption of coordinated bimanual movements. It remains to be determined whether these discrepancies between human and simian studies represent true species differences, or are related instead to possible differences in the precise locations of the damaged areas.

The importance of the "oculomotor" circuit in the control of eye movements has been brought into clearer focus by recent anatomical and physiological studies. Among other major clinical manifestations, patients with Huntington's disease—in whom there is profound degeneration of the caudate nucleus (and to a lesser extent the putamen)—often show severe disturbances in the initiation of voluntary saccades, and their saccades may be markedly slowed (Leigh et al 1983). These observations might be explained by an increase in the tonic, GABAergic nigrocollicular discharge (Hikosaka & Wurtz 1985b), which would be expected if degeneration of the "oculomotor" portion of the caudate nucleus resulted in decreased phasic disinhibition along the caudate-SNr-thalamic/collicular pathway.

While it is generally accepted that disturbances of motor function may result from basal ganglia damage, the precise role of the basal ganglia in disturbances of higher functions in humans is still controversial because of the often associated neuropathologic changes occurring in other structures. The cognitive deficits in patients with Huntington's disease, for example, might be accounted for, at least partially, by degeneration of the caudate nucleus with consequent interruption of the "dorsolateral prefrontal" or "lateral orbitofrontal" circuits. This suggestion would be difficult to prove, however, in view of the widespread cortical degeneration that is a frequent accompaniment of Huntington's disease. Nevertheless, it is noteworthy that recent positron emission tomographic studies have revealed that dementia in Huntington's disease is correlated with hypometabolism in the caudate nucleus rather than in the cerebral cortex (Kuhl et al 1982).

CONCLUDING REMARKS

Although the basal ganglia lack gross cytoarchitechtonic differentiation and were long viewed as lacking in specificity of connection and function, the anatomic and physiologic evidence accumulated over the past two decades has revealed a level of organization and functional specificity paralleling that of the cerebral cortex itself. We have reviewed the evidence relevant to our proposal that the basal ganglia be viewed as components of multiple parallel, segregated circuits. Each of the five proposed basal ganglia–thalamocortical circuits appears to receive input from several separate but functionally related cortical areas, traverse specific portions of the basal ganglia and thalamus, and project back upon one of the cortical areas providing input to the circuit, thus forming a partially "closed" loop (Figure 2).

Although we have placed emphasis on the segregation of functional circuits, this should not be construed as implying that integration is a minor function of these pathways. We have attempted to show that integration within each circuit appears to be of a highly specific nature, as exemplified by the retention of detailed place and modality specificity within the "motor" circuit despite progressive convergence of separate cortical influences along the pathways leading to the ventrolateral thalamus.

Multiple subsidiary circuits appear to modify and modulate the flow of information through the major basal ganglia–thalamocortical pathways and to provide additional routes for influences to be exerted on other structures. It is obvious that the basal ganglia should no longer be viewed as "centers" or structures having a role independent of the cerebral cortex and thalamus, with which they have intimate and highly specific afferent and efferent connections. Accordingly, from the functional standpoint it would seem more appropriate to attempt an appraisal of the distinctive functions of the individual basal ganglia–thalamocortical circuits than to try to assign functions to their component nuclei. Furthermore, in view of the apparently uniform organization of the five principal circuits, it would seem that the operations performed at the same levels of each circuit are likely to be quite comparable, even though the information transmitted through the individual circuits might differ considerably. It is possible, therefore, that knowledge about the operations performed at a particular level of one circuit, e.g. the pallidal portion of the "motor" circuit, might also be applicable to those performed at the same level of the other basal ganglia–thalamocortical circuits.

Future research may reveal functional subdivisions of the "oculomotor," "prefrontal," and "anterior cingulate" circuits comparable to the somato-topic channels within the "motor" circuit. For example, the "oculomotor" circuit might contain separate, parallel channels arranged according to a retinocentric (or, on the other hand, a spatial) coordinate system. For the present, the nature of the channels that might exist in the "prefrontal" and "anterior cingulate" circuits remains a matter for speculation, pending further clarification of the functional roles of these pathways. That there may be additional basal ganglia–thalamocortical circuits, beyond those proposed here, seems likely. This question can only be answered, however, by further anatomical and functional studies. Considering the enormous expansion of the frontal lobes in man and the selective targeting of basal ganglia influences on these frontal areas, it is even possible that circuits exist in man for which there are no counterparts in the monkey.

ACKNOWLEDGMENTS

Preparation of this review has been supported in part by United States Public Health Service grants NS00632 (G. E. A.), NS17678 (G. E. A.), NS16375 (M. R. D.), and NS02957 (P. L. S.), and through funds of the Veterans Administration Medical Research Service and a donation from the E. K. Dunn family. We are grateful to Drs. M. D. Crutcher, A. P. Georgopoulos, and S. J. Mitchell for their critical comments on the manuscript, and to N. Kauffman, K. Merrill, and D. Weaver for typing the manuscript and assisting with the preparation of the figures.

Literature Cited

Akert, K. 1964. Comparative anatomy of frontal cortex and thalamofrontal connections. In *The Frontal Granular Cortex and Behavior*, ed. J. M. Warren, K. Akert, pp. 372–96. New York: McGraw-Hill

Alexander, G. E. 1984. Instruction-dependent neuronal activity in primate putamen. *Soc. Neurosci. Abstr.* 10:515

Alexander, G. E., DeLong, M. R. 1985a. Microstimulation of the primate neostriatum: II. Somatotopic organization of striatal microexcitable zones and their relation to neuronal response properties. *J. Neurophysiol.* 53:1433–46

Alexander, G. E., DeLong, M. R. 1985b. Microstimulation of the primate neostriatum: I. Physiological properties of striatal microexcitable zones. *J. Neurophysiol.* 53:1417–32

Alexander, G. E., Witt, E. D., Goldman-Rakic, P. S. 1980. Neuronal activity in the prefrontal cortex, caudate nucleus and mediodorsal thalamic nucleus during delayed response performance of immature and adult rhesus monkeys. *Soc. Neurosci. Abstr.* 6:86

Allen, G. I., Tsukahara, N. 1974. Cerebrocerebellar communication systems. *Physiol. Rev.* 54:957–1006

Anderson, M., Yoshida, M. 1977. Electrophysiological evidence for branching nigral projections to the thalamus and the superior colliculus. *Brain Res.* 137:361–64

Baleydier, C., Mauguiere, F. 1980. The duality of the cingulate gyrus in monkey, neuroanatomical study and functional hypothesis. *Brain* 103:525–54

Barbas, H., Mesulam, M. M. 1981. Organization of afferent input to subdivisions of area 8 in the rhesus monkey. *J. Comp. Neurol.* 200:407–31

Beckstead, R. M., Edwards, S. B., Frankfurter, A. 1981. A comparison of the intranigral distribution of nigrotectal neurons labeled with horseradish peroxidase in the monkey, cat, and rat. *J. Neurosci.* 1:121–25

Biber, M. P., Kneisley, L. W., LaVail, J. H. 1978. Cortical neurons projecting to the cervical and lumbar enlargements of the spinal cord in young and adult rhesus monkeys. *Exp. Neurol.* 59:492–508

Bizzi, E., Schiller, P. H. 1970. Single unit activity in the frontal eye fields of unanesthetized monkeys during eye and head movement. *Exp. Brain Res.* 10:151–58

Brinkman, C. 1984. Supplementary motor area of the monkey's cerebral cortex: Short- and long-term deficits after unilateral ablation and the effects of subsequent callosal section. *J. Neurosci.* 4:918–29

Brinkman, C., Porter, R. 1979. Supplementary motor area in the monkey: Activity of neurons during performance of a learned motor task. *J. Neurophysiol.* 42:681–709

Carpenter, M. B., Nakano, K., Kim, R. 1976. Nigrothalamic projections in the monkey demonstrated by autoradiographic technics. *J. Comp. Neurol.* 165:401–16

Carpenter, M. B., Peter, P. 1972. Nigrostriatal and nigrothalamic fibers in the rhesus monkey. *J. Comp. Neurol.* 144:93–116

Cowan, W. M., Powell, T. P. S. 1966. Striopallidal projection in the monkey. *J. Neurol. Neurosurg. Psychiat.* 29:426–39

Crutcher, M. D., DeLong, M. R. 1984a. Single cell studies of the primate putamen.

I. Functional organization. *Exp. Brain Res.* 53:233-43

Crutcher, M. D., DeLong, M. R. 1984b. Single cell studies of the primate putamen. II. Relations to direction of movements and pattern of muscular activity. *Exp. Brain Res.* 53:244-58

Damasio, A. R., VanHoesen, G. W. 1980. Structure and function of the supplementary motor area. *Neurology* 30:359

DeLong, M. R. 1971. Activity of pallidal neurons during movement. *J. Neurophysiol.* 34:414-27

DeLong, M. R., Georgopoulos, A. P. 1979. Motor functions of the basal ganglia as revealed by studies of single cell activity in the behaving primate. *Adv. Neurol.* 24:131-40

DeLong, M. R., Georgopoulos, A. P. 1981. Motor functions of the basal ganglia. In *Handb. Physiol.*, Sect. 1, *The Nervous System*, Vol. 2, *Motor Control*, Part 2, ed. J. M. Brookhart, V. B. Mountcastle, V. B. Brooks, pp. 1017-61. Bethesda: Am. Physiol. Soc.

DeLong, M. R., Crutcher, M. D., Georgopoulos, A. P. 1983a. Relations between movement and single cell discharge in the substantia nigra of the behaving monkey. *J. Neurosci.* 3:1599-1606

DeLong, M. R., Georgopoulos, A. P., Crutcher, M. D. 1983b. Cortico-basal ganglia relations and coding of motor performance. *Exp. Brain Res. Suppl.* 7:30-40

DeLong, M. R., Alexander, G. E., Georgopoulos, A. P., Crutcher, M. D., Mitchell, S. J., Richardson, R. T. 1984. Role of basal ganglia in limb movements. *Human Neurobiol.* 2:235-44

DeLong, M. R., Crutcher, M. D., Georgopoulos, A. P. 1985. Primate globus pallidus and subthalamic nucleus: Functional organization. *J. Neurophysiol.* 53:530-43

DeVito, J. L., Anderson, M. E., Walsh, K. E. 1980. A horseradish peroxidase study of afferent connections of the globus pallidus in *Macaca mulatta*. *Exp. Brain Res.* 38:65-73

DeVito, J. L., Anderson, M. E. 1982. An autoradiogrâhic study of efferent connections of the globus pallidus in *Macaca mulatta*. *Exp. Brain Res.* 46:107-17

Divac, I., Rosvold, H. E., Szwarcbart, M. K. 1967. Behavioral effects of selective ablation of the caudate nucleus. *J. Comp. Physiol. Psychol.* 63:184-90

Evarts, E. V., Thach, W. T. 1969. Motor mechanism of the CNS: Cerebrocerebellar interrelations. *Ann. Rev. Physiol.* 31:451-98

Evarts, E. V., Wise, S. P. 1984. Basal ganglia outputs and motor control. *Functions of the Basal Ganglia, CIBA Found. Symp.* 107:83-96

Flowers, K. 1978. Lack of prediction in the motor behavior of parkinsonism. *Brain* 101:35-52

Fries, W. 1984. Cortical projections to the superior colliculus in the macaque monkey: A retrograde study using horseradish peroxidase. *J. Comp. Neurol.* 230:55-76

Fuster, J. M. 1981. Prefrontal cortex in motor control. See DeLong & Georgopoulos 1981, pp. 1149-78

Fuster, J. M., Alexander, G. E. 1973. Firing changes in cells of the nucleus medialis dorsalis associated with delayed response behavior. *Brain Res.* 61:79-91

Georgopoulos, A. P., DeLong, M. R., Crutcher, M. D. 1983. Relations between parameters of step-tracking movements and single cell discharge in the globus pallidus and subthalamic nucleus of the behaving monkey. *J. Neurosci.* 3:1586-98

Goldberg, M. E., Bushnell, M. C. 1981. Behavioral enhancement of visual responses in monkey cerebral cortex. II. Modulation in frontal eye fields specifically related to saccades. *J. Neurophysiol.* 46:773-87

Goldman, P. S., Nauta, W. J. H. 1976. Autoradiographic demonstration of a projection from prefrontal association cortex to the superior colliculus in the rhesus monkey. *Brain Res.* 116:145-49

Goldman, P. S., Nauta, W. J. H. 1977. An intricately patterned prefronto-caudate projection in the rhesus monkey. *J. Comp. Neurol.* 171:369-86

Goldman-Rakic, P. S. 1982. Cytoarchitectonic heterogeneity of the primate neostriatum: Subdivision into Island and Matrix cellular compartments. *J. Comp. Neurol.* 205:398-413

Goldman, P. S., Rosvold, H. E., Vest, B., Galkin, T. W. 1971. Analysis of the delayed-alternation deficit produced by dorsolateral prefrontal lesions in the rhesus monkey. *J. Comp. Physiol. Psychol.* 77:212-20

Graybiel, A. M., Ragsdale, C. W. 1978. Histochemically distinct compartments in the striatum of human being, monkey and cat demonstrated by the acetylthiocholinesterase staining method. *Proc. Natl. Acad. Sci. USA* 75:5723-26

Graybiel, A. M., Ragsdale, C. W., Edley, S. M. 1979. Compartments in the striatum of the cat observed by retrograde cell labeling. *Exp. Brain Res.* 34:189-95

Graybiel, A. M., Pickel, V. M., Joh, T. H., Reis, D. J., Ragsdale, C. W. Jr. 1981. Direct demonstration of a correspondence between the dopamine islands and acetylcholinesterase patches in the developing striatum. *Proc. Natl. Acad. Sci. USA* 78 : 5871–75

Hallett, M., Khoshbin, S. 1980. A physiological mechanism of bradykinesia. *Brain* 103 : 301–14

Hedreen, J., DeLong, M. R., Holm, G. 1980. Striatonigral relationship in macaques. *Soc. Neurosci. Abstr.* 6 : 272

Heimer, L. 1978. The olfactory cortex and the ventral striatum. In *Limbic Mechanisms*, ed. K. E. Livingston, O. Hornykiewicz, pp. 95–187. New York : Plenum

Heimer, L., VanHoesen, G. W., Rosene, D. L. 1977. The olfactory pathways and the anterior perforated substance in the primate brain. *Int. J. Neurol.* 12 : 42–52

Heimer, L., Wilson, R. D. 1975. The subcortical projections of the allocortex. Similarities in the neural associations of the hippocampus, the piriform cortex, and the neocortex. In *Golgi Centennial Symposium: Perspectives in Neurology*, ed. M. Santini, pp. 177–93. New York : Raven

Hemphill, M., Holm, G., Crutcher, M., DeLong, M. R., Hedreen, J. 1981. Afferent connections of the nucleus accumbens in the monkey. In *Neurobiology of the Nucleus Accumbens*, ed. R. Chronister, J. DeFrance, pp. 75–81. Brunswick, Maine : Haer Inst. Press

Herkenham, M., Pert, C. 1981. Mosaic distribution of opiate receptors, parafascicular projections and acetylcholinesterase in rat striatum. *Nature* 291 : 415–17

Hikosaka, O., Wurtz, R. H. 1983a. Visual and oculomotor functions of monkey substantia nigra pars reticulata. I. Relation of visual and auditory responses to saccades. *J. Neurophysiol.* 49 : 1230–53

Hikosaka, O., Wurtz, R. H. 1983b. Visual and oculomotor functions of monkey substantia nigra pars reticulata. II. Visual responses to fixation of gaze. *J. Neurophysiol.* 49 : 1254–67

Hikosaka, O., Wurtz, R. H. 1983c. Visual and oculomotor functions of monkey substantia nigra pars reticulata. III. Memory-contingent visual and saccade responses. *J. Neurophysiol.* 49 : 1268–84

Hikosaka, O., Wurtz, R. H. 1983d. Visual and oculomotor functions of monkey substantia nigra pars reticulata. IV. Relation of substantia nigra to superior colliculus. *J. Neurophysiol.* 49 : 1285–1301

Hikosaka, O., Wurtz, R. H. 1985a. Modification of saccadic eye movements by GABA-related substances. I. Effect of muscimol and bicuculline in monkey superior colliculus. *J. Neurophysiol.* 53 : 266–91

Hikosaka, O., Wurtz, R. H. 1985b. Modification of saccadic eye movements by GABA-related substances. II. Effects of muscimol in monkey substantia nigra pars reticulata. *J. Neurophysiol.* 53 : 292–308

Ilinsky, I. A., Jouandet, M. L., Goldman-Rakic, P. S. 1985. Organization of the nigrothalamocortical system in the rhesus monkey. *J. Comp. Neurol.* 236 : 315–30

Isseroff, A., Rosvold, H. E., Galkin, T. W., Goldman-Rakic, P. S. 1982. Spatial memory impairments following damage to the mediodorsal nucleus of the thalamus in rhesus monkeys. *Brain Res.* 232 : 97–113

Jacobson, S., Butters, N., Tovsky, N. J. 1978. Afferent and efferent subcortical projections of behaviorally defined sectors of prefrontal granular cortex. *Brain Res.* 159 : 279–96

Johnson, T. N., Rosvold, H. E. 1971. Topographic projections on the globus pallidus and the substantia nigra of selectively placed lesions in the precommissural caudate nucleus and putamen in the monkey. *Exp. Neurol.* 33 : 584–96

Jones, E. G., Coulter, J. D., Burton, H., Porter, R. 1977. Cells of origin and terminal distribution of corticostriatal fibers arising in the sensory-motor cortex of monkeys. *J. Comp. Neurol.* 173 : 53–80

Jurgens, U. 1983. Afferent fibers to the cingular vocalization region in the squirrel monkey. *Exp. Neurol.* 80 : 395–409

Kalil, K. 1978. Patch-like termination of thalamic fibers in the putamen of the rhesus monkey: An autoradiographic study. *Brain Res.* 140 : 333–39

Kelley, A. E., Domesick, V. B. 1982. The distribution of the projection from the hippocampal formation to the nucleus accumbens in the rat : An anterograde- and retrograde-horseradish peroxidase study. *Neuroscience* 7 : 2321–35

Kelley, A. E., Domesick, V. B., Nauta, W. J. H. 1982. The amygdalostriatal projection in the rat—an anatomical study by anterograde and retrograde tracing methods. *Neuroscience* 7 : 615–30

Kemp, J. M., Powell, T. P. S. 1971. The connections of the striatum and globus pallidus: Synthesis and speculation. *Philos. Trans. R. Soc. London Ser. B* 262 : 441–57

Kievit, J., Kuypers, H. G. J. M. 1977. Organization of the thalamocortical connections to the frontal lobe in the rhesus monkey. *Exp. Brain Res.* 29 : 299–322

Kim, R., Nakano, K., Jayaraman, A., Carpenter, M. B. 1976. Projections of the globus pallidus and adjacent structures : An autoradiographic study in the monkey. *J. Comp. Neurol.* 169 : 263–90

Krayniak, P. F., Meibach, R. C., Siegel, A.

1981. A projection from the entorhinal cortex to the nucleus accumbens in the rat. *Brain Res.* 209:427–31

Kuhl, D. E., Phelps, M. E., Markham, C. H., Metter, E. J., Riege, W. H., Winter, J. 1982. Cerebral metabolism and atrophy in Huntington's disease determined by 18FDG and computed topographic scan. *Ann. Neurol.* 12:425–34

Kunzle, H. 1975. Bilateral projections from precentral motor cortex to the putamen and other parts of the basal ganglia. An autoradiographic study in *Macaca fascicularis*. *Brain Res.* 88:195–209

Kunzle, H. 1976. Thalamic projections from the precentral motor cortex in *Macaca fascicularis*. *Brain Res.* 105:253–67

Kunzle, H. 1977. Projections from the primary somatosensory cortex to basal ganglia and thalamus in the monkey. *Exp. Brain Res.* 30:481–92

Kunzle, H. 1978. An autoradiographic analysis of the efferent connections from premotor and adjacent prefrontal regions areas 6 and 9) in *Macaca fascicularis*. *Brain Behav. Evol.* 15:185–234

Kunzle, H., Akert, K., Wurtz, R. H. 1976. Projection of area 8 (frontal eye field) to superior colliculus in the monkey. An autoradiographic study. *Brain Res.* 117:487–92

Kunzle, H., Akert, K. 1977. Efferent connections of cortical area 8 (frontal eye field) in *Macaca fascicularis*. A reinvestigation using the autoradiographic technique. *J. Comp. Neurol.* 173:147–64

Kuo, J. S., Carpenter, M. B. 1973. Organization of pallidothalamic projections in the rhesus monkey. *J. Comp. Neurol.* 151:201–36

LaPlane, D., Talairach, J., Meininger, V., Bancaud, J., Orgogozo, J. M. 1977. Clinical consequences of corticectomies involving the supplementary motor area in man. *J. Neurol. Sci.* 34:301–14

Leichnetz, G. R., Spencer, R. F., Hardy, S. G. P., Astruc, J. 1981. The prefrontal corticotectal projection in the monkey: An anterograde and retrograde horseradish peroxidase study. *Neuroscience* 6:1023–41

Leigh, R. J., Newman, S. A., Folstein, S. E., Lasker, A. G., Jensen, B. A. 1983. Abnormal ocular motor control in Huntington's disease. *Neurology* 33:1268–75

Liles, S. L. 1979. Topographic organization of neurons related to arm movement in the putamen. *Adv. Neurol.* 23:155–62

Liles, S. L. 1985. Activity of neurons in putamen during active and passive movements of wrist. *J. Neurophysiol.* 53:217–36

Lynch, J. C., Mountcastle, V. B., Talbot, W. H., Yin, T. C. T. 1977. Parietal lobe mechanisms for directed visual attention. *J. Neurophysiol.* 40:362–89

Macpherson, J. M., Marangoz, C., Miles, T. S., Wiesendanger, M. 1982. Microstimulation of the supplementary motor area (SMA) in the awake monkey. *Exp. Brain Res.* 45:410–16

Marsden, C. D. 1984. Which motor disorder in Parkinson's disease indicates the true motor function of the basal ganglia? *Functions of the Basal Ganglia, Ciba Found. Symp.* 107:225–41

Mishkin, M., Manning, F. J. 1978. Nonspatial memory after selective prefrontal lesions in monkeys. *Brain Res.* 143:313–23

Miyata, M., Sasaki, K. 1984. Horseradish peroxidase studies on thalamic and striatal connections of the mesial part of area 6 in the monkey. *Neurosci. Lett.* 49:127–33

Mohler, C. W., Goldberg, M. E., Wurtz, R. H. 1973. Visual receptive fields of frontal eye field neurons. *Brain Res.* 61:385–89

Monakow, K., Hartmann-von, Akert, K., Kunzle, H. 1978. Projections of the precentral motor cortex and other cortical areas of the frontal lobe to the subthalamic nucleus in the monkey. *Exp. Brain Res.* 33:395–403

Muakkassa, K. F., Strick, P. L. 1979. Frontal lobe inputs to primate motor cortex: Evidence for four somatotopically organized 'premotor' areas. *Brain Res.* 177:176–82

Murray, E. A., Coulter, J. D. 1981. Organization of corticospinal neurons in the monkey. *J. Comp. Neurol.* 195:339–65

Nauta, W. J. H. 1961. Fibre degeneration following lesions of the amygdaloid complex in the monkey. *J. Anat.* 95:515–31

Nauta, W. J. H. 1962. Neural associations of the amygdaloid complex in the monkey. *Brain* 85:505–20

Nauta, H. J. W. 1979. A proposed conceptual reorganization of the basal ganglia and telencephalon. *Neuroscience* 4:1875–81

Nauta, H. J. W., Cole, M. 1978. Efferent projections of the subthalamic nucleus: An autoradiographic study in monkey and cat. *J. Comp. Neurol.* 180:1–16

Nauta, W. J. H., Domesick, V. B. 1984. Afferent and efferent relationships of the basal ganglia. *Functions of the Basal Ganglia, Ciba Found. Symp.* 107:3–29

Nauta, W. J. H., Mehler, W. R. 1966. Projections of the lentiform nucleus in the monkey. *Brain Res.* 1:3–42

Nauta, W. J. H., Smith, G. P., Faull, R. L. M., Domesick, V. B. 1978. Efferent connections and nigral afferents of the nucleus accumbens septi in the rat. *Neuroscience* 3:385–401

Neafsy, E. J., Hull, C. D., Buchwald, N. A. 1978. Preparation of movement in the cat.

II. Unit activity in the basal ganglia and thalamus. *Electroenceph. Clin. Neurophysiol.* 44:714–23

Palmer, C., Schmidt, E. M., McIntosh, J. S. 1981. Corticospinal and corticorubral projections from the supplementary motor area in the monkey. *Brain Res.* 209:305–14

Pandya, D. N., Seltzer, B. 1982. Association areas of the cerebral cortex. *Trends Neurosci.* 5:386–90

Parent, A., Mackey, A., De Bellefeuille, L. 1983. The subcortical afferents to caudate nucleus and putamen in primate: A fluorescence retrograde double labeling study. *Neuroscience* 10:1137–50

Parent, A., Bouchard, C., Smith, Y. 1984a. The striatopallidal and striatonigral projections: Two distinct fiber systems in primate. *Brain Res.* 303:385–90

Parent, A., Smith, Y., Bellefeuille, L. 1984b. The output organization of the pallidum and substantia nigra in primate as revealed by a retrograde double-labeling method. In *The Basal Ganglia, Structure and Function*, eds. J. S. McKenzie, R. E. Kemm, L. N. Wilcock, pp. 147–60. New York: Plenum

Percheron, G., Yelnik, J., Francois, C. 1984. A golgi analysis of the primate globus pallidus. III. Spatial organization of the striatopallidal complex. *J. Comp. Neurol.* 227:214–27

Pickel, V. M., Sumal, K. K., Beckley, S. C., Miller, R. J., Reis, D. H. 1980. Immunocytochemical localization of enkephalin in the neostriatum of rat brain: A light and electron microscopic study. *J. Comp. Neurol.* 189:721–40

Powell, E. W., Leman, R. B. 1976. Connections of the nucleus accumbens. *Brain Res.* 105:389–403

Pribram, K. H., Chow, K. L., Semmes, J. 1953. Limit and organization of the cortical projection from the medial thalamic nucleus in monkey. *J. Comp. Neurol.* 98:433–48

Ragsdale, C. W., Graybiel, A. M. 1981. The fronto-striatal projection in the cat and monkey and its relationship to inhomogeneities established by acetylcholinesterase histochemistry. *Brain Res.* 208:259–66

Roland, P. E., Larsen, B., Lassen, N. A., Skinhoj, E. 1980. Supplementary motor area and other cortical areas in organization or voluntary movements in man. *J. Neurophysiol.* 43:118–36

Russchen, F. T., Bakst, I., Amaral, D. G., Price, J. L. 1985. The amygdalostriatal projections in the monkey. An anterograde tracing study. *Brain Res.* 329:241–57

Schell, G. R., Strick, P. L. 1984. The origin of thalamic inputs to the arcuate premotor and supplementary motor areas. *J. Neurosci.* 4:539–60

Selemon, L. D., Goldman-Rakic, P. S. 1985. Longitudinal topography and interdigitation of cortico-striatal projections in the rhesus monkey. *J. Neurosci.* 5:776–94

Szabo, J. 1962. Topical distribution of the striatal efferents in the monkey. *Exp. Neurol.* 5:21–36

Szabo, J. 1967. The efferent projections of the putamen in the monkey. *Exp. Neurol.* 19:463–76

Szabo, J. 1970. Projections from the body of the caudate nucleus in the rhesus monkey. *Exp. Neurol.* 27:1–15

Tanji, J., Evarts, E. V. 1976. Anticipatory activity of motor cortex neurons in relation to direction of movement. *J. Neurophysiol.* 39:1062–68

Tanji, J., Kurata, K. 1979. Neuronal activity in the cortical supplementary motor area related with distal and proximal forelimb movements' *Neurosci. Lett.* 12:201–6

Tanji, J., Taniguchi, K., Saga, T. 1980. Supplementary motor area: Neuronal response to motor instructions. *J. Neurophysiol.* 43:60–68

Tobias, T. J. 1975. Afferents to prefrontal cortex from the thalamic mediodorsal nucleus in the rhesus monkey. *Brain Res.* 83:191–212

Travis, A. M. 1955. Neurological deficiencies following supplementary motor area lesions in *macaca mulatta*. *Brain* 78:155–74

VanHoesen, G. W., Mesulam, M. M., Haaxma, R. 1976. Temporal cortical projections to the olfactory tubercle in the rhesus monkey. *Brain Res.* 109:375–81

VanHoesen, G. W., Yeterian, E. H., Lavizzo-Mourney, R. 1981. Widespread corticostriate projections from temporal cortex of the rhesis monkey. *J. Comp. Neurol.* 199:205–19

Vogt, B. A., Rosene, D. L., Pandya, D. N. 1979. Thalamic and cortical afferents differentiate anterior from posterior cingulate cortex in the monkey. *Science* 204:205–7

Weinrich, M., Wise, S. P. 1982. The premotor cortex of the monkey. *J. Neurosci.* 2:1329–45

Yeterian, E. H., VanHoesen, G. W. 1978. Cortico-striate projections in the rhesus monkey: The organization of certain cortico-caudate connections. *Brain Res.* 139:43–63

Ann. Rev. Neurosci. 1986. 9 : 383–413

THE MOLECULAR NEUROBIOLOGY OF THE ACETYLCHOLINE RECEPTOR

Michael P. McCarthy, Julie P. Earnest, Ellen F. Young, Seunghyon Choe, and Robert M. Stroud

Department of Biochemistry and Biophysics, University of California, San Francisco, California 94143

Introduction

The acetylcholine receptor (AChR) is the most thoroughly characterized component of the neuromuscular transduction process. Earlier reviews that summarize the structural and biochemical features of the AChR include Popot & Changeux (1984), Stroud (1983), Conti-Tronconi & Raftery (1982), and Karlin (1980). This receptor translates the binding of the neurotransmitter, acetylcholine (ACh), into a rapid increase and subsequent decrease in the permeability of the endplate membrane to the passage of cations. Inward flux of ions through the channel is passive, driven by electrochemical gradients across the receptor-containing membrane. The physiological effect is to temporarily depolarize the endplate, a response that is translated into muscular contraction in the case of a neuromuscular junction, or potentiation of electric tissue in the stacked asymmetric cells of electric organs in *Torpedo* (a marine elasmobranch) or *Electrophorus* (a freshwater teleost). The availability of acetylcholine receptors from electric tissue was a fundamental key to molecular characterization. The subunit stoichiometry of the four identified polypeptides has been unequivocally established as $\alpha_2\beta\gamma\delta$, and the funnel shape of the molecule has been well characterized with respect to position of the ion channel. Distribution of protein relative to the phospholipid bilayer and some aspects of the subunit arrangements in a quasipentameric structure around the ion channel have also been established. The genes for the four subunits that constitute the

383

0147–006X/86/0301–0383$02.00

minimal acetylcholine receptor from *T. californica* have been cloned and the amino acid sequences have been deduced. The amino acid sequence of the α subunits from *Torpedo*, bovine, and human sources shows the receptors to be close evolutionary homologues of one another. The $\alpha_2\beta\gamma\delta$ chains within any one species are more distantly related to each other than the α chain is between human and *Torpedo*, suggesting that the divergence of the four homologous chain types occurred early in the evolution of the synapse.

The neuromuscular nicotinic acetylcholine receptors are of interest as the target for autoimmune antibodies in myasthenia gravis, as the target for muscle relaxants used in surgical procedures, such as succinylcholine, and for studies related to development of the neural system. We here summarize what is known about the function of components in the system and the molecular characterization of the acetylcholine receptor, and the meaning of the structure for understanding the AChR at the molecular level.

Organization of the Postsynaptic Membrane

The postsynaptic membrane of the neuromuscular junction is predominantly characterized by its intricately folded structure and the high local concentration of AChR (Salpeter 1983). The high density of receptors is established in the postembryonic stage. Upon functional contact by a neuron at the nerve-muscle junction, previously diffuse AChRs in the postsynaptic membrane cluster at the junction, forming a tightly packed matrix of receptors (Ziskind-Conhaim et al 1984). Receptor molecules are localized at the crest of folds or invaginations of about 0.3–1.0 μm in depth, which occur across the face of the membrane. AChR density at the crest is about 10,000 AChR molecules per μm^2, or about 1 for every 100×100 angstrom area, in *T. californica* (Conti-Tronconi & Raftery 1982). An electron micrograph of a *T. californica* junction, demonstrating the relative orientation of the pre- and postsynaptic membranes and the basal lamina and the distribution of AChR molecules, is shown in Figure 1. Heuser & Salpeter (1979) observed the supramolecular organization of AChR molecules in rapidly frozen, deep-etched specimens. The advantage of this technique is that the specimen is frozen within the rotational relaxation time of a single molecule, and the structures revealed may be more biologically relevant than those seen with slower fixation procedures. Receptors occurred in strings of dimers, which associated to form four stranded strings and higher levels of organization in *T. californica*. In higher vertebrates, autoradiography and cytochemistry have shown that the AChR is concentrated at the crest (20,000–30,000 per μm^2; Conti-Tronconi & Raftery 1982), whereas acetylcholinesterase is more uniformly distributed down and throughout the fold (Hartzell et al 1976). This implies

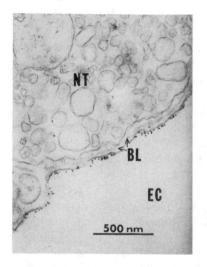

Figure 1 Fine structure of a *Torpedo californica* electroplaque synapse. The nerve terminal (NT), basal lamina (BL), and electrocyte (EC) are visible. The distribution of acetylcholine receptors was demonstrated by binding antibodies, conjugated with 5 nm gold beads (seen in the figure as black dots), to the cytoplasmic side of the AChR in the postsynaptic membrane. Glutaraldehyde-fixed and Epon-embedded tissue was stained with 1% uranyl acetate. Micrograph courtesy of E. Young.

that the invagination does not increase the receptive area, as was suggested earlier, but may function instead as a diffusion "sink" for excess ACh. Extrajunctional receptors are distributed at only 0.1–0.01 times the surface density of junctional receptors and are electrophysiologically distinct (Peper et al 1982).

In a mature synapse, the integral membrane protein subunits recognized as the minimal components of the AChR are intimately linked with other proteins. Although co-distribution of receptors and heparin sulfate proteoglycans in the basal lamina has been observed in cultured skeletal muscle cells (Bayne et al 1984), the predominant interactions appear to involve cytoplasmic membrane proteins, generally referred to as the 43 kD or v proteins. The distribution of these proteins, determined with immunoelectronmicroscopy, was seen to coincide with monoclonal antibody binding to AChR (Sealock et al 1984). The 43 kD proteins co-purify with the AChR, as seen by polyacrylamide gel electrophoresis, but can be removed by treatment with base ($\geq$ pH 10.5) (Neubig et al 1979), demonstrating that they are only peripherally associated with the membrane. As the cholinergic response can be reconstituted by purification of only the four polypeptides α, β, γ, and δ of the AChR, the function of the 43 kD proteins is not yet clear. There are apparently three types of 43 kD protein, one of which probably has phosphatase activity (Gordon et al 1983). The removal of these proteins alters the ultrastructure of the cytoskeleton (Cartaud et al 1981) and accelerates the rotational diffusion of the AChR (Rousselet et al 1982). In addition, a 43 kD protein can be chemically crosslinked to the β subunit of the AChR (Burden et al 1983). The above information is consistent with a model wherein the 43 kD proteins act largely to maintain the AChR in a closely-packed array at the synaptic crest.

X-ray diffraction reveals that the postsynaptic membrane has a lipid bilayer thickness of 40 ± 1 Å between phosphatidyl head groups (Ross et al 1977). The funnel-shaped receptor extends approximately 55 Å above the bilayer surface and provides an insulating environment for ions that enter the central ion channel (Klymkowsky & Stroud 1979). The lipid environment of *T. californica* electrocytes shows a fairly typical composition, approximately equimolar in cholesterol and phospholipids. It may be significant that the AChR-rich membranes contain more phosphatidyl ethanolamine, phosphatidic acid, and cholesterol, while the AChR-poor membranes have more phosphatidyl serine, sphingomyelin, and lysophosphatidyl choline (Gonzalez-Ros et al 1982). This study also demonstrated that the *T. californica* electrocyte lipid acyl chains contain an unusually high percentage of docosahexaenoic acid (22:6), although no significant differences were detected in the various fractions. *T. marmorata* electrocyte lipids have also been shown to contain high amounts of 22:6 fatty acids (Popot et al 1978).

Shape and Size of the AChR Molecule

The overall three-dimensional structure of the receptor has been analyzed by a number of methods. A three-dimensional model of the AChR is seen in Figure 2. The AChR was shown to be elongated by hydrodynamic measurement of the Stoke's radius of receptor-detergent complexes (Meunier et al 1972), assuming a receptor mass of 250,000 daltons. In neutron scattering studies, where the detergent component of solubilized AChR can be largely factored out, the receptor molecule was found to be cylindrical, with a radius of gyration of 4.6 nm (Wise et al 1979). The long axis of the receptor is perpendicular to the plane of the membrane. Small angle X-ray diffraction studies of oriented membranes determined that the overall length of the AChR was 11 nm, extending 5.5 nm beyond the extracellular surface of the membrane and about 1.5 nm into the cytoplasm (Ross et al 1977). The protrusion of the AChR molecule beyond the membrane was also seen in lateral views of receptor vesicles in the electron microscope (Klymkowsky & Stroud 1979). When viewed from above, the extracellular surface of the AChR appears as a rosette 80–90 Å in diameter. Each rosette contains a stained central pit (Cartaud et al 1978, Heuser & Salpeter 1979) of about 25 Å in diameter. The best evidence for this as the location of the ion channel is that it can be filled with stain to a depth of about 114 Å (Kistler et al 1982). This channel has been shown to conduct ions of less than 6.5 Å diameter at rates comparable to the rate of sodium ions (Maeno et al 1977, Dwyer et al 1980), defining a minimum diameter for the most constricted portion of the channel. In a series of experiments in which transport of different diameter mono- and divalent cations through the channel was compared to free diffusion rates in solution, it was seen that the channel provides an environment very similar to the bulk aqueous phase (Lewis & Stevens 1983), showing that the channel is water-filled in its open state.

Although little is known of the tertiary structure of the AChR, the overall secondary structure is fairly well characterized. The existence of long α-helices oriented perpendicularly to the membrane was deduced from small angle X-ray diffraction studies that showed characteristic 5.1 Å meridional and 10 Å equatorial peaks (Ross et al 1977). Circular dichroism studies of solubilized AChR molecules from *T. nobiliana* suggested a secondary structure that was 34% α-helix, 29% β-structure (including turns), and 37% random coil (Moore et al 1974), while comparable studies with *T. californica* indicated 20% α-helix, 50% β-structure, and 30% random coil (Mielke et al 1984). Resonance Raman spectroscopy of AChR molecules reconstituted into artificial membrane vesicles showed that the receptor from *T. marmorata* was 25% α-helix (plus 14% disordered α-helical ends) and 34% β-sheet (Aslanian et al 1983). Given the high degree of sequence

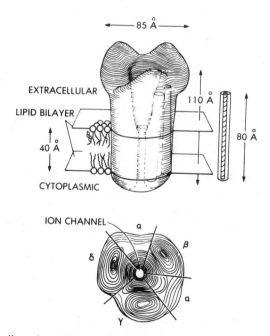

Figure 2 Three-dimensional model of AChR from *Torpedo californica*. The subunit locations around the central well are tentative assignments.

homology among these species, the secondary structure of the folded receptor is expected to be essentially the same for each of them as well. The membrane-spanning regions of the receptor are almost certainly α-helical, whereas most of the extra-membranous regions of the AChR are probably composed of β-sheet and random coil as shown by Fourier transform analysis of sequence amphipathicity (Finer-Moore & Stroud 1984).

The quarternary structure of the AChR is a quasi–five-fold pentamer composed of four subunit types with the stoichiometry $\alpha_2\beta\gamma\delta$. This arrangement is evolutionarily conserved from *Torpedo* and *Electrophorus* to mammalian AChR (Raftery et al 1983), although the molecular weights of the individual subunits vary slightly. The molecular weight of the *T. californica* AChR, calculated from the molecular weights of the subunits determined by gel electrophoresis and the known subunit stoichiometry, is 248,000 daltons (Weill et al 1974, Reynolds & Karlin 1978). This is apparently an underestimate (which can occur with glycoproteins), as a molecular weight of 268,000 daltons is suggested from the amino acid compositions of the subunits determined by cDNA analysis (Popot & Changeux 1984). In addition, the mature receptor is glycosylated (Vandlen et al 1979, Anderson & Blobel 1981) and phosphorylated (Vandlen et al 1979, Huganir et al 1984), and contains fatty acids linked to the α and β subunits (Olson et al 1984), suggesting a final molecular weight of at least 290,000 daltons, since the sugar residues are estimated to total 20,000 daltons (Popot & Changeux 1984). As each subunit was shown to be exposed on both sides of the membrane by protease susceptibility (Wennogle & Changeux 1980, Strader & Raftery 1980) and to be externally glycosylated (Anderson & Blobel 1981), and accessible to hydrophobic probes (Middlemas & Raftery 1983), it is now clearly established that they are quasisymmetrically positioned in the native receptor, rather as staves in a barrel, each occupying a $\sim 72°$ segment (Fairclough et al 1983). Although the subunits show 40–50% sequence homology with one another, they cannot substitute for one another. In in vitro expression systems, a significant response to ACh was seen only when the cDNA for all four subunits was available for expression (Mishina et al 1984). The arrangement of the individual subunits in the receptor is not known. The order $\alpha\gamma\alpha\beta\delta$ was suggested from the angles between β and δ subunits in artificially-generated trimers (Wise et al 1981), while the α subunit was visualized with avidin-binding to biotinylated toxin (Holtzman et al 1982). A different subunit arrangement of $\alpha\beta\alpha\gamma\delta$ was suggested on the basis of the binding of monoclonal antibodies, raised against the various subunits (Kistler et al 1982), and the position of α-toxin binding relative to dimer contacts (Zingsheim et al 1982a) as observed in the electron microscope. It is widely accepted that the α subunits are separated by one other subunit. A final state of association observed between native AChR molecules in *Torpedo* species is dimerization, maintained by a disulfide bond between δ subunits. The amino acid residues involved have not been explicitly identified, but are probably the penultimate cysteines located at the carboxy-terminal ends of each δ chain (Wennogle et al 1981). Dimers have been observed to form parallel (Kistler & Stroud 1981, Bon et al 1984) or anti-parallel (Fairclough et al 1983, Brisson & Unwin 1984) doublets when tightly packed in the membrane, but they may interact more flexibly when diluted in membrane patches (reviewed in Popot & Changeux 1984). Dimers are only found in the AChR from *Torpedo* and *Narcine*, and do not occur in *Electrophorus* or vertebrates. Their significance is unknown; a number of workers have shown monomers and dimers to be functionally identical (Anholt et al 1980, Boheim et al 1981, Fels et al 1982), while other workers have shown that they are not (Schindler et al 1984, Chang et al 1984).

Electron microscopic images of individual particles have been analyzed by image analysis and statistical correlation techniques that rely upon maximizing the overlap of reproducible structural features in the molecules. Summed and averaged views as seen in the electron microscope describe a receptor surface of varying density, with the ridges and depressions suggested to represent the contribution of individual subunits (Zingsheim et al 1982b, Bon et al 1984), and the binding of α-toxin to the surface of the receptor has been localized, presumably to the α subunit. Studies of two-dimensional receptor "crystal" lattices formed in the plane of the membrane (Ross et al 1977) or in tubular crystalline membranes (Kistler & Stroud 1981, Brisson & Unwin 1984) allow the image of the AChR to be enhanced by Fourier transformation and image filtration. The results of such studies look rather similar to those based on single receptors. In studies on the better ordered tubes from *T. marmorata* to a nominal resolution of 30 Å, the surface of the receptor appeared to be a regular pentagonal rosette, with an asymmetric distribution of mass around the periphery (Brisson & Unwin 1984). Similar results are seen in studies of thin three-dimensional AChR crystals analyzed in the electron microscope (M. P. McCarthy, A. K. Mitra, S. Choe, R. M. Stroud, manuscript in preparation), whereas earlier studies described a somewhat more horseshoe-shaped projection, probably reflecting the expected asymmetric mass distribution between the subunits. Although these studies map the surface of the receptor rather well, detailed analysis of the three-dimensional structure of the AChR must await the growth of large, well-ordered crystals and the completion of high resolution X-ray diffraction studies.

Electrophysiological Action of the AChR

Descriptions of the response of the AChR to ACh (or other agonists) follow two general forms: electrophysiological measurements, primarily of vertebrate neuromuscular junctions, or biochemical characterizations of purified receptor from the electric organs of certain fish. While we concentrate on biochemical approaches to defining AChR structure and action, the following is a summary of the relevant electrophysiological data.

A typical presynaptic action potential at a neuromuscular junction induces the release of $0.2–3.0 \times 10^6$ ACh molecules into the synaptic cleft. Following a 300 μsec lag, about 2.5×10^5 channels will transiently open in the postsynaptic membrane, generating an endplate current of around -400 nA (Peper et al 1982). Overall open times for channels vary [extrajunctional AChR molecules stay open longer (Neher & Sakmann 1976)] but average about 1 msec at a membrane holding potential of -100 mV in frog neuromuscular junction (Magleby & Stevens 1972). This duration is sufficient to allow the net conductance of 10,000 Na^+ ions through each channel. Opening of the channel is caused by the binding of two or more molecules of ACh per receptor, as channel opening varies roughly with the square of the agonist concentration (Adams 1975, Dionne et al 1978), with a Hill coefficient of 1.97 ± 0.06 (Neubig & Cohen 1980). Average open times are also influenced by temperature and membrane potential; hyperpolarization of the membrane lengthens average open times (Magleby & Stevens 1972). Channel conductance values, however, average about 25 pS regardless of conditions in a variety of vertebrate neuromuscular junctions (Adams 1981).

The advent of patchclamp techniques (pioneered by Neher & Sakmann 1976) allows a uniquely detailed view of receptor function in the form of single channel recordings. Mean open times of individual AChR may be separated into two general classes in frog muscle, long-term (average 10 msec) and short-term (average 0.15 msec) (Colquhoun & Sakmann 1981). The authors suggested that the brief open forms may correspond to mono-liganded AChR. In addition, short non-conducting gaps occur during bursts that are too brief to be caused by agonist dissociation and subsequent rebinding. This flickering behavior (or Nachschlag) varies among species. In frog muscle, at low agonist concentration, there is an average of 3 gaps/burst, with a mean gap duration of around 40–70 μsec (Colquhoun & Sakmann 1981), while in snake twitch muscle, gaps are rarer (0.3 gap/burst) but longer (average duration of 200 μsec) (Dionne & Leibowitz 1982). The number of gaps per burst and mean open times are largely unaffected by membrane potential, but gap duration is shortened with hyperpolarization (Leibowitz & Dionne 1984). These gaps have been claimed to describe excursions of doubly-liganded receptor into transiently nonconductive forms and may represent isomerizations that preceded the dissocation of bound ACh (Land et al 1984). Contrary to the findings of earlier noise-analysis studies (Colquhoun et al 1975), conductance does not vary with different agonists (Gardner et al 1984); this suggests a simple two-state, open-closed system. In this case, the energy barrier between the two forms may not be large, as spontaneous openings of mouse AChR channels were shown to occur at infrequent intervals (Jackson 1984). In addition, the binding energy from antagonists such as the curare analogue, tubocurarine, can induce channel opening (Trautman 1982). However, the occurrence of "sublevel" conductance states, typically at lower temperatures or hyperpolarizing conditions, in embryonic rat muscle (Hamill & Sakmann 1981) and chick myoballs (Aracava et al 1984) as well as in the presence of curare (Trautman 1982), demonstrates that the AChR is capable of adopting several open states.

Prolonged exposure of endplates to greater than micromolar concentrations of ACh results in a decrease of conductivity (Katz & Thesleff 1957). This phenomenon is known as desensitization and it varies in extent and duration with different agonists. It is not caused by direct blockage of the ion channel by agonist (Sakmann et al 1980), although channel blockage by agonists may occur (Sine & Steinbach 1984). Desensitization may involve agonist binding to sites different from those involved in agonist activation. In frog muscle, desensitization rates (at 20 μM ACh) have been estimated at about 2 sec^{-1}; recovery is slightly faster (Sakmann et al 1980). Much slower rates also correlated with desensitization have been observed, suggesting that the phenomenon associated with desensitization may involve more than one state or process (Adams 1981). Many drugs and local anesthetics initially suggested to increase desensitization rates have now been classed as noncompetitive channel blockers on the basis of electrophysiological analysis (Peper et al 1982). Recovery rates from drug-enhanced desensitization appear to be drug insensitive (Magazanik & Vyskocil 1973), implying that drug dissociation is followed by a slow isomerization back to the resting state.

Electrophysiological studies on *Torpedo* AChR have been limited. Single channel recordings of *T. marmorata* AChR reconstituted into planar lipid bilayers showed mean channel open times of about 3 msec, similar to those seen in vertebrate endplates, with about four-fold higher conductance. The latter may be due to the high experimental ionic strength (Boheim et al 1981). Analogous to observations at the vertebrate neuromuscular junction (Colquhoun & Sakmann 1981), *T. californica* AChR reconstituted into planar lipid bilayers exhibited two classes of open states, averaging 2.8 msec or 0.8 msec (Labarca et al 1984). Furthermore, a difference in conductance dependent upon the state of association of the AChR has recently been described (Schindler et al 1984); the conductance of the monomer $(20 \pm 2$ pS) falls within the range typical of vertebrate neuromuscular junctions, whereas the single channel conductance of a dimer appears two-fold higher. The authors suggested that the increased conductance of the associated species (which is not dependent upon covalent interactions) may reflect cooperative behavior important under physiological conditions. However, as noted above, a number of other investigators have failed to detect functional differences between *Torpedo* monomer and dimer preparations (Anholt et al 1980, Boheim et al 1981, Fels et al 1982).

Biochemistry of Agonist and Antagonist Binding

The number and importance of different agonist binding sites on the AChR remains a matter of some controversy. A majority of studies indicate the presence of two, high-affinity ACh binding sites per AChR monomer in *Torpedo* localized on the α subunit (reviewed in Conti-Tronconi & Raftery 1982, Popot & Changeux 1984). These sites may not be initially equivalent. The affinity reagents 4-(*N*-maleimido)benzyltrimethyl ammonium iodide (MBTA) and bromoacetylcholine (BAC) were found to label only one α subunit per AChR in *T. californica* AChR (Weill et al 1974, Damle et al 1978), whereas *p*-(trimethylammonium) benzenediazonium fluorobate labeled both α sites (Weiland et al 1979). In *T. marmorata* AChR, only one bromoacetylcholine binding site was reactive per monomer at 4°C, while both were labeled at 23°C (Wolosin et al 1980). Two classes of α subunits that differ in the extent of glycosylation have been observed in *T. californica* AChR (Conti-Tronconi et al 1984). These differences may contribute to the nonequivalence of the high affinity agonist binding sites. Although these sites may not be identical in structure, both are involved in channel opening. Even when one site is blocked by MBTA, ion flux could be induced by agonist binding to the second high-affinity site (Delegeane & McNamee 1980). ACh binding to these sites was shown to be weakly cooperative under desensitizing conditions (Fels et al 1982); models involving initially nonequivalent sites characterized by concave Scatchard plots (Prinz & Maelicke 1983) and long-lived, variable affinity states (Chang et al 1984) have been proposed. The Hill coefficient for channel opening is 1.97 ± 0.06 (Neubig & Cohen 1980), suggesting that the binding of two (or more) agonists is required for channel opening. The functionally significant number of binding sites is more difficult to determine, as most assays that correlate extent of binding with relevant operations such as ion flux are too slow, and thus probably describe the activities of largely desensitized receptor. One difficulty has been the contradictory values obtained for equilibrium carbamylcholine binding to the resting ($K_d = 30$ μM) and desensitized states of purified receptor (K_d = 10–100 nM) (Weiland et al 1977, Quast et al 1978) and those determined to induce half-maximal activity in electrophysiological studies of frog muscle (0.5–1 mM) (Dreyer et al 1978, Dionne et al 1978). Improved ion flux measurements of purified receptor, utilizing $^{22}Na^+$ (Neubig & Cohen 1980) and $^{86}Rb^+$ (Hess et al 1979) flux or Tl^+ quenching (Moore & Raftery 1980) yielded dose response curves similar to those determined by electrophysiological analysis. Complex kinetic schemes have been postulated to account for the differences observed in equilibrium and kinetic measurements (Neubig & Cohen 1980, Hess et al 1983), and the possibility that the high affinity binding sites identified by MBTA are involved with desensitization while low affinity sites located on other subunits are responsible for channel opening has been suggested (Dunn & Raftery 1982b, Dunn et al 1983). These researchers observed changes in the fluorescence of a bound, extrinsic probe indicative of a conformational change; this occurred with the time scale of channel opening even when the high-affinity sites were blocked by BAC and the receptor was in the desensitized state. Additional regulatory binding sites for agonists such as suberyldicholine on *Electrophorus* AChR (Pasquale et al 1983) and a voltage-dependent inhibitory ACh binding site in *Electrophorus* (Takeyasu et al 1983) and *T. californica* (Shiono et al 1984) have been described. The location of these sites is not identified, but direct channel blockage, which is consistent with some electrophysiological measurements (Sine & Steinbach 1984), was assumed not to occur on the basis of model considerations.

The relatively slow transformation from low to high affinity states of the AChR seen in biochemical studies is thought to parallel the desensitization process observed electrophysiologically. Desensitization rates determined by changes in intrinsic protein fluorescence differ for the agonists suberyldicholine, carbamylcholine, and ACh in *T. marmorata*, but the final fluorescent state in each case is the same, implying that the ultimate form is independent of the ligand (Barrantes 1978). Desensitization may proceed from monoliganded receptors, as seen in fluorescent studies of *Torpedo* and *Electrophorus* (Bonner et al 1976) and by kinetic analysis of ion flux measurements (Dunn et al 1980, Hess et al 1983). Two-step desensitization kinetics characterized by millisecond and second rates have been observed in reconstituted *T. californica* vesicles (Walker et al 1982), again in agreement with electrophysiological data (Adams 1981). Local anesthetics (Weiland et al 1977) and antagonists (Quast et al 1978) have been shown to accelerate agonist-induced rates of desensitization in *T. californica*, although Covarrubias et al (1984) did not see an enhancement in desensitization induced by competitive antagonists. Suberyldicholine binds more tightly to desensitized receptor than ACh, while carbamylcholine binds more weakly, a relationship that extends to the average open times sustained by these agonists (Spivak & Albuquerque 1982).

Competitive antagonists such as cobratoxin and α-bungarotoxin (αBgTx) bind noncooperatively to receptor in B3CH-1 cells, but with Hill coefficients less than one, implying nonequivalent sites (Sine & Taylor 1981). However, in *T. californica*, both sites appeared equivalent in membrane-bound and detergent-solubilized AChR (Ellena & McNamee 1980). Snake toxins were found to bind with high affinity (average K_d less than 0.1 nM; Weber & Changeux 1974) and their binding affinity seemed to be unaffected by desensitization (Weiland & Taylor 1979). They act by competitively blocking agonist binding and do not appear to interact with the ion channel. Curare, once considered a simple antagonist, appears to function as a partial agonist (similar to decamethonium)—both competing for (and activating) the agonist binding site and blocking open channel conductance (Trautman 1982).

The observed effect of agonist binding is channel opening and desensitization. On the molecular level, these processes involve conformational transitions of the AChR. ACh binding released 4–6 Ca^{2+} ions (Chang & Neumann 1976) or 6–12 terbium ions (Rubsamen et al 1978) from purified *Torpedo* AChR. Agonist binding quenched intrinsic receptor fluorescence, at rates that suggested desensitization (Barrantes 1978, Bonner et al 1976, Kaneda et al 1982). Conversely, the fluorescence of noncovalently bound ethidium (Schimerlik et al 1979) or the covalently bound fluorophore 5-(iodoacetamide) salicyclic acid (Dunn et al 1980) was enhanced by agonist binding. The sensitivity of a hydrophobic fluorophore, bound covalently to the β and γ chains of *T. californica*, to quenching by nitromethane was decreased by desensitizing concentrations of carbamylcholine, whereas α-bungarotoxin had the opposite effect (Gonzales-Ros et al 1983). These studies do not allow definite conclusions to be formed about the conformational states of the AChR, but are consistent with a simplistic model wherein the agonist-binding site becomes less exposed to solvent upon occupancy, while the rest of the receptor becomes more accessible except for the membrane-spanning regions. Fluorescent studies such as these do not allow the extent of conformational transitions (local or global) to be estimated, although agonist binding was shown to increase the β-structure contribution in AChR CD spectra (Mielke et al 1984). Tritium–hydrogen exchange experiments showed that the overall solvent accessibility of the resting and desensitized AChR were identical, whereas α-bungarotoxin binding restricted accessibility (M. P. McCarthy and R. M. Stroud, unpublished data). α-Bungarotoxin also induced the uptake of 4–6 Ca^{2+} upon binding to receptor (Chang & Neumann 1976), enhanced desensitization rates (Quast et al 1978), and modified the behavior of a number of agonists and noncompetitive blockers (Spivak & Albuquerque 1982). Whether these effects are due to a large-scale conformational change or to local changes involving blockage of an essential binding site is not known.

Location of Binding Sites

The location of agonist binding sites remains a topic of great interest. It has been known for several years that covalent modification of reduced α subunits by affinity reagents permanently blocks agonist (and some antagonist) binding sites (Damle & Karlin 1978, Damle et al 1978). Recently, the target cysteine residue(s) have been identified as Cys 208[1] and perhaps Cys 209 (Kao et al 1984), amino acids unique to the α subunit in terms of sequence homology (Noda et al 1983a). The requirement for prior reduction suggests the existence of a cystine disulfide near the agonist binding site in the native AChR. Because cysteines 208 and 209 follow one another in the sequence, Karlin and co-workers feel that these residues are unlikely candidates, as bridging between adjacent cysteines has never been observed in proteins; however, disulfide bonds are seen in crystal forms of a dipeptide, cysteinylcysteine (Capasso et al 1977), and are not otherwise energetically forbidden (Mitra & Chandrasekaran 1984). Instead, Kao et al (1984) suggested possible disulfide bonds with Cys 130 and 144 of the α subunit.

A sense of this binding site arises from analysis of the effects of agonists in terms of their structure. The bulky, quarternary amine head group of the agonist probably binds to a sterically-restricted, anionic region on the receptor, while hydrogen bond formation between ACh and AChR lend additional binding energy (reviewed in Spivak & Albuquerque 1982). Covalent cholinergic ligands have also been found to bind to the other subunits in unreduced AChR (Hucho et al 1976, Witzemann & Raftery 1977), perhaps reflecting low-affinity binding sites.

By definition, the competitive antagonist binding site overlaps the agonist binding site. In the case of snake neurotoxins (M.W. 7000–8000), this binding "site" is predicted to cover 20 × 30 Å of the receptor surface (Low 1979, Kistler et al 1982, Stroud 1983, Fairclough et al 1983). Binding of these toxins to the surface has been indirectly visualized in electron micrographs by complexing the toxins with gold-labeled antitoxin antibodies (Klymkowsky & Stroud 1979) and avidin binding to biotinylated toxin (Holtzman et al 1982), in the dimension perpendicular to the membrane by X-ray diffraction (Fairclough et al 1985), and in projection

---

[1] The numbering of amino acids in this review follows the consensus alignment of homologous residues of Fairclough et al (1983).

using difference image autocorrelation methods (Zingsheim et al 1982a). Snake toxins bind to the periphery of the receptor surface, away from the central pit, and are probably at least 50–60 Å apart from one another in AChR, with two toxins bound, as was seen in fluorescence energy transfer studies (Johnson et al 1984). As the snake toxins are known to competitively block agonist binding to the high affinity site on the α chain, this is presumably their binding site as well. However, covalent cross-linking studies have not produced a consistent picture. α-Bungarotoxin derivatives have been shown to label both the α and δ subunits in *T. californica* (Witzemann et al 1979) but in situ studies to crosslink all four subunits (Nathanson & Hall 1980). Given the size of the snake toxins, binding sites that involve more than one subunit are probable. Interaction of spin-labeled neurotoxin with the AChR suggested that Lys-27 of the toxin was in the vicinity of a disulfide on the AChR (Tsetlin et al 1982). Lys-27 has been shown to be about 16 Å from the high-affinity agonist binding site (Fairclough et al 1983), but which α subunit cysteines form the disulfide is not known. Detailed descriptions of agonist and antagonist binding to receptor will require X-ray crystallographic analysis of receptor-ligand co-crystals.

Sulfhydryl Groups

A disulfide bond has been identified that, upon reduction and subsequent covalent modification, modulates agonist-stimulated ion flux by altering the affinity of the agonist for its binding site (Damle & Karlin 1978, Walker et al 1984). Disulfide reduction by dithiothreitol (DTT) alone was found by Ben-Haim et al (1975) to affect ion channel conductance, and by Moore & Raftery (1979) to alter the transition from the low-affinity to high-affinity state. Sulphonation was shown by Steinacker (1979) to increase miniature end-plate potential frequency and amplitude in cutaneous pectoris muscle of *Rana pipiens*. On the basis of competition with ACh, Steinacker proposed that the sulphonation site was in the vicinity of the ACh binding site. Affinity-labeling experiments using MBTA (Karlin 1980) suggested that the distance between the anionic sites where the quaternary ammonium of acetylcholine binds and the sulfhydryl groups where MBTA attaches is about 1 nm. Walker et al (1984) demonstrated that reduction with DTT and subsequent alkylation with *N*-ethylmaleimide (NEM) or the more hydrophobic *N*-benzylmaleimide (NBM) dramatically decreased the flux response of asolectin-reconstituted AChR. In addition, alkylation by NEM was competitive with alkylation by the affinity reagent BAC, suggesting that NEM and NBM react with the same sulfhydryls as the affinity reagents MBTA and BAC. The above experiments confirm the existence of one or more sulfhydryl groups near the agonist-binding site. The cysteine residues that form disulfide bridges have for the most part not been identified. Two cysteines in the N-terminal hydrophilic domain of the receptor are conserved in all four subunits and in all species tested. These cysteines (α 130 and α 144) can easily be in close proximity (Noda et al 1983a), as revealed by the identified turns in the β-pleated sheet domain (Finer-Moore & Stroud 1984), supporting the contention that these two cysteines form a disulfide within each of the receptor subunits. It is thought that the two other cysteines on the α subunit hydrophilic domain participate in a disulfide linkage, based on the ability of the affinity labels MBTA and BAC to label residues 208 and 209 only after DTT reduction (Kao et al 1984). If Cys 130 and Cys 144 form a disulfide in the α subunit, the requirement for prior reduction suggests that either Cys 208 and Cys 209 form a disulfide between themselves, or that DTT treatment reduces other disulfides that block the accessibility of Cys 208 and 209 to affinity reagents.

Mishina et al (1985) have recently shown that mutation of *any* of the above-mentioned cysteines (130, 144, 208, or 209 on α) to serines completely eliminates the responsiveness of the expressed receptor to ACh. A mutation in either Cys 130 or Cys 144 completely prevents αBgTx binding, while a mutation in Cys 208 or Cys 209 reduces αBgtx binding to 28–39% of the control. These results may indicate that disulfide bonds involving residues 130 and 144 are critical for maintaining the tertiary structure of the hydrophilic domain that participates in toxin-binding.

There is evidence for one or more reactive sulfhydryl groups in the nonreduced receptor. Moore & Raftery (1979) found that para-chlormercuribenzoate (PCMB) modified the receptor in either the low-affinity or high-affinity state and blocked interconversion between states. This effect was seen in the absence of DTT treatment. Huganir & Racker (1982) demonstrated that the ability of various maleimides to inhibit agonist-stimulated ion transport was basically proportional to the hydrophobicity of the maleimide. This phenomenon was examined in detail by Walker et al (1984), who found that 0.4 mM NBM blocked 50% of the carbamylcholine-stimulated ^{86}Rb$^+$ influx in *Torpedo* AChR reconstituted into asolectin. The concentrations of NEM and iodoacetamide required to block 50% of the flux response were about 60 and 400 mM, respectively (based on their Figure 10). Toxin competition assays performed on NBM-modified membranes established that the inhibitory effect of NBM modification was not correlated with impaired ligand binding. These results, along with those of Huganir & Racker (1982), implicate a hydrophobic site associated with a sulfhydryl group that is capable of modulating the function of the ion channel.

Not only is there evidence that sulfhydryls can modulate channel activity, but recent evidence has shown that a conformational change in the receptor alters susceptibility of sulfhydryls to alkylation. Otero & Hamilton (1984) demonstrated that labeling of different subunits of AChR-rich membranes from *T. californica* with [^{3}H]NEM changes upon addition of cholinergic ligands. The native receptor, with or without carbamylcholine, was labeled only on the β subunit, whereas reduced receptor was labeled on the α and β subunits. The addition of carbamylcholine, and to a lesser extent choline, αBgTx and curare to the reduced receptor resulted in a decreased labeling of the α subunit, with no change in the labeling of the β subunit. Gallamine, which has been shown to block agonist-induced ion flux both as a competitive antagonist and in an alternate voltage-dependent fashion, increased NEM labeling of the α subunit up to twice the value of the reduced control, suggesting that gallamine makes additional disulfides accessible to DTT reduction. The small decrease in α-labeling caused by αBgTx, relative to the large decrease in α-labeling caused by carbamyl-choline, suggested to these researchers that steric hindrance of a reducible disulfide in the presence of cholinergic ligands is not responsible for the observed differences in NEM-labeling. Their observations are consistent with ligand-induced conformation changes altering accessibility of sulfhydryl groups.

Noncompetitive Blockers

Compounds that block or modulate the agonist-stimulated increase in cation permeability, but do not bind at the acetylcholine site (at pharmacologically relevant concentrations) include amine local anesthetics, histrionicotoxin (from the poison-dart frog of South America), the psychoactive tranquilizer phencyclidine, nonionic detergents, the antipsychotic chlorpromazine, and aliphatic alcohols (Spivak & Albuquerque 1982). These diverse compounds are not likely to have the same binding site or mechanism of action. There appear, however, to be several properties common to noncompetitive blockers.

Noncompetitive blockers stabilize the AChR in the desensitized state. The affinity of the AChR for some noncompetitive blockers is enhanced by binding of agonists and some antagonists; binding of [^{3}H]phencyclidine to *Torpedo* AChR-rich membranes was 10^3–10^4-fold faster when carbamyl-choline was added concurrently (Oswald et al 1984). Binding of local anesthetics to saturable site(s) on the AChR stabilized the receptor in a state of high affinity for agonist (Heidmann & Changeux 1979, Boyd & Cohen 1984). Agonists and local anesthetics (but not competitive antagonists) induced the transition of the AChR to the desensitized state (Covarrubias et al 1984). It is unclear whether the stabilization of the desensitized state is the primary pharmacological effect of noncompetitive blockers.

Noncompetitive blockers bind to multiple sites on AChR membranes. A saturable, high-affinity, allosteric binding site for noncompetitive blockers has been proposed (Heidmann et al 1983, Oswald et al 1983, Haring & Kloog 1984). Two classes of binding sites for [³H]phencyclidine on acetylcholine receptors from *T. ocellata* membranes have been identified: one high-affinity ($K_d = 6$–9 μM) site per receptor, and two low-affinity ($K_d = 85$ μM) sites per receptor (Haring & Kloog 1984). Binding of [³H]phencyclidine to the low-affinity sites could be completely blocked by the presence of [¹²⁵I] α-bungarotoxin. The results were consistent with the low-affinity site being the agonist-binding site, and evidence was presented to suggest that at close to millimolar concentrations, noncompetitive blockers can bind to the acetylcholine site and promote desensitization. Heidmann et al (1983) found that high-affinity binding of phencyclidine, meproadifen, and the detergent Triton X-100 could be competitively blocked by perhydrohistrionicotoxin, with a fixed stoichiometry of one site per receptor monomer, while binding of chlorpromazine and trimethisoquin was less sensitive to perhydrohistrionicotoxin, had lower affinity, and showed 10 to 30 sites per monomer. The number of low affinity sites was linearly dependent on the lipid-to-protein ratio in reconstituted membranes, suggesting that these low affinity sites are at the lipid-protein interface. The affinity of any noncompetitive blocker for the low-affinity site(s) is a function of the molecule's solubility in the lipid bilayer, or more importantly its solubility at the lipid-protein interface, and of the charge on the molecule. A third "nonsaturable" binding site in the bulk phase lipid was also proposed. The high lipid-solubility of several of the noncompetitive blockers suggests that even if there is one saturable high-affinity binding site per receptor monomer that is responsible for blocking agonist-stimulated ion flux, the lipid membrane provides the receptor with a locally high concentration of drug, which may have its own effects on receptor function through specific effects on the lipid-protein interface, or by affecting the local charge environment.

Recent experiments using inside-out cell patches and both tertiary and quaternary amine noncompetitive blockers demonstrated that a positive charge on the noncompetitive blocker is necessary for pharmacological effect at micromolar concentrations, and that the drug interacts with the AChR from the extracellular side of the membrane (Aracava et al 1984, Aracava & Albuquerque 1984). That the positively charged form of a tertiary amine local anesthetic binds with high affinity to reconstituted AChR and blocks ion flux has been corroborated by Earnest et al (1984) and Blickenstaff & Wang (1985).

Electrophysiological studies have shown that noncompetitive blockers do not affect single-channel conductance, but do decrease channel lifetime; the effect is generally voltage-dependent (Koblin & Lester 1979,

Albuquerque et al 1980, Lambert et al 1983), although voltage-independent blockage has been seen recently by Ribera et al (1985). The observation that the binding of some noncompetitive blockers to site(s) on the AChR is dependent upon the applied voltage has led to their further classification as "open channel blockers." Although some of the data can be explained most easily by direct occlusion of the channel by noncompetitive blockers, there is no direct evidence for binding site(s) for these compounds within the ion channel.

Attempts have been made to identify local anesthetic binding sites on the AChR using photoaffinity labels (Lester et al 1980, Oswald & Changeux 1981a,b, Heidemann & Changeux 1984, Muhn et al 1984) and alkylating derivatives of local anesthetics (Kaldany & Karlin 1983). Quinacrine mustard was shown by Kaldany & Karlin to block channel function at low concentrations (10 μM) without blocking agonist binding. Tritiated quinacrine mustard labeled the α and β chains of the receptor under these conditions. Kaldany & Karlin's results implicating the α and β chains as the sites of functionally significant local anesthetic binding differ from the results of Oswald & Changeux (1981a), which implicate the δ chain. In the latter the label was a radioactive photoaffinity derivative of the local anesthetic, trimethisoquin (5-azido[^{3}H]trimethisoquin), and its incorporation into the α chain was inhibited by cholinergic agonists and antagonists. Binding to the δ subunit was competitive with nonlabeled trimethisoquin and with [^{3}H]phencyclidine. [^{3}H]chlorpromazine irradiated with UV light showed an agonist-dependent rapid covalent incorporation into all five subunits, suggesting a site within the ion channel that becomes accessible when the channel is open (Oswald & Changeux 1981b, Heidmann & Changeux 1984). Photolabeling experiments of amphiphilic compounds such as local anesthetics have the intrinsic problem that rapid exchanges among lipids, proteins, and drugs in the lipid phase make identification of a specific binding site difficult. Muhn et al (1984) have attempted to resolve the interaction of AChR with [^{3}H]triphenylmethylphosphonium (a lipophilic cation thought to block the AChR ion channel) in millisecond-to-second time scales using a stopped-flow apparatus and a high-energy pulse laser. In the absence of cholinergic ligands, most of the label was incorporated into the α subunit. With cholinergic ligands, labeled anesthetic was incorporated into the δ and somewhat into the β subunits, although the α subunit still contained the majority of the label.

Discrepancies in labeling patterns could be due to differences in the time scale of interactions, or variability in the effects of the ligands on receptor function. (Functional effects of the label under investigation on agonist-stimulated ion flux were examined only by Kaldany & Karlin.) Moreover, it is possible that there are distinct binding sites for different noncompetitive inhibitors. Although binding sites for noncompetitive blockers may be identified, further research is required before the pharmacologically relevant binding site(s) is identified or understood.

Sequences of the Subunits

The full sequences of all four subunits in the AChR have now been determined and have proved uniquely important in providing structural detail and insights into function. Sequencing began at the amino acid level. Devillers-Thiery et al (1979) sequenced the first 20 amino acids at the N-terminus of the α chain of AChR from $T.$ $marmorata.$ Raftery et al (1980) determined the first 56 amino acids of each subunit from $T.$ $californica,$ and by quantitative comparison proved the stoichiometry of the chains as $\alpha_2\beta\gamma\delta.$ There is 100% correspondence between the first 20 amino acids of the α chain from $T.$ $marmorata$ and $T.$ $californica.$ Raftery et al (1980) further showed that all four subunit types have 35–50% sequence homology with one another in this region. Using synthetic oligonucleotide primers for short stretches of amino acids, Ballivet et al (1982) with the γ chain and Sumikawa et al (1982) with the α chain reported cloning and sequencing of the entire precursors of these chains, including the N-terminal signal sequences deduced from cDNA clones. Subsequently, Numa and his colleagues obtained the sequences for $T.$ $californica$ α subunit (Noda et al 1982), and then by similar techniques the β and δ subunit precursors (Noda et al 1983b). Sequencing of the γ subunit by Claudio et al (1983) yielded a proposed topological model for the chain that contained four very hydrophobic stretches about 26 amino acids in length, long enough to span the lipid bilayer in α helical conformation. Cloning of the $T.$ $marmorata$ α chain cDNA by Giraudat et al (1982) led to a similar topological model, and to the proposal that the most polar of the four hydrophobic helices within each chain could contribute to formation of an ion channel between all five subunits (Devillers-Thiery et al 1983). The orientation of the large, amino terminal region involved in agonist binding, the four hydrophobic membrane-spanning regions (M1-M4), the amphipathic channel forming region (A1), and the cytoplasmic regions of a consensus subunit are depicted in Figure 3. The evolutionary relatedness of the different chains was dramatically displayed by Noda et al (1983a) following their complete sequencing of all four chains. After alignment they were shown to be 19% identical at equivalent sites and 54% homologous, in that three (or four) chains with identical residues have a conservative substitution in the divergent chain. Comparison of $Torpedo,$ calf, and human subunit sequences led Numa and co-workers to propose that the α subunit has evolved more slowly than the other subunits, and that the basic $\alpha_2\beta\gamma\delta$ subunits stoichiometry was established 550–690 million years ago (Kubo et al 1985).

Determination of these sequences revolutionized thinking about the AChR. Noda et al (1983c) showed a 97% homology between α chains of human and calf, and 80–81% homology between human and *Torpedo* α chains. The functional AChR is highly conserved throughout evolution, as the recent determination of the sequences of mouse δ chains (La Polla et al 1984) and chick γ and δ chains (Nef et al 1984) has confirmed.

Examination of the splice junctions in both human genomic (Noda et al 1983c) and in *Torpedo* genomic sequences (Noda et al 1983a) showed that the 9 exons in human α gene and 12 exons in *Torpedo* γ and δ in some cases precisely corresponded to predicted structural domains of the receptor subunits; in all cases there are introns between the predicted hydrophobic membrane spanning regions M1 and M2, 15 amino acids beyond M3, and 13 amino acids prior to M4. In chicken and *Torpedo* another exon lies between M2 and M3. Thus the pattern of functional domains and specifically the transmembrane sequences being encoded in separate exons is strongly indicated in this membrane protein.

Analysis of the amphipathic character of secondary structures demonstrated one sequence (A1) just before M4 that, if in α helical

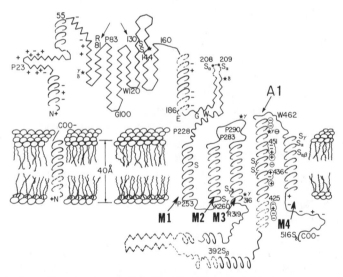

Figure 3 Secondary structure and topography of an acetylcholine receptor subunit. Hydrophobic, transmembrane α-helices are numbered M1–M4, and the amphipathic putative channel-forming α-helix is labeled A1. Possible glycosylation sites are marked with stars, and residues whose charge is conserved in all four subunits are depicted. Residue numbering follows the sequence alignment of homologous amino acids of Fairclough et al (1983).

conformation, would present one highly charged surface and one very hydrophobic surface (Stroud 1983, Fairclough et al 1983, Finer-Moore & Stroud 1984, Guy 1984), which suggested that the five subunits could each contribute one highly charged surface to the conformation of a central, water-filled, ion-conducting channel across the membrane.

Mishina et al (1984) subsequently obtained expression of their cDNA clones for all four subunits of the *Torpedo* AChR. The procedure involved production of mRNA under an SV40 promoter in sensitive COS monkey cells. The mRNA was subsequently injected into *Xenopus* oocytes and expressed. In 35 site-specific mutagenesis experiments aimed at alterations in the α chain of *Torpedo* AChR, Mishina et al (1985) showed that certain regions of the sequence are very important for different functions. Substitution of cysteine residues 130 or 144 for serine eliminated ACh-induced ion flux and reduced bungarotoxin binding to only 8%. The authors suggested that the extracellular β-sheet domain may be unstable without the putative S–S bond between these residues. Replacement of cysteine by serine at positions 208 or 209, shown by Kao et al (1984) to lie at the MBTA affinity-labeled site, destroyed ACh-induced ion flux sensitivity but only diminished α-BgTx binding to 28–39% relative to native receptor, consistent with the large binding surface predicted for α-BgTx binding (Low 1979, Kistler et al 1982). Substitution of 5–10 amino acids by 0–3 amino acids in M1, M2, or M3 resulted in no detectable ACh sensitivity and had a profound effect on bungarotoxin binding, possibly by affecting assembly of the AChR complex. Almost all substitutions in the region of the amphipathic helix (A1), which may contribute to ion channel formation, eliminated ACh sensitivity, while reducing toxin binding by 56–70%. The one exception was that deletion of residues 417–452, which comprise the *entire* amphipathic region, showed a slight ($\sim 3\%$) ion-fluxing capability in response to ACh. Mishina et al (1985) pointed out that the neighboring region, which may be dragged across the membrane during insertion, is also of amphipathic quality and could conceivably substitute for region A1. Evidently A1 is essential for ion flux, but is less crucial for inhibition of α-BgTx binding by carbamylcholine. Small substitutions within either the carboxy terminal sequence after M4, or within most of the cytoplasmic domain between residues 324–434, did not substantially impair functional properties.

Topography of the AChR in the Bilayer

With the amino acid sequences of all four subunits from *T. californica* known, a significant amount of topographical information can be deduced by analyzing the results of biochemical, immunological, and proteolytic studies. Each subunit contains a large amino terminal region ($M_r \sim 30{,}000$ daltons) that contains the sites for core glycosylation (Anderson & Blobel 1981, Anderson et al 1983). The most likely site for N-glycosylation is Asn 143, which is conserved in all four subunits (Noda et al 1983a). This hydrophilic domain is extracellular, as shown by protease susceptibility (Wennogle & Changeux 1980), and contains the high affinity agonist and antagonist bindings sites in the α subunit (Kao et al 1984). The amino terminal region corresponds to the portion of the receptor seen to extend 55 Å above the membrane on the synaptic side (Klymkowsky & Stroud 1979, Kistler et al 1982).

The first three of the four hydrophobic membrane-spanning regions (M1–M3) follow one another rather closely in the sequence, connected by short hydrophilic loops approximately 5 and 16 amino acids long, respectively (Claudio et al 1983, Devillers-Thiery et al 1983, Noda et al 1983a). The short loop between M1 and M2 is predicted to be cytoplasmic, and the longer loop connecting M2 and M3 should lie on the synaptic side of the membrane. The disposition of the long segment (approximately 155 amino acids in length) between M3 and M4 is largely cytoplasmic. The amphipathic helical segment A1 introduces a fifth membrane crossing, reduces the length of the large cytoplasmic segment to about 110 residues, introduces a short extracellular loop of about 20 amino acids between A1 and M4, and places the carboxy terminus of each subunit within the cytoplasm.

The initial portion of the link between M3 and M4 has been shown to be cytoplasmic by several groups. Wennogle et al (1981) generated a 16,000 dalton segment of the δ subunit, corresponding to the carboxy-terminal region, which included A1, M4, and some of the cytoplasmic linker, and showed that the site of cleavage was cytoplasmic by the disposition of phosphorylation sites. Huganir et al (1984) demonstrated the cytoplasmic location of the phosphorylation site, and have tentatively identified the specific phosphorylation sites as serines and tyrosines between residues 364–381 in all four subunits. Barkas et al (1984) generated antibodies to peptides corresponding to segments 153–184 and the carboxy-terminal segment 489–500 of the α chain, and showed that cleavage between the binding sites for these antibodies was enhanced by sonication, indicating that the cleavage site was cytoplasmic. More specifically, by using a colloidal-gold second antibody to visualize binding, antibodies raised against a peptide corresponding to residues 350–358 of the β chain (45 residues from the end of M3 and 32 residues from the start of A1) were shown to bind to the cytoplasmic side of the postsynaptic membrane in disrupted *T. californica* electroplaque (Young et al 1985).

The cytoplasmic location of the carboxy terminus of receptor subunits, which is consistent with the five-crossing model, has recently been demonstrated; Lindstrom and co-workers (Lindstrom et al 1984, Ratnam et al 1984) raised antibodies to peptides corresponding to the carboxy termini of all four subunits, and showed that detergent or lithium diiodosalicylate permeabilization of AChR vesicles was necessary for antibody binding. In vitro studies performed by Young et al (1985) using antibodies raised against the carboxy-terminal residues 501–516 of the δ subunit also showed that detergent solubilization of AChR vesicles exposed the carboxy terminus to antibody binding. In addition, the cytoplasmic location of the δ carboxy terminus was demonstrated in situ, in *Torpedo* electroplaque, by visualizing antibody binding in the electron microscope through use of colloidal-gold second antibodies. This proved that the carboxy terminus was cytoplasmic, and that detergent solubilization was not acting simply to expose an extracellular carboxy terminus that was buried in the protein. The cytoplasmic location of the δ carboxy terminus seems to require that the δ–δ disulfide bond in dimeric AChR must also be cytoplasmic.

Nature of the Suggested Ion Channel

The amphipathic sequence suggests that charged residues contributed by similar and homologous sequences from five subunits, present on one side of a 50 Å long α-helix within each subunit, could come together to form the lining of the ion channel. The five-fold symmetrical arrangement implies that similar residues will be brought together at similar heights relative to the bilayer, and that the channel lining would contain alternate layers of positively and negatively charged residues. While the net charge within this 40 Å transmembrane region would be neutral, the entrance would be predominantly negatively charged, and so provide specificity for cations, as anions would be repelled. The funnel-shaped structure of the molecule would also provide some insulation against the effects of membrane surface charge. Model-building experiments show that five α-helices brought together with five-fold symmetry and placed 11 Å apart as suggested by X-ray scattering (R. H. Fairclough and R. M. Stroud, unpublished results) would generate a central channel no less than 7.0 Å across. This diameter closely parallels the value of 6.5 Å estimated by electrophysiological measurements for conductivity of organic cations (Maeno et al 1977, Dwyer et al 1980). Such an arrangement of close-packed helices can accommodate passage of a hydrated sodium ion or an organic cation such as diethyldimethylammonium, simply by movement of side chains away from the center of the channel. Electrostatic energy calculations suggest that the barriers to passage of monovalent or divalent cations through this putative channel are about 2.5–3.7 kcal/mole, comparable to the barriers observed electrophysiologically. This arrangement of amphipathic helices also indicates at least three potential energy minima, or stable positions for cations within the channel. X-ray diffraction studies in which terbium ions, a calcium analogue, can be localized using anomalous dispersion (Fairclough et al 1985) clearly identified three major locations for terbium within the transmembrane region. These observations support the view that the ion channel is occupied by cations even in the resting state, at least in vitro.

Closing of the channel must involve a diameter decrease below 7.0 Å. The event that closes the ion channel is currently open to conjecture, and some models present themselves as testable candidates. For example, the aromatic Phe and Tyr residues at position 443 could serve as a possible gate near the cytoplasmic end of the channel. Terbium-binding experiments showed that the channel is filled with cations through most of its length (Fairclough et al 1985) and suggested that a site 17 Å from the cytoplasmic face, where essentially no terbium is bound, is a likely location for the gate.

Conclusions

The sequences of the subunit chains in the AChR provide the essential map for localization of functional determinants. When the three-dimensional structure at atomic resolution emerges, sites mapped in the structure can be related to their effect upon channel opening or desensitization. This correlation requires a profound understanding of the physiological function, the chemical change, and the electrophysiological consequences of modulation. Each facet of this analysis demands the ultimate of forefront technologies, but in the end, the synthesis of these perspectives will yield the best understanding of any element in the neuromuscular synapse and probably of any protein involved in cell-to-cell communication.

ACKNOWLEDGMENTS

We thank Geri R. Gilbert for typing the many versions of the manuscript, and Robert Love, Paul Bash, and Alok Mitra for their helpful suggestions and discussion. Related studies are supported by National Institute of Health Grant GM24485 and National Science Foundation Grant PCM83-16401 to R. M. S., National Institute of Health Grants GM09827 to M. P. M., and NS07241 to E. F. Y. Salary support for J. P. E. derives from National Institute of Health Grant NS13050 to Dr. Mark G. McNamee. S. C. was supported by a university fellowship from the University of California, Berkeley.

Literature Cited

Adams, P. R. 1981. Acetylcholine receptor kinetics. *J. Memb. Biol.* 58:161–74

Adams, P. R. 1975. An analysis of the dose-response curve at voltage-clamped frog endplates. *Pfleugers Arch.* 360:145–53

Albuquerque, E. X., Tsai, M.-C., Aronstam, R. S., Witkop, B., Eldefrawi, A. T., Eldefrawi, M. E. 1980. Phencyclidine interactions with the ionic channel of the acetylcholine receptor and electrogenic membrane. *Proc. Natl. Acad. Sci. USA* 77:1224–28

Anderson, D. J., Blobel, G. 1981. In vitro synthesis, glycosylation and membrane insertion of the four subunits of *Torpedo* acetylcholine receptor. *Proc. Natl. Acad. Sci. USA* 78:5598–5602

Anderson, D. J., Blobel, G., Tzartos, S., Gullick, W., Lindstrom, J. 1983. Transmembrane orientation of an early biosynthetic form of acetylcholine receptor δ subunit determined by proteolytic dissection in conjunction with monoclonal antibodies. *J. Neurosci.* 3:1773–84

Anholt, R., Lindstrom, J., Montal, M. 1980. Functional equivalence of monomeric and dimeric forms of purified acetylcholine receptors from *Torpedo californica* in reconstituted lipid vesicles. *Eur. J. Biochem.* 109:481–87

Aracava, Y., Albuquerque, E. X. 1984. Meproadifen enhances activation and desensitization of the acetylcholine receptor-ionic channel complex (AChR): Single channel studies. *FEBS Lett.* 174:267–74

Aracava, Y., Ikeda, S. R., Daly, J. W., Brookes, N. Albuquerque, E. X. 1984. Interactions of bupivacaine with ionic channels of the nicotinic receptor. Analysis of single-channel currents. *Mol. Pharmacol.* 26:304–13

Aslanian, D., Heidmann, T., Nigrerie, M., Changeux, J.-P. 1983. Raman spectroscopy of acetylcholine receptor-rich membranes from *Torpedo marmorata* and of their isolated components. *FEBS Lett.* 164:393–400

Ballivet, M., Patrick, J., Lee, J., Heinemann, S. 1982. Molecular cloning of cDNA coding for the γ subunit of *Torpedo* acetylcholine receptor in its membrane environment. *Proc. Natl. Acad. Sci. USA* 74:4460–70

Barkas, T., Juillerat, M., Kistler, J., Schwendimann, B., Moody, J. 1984. Antibodies to synthetic peptides as probes of acetylcholine receptor structure. *Eur. J. Biochem.* 143:309–14

Barrantes, F. J. 1978. Agonist-mediated changes of the acetylcholine receptor in its membrane environment. *J. Mol. Biol.* 124:1–26

Bayne, E. K., Anderson, M. J., Fambrough, D. M. 1984. Extracellular matrix organization in developing muscle: Correlation with acetylcholine receptor aggregates. *J. Cell Biol.* 99:1486–1501

Ben-Haim, D., Dreyer, F., Peper, K. 1975. Acetylcholine receptor: Modification of synaptic gating mechanism after treatment with a disulfide bond reducing agent. *Pfluegers Arch.* 355:19–26

Blickenstaff, G. D., Wang, H. H. 1985. The effects of spin-labeled local anesthetics on acetylcholine receptor-mediated ion flux. *Biophys. J.* 47:261a

Boheim, G., Hanke, W., Barrantes, F. S., Eibl, H., Sakmann, B., Fels, G., Maelicke, A. 1981. Agonist-activated ionic channels in acetylcholine receptor reconstituted into planar lipid bilayers. *Proc. Natl. Acad. Sci. USA* 78:3586–90

Bon, F., Lebrun, E., Gomel, J., van Rapenbusch, R., Cartaud, J., Popot, J.-L., Changeux, J.-P. 1984. Image analysis of the heavy form of the acetylcholine receptor from *Torpedo marmorata*. *J. Mol. Biol.* 176:205–37

Bonner, R., Barrantes, F. J., Jovin, T. M. 1976. Kinetics of agonist-induced intrinsic fluorescence changes in membrane-bound acetylcholine receptor. *Nature* 263:429–31

Boyd, N. D., Cohen, J. B. 1984. Desensitization of membrane-bound *Torpedo* acetylcholine receptor by amine noncompetitive antagonists and aliphatic alcohols: Studies of [^{3}H] acetylcholine binding and ^{22}Na$^+$ ion fluxes. *Biochemistry* 23:4023–33

Brisson, A., Unwin, P. N. T. 1984. Tubular crystals of acetylcholine receptor. *J. Cell Biol.* 99:1202–11

Burden, S. J., DePalma, R. L., Gottesman, G. S. 1983. Crosslinking of proteins in acetylcholine receptor-rich membranes: Association between the β-subunit and the 43 kD subsynaptic protein. *Cell* 35:687–92

Capasso, S., Mattia, C., Mazzarella, L., Puliti, R. 1977. Structure of a *cis*-peptide unit: Molecular conformation of the cyclic disulphide L-cysteinyl-L-cysteine. *Acta Crystallogr.* B33:2080–83

Cartaud, J., Benedetti, L., Sobel, A., Changeux, J.-P. 1978. A morphological study of the cholinergic receptor protein from *Torpedo marmorata* in its membrane environment and in its detergent-extracted purified form. *J. Cell Sci.* 29:313–37

Cartaud, J., Sobel, A., Rousselet, A., Devaux,

P. F., Changeux, J.-P. 1981. Consequences of alkaline treatment for the ultrastructure of the acetylcholine receptor-rich membranes from *Torpedo marmorata* electric organ. *J. Cell Biol.* 90:418–26

Chang, H. W., Neumann, E. 1976. Dynamic properties of isolated acetylcholine receptor proteins: Release of calcium ions caused by acetylcholine binding. *Proc. Natl. Acad. Sci. USA* 73:3364–68

Chang, H. W., Bock, E., Neumann, E. 1984. Long-lived metastable states and hysteresis in the binding of acetylcholine to *Torpedo californica* acetylcholine receptor. *Biochemistry* 23:4546–56

Claudio, T., Ballivet, M., Patrick, J., Heinemann, S. 1983. Nucleotide and deduced amino acid sequences of *Torpedo californica* acetylcholine receptor α subunit. *Proc. Natl. Acad. Sci. USA* 80:1111–15

Colquhoun, D., Sakmann, B. 1981. Fluctuations in the microsecond time range of the current through single acetylcholine receptor ion channels. *Nature* 294:464–66

Colquhoun, D., Dionne, V. E., Steinbach, J. H., Stevens, C. F. 1975. Conductance of channels opened by acetylcholine-like drugs in muscle end-plate. *Nature* 253:204–6

Conti-Tronconi, B. M., Hunkapiller, M. W., Raftery, M. A. 1984. Molecular weight and structural nonequivalence of the mature α subunits of *Torpedo californica* acetylcholine receptor. *Proc. Natl. Acad. Sci. USA* 81:2631–34

Conti-Tronconi, B. M., Raftery, M. A. 1982. The nicotinic cholinergic receptor: Correlation of molecular structure with functional properties. *Ann. Rev. Biochem.* 51:491–530

Covarrubias, M., Prinz, H., Maelicke, A. 1984. Ligand-specific state transitions of the membrane-bound acetylcholine receptor. *FEBS Lett.* 169:229–33

Damle, V. N., Karlin, A. 1978. Affinity labeling of one of two α-neurotoxin binding sites in acetylcholine receptor from *Torpedo californica*. *Biochemistry* 17:2039–45

Damle, V., McLaughlin, M., Karlin, A. 1978. Bromoacetylcholine as an affinity label of the acetylcholine receptor from *Torpedo californica*. *Biochem. Biophys. Res. Commun.* 84:845–51

Delegeane, A. M., McNamee, M. G. 1980. Independent activation of the acetylcholine receptor from *Torpedo californica* at two sites. *Biochemistry* 19:890–95

Devillers-Thiery, A., Changeux, J.-P., Parotaud, P., Strosberg, A. D. 1979. The amino terminal sequence of the 40 K subunit of the acetylcholine receptor protein from *Torpedo marmorata*. *FEBS Lett.* 104:99–105

Devillers-Thiery, A., Giraudat, J., Bentaboulet, M., Changeux, J.-P. 1983. Complete mRNA coding sequence of the acetylcholine binding α-subunit from *Torpedo marmorata*: A model for the transmembrane organization of the polypeptide chain. *Proc. Natl. Acad. Sci. USA* 80:2067–71

Dionne, V. E., Leibowitz, M. D. 1982. Acetylcholine receptor kinetics. A description from single-channel currents at snake neuromuscular junctions. *Biophys. J.* 39:253–61

Dionne, V. E., Steinbach, J. H., Stevens, C. F. 1978. An analysis of the dose-response relationship at voltage-clamped frog neuromuscular junctions. *J. Physiol. London* 281:421–44

Dreyer, F., Peper, K., Sterz, R. 1978. Determination of dose-response curves by quantitative ionophoresis at the frog neuromuscular junction. *J. Physiol. London* 281:395–419

Dunn, S. M. J., Blanchard, S. F., Raftery, M. A. 1980. Kinetics of carbamylcholine binding to membrane-bound acetylcholine receptor monitored by fluorescence changes of a covalently bound probe. *Biochemistry* 19:5645–52

Dunn, S. M. J., Conti-Tronconi, B. M., Raftery, M. A. 1983. Separate sites of low and high affinity for agonists on *Torpedo californica* acetylcholine receptor. *Biochemistry* 22:2512–18

Dunn, S. M. J., Raftery, M. A. 1982a. Multiple binding sites for agonists on *Torpedo californica* acetylcholine receptor. *Biochemistry* 21:6264–72

Dunn, S. J., Raftery, M. A. 1982b. Activation and desensitization of *Torpedo* acetylcholine receptor: Evidence for separate binding sites. *Proc. Natl. Acad. Sci. USA* 79:6757–61

Dwyer, T. M., Adams, D. J., Hille, B. 1980. The permeability of the endplate channel to organic cations in frog muscle. *J. Gen. Physiol.* 75:469–92

Earnest, J. P., Wang, H. H., McNamee, M. G. 1984. Multiple binding sites for local anesthetics on reconstituted acetylcholine receptor membranes. *Biochem. Biophys. Res. Commun.* 123:862–68

Ellena, J. F., McNamee, M. G. 1980. Interaction of spin-labeled *Naja naja siamensis* α-neurotoxin with acetylcholine receptor from *Torpedo californica*. *FEBS·Lett.* 110:301–4

Fairclough, R. H., Finer-Moore, J., Love, R. A., Kristofferson, D., Desmeules, P. J., Stroud, R. M. 1983. Subunit organization and structure of an acetylcholine receptor. *Cold Spring Harbor Symp. Quant. Biol.* 48: 9–20

Fairclough, R. H., Miake-Lye, R. C., Stroud, R. M., Hodgson, K. O., Doniach, S. 1985. Location of Tb (III) binding sites on acetylcholine receptor enriched membranes. *J. Mol. Biol.* Submitted

Fels, G., Wolff, E. K., Maelicke, A. 1982. Equilibrium binding of acetylcholine to the membrane-bound acetylcholine receptor. *Eur. J. Biochem.* 127: 31–38

Finer-Moore, J., Stroud, R. M. 1984. Amphipathic analysis and possible formation of the ion channel in an acetylcholine receptor. *Proc. Natl. Acad. Sci. USA* 81: 155–59

Gardner, P., Ogden, D. C., Colquhoun, D. 1984. Conductance of single ion channels opened by nicotinic agonists are indistinguishable. *Nature* 309: 160–62

Giraudat, J., Devillers-Thiery, A., Auffrey, C., Rougeon, F., Changeux, J.-P. 1982. Identification of a cDNA clone coding for the acetylcholine binding subunit of *Torpedo marmorata* acetylcholine receptor. *EMBO J.* 1: 713–17

Gonzalez-Ros, J. M., Farach, M. C., Martinez-Carrion, M. 1983. Ligand-induced effects at regions of acetylcholine receptor accessible to membrane lipids. *Biochemistry* 22: 3807–11

Gonzalez-Ros-, J. M., Llanillo, M., Paraschos, A., Martinez-Carrion, M. 1982. Lipid environment of acetylcholine receptor from *Torpedo californica*. *Biochemistry* 21: 3467–74

Gordon, A. S., Milfoy, D., Diamond, I. 1983. Identification of a molecular weight 43,000 protein kinase in acetylcholine receptor-enriched membranes. *Proc. Natl. Acad. Sci. USA* 80: 5862–65

Guy, H. R. 1984. A structural model of the acetylcholine receptor channel based on partition energy and helix packing calculations. *Biophys. J.* 45: 249–61

Hamill, O. P., Sakmann, B. 1981. Multiple conductance states of single acetylcholine receptor channels in embryonic muscle cells. *Nature* 294: 462–64

Haring, R., Kloog, Y. 1984. Multiple binding sites for phencyclidine on the nicotinic acetylcholine receptor from *Torpedo ocellata* electric organ. *Life Sci.* 34: 1047–55

Hartzell, H. C., Kuffler, S. W., Yoshikami, D. 1976. The number of acetylcholine molecules in a quantum and the interaction between quanta at the subsynaptic membrane of the skeletal neuromuscular synapse. *Cold Spring Harbor Symp. Quant. Biol.* 40: 175–86

Heidmann, T., Changeux, J.-P. 1979. Fast kinetic studies on the allosteric inter-actions between acetylcholine receptor and local anesthetic binding sites. *Eur. J. Biochem.* 94: 281–96

Heidmann, T., Changeux, J.-P. 1984. Time-resolved photolabeling by the noncompetitive blocker chlorpromazine of the acetylcholine receptor in its transiently open and closed ion channel conformation. *Proc. Natl. Acad. Sci. USA* 81: 1897–1901

Heidmann, T., Oswald, R. E., Changeux, J.-P. 1983. Multiple sites of action for noncompetitive blockers on acetylcholine receptor-rich membrane fragments from *Torpedo marmorata*. *Biochemistry* 22: 3112–27

Hess, G. P., Cash, D. J., Aoshima, H. 1979. Acetylcholine receptor-controlled ion fluxes in membrane vesicles investigated by fast reaction techniques. *Nature* 282: 329–31

Hess, G. P., Cash, D. J. Aoshima, H. 1983. Acetylcholine receptor-controlled ion translocation: Chemical kinetic investigations of the mechanism. *Ann. Rev. Biophys. Bioeng.* 12: 443–73

Heuser, J. E., Salpeter, S. R. 1979. Organization of acetylcholine receptors in quick-frozen deep-etched and rotary replicated *Torpedo* postsynaptic membrane. *J. Cell Biol.* 82: 150–73

Holtzman, E., Wise, D., Wall, J., Karlin, A. 1982. Electron microscopy of complexes of isolated acetylcholine receptor, biotinyl-toxin, and avidin. *Proc. Natl. Acad. Sci. USA* 79: 310–14

Hucho, F., Layer, P., Riefer, H. R., Bandini, G. 1976. Photoaffinity labeling and quarternary structure of the acetylcholine receptor from *Torpedo californica*. *Proc. Natl. Acad. Sci. USA* 73: 2624–28

Huganir, R. L., Racker, E. 1982. Properties of proteoliposomes reconstituted with acetylcholine receptor from *Torpedo californica*. *J. Biol. Chem.* 257: 9372–78

Huganir, R. L., Miles, K., Greengard, P. 1984. Phosphorylation of the nicotinic acetylcholine receptor by an endogenous tyrosine-specific protein kinase. *Proc. Natl. Acad. Sci. USA* 81: 6968–72

Jackson, M. B. 1984. Spontaneous openings of the acetylcholine receptor channel. *Proc. Natl. Acad. Sci. USA* 81: 3901–4

Johnson, D. A., Voet, J. G., Taylor, P. 1984. Fluorescence energy transfer between cobra α-toxin molecules bound to the acetylcholine receptor. *J. Biol. Chem.* 259: 5717–25

Kaldany, R.-R. J., Karlin, A. 1983. Reaction of quinacrine mustard with the acetylcholine receptor from *Torpedo californica*. *J. Biol. Chem.* 258: 6263–42

Kaneda, N., Tanaka, F., Kohno, M.,

410 McCARTHY ET AL

Hayashi, K., Yagi, R. 1982. Change in the intrinsic fluorescence of acetylcholine receptor purified from *Narke japonica* upon binding with cholinergic ligands. *Arch. Biochem. Biophys.* 218:376–83

Kao, P. N., Dwork, A. J., Kaldany, R.-R. J., Silver, M. L., Wideman, J., Stein, S., Karlin, A. 1984. Identification of the α-subunit half-cystine specifically labeled by an affinity reagent for the acetylcholine receptor binding site. *J. Biol. Chem.* 259:11662–65

Karlin, A. 1980. Molecular properties of nicotinic acetylcholine receptors. *Cell Surf. Rev.* 6:191

Katz, B., Thesleff, S. 1957. A study of the "desensitization" produced by acetylcholine on the motor endplate. *J. Physiol. London* 138:63–80

Kistler, J., Stroud, R. M. 1981. Crystalline arrays of membrane-bound acetylcholine receptor. *Proc. Natl. Acad. Sci. USA* 78:3678–82

Kistler, J., Stroud, R. M., Klymkowsky, M. W., Lalancette, R. A., Fairclough, R. H. 1982. Structure and function of an acetylcholine receptor. *Biophys. J.* 37:371–83

Klymkowsky, M. W., Stroud, R. M. 1979. Immunospecific identification and three-dimensional structure of a membrane-bound acetylcholine receptor *Torpedo californica*. *J. Mol. Biol.* 128:319–34

Koblin, D. D., Lester, H. A. 1979. Voltage-dependent and voltage-independent blockade of acetylcholine receptors by local anesthetics in *Electrophorus* electroplaques. *Mol. Pharmacol.* 15:559

Kubo, T., Noda, M., Takai, T., Tanuba, T., Kayano, T., et al. 1985. Primary structure of δ subunit precursor of calf muscle acetylcholine receptor deduced from cDNA sequence. *Eur. J. Biochem.* 149:5–13

Labarca, P., Lindstrom, J., Montal, M. 1984. The acetylcholine receptor channel from *Torpedo californica* has two open states. *J. Neurosci.* 4:502–7

Lambert, J. J., Durant, N. N., Henderson, E. B. 1983. Drug-induced modification of ionic conductance at the neuromuscular junction. *Ann. Rev. Pharmacol. Toxicol.* 23:505–39

Land, B. R., Harris, W. V., Salpeter, E. E., Salpeter, M. M. 1984. Diffusion and binding constants for acetylcholine derived from the falling phase of miniature endplate currents. *Proc. Natl. Acad. Sci. USA* 81:1594–98

La Polla, R. J., Mayne, K. M., Davidson, N. 1984. Isolation and characterization of cDNA clone for the complete protein coding region of the δ subunit of the mouse acetylcholine receptor. *Proc. Natl. Acad.*

Sci. USA 81:7970–74

Leibowitz, M. D., Dionne, V. E. 1984. Single-channel acetylcholine receptor kinetics. *Biophys. J.* 45:153–63

Lester, H. A., Nass, M. M., Krouse, M. E., Nerbonne, J. M. 1980. Electrophysiological experiments with photoisomerizable cholinergic compounds: Review and progress report. *Ann. NY Acad. Sci.* 346:475–90

Lewis, C. A., Stevens, C. F. 1983. Acetylcholine receptor channel ionic selectivity: Ions experience an aqueous environment. *Proc. Natl. Acad. Sci. USA* 80:6110–13

Lindstrom, J., Criado, M., Hochschwender, S., Fox, J. L., Sarin, V. 1984. Immunochemical tests of acetylcholine receptor subunit models. *Nature* 311:573–75

Low, B. W. 1979. The three-dimensional structure post-synaptic neurotoxins: Consideration of structure and function *Handb. Exp. Pharmacol.* 52:213–57

Maeno, T., Edwards, C., Ankaru, M. 1977. Permeability of the end-plate membrane activated by acetylcholine to some organic cations. *J. Neurobiol.* 8:173–84

Magazanik, L. G., Vyskocil, F. 1973. Desensitization of the motor end plate. In *Drug Receptors, A Symposium*, ed. H. P. Rang, pp. 105–19. London: Macmillan

Magleby, K. L., Stevens, C. F. 1972. The effect of voltage on the time course of end-plate currents. *J. Physiol. London* 223:151–71

Meunier, J. C., Olsen, R. W., Changeux, J.-P. 1972. Studies on the cholinergic receptor protein from *Electrophorus electricus*. III. Effect of detergent on some hydrodynamic properties of the receptor protein in solution. *FEBS Lett.* 24:63–68

Middlemass, D. S., Raftery, M. A. 1983. Exposure of acetylcholine receptor to the lipid bilayer. *Biochem. Biophys. Res. Commun.* 115:1075–82

Mielke, D. L., Kaldany, R.-R., Karlin, A., Wallace, B. A. 1984. Effector-induced changes in the secondary structure of the nicotinic acetylcholine receptor. *Biophys. J.* 45:205a

Mishina, M., Kurosaki, T., Tobimatsu, T., Marimoto, Y., Noda, M., et al. 1984. Expression of functional acetylcholine receptor from cloned cDNAs. *Nature* 307:604–8

Mishina, M., Tobimatsu, T., Imoto, K., Tanaka, K., Fujita, Y., et al. 1985. Location of functional regions of acetylcholine receptor α-subunit by site-directed mutagenesis. *Nature* 313:364–69

Mitra, A. K., Chandrasekaran, R. 1984. Conformational flexibilities in malformin A. *Biopolymers* 23:2513–24

Moore, H. P., Raftery, M. A. 1979. Ligand-induced interconversion of affinity states in membrane-bound acetylcholine receptor from *Torpedo californica*. Effects of sulfhydryl and disulfide reagents. *Biochemistry* 18:1907–11

Moore, H. P., Raftery, M. A. 1980. Direct spectroscopic studies of cation translocation by *Torpedo* acetylcholine receptor on a time scale of physiological relevance. *Proc. Natl. Acad. Sci. USA* 77:4509–13

Moore, W. M., Holliday, L. A., Puett, D., Brady, R. N. 1974. On the conformation of the acetylcholine receptor protein from *Torpedo nobiliana*. *FEBS Lett.* 45:145–49

Muhn, P., Fehr, A., Hucho, F. 1984. Photoaffinity labeling of acetylcholine receptor in millisecond time scale. *FEBS Lett.* 166:146–50

Nathanson, N. M., Hall, Z. W. 1980. *In situ* labeling of *Torpedo* and rat muscle acetylcholine receptor by a photoaffinity derivative of α-bungarotoxin. *J. Biol. Chem.* 255:1698–1703

Nef, P., Mauron, A., Stalder, R., Alliod, C., Ballivet, M. 1984. Structure, linkage and sequence of the two genes encoding the δ and γ subunits of the nicotinic acetylcholine receptor. *Proc. Natl. Acad. Sci. USA* 81:7975–79

Neher, E., Sakmann, B. 1976. Single-channel currents recorded from membrane of denervated frog muscle fibres. *Nature* 260:799–802

Neubig, R. R., Cohen, J. B. 1980. Permeability control by cholinergic receptors in *Torpedo* post-synaptic membranes: Agonist dose-response relations measured at second and millisecond times. *Biochemistry* 19:2770–79

Neubig, R. R., Krodel, E. K., Boyd, N. D., Cohen, J. B. 1979. Acetylcholine and local anesthetic binding to *Torpedo* nicotinic post-synaptic membranes after removal of non-receptor peptides. *Proc. Natl. Acad. Sci. USA* 76:690–94

Noda, M., Takahashi, H., Tanabe, T., Toyosato, M., Furutani, Y., et al. 1982. Primary structure of α-subunit precursor of *Torpedo californica* acetylcholine receptor deduced from a cDNA sequence. *Nature* 299:793–97

Noda, M., Takahashi, H., Tanabe, T., Toyosato, M., Kikyotani, S., et al. 1983a. Structural homology of *Torpedo californica* acetylcholine receptor subunits. *Nature* 302:528–32

Noda, M., Takahashi, H., Tanabe, T., Toyoto, M., Kikyotani, S., Hirose, T., Asai, M., Takashima, H., Inayama, S., Miyata, T., Numa, S. 1983b. Primary structures of β- and δ-subunit precursors of *Torpedo californica* acetylcholine receptor deduced from cDNA sequences. *Nature* 301:251–55

Noda, M., Furutani, Y., Takahashi, H., Toyosato, M., Tanabe, T., et al. 1983c. Cloning and sequence analysis of calf cDNA and human genomic DNA encoding α-subunit precursor of *Torpedo californica* acetylcholine receptor deduced from cDNA sequence. *Nature* 299:793–97

Olson, E. N., Glaser, L., Merlie, J. P. 1984. α and β subunits of the nicotinic acetylcholine receptor contain covalently bound lipid. *J. Biol. Chem.* 259:5364–67

Oswald, R. E., Changeux, J.-P. 1981a. Selective labeling of the δ-subunit of the acetylcholine receptor by a covalent local anesthetic. *Biochemistry* 20:7166–74

Oswald, R. E., Changeux, J.-P. 1981b. Ultraviolet light induced labeling by noncompetitive blockers of the acetylcholine receptor from *Torpedo marmorata*. *Proc. Natl. Acad. Sci. USA* 78:3925–29

Oswald, R. E., Heidmann, T., Changeux, J.-P. 1983. Multiple affinity states for noncompetitive blockers revealed by [³H] phencyclidine binding to acetylcholine receptor rich membranes from *Torpedo marmorata*. *Biochemistry* 22:3128–36

Oswald, R. E., Bamberger, M. J., McLaughlin, J. T. 1984. Mechanisms of phencyclidine binding to the acetylcholine receptor from *Torpedo* electroplaque. *Mol. Pharmacol.* 25:360–68

Otero, A. S., Hamilton, S. L. 1984. Ligand-induced variations in the reactivity of thiol groups of the α subunit of the acetylcholine receptor from *Torpedo californica*. *Biochemistry* 23:2321–28

Pasquale, E. B., Takeyasu, K., Udgaonkar, J. B., Cash, D. J., Severski, M. C., Hess, G. P. 1983. Acetylcholine receptor: Evidence for a regulatory binding site in investigations of suberyldicholine-induced transmembrane ion flux in *Electrophorus electricus* membrane vesicles. *Biochemistry* 22:5967–73

Peper, K., Bradley, R. J., Dreyer, F. 1982. The acetylcholine receptor at the neuromuscular junction. *Physiol. Rev.* 62:1271–1340

Popot, J.-L., Changeux, J.-P. 1984. Nicotinic receptor of acetylcholine: Structure of an oligomeric integral membrane protein. *Physiol. Rev.* 64:1162–1239

Popot, J.-L., Demel, R. A., Sobel, A., van Deenen, L. L. M., Changeux, J.-P. 1978. Interaction of the acetylcholine (nicotinic) receptor protein from *Torpedo marmorata* electric organs with monolayers of pure lipids. *Eur. J. Biochem.* 85:27–42

Prinz, H., Maelicke, A. 1983. Interaction of cholinergic ligands with the purified acetylcholine receptor protein. I. Equilibrium binding studies. *J. Biol. Chem.* 288:10263–71

412   McCARTHY ET AL

Quast, V., Schimerlik, M., Lee, T., Witzemann, V., Blanchard, S., Raftery, M. A. 1978. Ligand-induced conformation changes in *Torpedo californica* membrane-bound acetylcholine receptor. *Biochemistry* 17:2405–14

Raftery, M. A., Dunn, S. M. J., Conti-Tronconi, B. M., Middlemass, D. S., Crawford, R. D. 1983. The nicotinic acetylcholine receptor: Subunit structure, functional binding sites, and ion transport properties. *Cold Spring Harbor Symp. Quant. Biol.* 48:21–23

Raftery, M. A., Hunkapiller, M. W., Strader, C. D., Hood, L. E. 1980. Acetylcholine receptor: Complex of homologous subunits. *Science* 208:1454–57

Ratnam, M., Lindstrom, J. 1984. Structural features of the nicotinic acetylcholine receptor revealed by antibodies to synthetic peptides. *Biochem. Biophys. Res. Commun.* 122:1225–33

Reynolds, J., Karlin, A. 1978. Molecular weight in detergent solution of acetylcholine receptor from *Torpedo californica*. *Biochemistry* 17:2035–38

Ribera, A., Trautmann, A., Pinset, C., Changeux, J.-P. 1985. Chlorpromazine alters acetylcholine-activated channel kinetics. *Biophys. J.* 47:40a

Ross, M. J., Klymkowsky, M. W., Agard, D. A., Stroud, R. M. 1977. Structural studies of a membrane-bound acetylcholine receptor from *Torpedo californica*. *J. Mol. Biol.* 116:635–59

Rousselet, A., Cartaud, J., Devaux, P. F., Changeux, J.-P. 1982. The rotational diffusion of the acetylcholine receptor in *Torpedo marmorata* membrane fragments studied with a spin-labeled α-toxin: Importance of the 43K protein. *EMBO J.* 1:439–45

Rubsamen, H., Eldefrawi, A. T., Eldefrawi, M. E., Hess, G. P. 1978. Characterization of the calcium-binding sites of the purified acetylcholine receptor and identification of the calcium-binding subunit. *Biochemistry* 17:3818–25

Sakmann, B., Petlack, J., Neher, E. 1980. Single acetylcholine-activated channels show burst-kinetics in presence of desensitizing concentrations of agonist. *Nature* 286:71–73

Salpeter, M. M. 1983. Molecular organization of the neuromuscular synapse. In *Myasthenia Gravis*, ed. E. X. Albuquerque, A. T. Eldefrawi, pp. 105–30. London: Chapman & Hall

Schimerlik, M., Quast, V., Raftery, M. A. 1979. Ligand-induced changes in membrane-bound acetylcholine receptor observed by ethidium fluorescence. 1. Equilibrium studies. *Biochemistry.* 18:

1884–90

Schindler, H., Spillecke, F., Neumann, E. 1984. Different channel properties of *Torpedo* acetylcholine receptor monomers and dimers reconstituted in planar membranes. *Proc. Natl. Acad. Sci. USA* 81:6222–26

Sealock, R., Wray, B. E., Froehner, S. C. 1984. Ultrastructural localization of the M_r 43,000 protein and the acetylcholine receptor in *Torpedo* postsynaptic membranes using monoclonal antibodies. *J. Cell Biol.* 98:2239–44

Shiono, S., Takeyasu, K., Udgaonker, J. B., Delcour, A. H., Fujita, N., Hess, G. P. 1984. Regulatory properties of acetylcholine receptor: Evidence for two different inhibitory sites, one for acetylcholine and the other for a noncompetitive inhibitor of receptor function (Procaine). *Biochemistry* 23:6889–93

Sine, S. M., Steinbach, J. H. 1984. Agonists block currents through acetylcholine receptor channels. *Biophys. J.* 46:277–84

Sine, S. M., Taylor, P. 1981. Relationship between reversible antagonist occupancy and the functional capacity of the acetylcholine receptor. *J. Biol. Chem.* 256:6692–99

Spivak, C. E., Albuquerque, E. X. 1982. Dynamic properties of the nicotinic acetylcholine receptor ionic channel complex: Activation and blockade. In *Progress in Cholinergic Biology, Vol. 2*, ed. I. Hanin, A. M. Goldberg, pp. 323–57. New York: Raven

Steinacker, A. 1979. Sulphonation of cholinergic receptor disulphide bond increases response to ACh. *Nature* 278:358–60

Strader, C. D., Raftery, M. A. 1980. Topographic studies of *Torpedo* acetylcholine receptor subunits as a transmembrane complex. *Proc. Natl. Acad. Sci. USA* 77:5807–11

Stroud, R. M. 1983. Acetylcholine receptor structure. *Neurosci. Commun.* 1:124–38

Sumikawa, K., Houghton, J., Smith, J. G., Bell, L., Richards, B. M., Barnard, E. A. 1982. The molecular cloning and characterization of cDNA coding for the α subunit of the acetylcholine receptor. *Nucleic Acids Res.* 10:5809–22

Takeyasu, R., Udgaonker, J. P., Hess, G. P. 1983. Acetylcholine receptor: Evidence for a voltage-dependent regulatory site for acetylcholine. Chemical kinetic measurements in membrane vesicles using a voltage clamp. *Biochemistry* 22:5973–78

Trautman, A. 1982. Curare can open and block ionic channels associated with cholinergic receptors. *Nature* 298:272–75

Tsetlin, V. I., Karlsson, E., Utkin, Y. N., Pluzhnikov, K. A., Arseniev, et al. 1982.

Interacting surfaces of neurotoxins and acetylcholine receptor. *Toxicon* 20:83–93

Vandlen, R. L., Wu, W. C. S., Eisenach, J. C., Raftery, M. A. 1979. Studies of the composition of purified *Torpedo californica* acetylcholine receptor and of its subunits. *Biochemistry* 18:1845–54

Walker, J. W., Takeyasu, K., McNamee, M. G. 1982. Activation and inactivation kinetics of *Torpedo californica* acetylcholine receptor in reconstituted membranes. *Biochemistry* 21:5384–89

Walker, J., Richardson, C., McNamee, M. G. 1984. Effects of thiol-group modifications of *Torpedo californica* acetylcholine receptor on ion flux activation and inactivation kinetics. *Biochemistry* 23:2329–38

Weber, M., Changeux, J.-P. 1974. Binding of *Naja nicricollis* [³H]α-toxin to membrane fragments from *Electrophorus* and *Torpedo* electric organs. *Mol. Pharmacol.* 10:35–40

Weiland, G., Frismer, D., Taylor, P. 1979. Affinity labeling of the subunits of the membrane associated cholinergic receptor. *Mol. Pharmacol.* 15:213–26

Weiland, G., Taylor, P. 1979. Ligand specificity of state transitions in the cholinergic receptor—behavior of agonists and antagonists. *Mol. Pharmacol.* 15:197–212

Weiland, G., Georgia, B., Lappi, S., Choquell, C., Taylor, P. 1977. Kinetics of agonist-mediated transitions in state of the cholinergic receptor. *J. Biol. Chem.* 252:7648–56

Weill, C. L., McNamee, M. G., Karlin, A. 1974. Affinity labeling of purified acetylcholine receptor from *Torpedo californica*. *Biochem. Biophys. Res. Commun.* 61:997–1003

Wennogle, L. P., Changeux, J.-P. 1980. Transmembrane orientation of proteins present in acetylcholine receptor rich membranes from *Torpedo marmorata* studied by selective proteolysis. *Eur. J. Biochem.* 106:381–93

Wennogle, L. P., Oswald, R., Saitoh, T., Changeux, J.-P. 1981. Dissection of the 66,000 dalton subunit of the acetylcholine receptor. *Biochemistry* 20:2492–97

Wise, D. S., Schoenborn, B. P., Karlin, A. 1979. Analysis by low-angle neutron scattering of the structure of the acetylcholine receptor from *Torpedo californica* in detergent solution. *Biophys. J.* 28:473–96

Wise, D. S., Wall, J., Karlin, A. 1981. Relative locations of the β and δ chains of the acetylcholine receptor determined by electron microscopy of isolated receptor trimers. *J. Biol. Chem.* 256:12624–27

Witzemann, V., Raftery, M. A. 1977. Selective photoaffinity labeling of acetylcholine receptor using a cholinergic analogue. *Biochemistry* 16:5862–68

Witzemann, V., Muchmore, D., Raftery, M. A. 1979. Affinity-directed cross-linking of membrane-bound acetylcholine receptor polypeptides with photolabile α-bungarotoxin derivatives. *Biochemistry* 18:5515–18

Wolosin, J. M., Lyddiatt, A., Dolly, J. O., Barnard, E. A. 1980. Stoichiometry of the ligand-binding sites in the acetylcholine-receptor oligomer from muscle and from electric organ. Measurement by affinity alkylation with bromoacetylcholine. *Eur. J. Biochem.* 109:495–505

Young, E. F., Ralston, E., Blake, J., Ramachandran, J., Hall, Z. W., Stroud, R. M. 1985. Topological mapping of acetylcholine receptor: Evidence for a model with five transmembrane segments and a cytoplasmic COOH-terminal peptide. *Proc. Natl. Acad. Sci. USA* 82:626–30

Ziskind-Conhaim, L., Geffen, I., Hall, Z. 1984. Redistribution of acetylcholine receptors on developing rat-myotubes. *J. Neurosci.* 4:2341–49

Zingsheim, H. P., Barrantes, F. J., Frank, J., Hanicke, W., Neugebauer, D. C. 1982a. Direct structural localization of two toxin-recognition sites on an acetylcholine receptor protein. *Nature* 299:81–84

Zingsheim, H. P., Neugebauer, D. C., Frank, J., Hanicke, W., Barrantes, F. J. 1982b. Dimeric arrangement and structure of the membrane-bound acetylcholine receptor studied by electron microscopy. *EMBO J.* 1(5):541–47

Ann. Rev. Neurosci. 1986. 9 : 415–34

INACTIVATION AND METABOLISM OF NEUROPEPTIDES

Jeffrey F. McKelvy

Department of Neurobiology and Behavior, State University of New York at Stony Brook, Stony Brook, New York 11794

Shmaryahu Blumberg

Institute of Molecular Medicine, Sackler Faculty of Medicine, Tel Aviv University, Ramat-Aviv, Israel, and Bio-Technology General LTD, Kiryat Weizmann, Rehovot, Israel

INTRODUCTION

Over the past decade, neuropeptides, defined as peptides of known structure secreted by neurons, have received increasing attention as possible intercellular signals in both invertebrate and vertebrate nervous systems. (For a recent comprehensive treatise, see Krieger et al 1983.) Although the mechanisms of neuropeptide action are still mostly unknown, sufficient experimental evidence suggests that at least some neuropeptides have physiologically important actions on neural target cells [e.g. an LHRH-like peptide on amphibian sympathetic ganglionic neurons (Jan & Jan 1983); substance P on guinea pig inferior mesenteric ganglion cells (Dun & Minota 1981)] to warrant consideration of how these actions might be terminated. The inactivation of intercellular signals must be viewed as a complex process extending from the point of receptor activation by these molecules to their "clearance" from the area bearing their receptors. By this argument, the processes of diffusion away from the receptor, concentrative reuptake, extra and intracellular enzymatic hydrolysis or covalent modification, and internalization in association with receptor, or combinations of

0147–006X/86/0301–0415$02.00

these processes, are possible mechanisms by which neuropeptides could be inactivated. In our view, all of these possibilities must be entertained in the investigation of how neuropeptide actions are terminated. This is an important point because, the traditional notion has been that the neuropeptide is hydrolyzed by a highly specific peptidase situated adjacent to the receptor of the neuropeptide-responsive cell (Schwartz et al 1981). Although this pioneering model has been very useful in stimulating experimentation, it is founded on the assumptions that neuropeptide action requires rapid inactivation of the peptide and that enzymatic hydrolysis coupled to receptor activation is the necessary process to achieve this. As the preceding and ensuing discussions illustrate, these assumptions are not in accord with existing experimental evidence and the scope of possible theoretical models, and are thus too limited.

Virtually the entire published literature on the inactivation of neuropeptides has to do with their enzymatic breakdown. Although we feel other means of inactivation are likely, this review reflects the current state of the field by emphasizing enzymatic hydrolysis. The enzymatic hydrolysis literature is voluminous and we have not attempted to review all of it. Rather, we have chosen a point of view emphasizing functional considerations rather than biochemical characterization, and have focused on representative studies related to this point of view.

A PERSPECTIVE ON MODELS FOR THE ROLE OF PEPTIDASES IN THE BIOLOGICAL ACTIONS OF NEUROPEPTIDES

Sufficient data are still lacking on which to found a meaningful model—analogous to acetylcholinesterase action at the neuromuscular junction (Adams 1980)—for how peptidases might regulate the synaptic actions of neuropeptides. Such a model awaits the detailed analysis of good model systems for peptidergic synapses, such as the bullfrog sympathetic ganglion, for example, and hard electrophysiological, biochemical, and anatomical data on neuropeptide release, neuropeptide receptor localization and density, and, of course, neuropeptide hydrolysis. Our current level of understanding of the relationship between peptide hydrolysis and peptidergic neurotransmission somewhat parallels that of acetylcholinesterase activity and neuromuscular transmission during the 1940s and 1950s.

Interestingly, the limited quantitative data currently available for neuropeptides suggest parallels between neurotransmitter hydrolysis and cholinergic and peptidergic transmission. For example, the affinity of acetylcholine for its nicotinic receptor at the neuromuscular junction is of the order of 10^4–10^5 M^{-1}, with an average occupancy time in the 1 msec range. Electrophysiological analysis indicates that the durations of acetylcholine effects are also in the millisecond range, that is, of the order of agonist occupancy times. Although all of the information has not been obtained in the same experimental system, most neuropeptides exhibit affinities for their receptors in the 10^9–10^{10} M^{-1} range (e.g. Chang & Cuatrecasas 1979) and the duration of electrophysiological effects is in the one to several minute range (e.g. Jan & Jan 1983), i.e. again of the order of the average time neuropeptides occupy their receptors. Both acetylcholinesterase (Rosenberry 1975) and at least two purified neuropeptide-degrading enzymes [bovine brain aminopeptidase (Hersh & McKelvy 1981) and rat brain post proline cleaving enzyme (Blumberg & McKelvy 1985)] exhibit high first-order rate constants (k_{cat}/K_m) for hydrolysis. These observations suggest that, in both cases, diffusion of the transmitter away from its receptor followed by enzymatic hydrolysis underlies the regulation of transmission, even though the magnitudes of the number of molecules and the duration of effects are so disparate.

Even such a limited excursion into modeling the enzymatic inactivation of neuropeptides prompts the caveat that many more complexities are involved with neuropeptide than with cholinergic systems. The first complexity is sheer numbers: the large number of known biologically active neuropeptides, of the order of 30 or more in vertebrates by current count, the apparent participation of these neuropeptides in many neural subsystems, and the possibility that the inactivation of even a given neuropeptide may occur in more than one way, depending on the synapse under consideration. [The number of neuropeptides is certain to grow as results develop from studies of alternative RNA splicing (e.g. calcitonin gene related peptide, Rosenfeld et al. 1983), the sequencing of complementary DNA libraries from neural tissue (e.g. Sutcliffe et al. 1983), and the biological effects of the cryptic portions of neuropeptide precursor polyproteins (e.g. rat somatostatin cryptic peptide Lechan et al. 1983; synenkephalin, Liston et al. 1983).] The second complexity lies in our virtual ignorance of the number of types, the distribution, and the nature of neuropeptide receptors.

Given the collected uncertainties, even the surmise that dissociation of a neuropeptide from its receptor followed by diffusion and enzymatic hydrolysis may be important in regulating neuropeptide actions may not be in accord with two of the few examples of critically analyzed peptidergic actions. Thus, in the bullfrog sympathetic ganglion (see Jan & Jan 1983), two observations suggest that LHRH-like peptide regulation might not involve enzymatic hydrolysis analogous to that for acetylcholine at the neuromuscular junction:

1. The 5–10 min duration of the electrophysiological effect of "LHRH": the late slow excitatory postsynaptic potential (EPSP) would be significantly longer than the half time for dissociation of "LHRH" from its receptor if the "LHRH" receptor has binding characteristics similar to other neuropeptide receptors (a value of around 1 min). In 5–10 min "LHRH" molecules could bind to their receptors, initiate their cellular effects, and have enough time to diffuse away from receptors before the onset of the next episode of activation of "LHRH" target cells. Enzymatic hydrolysis could still be the ultimate fate of the peptide molecules if satellite cells, capsule cells, or other ganglionic cells that are the sources of peptidases are appropriately spatially segregated from the "LHRH" receptor bearing neuronal membranes.

2. A second reason for not thinking that enzymatic hydrolysis determines the time course of "LHRH" effects is the current view that "LHRH" molecules must diffuse tens of micrometers before they reach the B cells, one of the two ganglionic cells that exhibit the late slow EPSP (Jan & Jan 1983). Thus, while both B and C cells are activated by "LHRH," the peptidergic input to these cells is provided by preganglionic C fibers in synaptic contact only with C cells. If this system requires peptide diffusion over a long distance to initiate peptide action, then peptidases should be absent in the volume of the ganglion containing B and C cells.

Studies on neuropeptide inactivation should be pursued in a physiologically well-defined system. The preceding argument plausibly emphasizes diffusion and an absence of proximate peptidase action. However, other possibilities exist: for example, perhaps both "LHRH" receptors and "LHRH" hydrolyzing peptidases are present but are different on B and C cells, such that "LHRH" enzymatic hydrolysis is an important determinant of "LHRH" activity, but because of geometric and kinetic factors, enough "LHRH" gets to B cells to activate them. Another experimental system in which neuropeptide action has been critically studied is primary cultures of dorsal root ganglion, where the effects of enkephalin on substance P release by primary sensory neurons and on membrane properties of enkephalin-sensitive neurons have been studied (Mudge et al. 1979). In these studies both constitutive and exogenously added peptides were stable over a period of hours, suggesting the absence of peptidase action.

The above perspective is meant not to discourage the reader with the difficulties inherent in peptide inactivation studies, or to minimize the importance of peptidases, but rather to underscore the need for studies of neuropeptide inactivation at specified peptidergic synapses. The goal is to construct a molecular picture analogous to that for cholinergic transmission.

ENZYMATIC INACTIVATION OF SPECIFIED NEUROPEPTIDES

Overview

Although a large body of literature exists on the enzymatic hydrolysis of virtually all of the known neuropeptides by a variety of preparations, from crude homogenates to purified enzymes, most of it is largely biochemical in nature. In the case of enkephalin pentapeptides, however, some workers have attempted to explore physiological aspects of its inactivation as related to synaptic transmission. Likewise, for the decapeptide LHRH, some investigators have attempted to relate its inactivation to hypothalamic neuroendocrine regulation. Accordingly, we discuss inactivation studies on these two neuropeptides as representative examples.

Enkephalin

The opiate peptides, methionine^5-enkephalin (MENK: Tyr^1-Gly^2-Gly^3-Phe^4-Met^5) and leucine^5-enkephalin (LENK: Tyr^1-Gly^2-Gly^3-Phe^4-Leu^5), have been extensively studied as neurotransmitter candidates since their discovery in 1975. Coincident with interest in a neurotransmitter role for the enkephalins, and following a path taken in the early investigation of biogenic amine neurotransmitters, studies on the enzymatic breakdown of enkephalins began as soon as the primary structures of these molecules were known and synthetic peptides became available (Hughes et al. 1975). The enkephalins have been shown to be capable of undergoing enzymatic hydrolysis at three major sites within the pentapeptide molecule by a variety of preparations of nervous tissue. These sites are the Tyr^1-Gly^2, Gly^2-Gly^3 and Gly^3-Phe^4 bonds, the cleavage of any of which will generate products with little or no activity at opiate receptors. The enzymes catalyzing the hydrolysis of any of these bonds could therefore potentially be involved in the physiological regulation of the biological effects of the enkephalins.

It is instructive to analyze the attempts to ascribe a physiological significance to the enzymatic hydrolysis of the enkephalins, and to the several activities capable of acting on these peptides. Two observations on enzymatic modifications of MENK—early reports on the rapid removal of N-terminal tyrosine from MENK by rat brain homogenates (Hambrook et al. 1976) and the observation that substitution of a D-alanine residue in position two of MENK-enhanced enkephalin biological activity (Pert et al.

1976)—focused attention on aminopeptidase action on enkephalins. Subsequently, aminopeptidases were purified from the soluble fraction of whole brain from a number of species, by using MENK or LENK as a substrate (for a review, see Hersh 1982). Such biochemical characterization revealed these soluble aminopeptidases to be metalloenzymes of (in the majority of cases) molecular weight 92–100,000 and exhibiting K_m values for the enkephalins (for the most highly purified forms) in the 20–30 μM range. Of significance to physiological experimentation, these enzymes were found to be inhibited by several antibiotics, such as bacitracin, and puromycin, with potencies in the micromolar to submicromolar range. Early attempts to use these inhibitors to implicate aminopeptidase-catalyzed hydrolysis in the regulation of enkephalin actions met with little success. Thus, for example, puromycin concentrations as high as 100 μM were found to have little effect on the recovery from the medium of endogenous MENK pools liberated from brain slices by elevation of extracellular potassium (Patey et al. 1981).

This inability of puromycin to protect against MENK degradation in a tissue preparation while yet able to inhibit MENK hydrolysis by a purified soluble aminopeptidase led to the view that aminopeptidase activity was not of physiological relevance to enkephalin neuronal systems. Although a particulate locus of the aminopeptidase could be detected (e.g. Craves et al. 1978), this activity was also sensitive to inhibition by puromycin. In spite of this unpromising start, several recent observations suggest that membrane-associated aminopeptidase activity could indeed be involved in hydrolyzing the enkephalins in a physiologically relevant way. First, Hersh (1981) used detergent solubilization of a rat brain membrane fraction to isolate two forms of membrane-associated aminopeptidase activity. One of these forms was very similar in properties to the soluble, puromycin-sensitive aminopeptidase mentioned above. The second form, however, exhibited a much higher K_m for MENK (20–100 fold), but most significantly, a K_I for puromycin of about 1 mM, as opposed to a value of 1 μM for the first form of the enzyme. Aminopeptidase activity measured in a total crude membrane fraction was dominated by the puromycin-sensitive form. The puromycin-insensitive form, however, was sensitive to inhibition by another antibiotic, bestatin, which had been shown to be a potent inhibitor (micromolar to submicromolar IC_{50} or K_I values) of both soluble (Wagner & Dixon 1981) and particulate (Hudgin et al. 1981) aminopeptidase activities.

Although nothing is currently known about the cellular or subcellular disposition of these two membrane-associated forms of enkephalin-hydrolyzing aminopeptidase activity, recently reported results on en-kephalin degradation by whole-cell preparations incubated in the presence of antibiotic inhibitors support the possibility that two aminopeptidase pathways may be involved. Thus, de la Baume et al (1982, 1983) incubated rat striatal slices with exogenous [^{3}H-Tyr^1]-MENK in the presence and absence of puromycin or bestatin and analyzed the production of [^{3}H]-tyrosine as a measure of aminopeptidase action. The first and most noteworthy finding of this study was that about 80% of MENK degradation measured in this way was by aminopeptidase action, as opposed to cleavage of the other peptide bonds in the pentapeptide. Second, 0.1 mM puromycin protected against aminopeptidase degradation by about 40%, while 20 μM bestatin afforded about 80% protection. Similar results—i.e. the majority of degradation occurring via the aminopeptidase pathway (inferred by the application of inhibitors), and a relative insensitivity of the aminopeptidase action to puromycin as opposed to bestatin—were obtained in studies in which endogenous enkephalin-containing peptides, measured by MENK radioimmunoassay, were released from striatal slices by elevated extracellular potassium (de la Baume et al 1982, 1983). In these studies, peptidase action was probably occurring on the surfaces of the slices, since short exposure and sampling times were used and the extracellular space in the preparation, and the time required to equilibrate it with additions to the medium, were not determined. Several additional control experiments would thus be interesting: (a) equilibrate the slice preparation with peptidase inhibitors; (b) measure by radioimmunoassay the enkephalin content in the slices over longer incubation times with and without depolarization; (c) use pharmacological agents thought to modify enkephalin synaptic relationships in the striato-nigral pathway. However, any radioimmunoassay (RIA) studies involving enkephalins must take into account that, in addition to MENK pentapeptide, other MENK-containing oligopeptides are processed from the proenkephalin precursor, that LENK-containing peptides derived from prodynorphin are present in the corpus striatum, and that such C-terminally extended enkephalins can act as inhibitors of a membrane associated enkephalin-hydrolyzing aminopeptidase purified from rat brain (Hui et al 1983). Thus, the determination of enkephalin contents in tissue experiments involving the use of peptidase inhibitors must take care with the identification of these peptides [by high performance liquid chromatography (HPLC) RIA] and recognize that they may exhibit differences in the kinetics of their hydrolysis by aminopeptidases.

Other recent studies using cultures of neural cells suggest that aminopeptidases capable of hydrolyzing enkephalins may be situated on the external surfaces of neural cells and be highly active, relative to other enkephalin-metabolizing enzymes, in the inactivation of enkephalins. Thus, Lentzen & Palenker (1983) found that when [^{3}H-Tyr^1]-LENK was incubated with cell suspensions derived from rat brain primary astrocyte cultures, [^{3}H]-tyrosine was the major metabolite. [For example, of the total enkephalin degraded at 6 min (30%), 80% of that degraded was the result of aminopeptidase action, a magnitude comparable to that observed by de la Baume et al (1982, 1983) in striatal slices.] The production of [^{3}H]-tyrosine could be completely inhibited by bestatin, but, unfortunately, the relative abilities of puromycin and bestatin to inhibit glial aminopeptidase activity were not tested. Also unfortunate was the absence of data on the ability of media conditioned by these cultured cells to hydrolyze LENK. However, it is noteworthy that dispersed cell cultures grown on glass coverslips also gave rise to [^{3}H]-tyrosine as the main degradation product.

Also using cultured neural cells as a model system for investigating enkephalin hydrolysis, Bauer's laboratory has provided data that suggest a heterogeneous distribution of enkephalin-metabolizing enzymes between neuronal and glial cells (Horsthemke et al 1983). Using primary dispersed cell cultures enriched in neurons from rat brain and cultures enriched in astroblasts (which could be differentiated in vitro by addition of dibutyryl cyclic AMP) from mouse brain, these workers also found [^{3}H]-tyrosine to be the major product derived from [^{3}H-Tyr^1]-LENK when all three types of cultures (neuron-enriched, astroblast-enriched, and astrocyte-enriched) were incubated with the radiolabeled pentapeptide. The inclusion of puromycin or bestatin in the incubation medium (0.1 mM) gave results suggestive of multiple pathways of aminopeptidase hydrolysis of enkephalins. In neuronal cultures, bestatin inhibited [^{3}H]-tyrosine formation by 100%, while puromycin gave 86% inhibition; in astroblast and astrocytic cultures, these values were 94% and 26%, and 96% and 67%, respectively. These observations suggest a significant representation of bestatin-sensitive, puromycin-insensitive aminopeptidase activity on the external surface of glial cell membranes. However, a critical understanding of the nature of the aminopeptidases involved in these experiments awaits the purification of these enzymes, the generation of highly specific inhibitors of them, and the raising of antisera to them.

Collectively, the above evidence points to the possibility of a significant action of membrane-associated aminopeptidase activity in the breakdown of enkephalins, and warrants the further study of these enzymes to determine the precise physiological significance of their action. It is known that membrane-associated aminopeptidase activity against enkephalins can be inhibited by other neuropeptides (Hersh 1982, Hui et al 1983). Future studies should determine whether these aminopeptidases can use other neuropeptides as substrates, as the soluble forms do, and whether therefore they might be involved in regulating diverse neuropeptide systems. Of course, the ability of these aminopeptidases to be inhibited by other neuropeptides that are not substrates could be of functional significance in regulation at synapses where more than one neurotransmitter is involved. Also, the ability of the enkephalin-hydrolyzing aminopeptidases to cleave other neuropeptides at significant rates would not compromise a potential physiological role for them, since, as the following discussion of the most famous neuropeptidase, "enkephalinase," suggests, rigid biochemical specificity may be less important than neuroanatomical and geometrical specificity as determinants of the physiological roles of peptidases.

Early studies on enkephalin breakdown demonstrated that, in addition to the avid aminopeptidase activity exhibited by both soluble and particulate fractions of neural tissue, cleavage of the Gly^3-Phe^4 bond could be catalyzed by membrane fractions from brain (Malfroy et al 1978), as well as by purified angiotensin-converting enzyme (ACE) [a widely distributed dipeptidyl carboxypeptidase that catalyzes the conversion of angiotensin I to angiotensin II by removing the C-terminal dipeptide of the former] (Erdos et al 1978). Because analysis of the products derived from the hydrolysis of $[^3H$-$Tyr^1]$-LENK by a striatal membrane preparation did not reveal the presence of $[^3H$-$Tyr]$-Gly-Gly-Phe, and because of the observation that ACE, a known dipeptidyl carboxypeptidase, could cleave the Gly^3-Phe^4 bond of enkephalins, Malfroy et al (1978) inferred that the cleavage pattern generated by the striatal membranes was not due to sequential carboxypeptidase action, but to removal of the Phe-Met dipeptide, i.e. to a dipeptidyl carboxypeptidase. This Gly^3-Phe^4 cleaving activity in neural preparations was subsequently shown to be distinct from ACE (reviewed by Schwartz et al 1981, Hersh 1982), and was named "enkephalinase," presaging a body of work directed at defining a "neuropeptidase," analogous to acetylcholinesterase, active in regulating the synaptic actions of enkephalins. Thus, neurochemical and neuropharmacological studies were carried out that suggested a close relationship between a highly specific "enkephalinase" activity and opioid peptide systems (see Schwartz et al 1981):

1. A high affinity of striatal membrane "enkephalinase" for the enkephalins (i.e. K_m values in the micromolar range) was observed.
2. A likely metallopeptidase nature of the activity was inferred by virtue of "enkephalinase" inhibition by metal-chelating agents; by likening the active site of "enkephalinase" to that for the metalloenzyme carboxypeptidase A, the Phe^4 residue of enkephalins was assumed to fit in a Zn atom-containing enzyme active site so as to give rise to Gly^3-Phe^4 bond cleavage via a carboxypeptidase-like mechanism. The successful synthesis of a highly potent metal-chelating "enkephalinase" inhibitor,

thiorphan : 3-mercapto-2-benzylpropanoyl glycine (K_1 4 nM), was based on this model.

3. The use of thiorphan in in vitro striatal slice preparations protected against the degradation of both exogenous and released endogenous MENK, and exhibited potentiation of the naloxone-sensitive anti-nociceptive action of exogenously administered enkephalin analogs in vivo.

4. A parallel subcellular distribution, through sub-synaptosomal fractionation, of "enkephalinase" activity and opioid peptide binding activity was observed in rat cortex.

5. An increase in striatal membrane "enkephalinase" activity of the order of 25% was observed after four days of exposure to morphine from an implanted pellet.

This model postulates a highly enkephalin-specific dipeptidyl carboxypeptidase as being the most prominent membrane-associated enkephalin-degrading enzyme that could be involved in regulating the synaptic actions of these neuropeptides.

Two lines of investigation have led to the need to modify the model. The first line of investigation has used the highly potent aminopeptidase inhibitor, bestatin, in enkephalin-protection experiments in vitro that employed intact neural cells or tissue slices. These preparations have been described above in connection with studies on membrane-associated aminopeptidases (de la Baume et al 1982, 1983, Lentzen & Palenker 1983, Horsthemke et al 1983). In these same studies, "enkephalinase" activity—the production from [^{3}H-Tyr^1]-MENK or LENK of [^{3}H-Tyr^1]-Gly^2-Gly^3—and its inhibition by thiorphan were also investigated. In all of the cases studied—striatal slices, cultured astrocytes and astroblasts, and neuron-enriched cultures—enkephalinase activity was present, *but was minor relative to membrane-associated aminopeptidase activity*. Of interest in the study of Horsthemke et al was that the specific activity of "enkephalinase" was ten-fold higher in glial than in neuronal cultures.

The second line of recent investigation of "enkephalinase" activity has come from enzymological studies and has provided data that will have a major impact on future peptidase studies. The question at issue is the designation of "enkephalinase" as a dipeptidyl carboxypeptidase. The biochemical data supporting such a designation (reviewed by Schwartz et al 1981) did not directly test the possibility that this activity might be an endopeptidase. In 1981, Blumberg et al noted similarities between striatal membrane "enkephalinase" and the bacterial metalloendopeptidase, thermolysin, and showed that synthetic metal-chelating compounds (amino acid hydroxamates) were potent inhibitors of both of these enzyme activities. In other studies (Hudgin et al 1981) it was shown that other, nonopioid neuropeptides such as substance P and neurotensin, which possessed internal sequences that conformed to the specificity "rules" deduced from attempts to demonstrate a dipeptidyl carboxypeptidase nature for "enkephalinase," were excellent inhibitors of this activity. More critically, a metalloendopeptidase that exhibited a close similarity to "enkephalinase" activity was isolated from a pituitary membrane fraction (Almenoff et al 1981). Subsequently, work from Kenny's laboratory has provided a basis for a unifying hypothesis regarding "enkephalinase" activity. These workers had purified a metalloendopeptidase from pig kidney microvillar membranes that exhibits a preference for the hydrolysis of peptide bonds on the amino side of hydrophobic amino acid residues in oligopeptides (Kerr & Kenny 1974).

It has now been shown that this enzyme exhibits virtually identical hydrolysis of LENK and substance P to that catalyzed by a pig caudate nucleus synaptic membrane fraction in parallel experiments, in terms of products generated, response to inhibitors, and, most significantly, inhibition by a specific polyclonal antibody raised against the enzyme purified from pig microvillar membranes (Fulcher et al 1982, Matsas et al 1983). These observations—plus those from other studies showing that enkephalin-hydrolyzing metalloendopeptidases are also capable of hydrolyzing other neuropeptides such as bradykinin, neurotensin, and oxytocin (Almenoff et al 1981)—suggest that the activity in brain membranes originally termed "enkephalinase" is a metalloendopeptidase that has a broad tissue distribution and is capable of hydrolyzing a variety of neuropeptides. This suggests that the "synaptic neuropeptidase" model based on "enkephalinase" must be considerably revised. Does this mean that the neuropeptide-degrading metalloendopeptidase is "nonspecific" and cannot therefore play a physiologically important role in regulating the synaptic actions of neuropeptides, save perhaps a scavenger one? Based on the considerations presented in a preceding section on models for the roles of peptidases, we would answer "no." First, the ability of this enzyme to hydrolyze several biologically active peptides does not indicate a lack of specificity. In fact a chemical specificity to catalysis by the enzyme is manifest in a preference for cleaving an X–Y bond, where Y is one of seven hydrophobic amino acid residues in small to intermediate sized peptides. In addition, recent kinetic studies with the purified enzyme suggest that, among opioid peptides (derived from the three opioid peptide families: proenkephalin, proopiomelanocortin, prodynorphin), those containing the MENK sequence are more reactive than LENK-containing ones (Hersh 1984).

Since conformational factors probably play a major role in the susceptibility of oligopeptides to hydrolysis by peptidases (Hersh 1980, McKelvy 1983, Hui et al 1983), especially conformations directed by hydrophobicity forces as exemplified in the cleavage of the hydrophobic "signal peptides" from primary translation products by membrane-associated peptidases in the endoplasmic reticulum, the preferences exhibited by the metallo-endopeptidase may confer significant specificity on its action. This may be especially true when the quantitative differences (first order rate constants) in the ability of the enzyme to hydrolyze these substrates are well understood. Specificity defined in this way may be very important to regulation in synaptic fields in which a variety of neuropeptides are released, a phenomenon that seems likely to exist, given the abundance and wide distribution of neuropeptides. The properties of this neuropeptide-degrading metalloendopeptidase tell us something about the complexity of chemical signaling, certainly in the nervous system, but in the entire organism as well: a useful contribution to the regulation of signaling has been bound in the evolution of a small number of enzymes that are able to act on a wide variety of signals, perhaps on the basis of quantitative differences in activity due to the different "chemistries" of the signals. This view is in distinction to the existence of a large number of highly specific enzymes equivalent to the diversity of signals, a mechanism that does not appear superior to the more economical alternative. Interestingly, such a few-enzyme view gains some circumstantial support from our current understanding of the enzymes that "process" neuro- and hormonal peptides, where small families of tryptic, carboxypeptidase, and acylating enzymes can be inferred to be acting on many different, biologically active peptides (Loh et al 1984).

Luteinizing Hormone Releasing Hormone (LHRH)

The enzymatic inactivation of neuropeptides can be studied independently of their synaptic actions in a neuroendocrine projection system where the peptide-secreting neurons terminate on fenestrated capillary endothelial cells, into which the release peptides are borne by the blood to their target cells. This system affords the opportunity to investigate the role soluble peptidases might play in the regulation of intracellular peptide pools, and hence the amount of neuropeptide available for secretion, as opposed to the regulation of the lifetimes of extracellular peptides in synaptic space. In the laboratory of one of us (J. F. M.), the enzymatic hydrolysis in the hypothalamus of luteinizing hormone releasing hormone (LHRH: pGlu^1-His^2-Trp^3-Ser^4-Tyr^5-Gly^6-Leu^7-Arg^8-Pro^9-Gly^{10}-NH$_2$) has been studied to evaluate peptidase action in neuroendocrine regulation. We have investigated the regulation of gonadotropin secretion, which has the advantages of being a relatively well-defined physiological process—the integration of peripheral steroid and central neuronal regulatory signals by anatomically defined "final common pathway" hydrophysiotropic LHRH secreting neurons—and allowing for several kinds of correlations to be made between LHRH hydrolysis and physiological change. In spite of these advantages, however, until recently no clearly defined role had emerged for peptidases acting on LHRH at the hypothalamic level. This was probably due to several factors:

1. the presence of activities in hypothalamic extracts that could cleave LHRH at several potential sites ($pGlu^1$-His^2, Tyr^5-Gly^6, Gly^6-Leu^7, Pro^9-Gly^{10}-NH), without information about the intrahypothalamic distribution or compartmentation of these activities;
2. the use of inappropriate assays for hypothalamic LHRH degradation; for example, the use of synthetic arylamidase substrates or the use of LHRH radioimmunoassay in which cross-reactivities with possible LHRH degradation fragments were not assessed;
3. the use of nonoptimal conditions in physiological modelling.

Some recent studies have attempted to take all of these factors into account (Advis et al 1982, 1983, Krause et al 1982). The physiological model chosen was that of the positive feedback of ovarian steroids on rat hypothalamus to promote activation of LHRH neurons projecting from the medial preoptic area (POA) to the median eminence (ME) and a subsequent preovulatory surge of luteinizing hormone (LH) by the anterior pituitary gland during the first estrus cycle at puberty. The strategy for measuring peptidase activity was three-fold:

1. To carry out tissue sampling discretely, in individual animals, so that compartmentalization of the intrahypothalamic LHRH neuron system could be detected. Thus, the POA (site of cell bodies of origin of the hypophysiotropic LHRH system) was "punched" from coronal sections and the ME (site of hypophysiotropic LHRH nerve terminals) was discretely dissected.
2. To measure total LHRH degradation as initial rates of degradation of exogenous LHRH in crude homogenates of these sites without the addition of exogenous stabilizing agents, such as thiols or chelators, so that enzyme activity reflects the status of the tissue during the estrus cycle.
3. To carry out a *chemical* analysis of LHRH degradation; here we used high performance liquid chromatographic (HPLC) analysis of the loss of LHRH by measurement of the absorbance at 210 nm, combined with quantitation by digital integration.

Of particular importance was the need to use rapid HPLC analysis in order to process the large number of samples generated for correlations involving initial rate measurement to physiological status and yet unambiguously measure chemically the activity that gave rise to a loss of LHRH. This was achieved by use of an isocratic system based on the triethylammonium phosphate-acetonitrile system, which could separate the LHRH peak from all degradation products with a run time of six minutes per sample. The lack of cochromatography of any degradation product with the LHRH peak was assured by initially showing that collection of the latter peak from experimental samples gave amino acid analysis data consistent with only LHRH being present.

As additional aspects of the experimental design, we took advantage of the following:

1. The sensitivity of detection of LHRH loss enabled us to retain sufficient sample after isocratic measurements were complete to carry out gradient HPLC analysis of degradation products to define specific bond-cleaving activities that might be changing.

2. An enzyme cleaving LHRH at a specific bond—Pro^9-Gly^{10}-NH_2: postproline-cleaving enzyme (see McKelvy 1983), or proline endopeptidase—had been purified from brain and could be assayed in the remainder of our tissue sample if a sufficiently sensitive rapid assay could be developed.

Such an assay would detect changes in a defined peptidase capable of acting on LHRH. Accordingly, we developed a sensitive fluorometric assay for this activity based on its specificity for the carboxyl side of L-prolyl residues, the hydrolysis of carbobenzoxy-Gly-Pro-methylcoumarineamide to give the fluorophore 7-amino-3-methylcoumarine.

Using this experimental strategy, we made several findings that suggested that LHRH degradation at a specific bond cleavage site is regulated in the hypothalamus during positive feedback:

1. During the first estrus cycle at puberty, total LHRH hydrolysis in the median eminence was highly correlated with the ME content of LHRH throughout the estrus cycle, except during an interval 3 hr before the preovulatory surge of LH. At this time LHRH hydrolysis decreased and was dissociated from the still-rising ME content of LHRH, a situation expected as a prelude to active release of LHRH into the portal vessels.

2. There was no such correlation between LHRH degradation and content in the POA, the site of LHRH cell bodies, and neither was there any significant change in postproline-cleaving enzyme activity during the estrus cycle, suggesting that nonspecific changes were not contributing to the level of activity of LHRH degradation.

3. The same pattern of inhibition of LHRH hydrolysis and dissociation of level of degradation from LHRH content was observed following progesterone administration to estrogen-primed castrated animals.

4. The administration to ovariectomized, steroid-treated rats of diethyldithiocarbamate (DDC)—a norepinephrine synthesis inhibitor that can abolish the LH surge through blockade of its noradrenergic requirement at the hypothalamic level—resulted not only in failure of progesterone administration to elicit an LH surge but in the failure of progesterone administration to effect the transient inhibition of LHRH degradation prior to the LH surge seen in non-DDC-treated animals.

5. The analysis of LHRH degradation products in all of these physiological situations revealed that Tyr^5-Gly^6 cleavage accounted for the LHRH degradation observed, probably by a metalloendopeptidase similar to the partially purified pituitary "non-chymotrypsin-like" endopeptidase, described by Bauer and associates (Horsthemke & Bauer 1980), that cleaves LHRH. Moreover, this activity could be observed in both a soluble and a particulate fraction.

These observations suggest that degradation may be regulated in LHRH nerve terminals undergoing active and phasic LHRH release. The mechanism of this regulation is unknown, but could involve progesterone effects at the transcriptional level to decrease the rate of synthesis of the peptidase (if it turns over rapidly); alternatively the steroid could produce changes in the compartmentation of the enzyme in nerve terminals, resulting in its inability to gain access to LHRH, perhaps by inhibiting lysozome-secretory granule fusion. Analysis of the subcellular distribution of the enzyme in the median eminence during the time course of progesterone effects on peptidase activity would be of great interest in this regard.

INACTIVATION AND PROCESSING

The simplest definition of neuropeptide inactivation by enzymatic hydrolysis is that the action of the enzyme results in the generation of products that cannot effectively activate a particular receptor. But what if one of these hydrolysis products is now able to act on a different receptor, toward which the unhydrolyzed parent neuropeptide was inactive? This would be analogous to the "processing" reactions in neuropeptide biosynthesis in which biologically active peptides are derived, principally by proteolysis, from biologically inert precursors. In fact, several "mature" neuropeptides are acted on by (presumably catabolic) peptidases to give rise to new biological activities. Thus, for example, thyrotropin-releasing hormone (TRH: $pGlu^1$-His^2-Pro^3-NH_2) can be acted on by an exopeptidase that hydrolyzes the peptide bond between its cyclized N-terminal amino acid (pyroglutamic acid) and the histidyl residue to yield a dipeptide amide histidyl proline, amide, which rapidly and spontaneously undergoes cyclization to form a His-Pro diketopiperazine (DKP). The DKP is devoid of the pituitary thyroid-stimulating hormone (TSH) activity characteristic of the TRH molecule from which it was derived, but the DKP is able to inhibit the release of prolactin from the pituitary, whereas TRH cannot (Bauer et al 1978). Another example is the action of a homogeneous soluble enkephalin-hydrolyzing activity from bovine brain (Hersh & McKelvy 1981) on endorphins. Here, the aminopeptidase acts on γ-endorphin (β-lipotropin 61-77) so as to remove only N-terminal tyrosine (Hersh et al 1980). This modification abolishes agonist activity of γ-endorphin at opiate receptors, which requires N-terminal tyrosine, but generates a product, des-Tyr-γ endorphin, which is active in assays for certain adaptive behaviors (de Wied et al 1978).

The action of endopeptidases on neuropeptides have also been shown to generate new biological activities. For example, the action of a prolyl endopeptidase from rat brain on substance P (SP: Arg^1-Pro^2-Lys^3-Pro^4-Gln^5-Gln^6-Phe^7-Phe^8-Gly^9-Leu^{10}-Met^{11}-NH_2) results in cleavage of only the Pro^4-Gln^5 bond, yielding an N-terminally-derived basic tetra-peptide and a C-terminally-derived hydrophobic heptapeptide amide (Blumberg et al 1980). Subsequent biological studies demonstrated that these two peptide fragments exhibit distinct and non-overlapping biological activities: the N-terminal sequence enhances phagocytosis by macrophages and elicits histamine release from mast cells, whereas the C-terminal sequence causes smooth muscle to contract and depolar-izes motoneurons (reviewed by Blumberg & Teichberg 1982). These observations could be related to a contribution by primary sensory neurons to a response to peripheral injury.

Although other examples of this kind appear in the literature, informa-tion on which to judge the physiological significance of this phenomenon is still insufficient. Given our growing appreciation for the multipotent potential of neuropeptide polyproteins, however, molecules we now regard as "mature" neuropeptides might well be found to yield biologically significant species after being acted on by "degradative" peptidases. Perhaps peptide bond hydrolysis even to the level of single amino acids is of significance in some projection systems within the CNS.

UPTAKE OF NEUROPEPTIDES

This review reflects the current view that neuropeptides must be inactivated by enzymatic hydrolysis because high affinity uptake systems for peptides do not exist in nervous tissue. Nevertheless, neuropeptide uptake by slice preparations of neural tissue have been observed. In fact, the first such report came from S. Udenfriend's laboratory in 1964, before the great era of neuropeptides. Abraham et al (1964) observed that the dipeptide carnosine

(β-alanyl-histidine) was taken up by rat brain slices in a concentrative process (tissue to medium ratio = 22 at maximal uptake) which required glucose and oxygen. Furthermore, two studies on the uptake of TRH by CNS tissue have been reported. Pacheco et al (1981) showed that rat cerebellar slices accumulated [^{3}H-Pro]-TRH by a process that shared a number of characteristics with transport systems: (a) the accumulation saturated with time and was concentrative, with a maximal uptake at 60 min, when a tissue to medium ratio of 5:1 was achieved; (b) uptake saturated with increasing TRH concentration, exhibiting K_m values of 10 and 5 μM; (c) the process was temperature dependent, with a Q_{10} of 1.48; and (d) it was dependent on the presence of Na$^+$ in the medium. Charli et al (1984) studied the accumulation of [^{3}H]-TRH by rat hypothalamic slices, and found it to be time dependent, also with a maximal uptake at 60 min, but with a lower tissue to medium ratio, 1.3, than in the studies with cerebellar slices. This difference could be due to the much larger inulin space in the hypothalamic slices. The hypothalamic slices also exhibited a saturable accumulation of the tripeptide with respect to TRH concentration, with a K_m of 1 μM. The accumulation was also dependent on temperature and was inhibited by inhibitors of energy metabolism such as ouabain, dinitrophenol, and the absence of glucose. Nakata et al (1981) studied the accumulation of the C-terminal heptapeptide amide of substance P in slices of rat brain and rabbit spinal cord. They found that in both preparations of nervous tissue the accumulation process exhibited time, temperature, and energy dependence. The accumulation of the peptide was also saturable with respect to peptide concentration, exhibiting K_m values of 0.4 nM and 1 μM in the rat brain slices and a single K_m of 0.1 μM in the rabbit spinal cord slices.

Because we have so few studies of peptide uptake, and also because we know little about the role of peptidases these observations are difficult to relate directly to the regulation of neuropeptide activity at synapses. The long time courses (initial rates perhaps in the one to several minute range) and relatively high K_m values may suggest that uptake could be physiologically relevant, given the long duration of neuropeptide action and the half times and dissociation constants of neuropeptide-receptor interactions (*vide supra*); uptake, then, might operate in tandem with peptidases to regulate neuropeptide concentrations in the extracellular space. Clearly more studies are warranted on the uptake of peptides by nervous tissue.

FUTURE DIRECTIONS

Do we have sufficient critical evidence to assess how neuropeptide actions are terminated? On the basis of the findings summarized in this review, apparently we do not. However, research carried out on the enzymatic inactivation of neuropeptides by peptidases has made good progress in the biochemical definition of these enzymes, and the field stands ready to benefit from the application of biochemical tools, enzyme inhibitors and anti-enzyme antibodies, to neurobiological systems. Particularly promising is the use of such antibodies to answer the fundamental question of where are the various peptidases localized in nervous tissue at the cellular and subcellular levels. Use of such antibodies should also prove fruitful in studying the rates of biosynthesis, or the tissue levels at least, of the peptidases in defined neuronal projection systems that can be subjected to controlled physiological or pharmacological modulation of activity. Antipeptidase antibodies can be employed to screen cDNA expression libraries, with the ultimate goal of rapidly generating models for the primary sequences of these proteins. This is the most direct way to understand the relationships between the various aminopeptidases inferred to exist from inhibitor and enzyme purification studies, and the multiple forms of a given enzyme such as the neuropeptide-metabolizing metallo-endopeptidase from rat brain. Of course, cloning studies will permit the regulation of peptidase gene expression to be investigated.

Cellular neurophysiologists interested in neuropeptide actions at the synapse should, we feel, attempt to understand how the neuropeptide might undergo inactivation in their preparation. The application of peptidase inhibitors and antibodies to the enzymes could be especially helpful. Likewise, we urge investigators studying inactivation at the biochemical level to work on neuropeptide systems that are well characterized from the cellular neurophysiological perspective. A number of well-established cell culture systems could be profitably explored to answer fundamental questions about inactivation; for example, those of ciliary, dorsal root, and sympathetic ganglia. Finally, the possibility that peptide uptake could contribute to neuropeptide inactivation deserves further study, especially in simpler systems involving peptide-secreting neurons.

ACKNOWLEDGMENTS

J. F. M. gratefully acknowledges research support from the National Science Foundation and the National Institutes of Health. We thank Dr. Paul Adams for helpful discussions.

Literature Cited

Abraham, D., Pisano, J. J., Udenfriend, S. J. 1964. Uptake of carnosine and homo-carnosine by rat brain slices. Arch. Biochem. Biophys. 104 : 160–65
Adams, P. 1980. Aspects of synaptic potential

generation. In Information Processing in the Nervous System, ed. H. M. Pinsker, W. D. Willis Jr., pp. 109–24. New York: Raven
Advis, J. P., Krause, J. E., McKelvy, J. F.

1982. LHRH peptidase activities in discrete hypothalamic regions and anterior pituitary of the rat: Apparent regulation during the first estrus cycle at puberty. *Endocrinology* 110:1238–45

Advis, J. P., Krause, J. E., McKelvy, J. F. 1983. Evidence that endopeptidase-catalyzed LHRH cleavage contributes to median eminence LHRH levels during positive steroid feedback. *Endocrinology* 112:1147–50

Almenoff, J., Wilk, S., Orlowski, M. 1981. Membrane-bound pituitary metalloendopeptidase: apparent identity to enkephalinase. *Biochem. Biophys. Res. Commun.* 95:141–44

Bauer, K., Graf, K. J., Faivre-Bauman, A., Beier, S., Tixier-Vidal, A., Kleinkauf, H. 1978. Inhibition of prolactin secretion by histidyl-proline diketopiperazine. *Nature* 274:174–75

Blumberg, S., McKelvy, J. F. 1985. Submitted

Blumberg, S., Teichberg, V. 1982. The role of the N-terminal sequence in the biological activities of substance P. In *Regulatory Peptides: From Molecular Biology to Function, Adv. Biochem. Psychopharmacol.* 33:445–52

Blumberg, S., Teichberg, V. I., Charli, J. L., Hersh, L. B. McKelvy, J. F. 1980. Cleavage of substance P to an N-terminal tetrapeptide and a C-terminal heptapeptide by a postproline cleaving enzyme from bovine brain. *Brain Res.* 192:477–86

Blumberg, S., Vogel, Z., Alstein, M. 1981. Inhibition of enkephalin degrading enzymes from rat brain and of thermolysin by amino acid hydroxamates. *Life Sci.* 28:301–6

Chang, K. J., Cuatrecasas, P. 1979. Multiple opiate receptors: Enkephalins and morphine bind to receptors of different specificity. *J. Biol. Chem.* 254:2610–18

Charli, J. L., Ponce, G., McKelvy, J. F., Joseph-Bravo, P. 1984. Accumulation of thyrotropin releasing hormone by rat hypothalamic slices. *J. Neurochem.* 42:981–86

Craves, F. B., Law, P. Y., Hunt, C. A., Loh, H. H. 1978. The metabolic disposition of radiolabeled enkephalins *in vitro* and *in situ. J. Pharmacol. Exp. Ther.* 206:492–506

de la Baume, S., Gros, C., Yi, C. C., Chaillet, C., Marcais-Collado, H., Costentin, J., Schwartz, J.1982. Selective participation of both "enkephalinase" and aminopeptidase activities in the metabolism of endogenous enkephalins. *Life Sci.* 31:1753–56

de la Baume, S., Yi, C. C., Schwartz, J. C., Chaillet, P., Marcais-Collado, H., Costentin, J. 1983. Participation of both "enkephalinase" and aminopeptidase activities in the metabolism of endogenous enke-

phalins. *Neuroscience* 8:143–51

de Wied, D., Kovacs, G. L., Bohus, B., Van Ree, J. M., Greven, H. M. 1978. Neuroleptic activity of the neuropeptide beta-LPH$_{62-77}$. *Eur. J. Pharmacol.* 49:427–36

Dun, N. J., Minota, S. 1981. Effects of substance P on neurons of the inferior mesenteric ganglia of the guinea pig. *J. Physiol.* 321:259–71

Erdos, E. G., Johnson, A. R., Boyden, N. T. 1978. Hydrolysis of enkephalin by cultured human endothelial cells and by purified peptidylpeptidase. *Biochem. Pharmacol.* 27:843–48

Fulcher, I. S., Matsas, R., Turner, A. J., Kenny, A. J. 1982. Kidney neutral endopeptidase and the hydrolysis of enkephalin show similar sensitivity to inhibitors. *Biochem. J.* 203:519–22

Hambrook, J. M., Morgan, B. A., Rance, M. J., Smith, C. F. W. 1976. Mode of deactivation of the enkephalins by rat and human plasma and rat brain homogenates. *Nature* 262:782–83

Hersh, L. B. 1981. Solubilization and characterization of two rat brain membrane-bound aminopeptidases active on met-enkephalin. *Biochemistry* 20:2345–50

Hersh, L. B. 1982. Degradation of enkephalins: The search for an enkephalinase. *Mol. Cell. Biochem.* 47:35–43

Hersh, L. B. 1984. Reaction of opioid peptides with neutral endopeptidase. *J. Neurochem.* 43:487–93

Hersh, L. B., McKelvy, J. F. 1981. An aminopeptidase from bovine brain which catalyzes the hydrolysis of enkephalin. *J. Neurochem.* 36:171–78

Hersh, L. B., Smith, T. E., McKelvy, J. F. 1980. Cleavage of endorphins to des-tyr endorphins by homogeneous bovine brain aminopeptidase. *Nature* 286:160–62

Horsthemke, B., Bauer, K. 1980. Characterization of a non-chymotrypsin-like endopeptidase from anterior pituitary that hydrolyzes luteinizing hormone-releasing hormone at the tyrosyl-glycine and histidyl-tryptophan bonds. *Biochemistry* 19:2867–73

Horsthemke, B., Hamprecht, B., Bauer, K. 1983. Heterogeneous distribution of enkephalin-degrading peptidases between neuronal and glial cells. *Biochem. Biophys. Res. Commun.* 115:423–29

Hudgin, R. L., Charleson, S. E., Zimmerman, M., Mumford, R., Wood, P. L. 1981. Enkephalinase: Selective peptide inhibitors. *Life Sci.* 29:2593–2601

Hughes, J., Smith, T. W., Kosterlitz, H., Fothergill, I. A. Morgan, B. A., Morris, H. R. 1975. Identification of two related pentapeptides from the brain with potent opiate agonist activity. *Nature* 258:577–79

Hui, K. S., Wang, Y. J., Lajtha, A. 1983. Purification and characterization of an enkephalin aminopeptidase from rat brain membranes. *Biochemistry* 22: 1062–67

Jan, Y. N., Jan, Y. J. 1983. Coexistence and corelease of cholinergic and peptidergic transmitters in frog sympathetic ganglia. *Fed. Proc.* 42: 2929–33

Kerr, M. A., Kenny, A. J. 1974. The purification and specificity of a neutral endopeptidase from rabbit kidney brush border. *Biochem. J.* 137: 477–88

Krause, J. E., Advis, J. P., McKelvy, J. F. 1982. Characterization of the site of cleavage of LHRH under conditions of measurement in which LHRH degradation undergoes physiological change. *Biochem. Biophys. Res. Commun.* 108: 1475–81

Krieger, D. T., Brownstein, M. J., Martin, J. B., eds. 1983. *Brain Peptides.* New York: Wiley

Lechan, R. M., Goodman, R. H., Rosenblatt, M., Reichlin, S., Habener, J. 1983. Prosomatostatin-specific antigen in rat brain: Localization by immunohistochemical staining with an antiserum to a synthetic sequence of preprosomatostatin. *Proc. Natl. Acad. Sci. USA* 80: 2780–84

Lentzen, H., Palenker, J. 1983. Localization of the thiorphan-sensitive endopeptidase, termed enkephalinase A, on glial cells. *FEBS Lett.* 153: 93–97

Liston, D., Vanderhaegen, J., Rossier, J. 1983. Presence in brain of synenkephalin, a proenkephalin-immunoreactive protein which does not contain enkephalin. *Nature* 302: 62–65

Loh, Y. P., Brownstein, M. J., Gainer, H. 1984. Proteolysis in neuropeptide processing and other neural functions. *Ann. Rev. Neurosci.* 7: 189–222

Malfroy, B., Swerts, J. P., Guyon, A., Roques, B. P., Schwartz, J. C. 1978. High affinity enkephalin-degrading peptidase in brain is increased after morphine. *Nature* 276: 523–26

Matsas, R., Fulcher, I. S., Kenny, J., Turner, A. J. 1983. Substance P and leu-enkephalin are hydrolyzed by an enzyme in pig caudate synaptic membranes that is identical with the endopeptidase of kidney microvilli. *Proc. Natl. Acad. Sci. USA* 80: 3111–15

McKelvy, J. F. 1983. Enzymatic degradation of brain peptides. In *Brain Peptides*, ed. D. T. Krieger, M. J. Brownstein, J. B. Martin, pp. 118–33. New York: Wiley

Mudge, A. W., Leeman, S. E., Fischbach, G. D. 1979. Enkephalin inhibits release of substance P from sensory neurons in culture and decreases action potential duration. *Proc. Natl. Acad. Sci. USA* 76: 526–30

Nakata, Y., Kusaka, Y., Yajima, H., Segawa, T. 1981. Active uptake of substance P carboxy-terminal heptapeptide (5-11) into rat brain and rabbit spinal cord slices. *J. Neurochem.* 37: 1529–34

Pacheco, M. F., Woodward, D. J., McKelvy, J. F., Griffin, W. S. T. 1981. TRH in the rat cerebellum: II. Uptake by cerebellar slices. *Peptides* 2: 283–88

Patey, G., de la Baume, S., Schwartz, J. C., Gros, C., Roques, B., Fournie-Zaluski, M. C., Soroca-Lucas, E. 1981. Selective protection of methionine enkephalin released from brain slices by enkephalinase inhibition. *Science* 212: 1153–55

Pert, C. B., Pert, A., Chang, J. K., Fong, B. T. W. 1976. [D-Ala]-met-enkephalinamide: A potent, long-lasting synthetic pentapeptide analgesic. *Science* 194: 330–32

Rosenberry, T. L. 1975. Acetylcholinesterase. *Adv. Enzymol.* 43: 103–218

Rosenfeld, M. G., Mermod, J. J., Amara, S. G., Swanson, L. W., Sawchenko, P. E., Rivier, J., Vale, W., Evans, R. 1983. Production of a novel neuropeptide encoded by the calcitonin gene via tissue-specific RNA processing. *Nature* 304: 129–35

Schwartz, J. C., Malfroy, B., de la Baume, S. 1981. Biological inactivation of enkephalins and the role of enkephalin-dipeptidyl-carboxypeptidase ("enkephalinase") as neuropeptidase. *Life Sci.* 29: 1715–40

Sutcliffe, J. G., Milner, R. J., Bloom, F. E. 1983. Cellular localization and functions of the proteins encoded by brain specific mRNAs. *Cold Spring Harbor Symp. Quant. Biol.* 48: 477–84

Wagner, G. W., Dixon, J. E. 1981. Inhibitors of a rat brain enkephalin aminopeptidase. *J. Neurochem.* 37: 709–13

Ann. Rev. Neurosci. 1986. 9 : 435–87

INVERTEBRATE LEARNING AND MEMORY:
From Behavior to Molecules

Thomas J. Carew and Christie L. Sahley

Departments of Psychology and Biology, Yale University, New Haven,
CT 06520

INTRODUCTION

Classical ethologists have provided a long and rich tradition of studying the
learning and memory capabilities of a variety of invertebrate animals
(Schneirla 1929, von Frisch 1950, Lindauer 1960). In recent years, however,
interest in this field has been renewed because it has become clear that many
invertebrate animals can provide powerful model systems in which to
explore the cellular and molecular mechanisms of learning and memory.
Thus, a variety of disciplines such as neuroethology, psychology, and
modern neurobiology have merged in a concerted effort to develop
invertebrate preparations in which these combined approaches can be
focused on the general theme of the cellular basis of behavioral plasticity.
This overall strategy has come to be known as a "simple-systems" approach
to the study of learning and memory.

A "simple-systems" approach is not the exclusive domain of invertebrate
research. The general strategy of such an approach is to try to identify
neural circuits involved in a particular form of learning and then attempt to
specify the precise locus and nature of the cellular changes that are involved.
This approach has been successfully applied in a number of vertebrate
systems. These include eyeblink conditioning in the cat (Kim et al 1983,
Woody et al 1983, Woody 1984), cardiac conditioning in the pigeon (Cohen
1980, 1984), and conditioning of the nictitating membrane response in
rabbit (Thompson et al 1983, 1984, Moore et al 1982, Harvey et al 1985), as
well as cellular analog systems in cerebellum (Ito 1982), the red nucleus
(Tskahara, 1984), and hippocampus (McNaughton et al 1978, Levy &

435

Steward 1979, Lynch & Baudry 1984). Thus, there is currently a rapidly developing consolidation in vertebrate and invertebrate research, on both technical and conceptual levels, concentrating on the general question of the cellular mechanisms of learning and memory.

The purpose of this review is to survey recent progress made in the study of learning and memory in invertebrate animals. For reasons of both focus and constraints on length, we restrict the scope of the review in two ways. First, because of the recent major advances in the area, we primarily consider only instances of *associative learning* in invertebrates. Second, because the central theme of the review centers on neuronal mechanisms of learning, we only consider preparations in which some form of *cellular or molecular analysis* has been carried out, or is at least quite feasible. We divide the review into three parts. We first provide a brief overview of the major psychological principles and paradigms used in the modern study of learning and memory. We then consider how these principles and paradigms have been applied in a variety of invertebrate animals. We conclude with an attempt to synthesize some of the common themes and convergent ideas that have emerged from the study of learning and memory in invertebrates.

LEARNING: PRINCIPLES AND PARADIGMS

Learning is an inference made on the basis of an observed relationship between an animal's behavior and its past experience. Specifically, learning is a product of an animal's experience with one or more stimulus events and of the relationship between these events (Rescorla & Holland 1976, Sahley et al 1984). In making comparisons of learning and memory among invertebrates and vertebrates, it is particularly important to stress the similarity of *interevent relationships*. The kinds of stimuli that different species can learn about and the types of behaviors modified by learning can reflect unique evolutionary histories that may be very different, thus precluding meaningful comparisons of the stimuli responded to or the behavioral responses exhibited by different species in a learning experiment. Interevent relationships, however, are not species dependent; thus they permit direct and meaningful comparisons of learning across different species of animals.

Learning psychologists studying vertebrates have identified many of the interevent relationships that influence learned changes in behavior (Rescorla & Holland 1976, Macintosh 1974). Neuroethologists and neurobiologists studying invertebrates have looked to the vertebrate learning literature and have adapted many paradigms known to produce learning in vertebrates. Consequently it is now possible to determine whether a functional similarity exists between vertebrate and invertebrate learning processes. Before we turn to an assessment of invertebrate learning, we briefly discuss some key definitions and paradigms used in the modern study of learning.

Nonassociative Learning

Nonassociative learning refers to a change in behavior as a result of experience with either a single stimulus or two stimuli that are not temporally related. The two most common examples of nonassociative learning are habituation and sensitization. *Habituation* refers to a decrement in responding (which is not due to sensory adaptation or motor fatigue) as a result of repeated presentations of a single stimulus. *Sensitization* refers to an increase in responding to one stimulus due to one or more presentations of another intense or noxious stimulus.

Associative Learning

Associative learning implies the learning of a relationship between two or more events in the environment, and can be divided into two distinct types: *classical conditioning* and *instrumental conditioning or operant learning*. Classical conditioning is best illustrated by Pavlov's (1927) early experiments on dogs. A normal dog's reaction to meat powder (the unconditioned stimulus or US) is salivation (the unconditioned response). Following several presentations of a tone (the conditioned stimulus or CS) paired with meat powder, Pavlov found that the tone by itself came to elicit a conditioned response (CR), which was similar to the normal response to the meat powder: salivation. Procedurally, in classical conditioning the CS and US are presented to the animal with no dependence on the animal's response. The animal learns an association between these two stimuli and this association can modify many behaviors (Domjan & Burkhard 1982, Mackintosh 1974).

Instrumental learning or operant conditioning, in contrast, is response-contingent. This means that the presentation of the US is dependent upon the animal exhibiting a specific behavioral response. Thus, the animal learns an association between the response it emits and the consequence of that response (Domjan & Burkhard 1982, Mackintosh 1974). The first example of instrumental learning was Thorndike's report (1898) of cats learning to escape from a puzzle box. When initially placed in the box, cats had great difficulty freeing themselves. However, after many trial and error sequences they learned which specific response (conditioned response) allowed them to escape (reinforcement).

Temporal Pairing

Learning is critically dependent upon the time interval separating the CS and the US (the interstimulus interval or ISI). Depending on particular associations to be learned and the response that is measured, the optimal time interval for learning varies from milliseconds (e.g. the rabbit nictitating membrane response; Gormezano 1972) to hours (taste aversion; Garcia et al 1966, Garcia & Koelling 1966, Revusky & Garcia 1971).

First-Order Conditioning

The examples of classical and instrumental conditioning (from Pavlov and Thorndike respectively) described above are illustrations of first-order conditioning. First-order conditioning implies that the learning is due to the direct pairing of the CS and US (in classical conditioning), or the response and reinforcement (in operant conditioning).

Second-Order Conditioning

The defining feature of second-order conditioning is that a stimulus (CS) can come to produce a conditioned response without ever being directly paired with the US (Rescorla 1980). In this paradigm, two CSs are used (S1 and S2). For example, after an association is first acquired between a tone (S1) and shock (US), the tone can be paired with a new stimulus, say a light (S2). Following this training, the animal will show conditioned responding to the light (S2) even though the light had never been paired with the shock.

Blocking

A classic experiment by Kamin introduced a phenomenon called *blocking* that emphasized the importance of predictive relationships on the acquisition of learning (Kamin 1969). Kamin found that animals trained with compound conditioned stimuli paired with a US (i.e. S1*S2-US) subsequently showed a conditioned response to both S1 and S2. However, if animals received prior training to only one element (S1) of the compound stimulus, they failed to show learning to the second element (S2) following repeated presentations of the compound paired with the US (S1*S2-US). That is, the added CS (S2) failed to acquire conditioning even though it had been repeatedly paired with the US. The prior training to S1 blocked conditioning to S2. This suggests that when a US is well predicted by the presence of a CS with which it had been previously paired, that CS will interfere with or block conditioning to other CSs (Kamin 1969).

Contingency vs Contiguity

In another demonstration of predictive relationships, Rescorla (1968) varied the predictive value of the CS by systematically altering the degree to which US occurrence was *contingent* upon CS occurrence. This was done by giving different groups of animals the same number of CS-US pairings, but by giving some groups extra numbers of US presentations. Thus all groups had the same amount of CS-US *contiguity* (pairings), but differing amounts of CS-US *contingency* (correlation). The point is that when the US was as likely to occur in the absence of the CS as in its presence, the CS no longer *predicted* the US, even though it may have had a strong temporal relation with it. Rescorla found that as the predictive value of the CS decreased, so did its ability to evoke conditioned responding.

Conditioned Inhibition

Conditioned inhibition is the result of a special kind of predictive relationship. As first described by Pavlov (1927) and more recently extended by Rescorla (1969), animals can actually learn that a CS predicts the absence of the US. When this occurs, inhibitory learning takes place. Procedurally, conditioned inhibition can be obtained by several experimental manipulations. For example, prior to CS-US pairings, animals can be given explicitly unpaired CS-US presentations. Thus animals first learn a negative correlation between CS and US, endowing the CS with inhibitory properties in subsequent learning tasks.

Overshadowing

The salience of each element in a compound stimulus greatly influences the amount of learning to that stimulus. When two stimuli are presented as a compound, the particular stimulus that is more strongly perceived by the animal will tend to *overshadow* the weaker stimulus, i.e. less conditioning will occur to the weaker stimulus (Pavlov 1927). Rescorla & Wagner (Rescorla 1969, Rescorla & Wagner 1972) suggest that the separate stimuli within a compound compete for a fixed amount of associative strength, with the most salient stimulus getting the larger amount. Mackintosh suggests that each stimulus competes for attention (1974).

Conditioned Emotional Response (CER)

The CER procedure was first introduced as an indirect measure of conditioned fear by Estes & Skinner (1941). An "emotional response" (fear) is thought to become associated with stimuli that are paired with shock. Thus when the conditioned stimulus (say a light) is paired with shock, the subsequent presentation of the light will make the animal become fearful. The conditioned fear can have dramatic effects on ongoing behavior. For example, presentation of the light can disrupt appetitively motivated behaviors such as feeding or bar-pressing for food, and, in contrast, it can facilitate aversively motivated behaviors such as escape or avoidance responses. Thus there is a motivationally consistent response to the fear-eliciting conditioned stimulus (for review see Macintosh 1974).

The preceding brief overview of learning principles and paradigms illustrates that animals are capable of forming a variety of associations between events occurring on their world. In the sections that follow we examine the way these paradigms have been applied to the study of learning and memory in a number of invertebrate animals. In each section we first describe behavioral experiments and then discuss cellular and, where relevant, molecular studies.

APIS MELLIFERA

Learning in honey bees has been observed since Aristotole, but systematic investigation began with the classic studies of von Frisch (1921, 1950). Two distinct methods have been used to study learning in bees: the free-flying procedure and the proboscis extension paradigm.

Free-Flying Procedure

Menzel and his colleagues trained bees to fly from their hive to an experimental testing station on which spectral colors were projected onto ground glass disks. Individual bees were exposed to two complementary stimulus colors, and an initial baseline preference was recorded. The bee was then rewarded on the nonpreferred color (the CS) by allowing her to drink sugar water (the US) after she has landed on the color (no sugar water was given when the bee landed on the other color). With this procedure Menzel and co-workers demonstrated that bees can learn to approach a number of colors that have each been paired with the US. Approaches to violet were learned most readily and to bluish green least readily (Menzel 1967, 1968, Menzel & Erber 1972, Menzel et al 1974). Temporal characteristics are especially important for the bee in learning the color–sugar water association. Menzel reports that the bee is able to learn the association with an interstimulus interval of up to 2–3 sec, if the CS is presented before the US. However, when the color CS is presented only following the bee's landing and feeding, that color is not associated with the US (that is, backward conditioning is not effective) (Grossmann 1970, Menzel 1968, Opfinger 1931). This is consistent with earlier findings that a bee landing on one color and then feeding on another will prefer the landing color to the feeding color (Opfinger 1931).

In contrast to vertebrates and some other invertebrates (Hawkins et al 1983b, Farley & Kern 1984, Sahley et al 1981b), the predictive value of the CS does not appear to be important for associative learning in the bee. To examine this question, Couvillon et al (1983) used a blocking paradigm with two groups of bees: One group was first trained with a single conditioned stimulus, S1-US, and subsequently trained with a compound stimulus, S1∗S2-US (Blocking Group); a second group was trained first with the compound and then with a single stimulus (Control Group). The groups were then compared in their performance to the second conditioned stimulus (S2). If predictability of the US were important to stimulus selection, one would expect less response to S2 in the Blocking Group than in the Control Group. This was not observed. In fact, Couvillon et al (1983) report relatively more response to S2 in the Blocking Group.

When two or more stimuli are presented together as a compound stimulus and paired with a US, the response often is associated with only one of the two stimulus elements. This is termed *stimulus selection*. The salience of an element within a compound greatly influences learning to each element (i.e. one element can "overshadow" another—see previous section). In several experiments bees show overshadowing. Von Frisch (1921) found that after exposure to both an odor and a color in a home hive, bees would respond more to a hive marked with odor than to one marked with color. Similarly, bees learn a discrimination more quickly when the odor is the relevant cue than when color is important (Klosterhalfen et al 1978). Finally, odor also appears to overshadow color when bees are presented odor-color compounds in the free flying learning procedure (Couvillon et al 1983).

The retention of learning in bees appears to be a biphasic function (Menzel 1979). That is, after a single training trial, retention is highest immediately after learning and declines within the next 2–3 min. However, retention again increases 10–20 min following conditioning and then slowly decays over the next 3–5 days. Interestingly, following only 3 CS-US pairings there appears to be no memory loss over the lifetime of the bee (2 wk). Three training trials appear to produce maximal learning: groups of bees given additional training show no differences in retention (Menzel 1979). Menzel postulates short- and long-term memory phases to account for the biphasic nature of retention. Consistent with this hypothesis, memory is more easily interfered with during a transition phase than it is after the memory has become stable. Bees experiencing electroconvulsive shock (ECS), cooling to $+1°C$, or CO_2 or N_2 narcosis within the first 2 min following conditioning show a dramatic decrease in learning, compared to bees experiencing the same treatment 4–7 min following training (Menzel 1968, Erber 1976, Menzel et al 1974).

The nature of the information used by bees in remembering flower shapes has been extensively investigated (Wehner 1972, Collett & Cartwright 1982, Cartwright & Collett 1983, Gould 1985). From these studies it appears that bees can remember eidetic images of landmarks. For example, Gould (1985) found that bees were able to learn to distinguish between a variety of patterns that differed primarily or exclusively in the spatial relations among the elements. He concludes that bees can store flower patterns as a low-resolution photograph in their brains.

Proboscis Extension

The proboscis extension paradigm was originally developed by Kuwabara (1957) and has recently been used extensively. In this procedure, forager bees are caught at the entrance to the hive and harnessed individually in small tubes. A sugar solution (the US) presented to the antennae elicits a reflex response in which the antennae move in a forward direction, the mandibles open, and the proboscis is extended. If an olfactory stimulus (the CS) such as a flower odor is paired with sugar water to the antennae, bees quickly (one trial) develop a conditioned response to the odor that was paired with the sugar water (US) (Erber 1980, 1981, Vareschi 1971, Masuhr & Menzel 1972, Menzel et al 1974). Color-sugar water associations are learned more slowly (Kuwabara 1957, Masuhr & Menzel 1972).

This paradigm has been very useful for studying higher-order learning phenomena in the bee. For example, Bitterman and his colleagues have demonstrated conditioned inhibition (Bitterman et al 1983, Couvillon & Bitterman 1980). They found that bees that had experienced explicitly unpaired presentations of the odor and sugar water extended their proboscises when placed in the training situation with no odor present, and then retracted them when the odor was presented. This is in contrast to bees that had experienced odor-sugar water pairings. Animals trained with paired CS-US presentations extend their proboscises when the odor is presented and not when it is absent. Moreover, bees show a dramatic retardation of learning to the formerly unpaired CS when it is subsequently paired with sugar water. These two findings taken together provide good evidence of conditioned inhibition in the bee. In addition, Menzel & Bitterman (1983) have shown that the variables that produce second-order conditioning in vertebrates also result in second-order conditioning in the bee.

In addition to proboscis extension, bees respond to sugar-water stimuli presented to the antennae by moving the antennae toward the drop of sugar water. This response can also be conditioned. After five odor-sugar water pairings the bee responds with antennal movements directed to the odor (Erber 1982). Likewise, pairings of a moving striped pattern with sugar to the antennae result in conditioned antennal movements to the moving pattern (Erber & Schildberger 1980).

Neural Studies of Bee Learning

The bee brain (supraesophageal ganglion) is approximately 1 mm^3 and contains about 900,000 neurons (Witthoft 1967). The midbrain or mushroom bodies of the brain have been thought for anatomical reasons to be important for learning in the bee (Mobbs 1982). Moreover, cooling of the antennal lobes and the mushroom bodies shortly after learning appears to block memory formation (Erber et al 1980, Menzel 1984). The mushroom bodies receive input from both the visual and olfactory centers in the brain and have an extremely high density of fibers and synaptic connections (von Alten 1919, Mobbs 1982). Intracellular recording from neurons in the bee brain is quite difficult because individual neurons are very small and their cell bodies are electrically remote from sites of neuronal activity, necessitating recording from axons or dendrites. Axon diameters are typically less than 2 μm (Erber 1984).

Despite these difficulties, one class of interneurons located in the central part of the bee brain has been identified that shows response changes during learning. These neurons are characterized by their multimodal response. They respond both to odors and to sugar water presented to the antennae (Erber 1978, 1980, 1981). During learning, the sugar water response ceases and the neurons only respond to the odor CS. The odor-response is not selective, however; these cells continue to respond to many odors (Erber 1978, 1981). Interestingly, these neurons are also responsive to moving patterns, and, following antennal stimulation with sugar water, their responsiveness to a given moving pattern is significantly altered (Erber 1981).

LIMAX MAXIMUS

The garden slug, Limax maximus, is a generalized herbivore whose preferred diet is made up of foods such as carrot, potato, and mushroom. Gelperin (1975) first showed that consumption of one of these foods could be dramatically decreased in a single trial by pairing feeding to the food with CO_2 toxicosis. Two classes of stimuli dominate the feeding behavior of Limax, olfactory and gustatory. Slugs are attracted by odor to locomote toward potential food sources. When the potential food source is located, Limax everts its lips to taste the food. If a bitter taste is encountered the food is rejected. A series of experiments has demonstrated that associative learning can clearly modify a slug's preference for a particular odor. For example, Sahley et al (1981a) found that slugs quickly learn to avoid selectively odors paired with the bitter taste of quinidine sulfate (when given a choice between the odor paired with the quinine and their normal diet odor) compared both to naive slugs and to slugs experiencing the odor and taste in an explicitly unpaired relationship. Animals in both of these control groups showed no decrease in their preference for the odor. The learned aversion was retained for approximately 3 days, and the maximum CS-US interval was approximately 60 sec (Sahley et al 1981a).

As discussed above, if a CS is to become associated with a US, the CS must be temporally related to the US, and it is thought that the CS must also predict US occurrence (Kamin 1969, Rescorla 1968, Wagner 1969). Experiments in slugs have assessed the importance of predictability by using procedures modeled after Kamin's (1969) blocking design (Sahley et al 1981b). Kamin observed that prior conditioning to one element, S1, of a compound stimulus, S1*S2, blocked or prevented conditioning to the second element, S2. In slugs, prior conditioning to carrot odor (S1) dramatically reduced or blocked learning to potato odor (S2) when a compound presentation of carrot and potato odor was paired with a quinine US. In contrast, slugs without prior training to carrot learned to avoid both carrot odor and potato odor when the compound was paired with the quinine. In addition, neither slugs that experienced the compound S1*S2-quinine pairing preceded by backward pairings of quinidine and carrot odor, nor slugs that experienced the carrot odor-quinine pairings followed by potato odor-quinine pairings showed any evidence of blocking. All slugs showed conditioning to the potato odor. The temporal pairing of potato odor and quinine thus was not a sufficient condition for associative learning. Rather, the predictive relationship between the odor and the quinine is important, much as Kamin found in rats.

Experiments with slugs also indicate that the associative processes of *Limax* can be influenced by second-order conditioning. Slugs first presented with carrot-quinine (Phase 1) and then given potato-carrot pairings (Phase 2) dramatically reduced their preference for potato odor even though potato had never directly been paired with the quinine (Sahley et al 1981b). To determine whether this finding was truly second-order conditioning (Rizley & Rescorla 1972) the performance of the slugs receiving second-order training was compared to slugs in control groups not experiencing both pairings. Slugs in the control groups showed no change in their preference for potato odor.

Many animals are capable of using post-ingestive consequences of eating a specific food to modify the extent to which the food will be subsequently ingested. Post-ingestive cues that are most often utilized include nutritional deficiencies (Rozin 1969, Zahler & Harper 1972) and gastrointestinal distress (Garcia & Koelling 1966). When animals experience either of these consequences they show a profound aversion to the specific food preceding the consequence (Garcia & Koelling 1966, Rozin & Kalat 1973). It is commonly thought that the food aversion is an example of associative learning (Zahorik 1977, Domjan & Burkhard 1982). Recent experiments show that slugs use post-ingestional consequences to modify their feeding. Delaney & Gelperin (1985) found that slugs will selectively avoid eating a diet after a single meal when an essential amino acid (methionine) is removed from the diet. In contrast, the removal of a nonessential amino acid (alanine) does not produce a change in feeding.

Finally, in addition to food avoidance learning, *Limax* has been shown to be capable of appetitive learning. Slugs can be trained to approach an odor that has been paired with either the taste of a fructose solution or an opportunity to feed on rat chow extract (Sahley et al 1982).

To study the cellular events underlying associative learning in *Limax*, Gelperin and colleagues developed a semi-intact preparation consisting of cerebral ganglion, buccal ganglion, and lips (Prior & Gelperin 1977, Gelperin et al 1978). Food extract applied to the lips while recording from buccal ganglion nerve roots reliably elicited a coordinated rhythmic pattern of motoneuron activity (feeding motor program); this has been demonstrated to be the neural correlate of feeding (Gelperin et al 1978) and to correspond to behavior in the intact animal (Reingold & Gelperin 1977). Training procedures similar to those originally used by Gelperin (1975) in the intact slug have been successfully used to demonstrate associative learning in the semi-intact preparation (Chang & Gelperin 1980). Paired presentations of a food stimulus (CS) and a bitter taste (US) to the lips of the preparation result in a rapid and selective suppression of the feeding motor program to the taste paired with the US. The neurons mediating this learning appear to be centrally rather than peripherally located, since learning can be obtained when training is done on one lip and testing is done on the other (Culligan & Gelperin 1983).

Several nerve cells have been identified that appear to have a role in the sequence of actions associated with the ingestion of food, the "feeding motor program." These include several motorneurons within the buccal ganglion (Gelperin & Forsythe 1976, Gelperin 1981) and salivary neurons within the cerebral ganglion (Prior & Gelperin 1977, Copeland & Gelperin 1983). A serotonergic interneuron, the metacerebral giant cell, has a modulatory influence on the feeding motor program (Gelperin 1981). Retrograde filling of the connectives from the buccal ganglia to the cerebral ganglion reveals a cluster of about 12 neurons. These neurons are of special interest because similar neurons have been identified in *Pleurobranchaea* (see below) and have a command-like effect on feeding (Gillette et al 1982c). Wieland & Gelperin (1983) suggest that command-like neurons in *Limax* may be dopaminergic, since bath-applied dopamine triggers the feeding motor program in a dose-dependent manner.

Although associative learning has been demonstrated in *Limax* in both the whole animal and in the isolated nervous system, there are several important differences between the two preparations. In the whole animal an odor-taste association modifies an appetitive behavior (locomotion). In contrast, in the semi-intact preparation a taste-taste association modifies consummatory behavior (feeding). Results from several experiments suggest that both the whole-animal and the isolated-brain learning involve similar neural events. Gelperin & Culligan (1982) and subsequently

Culligen & Gelperin (1983) report that following food taste-quinine pairings in the whole animal, which result in odor-avoidance, selective suppression of the feeding motor program is evident in the isolated brain. In addition, appetitive conditioning procedures produce both increased approach responses to the specific odor paired with sugar and an odor-induced feeding response to the same odor (Sahley et al 1982). This suggests that odor-taste associations can modify the olfactory-guided search for food as well as the taste-guided consummatory response.

PLEUROBRANCHAEA CALIFORNICA

The marine mollusc, *Pleurobranchaea*, responds to food with a characteristic proboscis extension followed by a "bite-strike" response in which the animal fully extends its proboscis and strikes the food with a biting motion. Mpitsos & Davis (1973) were the first to describe modification of this behavior by associative learning. They found that animals that experienced paired food presentation (the CS) and a strong electric shock (the US) would quickly learn to cease responding to the food CS. Animals that had experienced both the CS and the US but in an explicitly unpaired relationship did not show an acquired avoidance to the food (Mpitsos & Davis 1973). In later experiments the training procedures were modified to a more traditional avoidance learning paradigm. Animals were usually given ten conditioning trials consisting of a presentation of the food stimulus paired with an electric shock to the oral veil. Trials were separated by an intertrial interval of 1 hr. If the animals made an avoidance response to the food, the shock was omitted. Learning was indicated by an active avoidance of the food (Mpitsos & Collins 1975, Mpitsos et al 1978). Nonassociative factors were assessed with explicitly unpaired and random control groups. Animals in the explicitly unpaired group showed significantly less food avoidance than animals in the paired group. Since these two groups were identical except for the temporal pairing of the CS and US, the food avoidance response was due to a learned association between the food and the shock. Interestingly, the paired effect did not emerge immediately: Avoidance responses for animals experiencing paired CS-US presentations did not exceed those experiencing random CS and US presentations until 12 hr following training. In addition, Gillette et al (1984) have recently shown that the performance of this learning task is dependent on the motivational state of the animal.

The stimulus specificity of the learned avoidance has been assessed by Davis and colleagues (Davis et al 1974a, 1980). They found that animals conditioned to avoid squid homogenate showed no generalization of avoidance behavior to sea anemone (Davis et al 1974a). Moreover, animals receiving differential conditioning (two different CSs counterbalanced across groups of animals with only one stimulus paired with the US) avoided only the specific food that was paired with the US and showed no change in responsiveness to the unpaired stimulus. If nonassociative factors contributed to the learned change in behavior, one would expect these factors to affect each odor similarly (Davis et al 1980).

The neural circuit underlying feeding has been extensively characterized. The oral veil with fused lateral tentacles and the rhinophores are the major chemosensory organs involved in food detection (Matera & Davis 1982, Davis & Matera 1982). Sensory receptors are located within the sensory epithelia and are thought to be responsible for detection of food stimuli (Bicker et al 1982a,b). Axons from these receptors project to the tentacular and rhinophore ganglia, which are located peripherally at the base of the tentacles and rhinophores, respectively (Bicker 1982a). A population of about 100 neurons has been identified within each peripheral ganglion that projects to the central cerebropleural ganglion or brain. These peripheral neurons appear to have chemosensory, mechanosensory, or bimodal response characteristics (Bicker 1982a).

The central control of feeding has been analyzed in considerable detail (Davis et al 1973, 1974a,b, 1984, Gillette & Davis 1977, Gillette et al 1980, 1982a–c, Kovac et al 1982, 1983a,b, Siegler 1977, Siegler et al 1974). Two central ganglia appear to be involved in feeding; the cerebropleural and the buccal ganglia. The peripheral ganglia are directly connected to the cerebropleural ganglia, and the buccal ganglion is also directly attached to the cerebropleural ganglia via the cerebropedal connectives. Many neurons involved in feeding have been characterized in *Pleurobranchaea*. We limit our discussion to neurons that appear to have a role in learning. A group of "feeding command neurons," the paracerebral neurons (PCNs) (Gillette et al 1978, 1982c) have been studied in naive and conditioned animals. Normally the PCNs are excited by food to the lips in hungry animals. In contrast, in trained animals food stimulation produces an intensive barrage of inhibitory postsynaptic potentials (IPSPs) in the PCNs; this same change (increased inhibition) occurs in satiated animals as well (Davis & Gillette 1978, Davis et al 1983). A cluster of interneurons, the INT-2s, receive direct chemosensory excitation and in turn make excitatory synapses on the retractor motoneurons and inhibitory synapses on the protector motoneurons. The INT-2s appear to be the major source of IPSPs to the paracerebral neurons (Kovac et al 1983a,b). As a result of either conditioning or food satiation, the INT-2s respond to food stimuli with a prolonged barrage of action potentials, resulting in suppression of feeding (London & Gillette 1985). Changes in the INT-2s appear to reflect a general decrease in feeding motivation rather than a specific mechanism of associative learning (Davis & Gillette 1978, Davis et al 1983, Kovac et al 1985). Interestingly, when the nervous system is removed from well-trained animals, the PCN neurons still show a barrage of IPSPs in response to food stimuli delivered to the attached oral veil; however, unlike in the semi-intact animal, PCNs from brains removed from naive, control, or satiated animals showed indistinguishable *excitatory* responses to food stimuli. Presumably, surgical isolation of the CNS removed the inhibitory signals arising from satiation, but left intact the inhibitory signals arising from conditioning. This preparation thus offers considerable promise for studying cellular mechanisms of learning that can be distinguished from more general changes due to altered motivational state.

LYMNAEA STAGNALIS

Recently, the pond snail, *Lymnaea stagnalis*, has been shown to be capable of appetitive classical conditioning. Audesirk et al (1982) found that *Lymnaea* are able to learn an association between a neutral chemostimulus CS (amyl acetate) and an appetitive US (a mixture of sucrose and casein digest). After 15 training trials, snails that experienced paired presentations of the CS and US showed significantly more feeding movements to the CS than snails that experienced either random or explicitly unpaired presentations of the CS and US, or presentations of the CS or US alone. Moreover, snails that have experienced spaced training (five trials per day for three consecutive days; ITI = 90 min) showed greater retention of the learned association than snails that had experienced massed training trials (15 trials on a single day; ITI = 45 min). Interestingly, in contrast to other molluscs (Sahley et al 1981b, Hawkins et al 1983b), *Lymnaea* showed backward as well as forward conditioning; backward conditioning effects were shorter-lived (2 days) compared to forward-conditioning effects (4–6 days). Alexander et al (1982) extended these studies in an examination of the effects of age and motivation of the learning. They reported that young snails (aged 8 wk) learned the task in both food-deprived and sated conditions, while older snails (aged 32 wk) learned the task only under food-deprived conditions.

There have been no reports of cellular changes that might underly the learned change in behavior. However, the neuronal circuitry mediating the feeding response has been described in considerable detail (Benjamin & Rose 1979, 1980, McCrohan & Benjamin 1980, Rose & Benjamin 1979, 1981a,b). Thus *Lymnaea* holds clear promise for future neurophysiological studies of appetitive learning.

ACHATINA FULICA

The land snail, *Achatina fulica*, shows a dramatic change in preference of food odors as a result of experience. Croll & Chase (1977, 1980) found that, as a result of restricted feeding to a specific food, both juvenile and adult snails subsequently preferred the odor of the food they had fed on in comparison to a novel food odor. The preference was indicated by preferential orientation toward the odor, decreased latency to begin locomotion, and more rapid locomotion toward the experienced odor than toward a novel odor. Snails that were exposed to the odor but not allowed to feed did not show a preference for the odor. Moreover, increased levels of food deprivation enhanced the learning for both juveniles and adults. The learning was extremely long-lived: After a 48-hr feeding exposure the learned preference was observed up to 3 wk. In addition, the age of the snail influenced the acquisition of learning. Learning in the adult snails required a longer feeding experience with the food than did learning in juveniles. Thus, as in *Lymnaea* (Alexander et al 1982), both motivational level and age can modulate the acquisition and retention of this form of appetitive learning.

HIRUDO MEDICINALIS

The medicinal leech, *Hirudo*, has only recently been the focus of learning studies. The behavior receiving the most experimental attention is the shortening reflex, which can be triggered either by light or by a tactile stimulus. The first evidence of associative learning in leeches was provided by Henderson & Strong (1972), who found that animals experiencing paired presentations of a photic stimulus (CS) and an electric shock (US) showed an enhanced shortening response to the light (which developed over training trials) compared to leeches receiving either random or unpaired presentations of the CS and US, or presentations of the CS or US alone. More recently, Sahley & Ready (1985) found that the shortening reflex elicited by a tactile stimulus in the leech could also be modulated by associative learning: The shortening response was enhanced for animals experiencing paired presentations of a tactile CS and an aversive shock US. In contrast, leeches experiencing either the US alone or explicitly unpaired or random CS and US presentations showed little or no increase in the shortening response, and leeches experiencing CS alone presentations showed a significant decrease in responding to the CS.

    In addition to the associative changes, the shortening response is clearly modulated by nonassociative factors. Habituation of the photic-elicited shortening reflex (Lockery & Kristan 1985, Lockery et al 1985) and the touch-elicited shortening reflex (E. Bird, personal communication; Stoller & Sahley 1985) have been observed in several laboratories. For example, Stoller & Sahley (1985) showed that the reflex recovers from habituation after a 20 min rest period (spontaneous recovery) and is facilitated if electric shock is applied to the tail of the animal (dishabituation). Finally, shock applied prior to habituation training produces retarded habituation of the reflex.

In addition to the shortening reflex, habituation has been observed to modulate more complex behaviors, such as the shock-elicited swimming response (Debski & Friesen 1985) and the water current–elicited motor response (Ratner 1972).

To investigate cellular events underlying response decrement and facilitation in the leech, a neural analog system has been developed. The fast-conducting system (FCS) is a chain of electrically coupled neurons, one to each ganglion, that mediates nerve cord shortening (Bagnoli et al 1975). Repeated tactile stimulation of the skin produces habituation of the FCS response and noxious stimulation produces dishabituation. Bath-applied serotonin or dopamine also dishabituates the FCS response, and the facilitation is blocked both by preincubation with methysergide, a serotonin antagonist, and by preincubation with imidaxole, a cAMP-phosphodiesterase activator (Belardetti et al 1982). Consistent with the pharmacological effects, increase in the synthesis of cAMP in the segmental ganglia of the leech has been observed following sensitization (Biondi et al 1982). These findings, together with the lack of changes in the sensory neurons thought to mediate the tactile stimulation, suggest that (similar to sensitization in *Aplysia*, see below) the observed facilitatory cellular changes within the FCS may be due to a serotonin-mediated increase of cAMP (Biondi et al 1982).

Less is known about the cellular mechanisms mediating habituation and sensitization of the shortening response of the intact animal. Training parameters identical to those employed in the behavioral experiments have recently been used in a surgically reduced body wall preparation first introduced by Baylor & Nicholls (1969). Apparently no change occurs in the one class of primary mechanosensory cells (P cells) mediating the touch-elicited shortening response. Rather, changes appear to occur in the motoneuron (L) mediating the response (Stoller & Sahley 1985).

SCHISTOCERCA GREGARIA

In 1962, Horridge (1962a,b) introduced a clever, simple operant conditioning paradigm in insects: Headless locusts (*Schistocerca*) or cockroaches were suspended above a dish of saline solution and received electric shocks (through an electrode attached to their leg) when they lowered their leg and made contact with the saline. Animals quickly learned to hold the leg that had been shocked away from the saline. This basic paradigm has been modified and refined over the years in a number of ways, both to make the preparation amenable to cellular studies and to explore the kinds of reinforcement that are effective in producing the learning. For example, Forman & Hoyle (1978) showed that locusts could learn the leg position task to obtain heat reinforcement in a cold room. Hoyle (1980) subsequently found that locusts or grasshoppers could learn the task either to terminate an aversive stimulus (a loud sound/vibration) or to obtain a food reward (sugar water or grass). Pointing out problems with the original Horridge paradigm, Forman (1984) further refined the procedure. Locusts were required to maintain a particular femero-tibial joint angle to avoid aversive heating of the head. By examining leg muscle myographic activity patterns during the learning, Forman & Zill (1984) were able to identify three distinct motor strategies (repeated flexion-extension, changes in muscle tonus, and tonic slow excitor motor neuron activity) that were selectively employed during the learning, depending upon the range of joint angle required. A fascinating variant on this form of learning has recently been reported by Hoyle & Field (1983), who found that a primitive New Zealand insect (the "Weta") could exhibit leg-position learning in the absence of electrical activity in the leg muscles generating the force, suggesting that the animals use a peripheral catch-like mechanism to generate the muscle tension necessary for the expression of the operant response.

First steps toward a neuronal analysis were made by Eisenstein & Cohen (1965), who showed that the prothoracic ganglion (which controls the leg) is both necessary and sufficient for the operant learning. In that same year, Hoyle (1965) recorded the activity of the excitor motoneuron to a leg muscle involved in the learning (the adductor of the coxa) and found that he could directly condition the firing rate of the motoneuron: When foot shock was delivered to the animal contingent upon a decline in firing rate, the neuron would increase its rate as many as four times the original rate. Similar results were obtained in an isolated ganglion-nerve-muscle preparation (Hoyle 1966). An important observation in these studies was that the learning did not depend on feedback from leg position, since it occurred if the leg was immobilized or if the tendon for the leg-lift muscle was cut. Subsequently, Tosney & Hoyle (1977) developed an automated training regime controlled by computer and found that they could condition both increases ("up-training") or decreases ("down-training") in motoneuron firing rate.

To explore the mechanism of the conditioned change in firing frequency, Woolacott & Hoyle (1977) first blocked synaptic transmission (with high Mg^{2+}, low Ca^{2+} saline) to assess the intrinsic pacemaker frequency of the motoneurons, then conditioned the cells in normal saline. To test for pacemaker shifts produced by the learning, they once again blocked synaptic transmission after training and found that the intrinsic firing rate of the motoneurons was increased following up-training and decreased following down-training. Thus, conditioning altered the pacemaker rhythm of the cells. However, alterations in synaptic input also were implicated, since the frequency shifts with transmission blocked were not as great as in normal saline solution. Clues to the nature of the change in the motoneurons were obtained by Woolacott & Hoyle (1976; see also Hoyle 1982a,b), who found (a) that the input resistance of the motoneuron increases with up-training and decreases with down-training, and (b) that the hyperpolarizing afterpotentials were large in amplitude with low frequencies (down-training) and reduced with high frequencies (up-training). These results are consistent with the hypothesis that learning produced a reduction in a K^+ conductance in the motoneurons. Based on these and related observations, a model has been proposed (Woolacott & Hoyle 1976, Hoyle, 1979, 1982a,b) postulating that a prolonged modulation of a voltage-dependent K^+ conductance in the motoneurons shifts the intrinsic pacemaker frequency in these cells to produce the increase and decrease in firing rates observed during up-learning and down-learning, respectively. Hoyle (1982b) further proposed that an efference copy mechanism is involved in providing feedback to a "selector" of a range of K^+ conductance values that ultimately produces the appropriate conductance to achieve goal-related frequency modulation. Certainly much of this model is speculative, but it provides a creative framework for important future research.

HERMISSENDA CRASSICORNIS

The nudibranch mollusc, Hermissenda, shows a reliable form of positive phototactic behavior: Both in an open field and in enclosed experimental tubes animals readily locomote toward a light source (Alkon 1974, Crow & Alkon 1978, Crow 1985a). This positive phototaxis can be suppressed if the animal receives prior exposure to light in conjunction with rotation. Alkon (1974) first described this basic phenomenon, and Crow & Alkon (1978) provided the first demonstration of an associative modification of this behavior. Using an automated training and testing procedure, they showed that three days of training, consisting of 50 trials per day of light (the CS) paired with rotation (the US), led to a significant increase in the animals' response latency to enter a test light. Groups receiving this paired training showed significantly greater suppression of phototaxis than a number of control groups, including those that received random or unpaired CS-US presentations, or random CS or US alone. The learning was retained for several days and, upon retraining, the paired group showed savings in their reacquisition performance. Subsequent to this first description of associative learning in *Hermissenda*, which demonstrated temporal specificity, prolonged retention and savings, other properties of learning have also been examined. For example, stimulus specificity has been reported by Farley & Alkon (1980, 1982a), who examined the effects of orientation on the learned response; and response specificity has been demonstrated by Crow & Offenbach (1979, 1983), who found that increased response latencies following learning were not due to a decrease in general activity, but to a significant increase in latency in conditioned animals to initiate locomotion toward light. More recently, extinction of the conditioning (Richards et al 1983) and contingency effects (Farley & Kern 1985) have also been described. In addition, Lederhendler et al (1983) have provided evidence that, following pairing of light and rotation, light produces a contraction of the foot that is similar to the contractions produced by rotation alone. Finally, using a different behavioral measurement, the shadow response (in which animals will avoid darkness by turning at light-dark borders), Lederhendler & Alkon (1984) have suggested that conditioning changes *Hermissenda*'s ability to detect light-dark intensity differences.

A cellular analysis of associative learning in *Hermissenda* has been made possible by the relative simplicity and accessibility of the sensory structures involved in the visual (CS pathway) and vestibular (US pathway) systems, which have been examined in considerable detail (for review see Alkon 1983, 1984). The eyes of *Hermissenda* contain only five photoreceptors each (two Type A and three Type B), all of which respond to light with a depolarizing generator potential that gives rise to propagated action potentials (Dennis 1967, Alkon & Fuortes 1972). Particular attention has been focused on Type B photoreceptors, which show a long-lasting depolarization (LLD) that outlasts a discrete presentation of light for many seconds and is accompanied by an increase in membrane resistance (Alkon & Grossman 1978). The ionic conductances in the photoreceptors (in the dark) have been shown to include three outward K^+ currents [early (I_A), late (I_K), and Ca^{2+}-dependent (I_C)], an inward Ca^{2+} conductance, and (in response to light) an early inward Na^+ conductance and an outward Ca^{2+}-dependent K^+ conductance (Alkon 1984). The next stage in the visual system, after the photoreceptors, is the optic ganglion, which has 13 second-order visual interneurons. The vestibular apparatus (the statocysts) contain 13 hair cells, which show increased impulse activity in response to rotation. The synaptic interactions among these neural elements have been characterized and can be summarized as follows: Type A and B photoreceptors inhibit each other; caudal hair cells and Type B cells inhibit cells in the optic ganglion (S/E cells); and S/E cells excite Type B cells and inhibit caudal hair cells (Alkon 1979, 1983, 1984, West et al 1982). More recently, progress has been made in identifying some motoneurons possibly involved in phototactic behavior (Crow 1981, Jerussi & Alkon 1981, Takeda & Alkon 1982, Goh & Alkon 1984).

Evidence implicating a primary change in the Type B photoreceptors as a result of the learning was first obtained by Crow & Alkon (1980), who examined B cell activity several hours after training and found that spontaneous activity of dark-adapted B photoreceptors was significantly increased in animals that received previous paired training compared to randomly trained controls. Moreover, these "paired" B photoreceptors had more depolarized resting potentials and higher input resistances, even when axotomized, indicating that these changes were intrinsic to the B cells. Further studies by Farley & Alkon (1982a) and West et al (1982) have confirmed and extended these observations: 1–2 days following training, Type B cells from paired animals were no longer depolarized, but did show increased light responses and LLDs and elevated input resistances. Changes in putative motoneurons correlated with training have also been reported (Crow 1981, Takeda & Alkon 1982, Goh et al 1985). For example, Crow (1981) showed that the putative pedal motoneuron (P5) exhibited significantly higher inhibitory synaptic input in response to light in paired compared to randomly trained animals. More recently, Goh et al (1985) reported that another putative motoneuron (MN1) showed a smaller increase in impulse frequency in response to a light step in animals receiving paired training compared to randomly trained controls, up to 54 hr after training, and that MN1's response to light was positively correlated with the animals' behavioral retention latency and inversely correlated with B photoreceptor input resistance. Finally, in addition to training-induced changes' leading to increased excitability in B photoreceptors, recent evidence suggests that training produces a concomitant decrease in excitability of Type A photoreceptors (Richards & Farley 1984). Taken collectively, these data have led to the hypothesis that the associative change in phototactic behavior is due to training-produced *enhancement* of the Type B photoreceptor's light-induced generator potential and impulse activity (an alternative mechanism involving a *decrease* in the B cell's response has been proposed by Crow 1985b, see below). This greater B-cell activity is thought to increase B-cell inhibition onto Type A cells, which are hypothesized to be responsible for mediating phototaxis through their connections to interneurons and motoneurons (Alkon 1983, 1984).

At the next level of analysis, Alkon and colleagues examined the acquisition and retention of the conditioned changes in the B cells' photoresponse.

Acquisition

The acquisition process has been postulated to involve a pairing-specific cumulative depolarization in the B photoreceptors that occurs across training trials (Alkon 1979, 1980a,b). This model is partly predicted on the observation that the depolarizing tail of the generator potential is enhanced in animals receiving paired training (Crow & Alkon 1980). Moreover, this enhancement can be mimicked by pairing light and rotation in a semi-intact preparation (Alkon 1980a,b) or light and caudal hair cell stimulation in an isolated nervous system (Farley & Alkon 1982b). It has also been suggested that membrane changes in the B cells can be causally related to associative changes (Farley et al 1983). The cumulative depolarization is hypothesized to be derived from two sources: (a) rebound excitation in the B cells at the termination of rotation, due to release from inhibition from caudal hairs; and (b) excitatory input from S/E optic ganglion cells, which themselves rebound (at the offset of paired light and rotation) from inhibition derived from the B cells and hair cells (Alkon 1979, 1980a,b).

The synaptic excitation following stimulus pairing is thought to be enhanced by means of a positive feedback cycle: Synaptic input produces membrane depolarization, which causes both an increase of a voltage-dependent Ca^{2+} current and a transient activation, followed by a prolonged inactivation, of two K^+ currents (I_A and I_C), which further increases the input resistance of the B cell (Alkon et al 1982a,b, Alkon 1984). This increase in input resistance will in turn enhance subsequent synaptic input, giving rise to greater depolarization, and so on. Pairing and stimulus specificity are thus thought to result from enhanced cumulative depolarization that arises out of a positive feedback cycle between synaptic and light-induced depolarization. An essential feature of this pairing model is that the CS and US show simultaneous *offset*. Thus, this model, in principle, would predict that US onset, followed some time later by CS onset (i.e. backward conditioning), would still produce associative changes if the stimuli terminated simultaneously.

The net result of cumulative depolarization is thought to be an increase in Ca^{2+} accumulation in the Type B photoreceptors (Alkon 1979). Recent experiments by Connor & Alkon (1984) are consistent with this hypothesis. Using the Ca^{2+} indicator dye, Arsenazo III, they found that pairing light and rotation increased intracellular Ca^{2+} in the B photoreceptors over that produced by light alone. However, the necessary role of reduced K^+ (I_A and I_C) as well as the role of Ca^{2+} has recently been brought into question by Grover & Farley (1984), who showed that B cells still exhibit pairing-specific cumulative depolarization when I_A and I_C are blocked and when there is no Ca^{2+} in the bathing solution.

In addition to intrinsic factors such as cumulative depolarization and Ca^{2+} accumulation, synaptic modulatory effects have also recently been implicated in *Hermissenda*. For example, Wu & Farley (1984) reported that, similar to *Aplysia* sensory neurons (Klein & Kandel 1980, Klein et al 1982), serotonin increases input resistance, reduces K^+ currents, and enhances

Ca^{2+} currents in B photoreceptors. Moreover, extracts of optic ganglia have been shown to increase B cell input resistance and LLD (McElearney & Farley 1983). The optic ganglia contain neurons that exhibit green fluorescence with a Falk-Hillarp stain, thus suggesting the presence of catecholamines. Norepinephrine and clonidine (α_2 adrenergic receptor agonists) have recently been shown to reduce both I_A and I_C in B cells (Sakakibara et al 1984). This has prompted the suggestion that the release of an α_2 receptor agonist from cells in the optic ganglion may facilitate the reduction of I_A and I_C during acquisition (Sakakibara et al 1984, Alkon 1984).

Retention

Retention of the learning is thought to involve a Ca^{2+}-induced long-term reduction of at least two K^+ currents (I_A and I_C). Several lines of evidence support this hypothesis (for review see Alkon 1984). For example, Alkon et al (1982a,b) found that I_A was significantly reduced in the B cells 1–2 days after paired training, and they (1982b) showed that direct elevation of intracellular Ca^{2+} reduces I_A, but does not affect another K^+ current, I_B (which is an aggregate current consisting of at least the late K^+ current, I_K, and the calcium-dependent K^+, I_C). In contrast to these earlier studies, more recent experiments also implicate the Ca^{2+}-dependent K^+ current, I_C (Farley & Alkon 1983, Forman et al 1984, Farley et al 1984). The model for retention postulates that over the period of days that the learning is retained, the membrane potential of the B cells returns to pretraining levels, but I_A and I_C remain reduced. The magnitude of this reduction (30–40% compared to controls) is thought to be sufficient to account for the enhanced generator potential observed in B cells on days after training (Alkon 1984).

On a biochemical level, an important clue concerning possible molecular mechanisms involved in the retention of the learning came from a study by Neary et al (1981), who showed that conditioning induced a change in the level of incorporation of ^{32}P in a 20,000-molecular weight phosphoprotein in the eyes of *Hermissenda*, suggesting that protein phosphorylation might contribute to retention. Moreover, Neary & Alkon (1983) obtained evidence that this protein phosphorylation was related to inactivation of I_A, and Neary et al (1984) showed that the phosphorylation was regulated by both Ca^{2+} and cAMP (for review see Neary 1984). Further study of protein phosphorylation was carried out by Alkon et al (1983a), who injected the catalytic subunit of the cAMP-dependent protein kinase into the B cells and found that it reduced the late K^+ currents (I_B) more than the early I_A. Since reduction of I_A was thought to be more involved in the learning (Alkon et al 1982a), cAMP-dependent phophorylation was not considered to be a primary mechanism (Alkon et al 1983a). It was known that Ca^{2+} injection produces suppression of both I_A (Alkon et al 1982b) and I_C (Alkon et al 1983b). This led to a focus on phosphorylation by means of a Ca^{2+}-calmodulin mechanism. Recently, Acosta-Urquidi et al (1984) have reported that injection of a species of Ca^{2+}-calmodulin-dependent protein kinase can reduce both I_A and I_B in B photoreceptors. They propose that intracellular Ca^{2+} accumulates during pairing, giving rise to a Ca^{2+}-calmodulin mediated phosphorylation of I_A and I_B channels, and that sustained phosphorylation of these K^+ channels could produce long-term retention of the associative learning.

An Alternative Proposal

Recent work in *Hermissenda* by Crow has raised two important issues concerning the behavioral results and the cellular model proposed by Alkon and co-workers: The first issue addresses the question of the associative and nonassociative components of the conditioning, and the second raises the question of whether complex circuit interactions between Type B and A photoreceptors are necessary to explain the conditioned behavior.

In a series of behavioral experiments, Crow (1983a) examined the contribution of associative and nonassociative factors to the conditioned suppression of phototactic behavior of *Hermissenda* by varying the number of conditioning trials and the time between training and testing. Few trials (five or ten) in a single session produced no changes in behavior when tested immediately after training and showed only nonassociative effects (i.e. paired performance comparable to random controls) when tested 15 min or longer. Consistent with previous observations (Crow & Alkon 1978, Crow & Harrigan 1979), with multiple sessions (3 days, 50 trials each) significant nonassociative changes were first observed, followed within 30 min of training by significant associative effects. Crow (1983a) emphasized that, since training with only a few trials produced no associative behavioral changes, cellular studies that use very few trials (e.g. Alkon 1980b, Farley & Alkon 1982b) may not yield neural correlates that can be directly related to the behavioral conditioning.

In a recent study Crow (1985a,b) addressed an apparent paradox in the results of the behavioral and cellular studies of Alkon and co-workers: *Hermissenda* are positively phototactic and conditioning results in a suppression of that response, yet it has been suggested that an enhanced photoresponse in B photoreceptors is critical for producing the learning (see Alkon 1983, 1984). How can an *enhanced* photoresponse produced by conditioning *decrease* phototactic behavior? As described above, Alkon and colleagues propose complex circuit interactions among Type B and A photoreceptors and their related interneurons and motoneurons to account for the decreased behavior (Alkon 1983). Crow (1985a,b) proposes an alternative model. In a psychophysical study, Crow (1985a) found that *Hermissenda* show a graded increase in positive phototaxis in response to an increasing series of test light intensities. Conditioning resulted in a significant suppression of phototactic behavior at all light intensities examined, including intensities in which the LLD in B photoreceptors would be expected to be either absent or not affected by conditioning (Farley & Alkon 1982a). Crow suggests that some other change in the B cell might account for the conditioning.

In a second paper Crow (1985b) addressed this question directly. He reasoned that the appropriate condition in which to measure changes in the B photoreceptor should correspond to those in which changes in phototactic behavior of *Hermissenda* are actually expressed during behavioral testing. Specifically, during behavioral testing animals are subject to light adaptation and do not begin to emit a phototactic response for several minutes. Therefore Crow examined the B-cell photoreceptor responses (in previously conditioned animals) both immediately after light onset and after 5 min of continuous light. When examined in continuous light 5 min after light onset, the B-cell response was significantly reduced in animals that received paired training compared to random controls, both in terms of B-cell activity and (in axotomized cells) in the amplitude of the steady-state generator potential. Thus when activity of B photoreceptors is examined under conditions of illumination that resemble conditions under which conditioned behavior is expressed, conditioning produces a *reduction* in the B cell response.

This finding is the opposite of that observed by Alkon and co-workers, who found *increased* B cell activity after conditioning (Alkon 1983). In the previously published reports by Alkon and co-workers (see Alkon 1983, 1984), cellular correlates have been based upon analysis of a very early phase of the B-cell photoresponse. Crow argues that this is not the behaviorally relevant phase of the photoreceptor response to examine, since the behavior is generated only after the animals have been light-adapted for many minutes. As an alternative to the network hypothesis involving B-cell inhibition of A cells (Alkon 1983, 1984), Crow (1985a) suggests that conditioning produces *diminished* B-cell activity that is directly translated into diminished phototaxis. As a possible mechanism for the long-term reduction of the B cells' photoresponse, Crow reported recent studies that show that conditioning produces two changes in the adaptation properties of B photoreceptors: (*a*) a significant increase in light-evoked photoreceptor desensitization that cannot be mimicked by blocking I_A, and (*b*) a significant delay in recovery of sensitivity following an adapting light (Crow 1982, 1983b). Since Ca^{2+} levels are thought to be modulated by conditioning (Alkon 1984) and Ca^{2+} is known to play an important role in receptor adaptation, regulation of intracellular Ca^{2+} may also contribute importantly to conditioned alteration of the adaptation properties of Type B photoreceptors.

The proposals by Alkon and colleagues and by Crow are not necessarily mutually exclusive; both mechanisms could contribute to learning in *Hermissenda*. Further work will undoubtedly clarify this question.

APLYSIA CALIFORNICA

Aplysia californica is a marine mollusc that has been used extensively to study nonassociative as well as associative forms of learning. Although the primary focus of this chapter is on associative learning, a brief review of nonassociative learning, specifically habituation and sensitization, is important as background for the associative work. Several more detailed reviews have recently been published (Kandel & Schwartz 1982, Kandel et al 1983, Kandel 1984). Many response systems in *Aplysia* exhibit habituation and sensitization. These include the gill and siphon withdrawal reflex (Pinsker et al 1970), the tail withdrawal reflex (Walters et al 1983b, see below), inking (Carew & Kandel 1977), escape locomotion (Walters et al 1978), and the feeding system (Kupfermann & Pinsker 1968). We restrict our discussion to studies of the gill and siphon withdrawal reflex, as this has been most extensively analyzed. We divide our discussion into two parts, first considering behavioral and then cellular studies.

Behavioral Studies of Habituation and Sensitization

Because of its simplicity and relatively well-delinated neural circuit, the behavior that has received the most experimental attention in *Aplysia* has been the defensive withdrawal reflex of the mantle organs (the gill, siphon, and mantle shelf). This reflex can be elicited by a light tactile stimulus to the siphon, which evokes a withdrawal of these organs into the mantle cavity. If the same stimulus is repeatedly delivered, the reflex readily habituates (Pinsker et al 1970) and remains so for several minutes to a few hours. However, reflex amplitude can be immediately restored by delivering a strong sensitizing stimulus to another site on the body, such as the head or tail, which produces dishabituation. Dishabituation is simply a specialized case of a more general facilitatory process, sensitization (Carew et al 1971). Both habituation and sensitization can be transformed into a long-term form lasting weeks by giving repeated training sessions over several days (Carew et al 1972, Pinsker et al 1973).

Cellular Studies of Habituation and Sensitization

The investigation of habituation and sensitization of the gill withdrawal reflex in *Aplysia* has been carried out on three levels: synaptic, biophysical, and molecular.

SYNAPTIC ANALYSIS   The neural circuit for the gill and siphon withdrawal reflex has been extensively studied and is relatively well understood. It consists of a small population of about 24 identified primary afferent neurons, siphon mechanoreceptors, which make monosynaptic excitatory connections onto identified gill and siphon motoneurons as well as several interneurons (Byrne et al 1974, Castellucci et al 1970, Kupfermann et al 1974, Perlman 1979, Hawkins et al 1981a,b). Several lines of evidence indicate that habituation and sensitization are due in large part to homosynaptic depression and heterosynaptic facilitation, respectively, at the synapses from the siphon sensory neurons onto their targets (Castellucci et al 1970, Byrne et al 1978). Moreover, a quantal analysis of the sensory neuron synapses during homosynaptic depression and hemosynaptic facilitation showed that both of these processes are presynaptic in origin (Castellucci & Kandel 1974, 1976). In addition to short-term changes, these same synaptic connections have been shown to be chronically depressed or enhanced in animals receiving long-term habituation or sensitization training (Castellucci et al 1978, Carew et al 1979, Frost & Kandel 1984). An important anatomical correlate of these observations has been described by Bailey & Chen (1983), who found that the number, size, and vesicle complement of active zones (synaptic release sites) in the sensory neuron terminals are significantly reduced in long-term habituated animals and significantly increased in long-term sensitized animals. Moreover, in a recent study, Bailey & Chen (1984) found that long-term habituated animals have significantly fewer varicosities (presynaptic terminals) per sensory neuron than controls. Hence these simple forms of learning produce profound ultrastructural as well as functional alterations of the sensory neuron synapses of the gill withdrawal reflex.

BIOPHYSICAL ANALYSIS   To analyze the ionic currents modulated in the sensory neurons during habituation and sensitization, Klein & Kandel (1978, 1980) examined the duration of the sensory neuron action potential and found that it was reduced in duration during repeated activation and dramatically prolonged during heterosynaptic facilitation. Accompanying these alterations in spike duration, synaptic transmission was also reduced or enhanced. Voltage clamp studies showed that habituation was accompanied by a progressive decrease in a voltage-sensitive inward Ca^{2+} current (Klein et al 1980). Recent quantitative simulation studies by

Gingrich & Byrne (1985) suggest, however, that Ca^{2+} inactivation may not fully account for habituation and that depletion of neurotransmitter may also play a significant role. Klein & Kandel (1980) further found that heterosynaptic facilitation was produced by a decrease in an outward K^+ current that prolonged the depolarization of the spike and thus enhanced the voltage-sensitive Ca^{2+} current, which in turn enhanced synaptic release. In addition to modulation of Ca^{2+} due to changes in duration of the action potential, a component of Ca^{2+} accumulation independent of spike shape has recently been identified. Boyle et al (1984) injected sensory neuron cell bodies with the Ca^{2+}-sensitive dye, Arsenazo III, and found that serotonin enhanced free Ca^{2+} accumulation in response to constant-duration voltage-clamp depolarizations, indicating that the Ca^{2+} changes were mediated by some intrinsic change in the handling of Ca^{2+} by the sensory neurons. This intrinsic enhancement of Ca^{2+} accumulation is thought to act synergistically with the reduction in K^+ current to enhance transmitter release. Consistent with this observation, Gingrich & Byrne (1985) suggest that in addition to spike broadening, transmitter mobilization may also contribute to heterosynaptic facilitation.

Several lines of evidence suggest that serotonin is one of the facilitating transmitters in *Aplysia* (Brunelli et al 1976, Bailey et al 1981, Kistler et al 1983, 1985). Immunocytochemical studies indicate that serotonin terminals contact the sensory neurons (Kistler et al 1985). Serotonin mimics heterosynaptic facilitation by broadening the action potential, decreasing a K^+ conductance, and enhancing synaptic release from the sensory neurons (Klein & Kandel 1978, 1980). A voltage-clamp analysis showed that the K^+ current modulated by serotonin differs from previously described K^+ currents (the delayed K^+, fast K^+, Ca^{2+}-dependent K^+, and muscarine-sensitive K^+) (Klein et al 1982). A patch clamp analysis by Siegelbaum and colleagues (1982) further characterized the serotonin-sensitive K^+ channel (the "S" channel, which gives rise to a unique K^+ current, I_S) in the sensory neurons: It is active at the resting potential, is moderately voltage dependent, and is independent of intracellular Ca^{2+}. Closure of this channel can account for increases in spike duration, Ca^{2+} influx, and enhanced transmitter release responsible for sensitization (Siegelbaum et al 1982).

MOLECULAR ANALYSIS   In investigations at a molecular level most progress has been made in the analysis of short-term sensitization. Based on several independent lines of evidence (see Kandel & Schwartz 1982 for review), Klein & Kandel (1980) proposed a molecular model for sensitization postulating that serotonin increases adenylate cyclase activity in sensory neuron terminals, thereby increasing intracellular cyclic adenosine monophosphate (cAMP). cAMP in turn stimulates the activity of enzymes (protein phosphokinases) responsible for phosphorylating substrate proteins, one of which is the K^+ channel or a protein closely associated with it. Phosphorylation of the K^+ channel closes it, giving rise to increased duration of the spike, more Ca^{2+} influx, and thus more transmitter release. Consistent with this suggestion, Siegelbaum et al (1982) found that cAMP closes the S channel in cell-attached membrane patches, and more recently Shuster et al (1985) have found that the catalytic subunit of the cAMP-dependent protein kinase also closes the S channel in cell-free membrane patches, suggesting that the kinase acts on the internal membrane surface to phosphorylate either the S channel itself or an associated protein that regulates the channel.

Support for the molecular model was obtained by Castellucci et al (1980), who found that injection of the catalytic subunit of the cAMP-dependent protein kinase into the sensory neuron broadened its spike and increased its transmitter release. Moreover, facilitation set in motion by serotonin can be blocked by injection of a specific protein inhibitor of the kinase (Castellucci et al 1982), suggesting that the time course of sensitization is due to the persistent activity of the cAMP-dependent kinase (and not, for example, to the stable phosphorylation of the substrate protein). Consistent with this hypothesis, Bernier et al (1982) have shown that cAMP is elevated three- to four-fold in single sensory neurons in response to serotonin or stimulation of facilitatory neural pathways and that the time course of its elevation parallels that of the presynaptic facilitation. Experiments involving the injection of a blocker of adenylate cyclase (GDP β-S) indicate that the elevation of cAMP during sensitization appears to be due to persistent activity in the adenylate cyclase complex (which has at least three components: a receptor, a regulatory G protein, and a catalytic subunit); this is thought to be the time-keeping step in short-term sensitization (Castellucci et al 1983, Schwartz et al 1983).

Recent evidence suggests that multiple facilitatory transmitters may converge on the adenylate cyclase complex in the sensory neurons. Abrams et al (1984b) have shown that in addition to serotonin, two endogenous neuropeptides in *Aplysia*, the small cardioactive peptides SCP_A and SCP_B, facilitate synaptic transmission from the sensory neurons by a cAMP-dependent closure of the S channel. Moreover, recent immunocytochemical studies by Ono & McCaman (1984) and Kistler et al (1983, 1985) have shown that one class of identified interneurons in *Aplysia*, the L_{29} cells, which produce presynaptic facilitation of the sensory neurons and which were originally thought to be serotonergic (Hawkins et al 1981b), do not use serotonin as their transmitter. Thus it appears that there are at least three agents that mimic heterosynaptic facilitation, as well as the natural transmitter of L_{29}, that may converge via different receptors on a common molecular cascade in the siphon sensory neurons (Abrams et al 1984b).

In recent years associative learning has been described in a number of behavioral systems in *Aplysia*, including (*a*) the defensive arousal system, (*b*) the withdrawal reflex of the mantle organs, (*c*) the tail withdrawal reflex, (*d*) the feeding system, and (*e*) the head-waving system. We discuss each of these in turn.

Conditioned Defensive Arousal

In one of the earliest studies of classical conditioning in *Aplysia*, Walters et al (1979) employed a CER paradigm to examine whether *Aplysia* could learn to associate a chemosensory CS (a shrimp extract) with an aversive US (head shock). The behavioral index of conditioning was modulation of escape locomotion by the chemosensory CS. Controls received unpaired CS-US presentations, the CS alone, or the US alone. Training lasted 2 days (three trials per day), followed by testing the next day. During testing all animals showed comparable escape locomotion in the absence of the CS. However, in the presence of the CS, paired animals showed significantly more escape than all controls. Interestingly, the CS by itself did not elicit an obvious locomotor conditioned response after training, leading Walters et al (1979) to propose that the CS had come to elicit a central defensive state (analogous to conditioned "fear") that modulated escape locomotion. To further test this notion, Walters et al (1981) examined a variety of other behavioral responses in the presence of the shrimp CS following similar training and found these responses modulated by the CS in a motivationally consistent fashion: Four defensive responses (head and siphon withdrawal, inking and escape locomotion) were significantly facilitated in the presence of the CS, while an appetitive response (feeding) was significantly suppressed. These behavioral results thus supported the conditioned-fear hypothesis. Further support was obtained by Carew et al (1981a), who recorded intracellularly from motoneurons in the escape-locomotion, siphon-withdrawal, and inking systems from previously trained and control animals. In the presence of the CS, animals that received paired training showed significant facilitation of synaptic input to all these motor systems, whereas controls showed no facilitation. The CS produced no change in resting potential or input resistance of motoneurons in trained animals, suggesting that the critical changes produced by the CS were presynaptic to the motoneurons. These results, together with the behavioral data, led these authors to propose that the conditioning exhibited by *Aplysia* in this CER paradigm reflects an association between the CS and a central defensive arousal state resembling conditioned fear in mammals.

Conditioned Withdrawal of the Gill and Siphon

Classical conditioning of gill and siphon withdrawal has been studied in two preparations: surgically reduced in vitro preparations and intact animal preparations.

IN VITRO CONDITIONING   Lukowiak & Sahley (1981) paired a photic stimulus to the siphon as the CS with a strong tactile stimulus directly to the gill as the US. Normally the photic stimulus does not produce gill withdrawal, but after repeated pairing with the US, the CS came to elicit gill withdrawal in seven out of ten preparations, whereas it never elicited withdrawal in control preparations that received random CS-US presentations. Subsequently, Lukowiak (1982) described a new paradigm that was more amenable to cellular investigation: a weak tactile stimulus to the siphon served as the CS, while the US remained a strong tactile stimulus to the gill. After repeated CS-US pairing, the weak siphon CS evoked an enhanced siphon and gill withdrawal response, while a number of controls (random CS-US pairing, CS alone, and US alone) showed no enhancement. Furthermore, Lukowiak (1983) recorded intracellularly from gill motor neurons during paired training and found that the siphon CS evoked larger synaptic responses in these cells as training progressed. Further experiments indicated that increased synaptic transmission from siphon sensory neurons contributed to the facilitated input in the gill motoneurons. This in vitro preparation provides the opportunity to analyze cellular events in the CNS, during the acquisition of a learned response.

INTACT ANIMALS   Carew et al (1981b) examined whether the gill and siphon reflex, in addition to sensitization, might also show associative learning. They paired a weak tactile stimulus to the siphon as the CS with a strong electric shock to the tail as the US. After repeated pairing trials, the CS came to elicit significantly enhanced gill and siphon withdrawal compared to controls, which received random or unpaired CS-US presentations, or the CS or US alone. Conditioning was acquired within 15 trials and was retained for several days. More recently, using a related paradigm, Carew et al (1983) have demonstrated differential classical conditioning of the siphon withdrawal component of this reflex. In a differential conditioning two conditioned stimuli are used in the same animal, one (CS+) is paired with the US, the other (CS−) is specifically unpaired. Thus each animal can serve as its own control. Weak tactile stimuli applied to the siphon or mantle shelf served as discriminative stimuli (CS+ and CS−), and tail shock served as the US. Differential conditioning could be acquired in a single trial, was retained for more than 24 hr, and increased in strength with increased number of trials. Moreover, conditioning could be produced with two separate sites on the siphon skin as discriminative stimuli, permitting a cellular analysis of differential conditioning in sensory neurons from the siphon (see below). Finally, Hawkins et al (1983b) found that the interstimulus interval function (which describes the CS-US intervals that produce effective conditioning) was quite steep; reliable learning was produced only with a CS-US interval of 0.5 sec; learning was less at 1.0 sec intervals. No conditioning occurred with forward intervals of 2, 5, and 10 sec, or with backward (US-CS) pairings. Moreover, they found that the conditioning depended upon the contingency between the CS and US, that is, upon the degree to which the CS predicted the US.

To investigate the cellular mechanisms of classical conditioning, Hawkins et al (1983a) used a differential training procedure in the neural circuit for siphon withdrawal. They recorded from two siphon sensory neurons that both made excitatory monosynaptic connections onto a siphon motoneuron. Tail shock produced significantly greater facilitation of synaptic transmission from a sensory neuron if the shock was immediately preceded by spike activity (produced by intracellular current in the sensory neuron) than if the shock and spike activity were specifically unpaired or if the shock occurred alone. These results suggested that the temporal specificity of the conditioning might be due to an activity-dependent enhancement of facilitation. Further evidence suggested that this enhancement is presynaptic in origin: Hawkins et al (1983a) examined the duration of the action potential in the sensory neurons (in 50 mM tetraethylammonium) and found that the differential training produced a significantly greater enhancement of spike duration in paired compared to unpaired sensory neurons. Extending these findings, recent voltage clamp experiments by Hawkins & Abrams (1984) suggest that the activity-dependent facilitation involves modulation of the same ionic channel in the sensory neurons (the S channel) that is involved in sensitization.

These experiments suggest that the mechanism of classical conditioning in *Aplysia* may involve a temporally specific, activity-dependent enhancement of the same cellular process (presynaptic facilitation) that is involved in sensitization. Further evidence that this mechanism can account for the behavioral conditioning has been obtained by Clark (1984), who found that just as backward (US-CS) pairing does not produce conditioning (Hawkins et al 1983b), neither does it produce activity-dependent facilitation of synaptic transmission. Forward pairing produces significantly greater facilitation than backward pairing.

Another model of conditioning emphasizing the importance of activity is that proposed by D. O. Hebb (1949), who postulated a critical role for postsynaptic activity in synaptic facilitation. Carew et al (1984) directly tested this hypothesis in *Aplysia* and found that postsynaptic activity (in siphon motoneurons that mediate the conditioned response) was neither necessary nor sufficient for associative synaptic changes, indicating that at least in this instance, conditioning was not due to a Hebb-type mechanism but rather to activity-dependent enhancement of presynaptic facilitation.

The activity-dependent model discussed above emphasizes the relationship between sensitization and its amplified form, classical conditioning. This model thus raises two important questions: How does activity affect the molecular machinery involved in sensitization (specifically, the cAMP cascade)? What aspect of activity is important for the amplification? Preliminary evidence addressing the first question suggests that the cAMP cascade is directly modulated by spike activity: Sensory neurons receiving a brief application of serotonin (as the US) immediately after a train of spikes show a four-fold greater level of cAMP than cells exposed to 5-HT alone (Kandel et al 1983, Abrams et al 1984a). With regard to the second question, two lines of evidence suggest that a brief elevation of Ca^{2+} produced by spike activity in the sensory neurons might serve as the signal for temporal specificity by increasing the cell's normal cAMP response to serotonin (Abrams et al 1983):

1. Action potentials paired with serotonin normally enhance the facilitatory (spike-broadening) response to serotonin in sensory neurons, but this activity-dependent enhancement does not occur in the *absence* of Ca^{2+} influx (Kandel et al 1983).
2. Stimulation of adenylate cyclase from *Aplysia* nervous tissue by serotonin is enhanced substantially by Ca^{2+}/calmodulin. The ability of serotonin to activate the cyclase is markedly reduced following depletion of calmodulin or following the addition of a calmodulin inhibitor (T. W. Abrams et al, in preparation).

Although these results are preliminary, they suggest that classical conditioning and sensitization in *Aplysia* share at least part of a common molecular cascade, and that Ca^{2+} influx during activity may be a critical priming factor for enhancement of that cascade during conditioning.

Classical Conditioning in the Tail Withdrawal Reflex

The tail withdrawal reflex in *Aplysia* is mediated by an identified cluster of mechanoafferent sensory neurons that synapse monosynaptically on tail motoneurons (Walters et al 1983a). Sensitization of the reflex as well as serotonin produces an increase in synaptic transmission from these neurons (Walters et al 1983b). In addition, serotonin produces an increase in input resistance and a concomitant elevation of cAMP (Byrne & Walters 1982, Ocorr et al 1983, Pollack et al 1982). More recently, Walsh & Byrne (1984) have shown that the cyclase activator, forskolin, mimics the 5-HT response in the sensory neurons and blocks subsequent 5-HT responses, suggesting that these agents act through a common saturable mechanism. Furthermore, Ocorr & Byrne (1985) have shown that, in addition to 5-HT, SCP_B (but not tryptamine or FMRF amide) elevates cAMP levels in these neurons, and that 5-HT and SCP_B produce an inward current associated with a increase in input resistance. Taken collectively, these data suggest that sensitization in the tail withdrawal reflex closely resembles that in the gill and siphon withdrawal reflex: Both may involve a serotonin-sensitive, cAMP-mediated decrease in K^+ conductance.

In addition to the nonassociative effects described above, the sensory neurons of the tail withdrawal reflex also show clear associative plasticity. Walters & Byrne (1983a) applied a cellular analogue of a differential conditioning procedure, recording simultaneously from three sensory neurons and a tail motoneuron. One sensory neuron (CS +) was activated (by intracellular current) paired with tail shock (the US), another (CS −) was activated unpaired with the US, and the third (sensitization control) was not activated during training. Following five trials, the EPSP from the paired sensory neurons (CS +) was significantly facilitated compared to both the CS − and sensitization controls. Walters & Byrne (1983b) further observed that CS + neurons show a slow depolarization after the US, while CS − or sensitization control neurons show a hyperpolarization. They suggest that, as a function of this depolarization, a voltage-sensitive Ca^{2+} conductance activated near the resting potential may be modulated by associative training. Based on these results, Walters & Byrne (1983a,b) propose that classical conditioning is produced by activity-dependent neuromodulation of synaptic transmission, and that Ca^{2+} could be a major intracellular messenger involved in this process. In a series of recent experiments, Walters & Byrne (1983c, 1985) have also suggested that activity-dependent neuromodulation may be a mechanism of long-term synaptic enhancement similar to long-term potentiation commonly observed in the mammalian CNS.

Recent studies by Ocorr et al (1983, 1985) have extended this analysis to a molecular level. They used bilateral clusters of the tail sensory neurons in a biochemical analogue of the classical conditioning paradigm: Depolarization of an entire cluster by means of a 5 sec exposure to high K^+ sea water served as the CS, and a 15 sec exposure to 5-HT served as the US. One cluster received paired, the other specifically unpaired, CS-US presentations. Following only one trial the paired cluster showed a significantly elevated cAMP content compared to the unpaired cluster. CS alone and US alone produced no significant changes. Preliminary studies of phosphodiesterase (PDE) activity by Ocorr & Byrne (1984) suggest that the increase in cAMP is due at least partly to stimulation of cyclase activity rather than inhibition of PDE. Taken together, these results led Ocorr et al (1985) to suggest that a pairing-specific enhancement of cAMP levels, perhaps by means of a Ca^{2+}-calmodulin dependent step, may be a biochemical mechanism for associative learning. This hypothesis can be directly tested in view of the recent report by Ingram & Walters (1984) that the tail withdrawal reflex is capable of differential classical conditioning.

The similarities of the behavioral, cellular, and molecular results of studies of associative conditioning from two systems in *Aplysia* that have been independently analyzed (the reflex withdrawal systems of the mantle and tail) are striking. Results from both systems emphasize the importance of neuronal activity, Ca^{2+} entry, cyclase activation, and reduction of a K^+ conductance for a pairing mechanism.

Conditioned Cessation of Feeding Responses

Aplysia are herbivores that normally feed on a variety of seaweeds. Susswein & Schwarz (1983) have described an interesting form of associative learning in this feeding behavior. They presented animals with seaweed wrapped in a plastic net, allowing the animals to taste the food and attempt to eat it, but not swallow it. Thus, the net-enclosed food became lodged in the buccal cavity. Animals receiving this treatment rapidly learned to cease responding to this inedible food, and memory was maintained for at least 24 hr. An essential component of the learning is that the food become stuck within the buccal activity; this the authors suggest gives rise to stimuli that become paired with lip stimuli from the taste of the food. The learning also displays stimulus specificity: lip stimuli occurring sequentially (rather than paired) with food stuck in the buccal cavity do not produce the conditioned cessation of feeding. Recently M. Schwarz, S. Markovich, and A. J. Susswein (in preparation) have further specified the parametric features of the learning and show them to be clearly dissociable from adaptation and satiation. Finally, Schwarz & Susswein (1984) have found that animals whose esophageal nerves are cut take twice as long to learn as sham controls, suggesting that these nerves carry information for learning that food is inedible. They suggest that gut inputs in these nerves can act as positive or negative reinforcers of feeding. Susswein & Schwarz (1983) suggest that the learning in their paradigm resembles operant conditioning, since the change in behavior is contingent upon the consequences of the animal's response. Since the neural circuit of feeding has been examined in detail (Cohen et al 1978, Rosen et al 1979, 1982), this preparation offers considerable promise for a cellular analysis of an interesting form of associative learning.

Operant Conditioning of the Head-Waving Response

Aplysia readily show side-to-side head-waving behavior in a number of behavioral contexts, including searching for a hold-fast, egg laying, and in "spontaneously" exploring their environment. Cook & Carew (1985a,b) have recently shown that this response can be operantly conditioned. Specifically, *Aplysia* can readily be operantly trained (within 10 min) to modify their head-waving response, increasing its frequency to one side of the body in order to terminate the presentation of a strong light that the animals find aversive. Yoked controls do not acquire the operant response. Two lines of evidence suggest that the operant responding is under the control of reinforcement contingencies:

1. When the contingencies are reversed (the original positive side now punished and vice versa), contingent-trained *Aplysia* (but not yoked controls) show a significant reduction in their responding, and when the original contingencies are reinstated, contingent animals once again show significant instrumental responding to the original side.
2. Yoked controls do not learn the operant task, yet these same animals readily learn when reinforcement is made contingent upon their responding.

Finally, contingent trained animals do not appear to use external cues (such as visual stimuli) to determine their body position during acquisition of the operant response. Rather, internally derived cues (e.g. proprioceptive or reafferent signals) appear to provide important feedback during acquisition of the operant task.

Since some of the cellular elements involved in both the operant response and the reinforcement pathways are known (Hening et al 1979, Jahn-Parwar & Fredman 1978, Olson & Jacklett 1984), this form of operant conditioning in *Aplysia* may prove to be quite useful in studying the cellular mechanisms of instrumental learning. Moreover, since mechanisms of classical conditioning have been elucidated in two other response systems in *Aplysia*, it may also be possible to address the theoretical question of the relationship between classical and operant conditioning in mechanistic terms.

DROSOPHILA MELANOGASTER

Because of its relatively small genome and extremely well-developed genetic methodology, the fruit fly, *Drosophila melanogaster*, has been the focus of a novel and creative strategy in the study of learning: Single gene lesions are used to disrupt the components of learning one by one. The goal is to find genes that produce specific lesions in the learning or memory apparatus, and then identify the gene product affected by the lesion to pinpoint a specific molecular component of the learning or memory process (Quinn 1984, Dudai 1985). The value of this approach is that is uses a totally different methodology from the physiological studies described in previous sections, but addresses common basic questions.

The first demonstration of associative learning in the fly was reported in *Phormia regina* (the blowfly) by Nelson (1971). Flies normally extend their proboscis in response to sugar applied to their legs. In Nelson's experiments a chemostimulus (the CS) applied to the leg was paired with sugar (the US) and resulted in conditioned proboscis extension to the CS. Since this original report, learning has been observed with a number of procedures. Adult and larvae *Drosophila* quickly learn to avoid a specific odor after it has been paired with shock (Quinn et al 1974, Dudai 1977, Aceves-Pina & Quinn 1979) or to approach an odor that has previously been paired with sugar (Tempel et al 1983). In addition, flies learn to avoid a color that was paired with mechanical shaking (Menne & Spatz 1977). Flies can also learn to either extend or flex their legs to avoid shock (Booker & Quinn 1981).

Most of the genetic and biochemical analysis of associative learning in fruit flies centers on the flies' ability to learn and remember an odor-shock association—which leads to a decrease in approach to the paired odor (Quinn et al 1974), or an odor-sucrose association—which leads to a significant increase in approach behavior to the paired odor (Tempel et al 1983). The procedures are straightforward. In a strategy that takes advantage of the flies' normal phototaxis, flies are allowed to enter plexiglass tubes illuminated by a fluorescent lamp. Small printed circuits are used as grids to deliver shock (ca. 90 V, AC) for the aversively motivated experiments, and the same grids are covered with sucrose (applied directly to the grid) in the appetitively motivated experiments. The grids are painted with odorants and then placed in each tube, so flies that phototax into the tubes are simultaneously exposed to either an odorant and electric shock or an odorant and sucrose. In a subsequent test without reinforcement, most conveniently done with a T-maze, normal flies tend to avoid the odorant previously paired with shock and to approach the odorant previously paired with sucrose. The data are usually presented as the fraction of flies avoiding or approaching the shock- or sugar-associated odor during a test trial, minus the fraction avoiding or approaching the control odor during the test trial.

The odor-shock learning paradigm has been used to isolate mutant flies that are deficient in the learning task. Mutations in six X-genes in *Drosophila* now appear to cause learning or memory deficits. These include *Dunce* (*dnc*), *rutabaga* (*rut*), *cabbage* (*cab*), *turnip* (*tur*), *radish* (*rad*), and *amnesiac* (*amn*) (Dudai et al 1976, Aceves-Pina & Quinn 1979, Quinn et al 1979, Tempel et al 1983). Each mutant has a characteristic deficit in learning and/or memory. For example, both *dnc* and *rut* flies show appreciable learning when tested within 30 sec following training, but memory in these flies decays rapidly. *Amn* also shows normal learning but aberrant memory. An additional mutant isolated by Wright (1977), the *Dopadecarboxylase* (*Ddc*) mutation has a deficiency in neurotransmitter synthesis (Livingstone & Tempel 1983) and also fails to learn both odor-shock associations as well as odor-sucrose associations. In this case, one can use mild mutations and show that it is acquisition that is blocked; rates of forgetting of the subsequent memory are normal (Tempel et al 1984). In the following sections we briefly discuss some of the salient features of each of these mutants.

Dunce

The memory mutant *Dunce* was first isolated by Benzer and colleagues in 1976. This mutant has subsequently been the focus of intense study. *Dunce* was initially thought to have practically no short-term memory, but later experiments showed evidence of an initial memory that decays very rapidly (Dudai 1979, 1985 Tempel et al 1983). It appears that the locus of the *dnc* gene is in chromere 3D4 on the X chromosome, and it is thought that this gene codes for an isozyme of the cAMP-phosphodiesterase (PdE II), one of two enzymes in *Drosophila* that hydrolyze cyclic AMP (Byers et al 1981, Kauvar 1982, Kiger 1979, Kiger & Golanty 1977, Shotwell 1983). A 35 kD high-cAMP affinity, low-cGMP affinity form of the enzyme is defective in the *dnc* mutant. Moreover, correlations among the *dnc* gene dose, reduced activity of the enzyme, and reduced memory have been reported (Shotwell 1983).

Rutabaga

Rutabaga (*rut*), also a memory mutant, has been reported to be defective in a limited subpopulation of membrane-bound adenylate cyclase. (Aceves-Pina et al 1983, Dudai & Zvi 1984, Dudai et al 1983, Livingstone & Tempel 1983, Livingstone et al 1984). The behavioral deficit and the biochemical deficit map to the same small segment of the X-chromosome, 12E1-13A5 (Livingstone et al 1984). Biochemical analyses indicate that the *rutabaga* mutant is specifically deficient in calcium calmodulin-activated cyclase activity and normal in the cyclase regulatory proteins (Livingstone et al 1982, 1984).

Turnip

The mutant, *turnip*, is severely debilitated in olfactory learning (Aceves-Pina & Quinn 1979), and has been characterized as a rapidly-forgetting mutant (Quinn et al 1979). Interestingly, *turnip* shows only a slight deficit in learning visual color discrimination (Dudai & Bicker 1978, Folkers 1982). The biochemical deficit in *turnip* appears to affect serotonin binding. High affinity binding is characteristic of normal flies, but serotonin binding in *turnip* flies shows a reduction in high-affinity binding different kinetics (Aceves-Pina et al 1983). Recent experiments indicate that the decrease in high-affinity binding seen in *turnip* may be due to abnormalities in membrane GTP-binding proteins and in GTP hydrolysis (Aceves-Pina 1983, Quinn 1984).

Dopadecarboxylase

The enzyme, dopa-decarboxylase, is necessary for synthesizing precursors to melanin and other compounds in the cuticle. The *Ddc* mutation was isolated by Wright (1977) for these functions before its effect on CNS neurotransmitters or fly behavior became known. Flies with severe lesions in the *Ddc* gene show no evidence of learning (Tempel et al 1983, 1984, Tempel & Quinn 1982). However, if flies are bred to have moderate lesions in the gene, some learning is observed (more learning than in full *Ddc* mutation but less learning than in wild type). More importantly, memory retention appears to be normal (Tempel et al 1984). The dopadecarboxylase enzyme is necessary for the synthesis of dopamine (Dewhurst et al 1972) and serotonin (Livingstone & Tempel 1983). Since mutant flies are deficient in these transmitters (Wright 1977) and since *Ddc* shows an acquisition deficit (unlike other *Drosophila* mutants that show memory deficits), it has been suggested that serotonin, dopamine, or both may be essential for learning but not for memory in *Drosophila* (Aceves-Pina et al 1983, Tempel et al 1984).

Amnesiac

Flies with this mutation show normal learning but abnormally rapid forgetting with the olfactory task, as compared to normal flies (Quinn et al 1979). Learning of color discrimination appears to be only slightly disrupted by this lesion (Dudai & Bicker 1978, Folkers 1982). The nature of the biochemical deficit has not been identified as yet, but the mutation has been mapped to chromomere 18F-19A on the X chromosome (Tully 1984).

In addition to approach and avoidance responses, *Drosophila* courtship behavior is also modified by learning. If males are first placed with sexually unreceptive females that reject the males' courtship attempts and then subsequently are placed with receptive females, the males' courtship of the receptive females is depressed for about 2 hr (Siegel & Hall 1979, Tompkins et al 1983). Interestingly, all the learning mutants (*amnesiac, cabbage, rutabaga, turnip, dunce*, and heat-treated *Ddc* male flies) recovery from this rejection experience much more quickly than wild-type flies (Gailey et al 1984, Tempel et al 1984). The modification of courtship behavior in the learning mutants is especially interesting, since this reflects a deficit in a presumably naturally occurring form of associative learning.

Some of the learning mutants also show behavioral deficits on other simpler, nonassociative learning tests. Normal flies show a characteristic habituation of proboscis extension to repeated leg stimulation with a dilute sugar solution. *Dunce* flies show a much slower rate of habituation than normal flies (Duerr & Quinn 1982). In addition to habituation, normal flies show sensitization, another form of nonassociative learning. Concentrated sugar solution applied to the proboscis results in proboscis extension to a number of stimuli that do not usually evoke the response. Both *dunce* and *turnip* flies show significantly less sensitization than normal wild-type flies (Duerr & Quinn 1982). Thus *Drosophila* mutants that block associative learning also block nonassociative processes, suggesting that, as in *Aplysia*, nonassociative and associative learning may use aspects of common molecular machinery.

CONCLUDING REMARKS

The progress achieved over the last 10–15 years in studying a wide variety of forms of learning in simple invertebrate animals is quite striking. There is now no question, for example, that associative learning is a common capacity in several invertebrate species. In fact, the higher-order features of learning seen in some invertebrates (notably bees and *Limax*) rivals that commonly observed in such star performers in the vertebrate laboratory as pigeons, rats, and rabbits. This is not to imply that invertebrates can learn anything that vertebrates can—there undoubtedly are phylogenetic constraints on the nature of associations that animals are capable of making. It is meant to indicate however, that several invertebrate preparations are now at hand that provide quite reasonable simple model systems in which to explore mechanisms of complex learning and memory phenomena on behavioral, cellular, and molecular levels.

Because of the diverse approaches used to study invertebrate learning and because neuronal mechanisms have been investigated in so few preparations, an attempt to derive "general principles" from work in invertebrates is premature; nevertheless, common themes and convergent ideas provide a framework for further investigations. One such theme is that, in all cases examined thus far, learning (both nonassociative and associative) can be localized to individual neurons and involves alterations of either previously existing synaptic connections or intrinsic cellular properties; in no case have novel synapses or new biophysical properties been induced by learning. The tentative principle then appears to be that certain neurons or groups of neurons are endowed with the capacity for plastic change, and experience then promotes that change in previously existing circuits.

A second theme that emerges from work in at least three different preparations (locust, *Hermissenda*, and *Aplysia*) is that modulation of K^+ conductance is involved in memory storage. Both the particular species of K^+ channel and the location of that channel in the neuron can vary: In locust, I_K is thought to change at a pacemaker region of a motoneuron, modulating firing frequency; in *Hermissenda* I_A and I_C are thought to be reduced in a class of photoreceptors, increasing their excitability; in *Aplysia*, I_S is thought to be reduced in cell bodies and terminals of sensory neurons, increasing transmitter release by broadening the action potential. Despite the differences in cellular detail, it is striking that three very different forms of learning appear to use alteration of K^+ conductance as a storage mechanism. The results from these studies underscore an important principle: The *location* of a conductance change in a neuron can have very different functional consequences. Consider a reduction in K^+ conductance. If localized to dendrites this could enhance synaptic input to a cell (since the same synaptic current would produce a larger potential change); if localized to a trigger zone it could increase excitability by reducing the membrane potential (it could also have direct effects on voltage threshold); if localized to a pacemaker region it could modulate firing frequency; and if localized to terminals it could broaden action potentials and thus enhance transmitter release. Of course these effects are not mutually exclusive. Cell-wide changes in K^+ conductance could produce all of these effects in the same neurons. Thus, a common general mechanism such as a reduction of K^+ conductance, when localized to different neuronal regions, can have widespread and diverse effects that can modulate both the way a neuron receives and transmits information.

A third recurrent theme concerns the molecular mechanisms involved in learning. In both *Hermissenda* and *Aplysia*, "second-messengers" are implicated as important for memory because of their role in mediating phosphorylation of substrate proteins (in both cases perhaps K^+ channels). In *Hermissenda*, both cAMP and Ca^{2+}-calmodulin are implicated in phototactic conditioning (with Ca^{2+}-calmodulin suggested as the more likely candidate), whereas in *Aplysia*, cAMP is thought to be the primary second messenger both in sensitization and (in amplified form) in classical conditioning of both gill and siphon withdrawal and tail withdrawal. Interesting parallels with *Aplysia* are seen in *Drosophila*, since learning mutants with impaired adenylate cyclase activity (*rutabaga*) and impaired cAMP phosphodiesterase (*dunce*) also show both impaired sensitization and classical conditioning. The mutant, *rutabaga*, which is defective in a

Ca^{2+}-activated adenylate cyclase, is of further relevance to both *Hermissenda* and *Aplysia*, since Ca^{2+} influx during a cellular response is thought to be important in each system. In *Hermissenda*, Ca^{2+} influx is thought to play a key role in both cumulative depolarization and in subsequent protein phosphorylation, and in *Aplysia*, Ca^{2+} influx with each action potential is thought to be the pairing-specific signal that amplifies the cAMP response during conditioning. Although clearly speculative, further characterization of the gene product affected in *rutabaga* may provide insights into specific molecules involved in forming the temporal associations observed in both *Hermissenda* and *Aplysia*. Finally, the suggestion that second messengers such as Ca^{2+} or cAMP are involved in several invertebrate (as well as vertebrate) systems raises the interesting question: What is the utility of second messengers in learning? Two possible suggestions are (*a*) elevation of the concentration of second messengers and their subsequent decline may provide the substrate for short-term memory; and (*b*) second messengers provide an excellent means of communicating a locally-produced event (e.g. stimulus pairing occurring at a particular synaptic site) with other regions of the neuron, such as other synaptic sites where input or output might be modulated (perhaps as described above by decreased K^+ conductance), or to the nucleus where genomic changes underlying long-term memory could be produced.

A fourth consideration that emerges is that, although *storage* mechanisms (such as a reduced K^+ conductance mediated by a second messenger) may be common, *acquisition* mechanisms may be different. This suggestion is based on only two examples in which models for acquisition have been advanced (*Hermissenda* and *Aplysia*) and should thus be considered tentative at best. Acquisition in *Hermissenda* is thought to involve cumulative depolarization in photoreceptors induced directly by light and synaptically by the simultaneous offset of light and rotation. Acquisition in *Aplysia* is thought to involve activity-dependent enhancement of pre-synaptic facilitation mediated by biogenic amines (and possibly peptides as well). The postulated acquisition mechanisms therefore appear quite different in these two systems. One (*Hermissenda*) involves primarily intrinsic (nonsynaptic) changes, while the other (*Aplysia*) involves primarily heterosynaptic (neuromodulatory) changes. It is possible, as further insights into these and other acquisition mechanisms are obtained, that aspects of common mechanisms will be uncovered. Alternatively, it is possible that nature uses different acquisition mechanisms for learning, but common storage mechanisms for memory. It is too soon to tell.

In conclusion, it is heartening that the possible mechanisms of learning and memory suggested from work in the simple invertebrate systems we have reviewed are not highly specialized or esoteric physiological or biochemical events. On the contrary, neuromodulation by biogenic amines, second messenger–mediated intracellular events, and modulation of ionic conductances by protein phosphorylation are highly conserved mechanisms that are widespread throughout the animal kingdom. Thus we have reason to hope that the distinction between vertebrate and invertebrate learning and memory is one that will diminish as our understanding of underlying mechanisms increases.

ACKNOWLEDGMENTS

We are indebted to Emilie Marcus for her expert and untiring assistance in all phases of preparing this manuscript. Research in TJC's laboratory is supported by NIH Career Development Award 7-KOZ-MH00081-09, NIH BRSG Grant 507-RR-07015, and NSF Grant BSN 8311300, and in CLS's laboratory by NIH BRSG Grant 2-507-RR-07015, NIH Grant 5RO1-MH379, and Sloan Foundation Fellowship BR2486.

Literature Cited

Abrams, T. W., Bernier, L., Hawkins, R. D., Kandel, E. R. 1984a. Possible roles of Ca^{++} and cAMP in activity-dependent facilitation, a mechanism for associative learning in *Aplysia*. *Soc. Neurosci. Abstr.* 10:269

Abrams, T. W., Carew, T. J., Hawkins, R. D., Kandel, E. R. 1983. Aspects of the cellular mechanism of temporal specificity in conditioning in *Aplysia*: Preliminary evidence for Ca^{++} influx as a signal of activity. *Soc. Neurosci. Abstr.* 9:168

Abrams, T. W., Castellucci, V. F., Camardo, J. S., Kandel, E. R., Lloyd, P. E. 1984b. Two endogenous neuropeptides modulate the gill and siphon withdrawal reflex in *Aplysia* by presynaptic facilitation involving cAMP-dependent closure of a serotonin-sensitive potassium channel. *Proc. Natl. Acad. Sci. USA* 81:7956–60

Aceves, Pina, E. O., Quinn, W. G. 1979. Learning in normal and mutant *Drosophila* larvae. *Science* 206:93–95

Aceves-Pina, E. O., Booker, R., Duerr, J. S., Livingstone, M. S., Quinn, W. G., Smith, R. F., Sziber, P. P., Tempel, B. L., Tully, T. P. 1983. Learning and memory in *Drosophila*, studied with mutants. *Cold Spring Harbor Symp. Quant. Biol.* 48:831–40

Acota-Urquidi, J., Alkon, D. L., Neary, J. T. 1984. Ca^{++}-dependent protein kinase injection in a photoreceptor mimics biophysical effects of associative learning. *Science* 224:1254–57

Alexander, J. E. Jr., Audesirk, T. E., Audesirk, G. J. 1982. Rapid, nonaversive conditioning in a freshwater gastropod. II. Effects of temporal relationships on learning. *Behav. Neural Biol.* 36:391–402

Alkon, D. L. 1974. Associative training of *Hermissenda*. *J. Gen Physiol.* 64:70–84

Alkon, D. L. 1979. Voltage-dependent calcium and potassium ion conductances: A contingency mechanism for an associative learning model. *Science* 205:810–16

Alkon, D. L. 1980a. Cellular analysis of a gastropod (*Hermissenda crassicornis*) model of associative learning. *Biol. Bull.* 159:505–60

Alkon, D. L. 1980b. Membrane depolarization accumulates during acquisition of an associative behavioral change. *Science* 210:1375–76

Alkon, D. L. 1983. Learning in a marine snail. *Sci. Am.* 249:70–84

Alkon, D. L. 1984. Calcium-mediated reduction of ionic currents: A biophysical memory trace. *Science* 226:1037–45

Alkon, D. L., Acosta-Urquidi, J., Olds, J., Kuzma, G., Neary, J. T. 1983a. Protein kinase injection reduces voltage-dependent potassium currents. *Science* 219:303–6

Alkon, D. L., Farley, J., Hay, B., Shoukimas, J. J. 1983b. Inactivation of Ca^{++}-dependent K^+ current can occur without significant Ca^{++}-current inactivation. *Soc. Neurosci. Abstr.* 9:1188

Alkon, D. L., Fuortes, G. F. 1972. Responses of photoreceptors in *Hermissenda*. *J. Gen. Physiol.* 60:631–49

Alkon, D. L., Grossman, Y. 1978. Long-lasting depolarization and hyperpolarization in eye of *Hermissenda*. *J. Neurophysiol.* 41:1328–42

Alkon, D. L., Lederhendler, I., Shoukimas, J. J. 1982a. Primary changes of membrane currents during retention of associative learning. *Science* 215:693–95

Alkon, D. L., Shoukimas, J. J., Heldman, E. 1982b. Calcium-mediated decrease of a voltage-dependent potassium current. *Biophys. J.* 40:245–50

Audesirk, T. E., Alexander, J. E. Jr., Audesirk, G. J., Moyer, C. M. 1982. Rapid, nonaversive conditioning in a freshwater gastropod. I. Effects of age and motivation. *Behav. Neural Biol.* 36:379–90

Bagnoli, P., Brunelli, M., Magni, P., Pelligrino, M. 1975. The neuron of the fast conducting system in *Hirudo medicinalis*: Identification and synaptic connections with primary afferent neurons. *Arch. Ital. Biol.* 113:21–43

Bailey, C. H., Chen, M. 1983. Morphological basis of long-term habituation and sensitization in *Aplysia*. *Science* 220:91–93

Bailey, C. H., Chen, M. 1984. Further studies on the morphological basis of long-term habituation in *Aplysia*. *Soc. Neurosci. Abstr.* 10:131

Bailey, C. H., Hawkins, R. D., Chen, M., Kandel, E. R. 1981. Interneurons involved in mediation and modulation of gill-withdrawal reflex in *Aplysia*. IV. Morphological basis of presynaptic facilitation. *J. Neurophysiol.* 45:340–60

Baylor, D. A., Nicholls, J. 1969. Chemical and electrical synapses between cutaneous mechanoreceptors in the CNS of the leech. *J. Physiol.* 203:591–609

Belardetti, P., Biondi, C., Colombaioni, L., Brunelli, M., Trevisani, A., Zavagno, C. 1982. Role of serotonin and cyclic AMP on facilitation of the Fast Conducting System activity in the leech, *Hirudo medicinalis*. *Brain Res.* 246:89–103

Benjamin, P. R., Rose, R. M. 1979. Central generation of bursting in the feeding system of the snail *Lymnaea stagnalis*. *J. Exp. Biol.* 80:93–118

Benjamin, P. R., Rose, R. M. 1980. Interneuronal circuitry underlying cyclical feeding in gastropod molluscs. *Trends Neurosci.* 3:272–74

Bernier, L., Castellucci, V. F., Kandel, E. R., Schwartz, J. H. 1982. Facilitatory transmitter causes a selective and prolonged increase in adenosine 3′:5′-monophosphate in sensory neurons mediating the gill and siphon withdrawal reflex in *Aplysia*. *J. Neurosci.* 2:1682–91

Bicker, G., Davis, W. J., Matera, E. M. 1982a. Mechano- and chemoreception in *Pleurobranchaea californica*. II. Neuroanatomical and intracellular analysis pf centripetal pathways. *J. Comp. Physiol.* 149:235–50

Bicker, G., Davis, W. J., Matera, E. M., Kovac, M. P., Stormo-Gipson, J. 1982b. Mechano- and chemoreception in *Pleurobranchaea californica*. I. Extracellular analysis of afferent responses. *J. Comp. Physiol.* 149:221–34

Biondi, C., Berladetti, F., Brunelli, M., Portolan, A., Trevisani, A. 1982. Increased synthesis of cyclic AMP and short-term plastic changes in the segmental ganglia of the leech, *Hirudo medicinalis*. *Cell. Molec. Neurobiol.* 2:81–91

Bitterman, M. E., Menzel, R., Fietz, A., Schader, S. 1983. Classical conditioning of proboscis-extension in honeybees. *J. Comp. Physiol. Psychol.* 97:107–19

Booker, R., Quinn, W. G. 1981. Conditioning of leg position in normal and mutant *Drosophila*. *Proc. Natl. Acad. Sci. USA* 78:3940–44

Boyle, M. B., Klein, M., Smith, S. J., Kandel, E. R. 1984. Serotonin increases intracellular Ca^{++} transients in voltage-clamped sensory neurons of *Aplysia californica*. *Proc. Natl. Acad. Sci. USA* 81:7642–46

Brunelli, M., Castellucci, V. F., Kandel, E. R. 1976. Synaptic facilitation and behavioral sensitization in *Aplysia*: Possible role of serotonin and cAMP. *Science* 194:1178–81

Byers, D., Davis, R. L., Kiger, J. A. 1981. Defect in cyclic AMP phosphodiesterase due to the dunce mutation of learning in *Drosophila melanogaster*. *Nature* 289:79–81

Byrne, J. H., Castellucci, V. F., Kandel, E. R. 1978. Contribution of individual mechanoreceptor neurons mediating defensive gill-withdrawal in *Aplysia*. *J. Neurophysiol.* 41:418–31

Byrne, J., Castellucci, V., Kandel, E. R. 1974. Receptive fields and response properties of mechanoreceptor neurons in innervating siphon skin and mantle shelf in *Aplysia*. *J. Neurophysiol.* 37:1041–64

Byrne, J. H., Walters, E. T. 1982. Associative conditioning of single sensory neurons in *Aplysia*: II. Activity-dependent modulation of membrane responses. *Soc. Neurosci. Abstr.* 8:386

Carew, T. J., Castellucci, V. F., Kandel, E. R. 1971. An analysis of dishabituation and sensitization of the gill-withdrawal reflex in *Aplysia*. *Int. J. Neurosci.* 2:79–98

Carew, T. J., Castellucci, V. F., Kandel, E. R. 1979. Sensitization in *Aplysia*: restoration of transmission in synapses inactivated by long-term habituation. *Science* 205:417–19

Carew, T. J., Hawkins, R. D., Abrams, T. W., Kandel, E. R. 1984. A test of Hebb's postulate at identified synapses which mediate classical conditioning in *Aplysia*. *J. Neurosci.* 4:1217–24

Carew, T. J., Hawkins, R. D., Kandel, E. R. 1983. Differential classical conditioning of a defensive withdrawal reflex in *Aplysia californica*. *Science* 219:397–400

Carew, T. J., Kandel, E. R. 1977. Inking in *Aplysia californica*: III. Two different synaptic conductance mechanisms for triggering the central program for inking. *J. Neurophysiol.* 40:721–34

Carew, T. J., Pinsker, H. M., Kandel, E. R. 1972. Long-term habituation of a defensive withdrawal reflex in *Aplysia*. *Science* 175:451–54

Carew, T. J., Walters, E. T., Kandel, E. R. 1981a. Associative learning in *Aplysia*: Cellular correlates supporting a conditioned fear hypothesis. *Science* 211:501–4

Carew, T. J., Walters, E. T., Kandel, E. R. 1981b. Classical conditioning in a simple withdrawal reflex in *Aplysia californica*. *J. Neurosci.* 1:1426–37

Cartwright, B. A., Collett, T. S. 1983. Landmark learning in bees: Experiments and models. *J. Comp. Physiol. A* 151:521–43

Castellucci, E. F., Bernier, L., Schwartz, J. H., Kandel, E. R. 1983. Persistent activation of adenylate cyclase underlies the time course of short-term sensitization in *Aplysia*. *Soc. Neurosci. Abstr.* 9:169

Castellucci, V. F., Carew, T. J., Kandel, E. R. 1978. Cellular analysis of long-term habituation of the gill-withdrawal reflex of *Aplysia californica*. *Science* 202:1306–8

Castellucci, V. K., Kandel, E. R. 1974. A quantal analysis of the synaptic depression underlying habituation of the gill-withdrawal reflex in *Aplysia*. *Proc. Natl. Acad. Sci. USA* 71:5004–8

Castellucci, V. F., Kandel, E. R. 1976. Presynaptic facilitation as a mechanism for behavioral sensitization in *Aplysia*. *Science* 194:1176–78

Castellucci, V. F., Kandel, E. R., Schwartz, J. H., Wilson, F. D., Nairn, A. C., Greengard, P. 1980. Intracellular injection of the catalytic subunit of cyclic AMP-dependent protein kinase simulates facilitation of transmitter release underlying behavioral sensitization in *Aplysia*. *Proc. Natl. Acad. Sci. USA* 77:7492–96

Castellucci, V. F., Nairn, A., Greengard, P., Schwartz, J. H., Kandel, E. R. 1982. Inhibitor of adenosine 3′:5′-monophosphate-dependent protein kinase blocks presynaptic facilitation in *Aplysia*. *J. Neurosci.* 2:1673–81

Castellucci, V., Pinsker, H., Kupfermann, I., Kandel, E. R. 1970. Neuronal mechanisms of habituation and dishabituation of the gill-withdrawal reflex in *Aplysia*. *Science* 167:1745–48

Chang, J. J., Gelperin, A. 1980. Rapid taste-aversion learning by an isolated molluscan central nervous system. *Proc. Natl. Acad. Sci. USA* 77:6204–6

Clark, G. A. 1984. A cellular mechanism for the temporal specificity of classical conditioning of the siphon-withdrawal response in *Aplysia*. *Soc. Neurosci. Abstr.* 10:268

Cohen, D. H. 1980. The functional neuroanatomy of the conditioned response. In *Neural Mechanisms of Goal-Directed Behavior and Learning*, ed. R. F. Thompson, L. H. Hicks, V. B. Shvyrkov, pp. 283–302. New York: Academic

Cohen, D. H. 1984. Identification of vertebrate neurons modified during learning: Analysis of sensory pathways. In *Primary Neural Substrates of Learning and Behavioral Change*, ed. D. L. Alkon, J. Farley, pp. 129–54. Cambridge: Cambridge Univ. Press

Cohen, J. L., Weiss, K. R., Kupfermann, I. 1978. Motor control of buccal muscles of *Aplysia*. *J. Neurophysiol.* 41:157–80

Collett, T. S., Cartwright, B. A. 1982. How honey bees use landmarks to guide their return to a food source. *Nature* 295:560–64

Connor, J., Alkon, D. L. 1984. Light- and voltage-dependent increases of calcium ion concentration in molluscan photoreceptors. *J. Neurophysiol.* 51:745–52

Cook, D. G., Carew, T. J. 1985a. Operant conditioning of head-waving in *Aplysia*. *Soc. Neurosci. Abstr.* 11:796

Cook, D. G., Carew, T. J. 1985b. Operant conditioning of head-waving in *Aplysia*. *Proc. Natl. Acad. Sci. USA*. In press

Copeland, J., Gelperin, A. 1983. Feeding and a serotonergic interneuron activate an identified autoactive salivary neuron in *Limax maximus*. *Comp. Biochem. Physiol.* 76:21–30

Couvillon, P. A., Bitterman, M. E. 1980. Some phenomena of associative learning in honeybees. *J. Comp. Physiol. Psychol.* 94:878–85

Couvillon, P. A., Bitterman, M. E. 1982. Compound conditioning in honeybees. *J. Comp. Physiol. Psychol.* 96:192–99

Couvillon, P. A., Klosterhalfen, S., Bitterman, M. E. 1983. Analysis of overshadowing in honeybees. *J. Comp. Psychol.* 97:154–66

Croll, R. P., Chase, R. 1977. A long-term memory for food odors in the land snail, *Achatina fulica*. *Behav. Biol.* 19:261–68

Croll, R. P., Chase, R. 1980. Plasticity of olfactory orientation to foods in the snail *Achatina fulica*. *J. Comp. Physiol. A* 136:267–77

Crow, T. 1981. Neurophysiological correlates of conditioning in identified putative motor neurons in *hermissenda*. *Soc. Neurosci. Abstr.* 7:352

Crow, T. 1982. Sensory neuronal correlates of associative learning in *Hermissenda*. *Soc. Neurosci. Abstr.* 8:824

Crow, T. 1983a. Conditioned modification of locomotion in *Hermissenda crassicornis*: Analysis of time-dependent associative and nonassociative components. *J. Neurosci.* 3:2621–28

Crow, T. 1983b. Modification of B-photo-receptor adaptation predicts differential light responding following conditioning in *Hermissenda*. *Soc. Neurosci. Abstr.* 9:167

Crow, T. 1985a. Conditioned modification of phototactic behavior in *Hermissenda*. I. Analysis of light intensity. *J. Neurosci.* 5:209–14

Crow, T. 1985b. Conditioned modification of phototactic behavior in *Hermissenda*. II. Differential adaptation of B photoreceptors. *J. Neurosci.* 5:215–23

Crow, T. J., Alkon, D. L. 1978. Retention of an associative behavioral change in *Hermissenda*. *Science* 201:1239–41

Crow, T. J., Alkon, D. L. 1980. Associative behavioral modification in *Hermissenda*: Cellular correlates. *Science* 209:412–14

Crow, T., Harrigan, J. F. 1979. Reduced behavioral variability in laboratory-reared *Hermissenda crassicornis*. *Brain Res.* 173:179–84

Crow, T., Offenbach, N. 1979. Response-specificity following behavioral training in the nudibranch mollusc *Hermissenda crasscornis*. *Biol. Bull.* 157:364

Crow, T., Offenbach, N. 1983. Modification of the initiation of locomotion in *Hermissenda*. *Brain Res.* 271:301–10

Culligan, N., Gelperin, A. 1983. One-trial associative learning by an isolated molluscan CNS: Use of different chemoreceptors for training and testing. *Brain Res.* 266:319–27

Davis, W. J. 1984. Neural consequences of experience in *Pleurobranchaea californica*. *J. Physiol. Paris* 78:793–98

Davis, W. J., Gillette, R. 1978. Neural correlate of behavioral plasticity in command neurons of *Pleurobranchaea*. *Science* 199:801–4

Davis, W. J., Gillette, R., Kovac, M. P., Croll, R. P., Matera, E. M. 1983. Organization of synaptic inputs to paracerebral feeding command interneurons of *Pleurobranchaea californica*. III. Modifications induced by experience. *J. Neurophysiol.* 49:1557–72

Davis, W. J., Kovac, M. P., Croll, R. P., Matera, E. M. 1984. Brain oscillator(s) underlying rhythmic cerebral and buccal motor output in the mollusc, *Pleurobran-*

chaea californica. *J. Exp. Biol.* 110:1–16

Davis, W. J., Matera, E. M. 1982. Chemoreception in gastropod mulluscs: Electron microscopy of putative receptor cells. *J. Neurobiol.* 13:79–84

Davis, W. J., Mpitsos, G. J., Pinneo, J. M. 1974a. The behavioral hierarchy of the mollusk *Pleurobranchaea*. I. The dominant position of the feeding behavior. *J. Comp. Physiol.* 90:207–24

Davis, W. J., Siegler, M. V. S., Mpitsos, G. J. 1973. Distributed neuronal oscillators and efference copy in the feeding system of *Pleurobranchea*. *J. Neurophysiol.* 36:258–74

Davis, W. J., Mpitsos, G. J., Siegler, M. V. S., Pinneo, J. M., Davis, K. B. 1974b. Neuronal substrates of behavioral hierarchies and associative learning in the mollusk *Pleurobranchaea*. *Am. Zool.* 14:1037–50

Davis, W. J., Villet, J., Lee, D., Rigler, M., Gillette, R., Prince, E. 1980. Selective and differential avoidance learning in the feeding and withdrawal behavior of *Pleurobranchaea californica*. *J. Comp. Physiol.* 138:157–65

Debski, E. A., Friesen, W. O. 1985. Habituation of swimming activity in the medicinal leach. *J. Exp. Biol.* In press

Delaney, K., Gelperin, A. 1985. Post ingestive food-aversion learning to amino acid deficient diets by the terrestrial slug *Limax maximus*. In press

Dennis, M. J. 1967. Electrophysiology of the visual system in a nudibranch mollusc. *J. Neurophysiol.* 30:1439–65

Dewhurst, S. A., Croker, S. G., Ikeda, K., McCaman, R. E. 1972. Metabolism of the biogenic amines in *Drosophila* nervous tissue. *Comp. Biochem. Physiol.* 43B:975–84

Domjan, M., Burkhard, B. 1982. *The Principles of Learning and Behavior*. Monterey, Calif.: Brooks/Cole

Dudai, Y. 1977. Properties of learning and memory in Drosophila melanogaster. *J. Comp. Physiol.* 114:69–81

Dudai, Y. 1979. Behavioral plasticity in a *Drosophila* mutant, *dunce*. *J. Comp. Physiol.* 130:271–85

Dudai, Y. 1985. Genes, enzymes and learning in *Drosophila*. *Trends Neurosci.* 8:18–22

Dudai, Y., Bicker, G. 1978. Comparison of visual and olfactory learning in *Drosophila*. *Naturwissenschaften* 65:495–96

Dudai, Y., Jan, Y.-N., Byers, D., Quinn, W. G., Benzer, S. 1976. *Dunce*, a mutant of *Drosophila* deficient in learning. *Proc. Natl. Acad. Sci. USA* 73:1684–88

Dudai, Y., Uzzan, A., Zvi, S. 1983. Abnormal activity of adenylate cyclase in the *Drosophila* memory mutant *rutabaga*. *Neurosci. Lett.* 42:207–12

Dudai, Y., Zvi, S. 1984. Adenylate cyclase in *Drosophila* memory mutant rutabaga displays an altered Ca^{++} sensitivity. *Neurosci. Lett.* 47:119–24

Duerr, J. S., Quinn, W. G. 1982. Three *Drosophila* mutants that block associative learning also affect habituation and sensitization. *Proc. Natl. Acad. Sci. USA* 79:3646–50

Eisenstein, E. M., Cohen, M. J. 1965. Learning in an isolated insect ganglion. *Anim. Behav.* 13:104–8

Erber, J. 1976. Retrograde amnesia in honeybees *(Apis mellifera carnica)*. *J. Comp. Physiol. Psychol.* 90:41–46

Erber, J. 1978. Response characteristics and after effects of multimodal neurons in the mushroom body area of the honey bee. *Physiol. Entomol.* 3:77–89

Erber, J. 1980. Neural correlates of nonassociative and associative learning in the honey bee. *Verb. Dtsch. Zool. Ges.* 250–61

Erber, J. 1981. Neural correlates of learning in the honeybee. *Trends Neurosci.* 4:270–73

Erber, J. 1982. Movement learning of freeflying honeybees. *J. Comp. Physiol.* 146:273–82

Erber, J. 1984. Response changes of single neurons during learning in the honeybee. See Cohen 1984, pp. 275–85

Erber, J., Schildberger, K. 1980. Conditioning of an antennal reflex to visual stimuli in bees *(Apis mellifera L.)*. *J. Comp. Physiol.* 135:217–25

Erber, J., Masuhr, T. H., Menzel, R. 1980. Localization of short-term memory in the brain of the bee, *Apis mellifera*. *Physiol. Entomol.* 5:343–58

Estes, W. K., Skinner, B. F. 1941. Some quantitative properties of anxiety. *J. Exp. Biol.* 29:390–400

Farley, J., Alkon, D. L. 1980. Neural organization predicts stimulus specificity for a retained associative behavioral change. *Science* 210:1373–75

Farley, J., Alkon, D. L. 1982a. Associative neural and behavioral change in *Hermissenda*: Consequences of nervous system orientation for light- and pairing-specificity. *J. Neurophysiol.* 48:785–807

Farley, J., Alkon, D. L. 1982b. Cumulative depolarization and short-term associated conditioning in *Hermissenda*. *Soc. Neurosci. Abstr.* 8:825

Farley, J., Alkon, D. L. 1983. Changes in *Hermissenda* type B photoreceptors involving a voltage-dependent Ca^{++} current and Ca^{++}-depe·dent K^+ current during retention of associative learning. *Soc. Neurosci. Abstr.* 9:167

Farley, J., Kern, G. 1985. Contingency-sensitive phototaxic behavioral changes in *Her-*

missenda: Temporally-specific attenuation of conditioning. *Animal Learn. Behav.* In press

Farley, J., Richards, W. G., Ling, L. J., Liman, E., Alkon, D. L. 1983. Membrane changes in a single photoreceptor cause associative learning in *Hermissenda*. *Science* 221:1201–3

Farley, J., Sakakibara, M., Alkon, D. L. 1984. Associative-training correlated changes in I_C and I_{Ca} in *hermissenda* Type B photoreceptors. *Soc. Neurosci. Abstr.* 10:270

Folkers, E. 1982. Visual learning and memory of *Drosophila melanogaster* wild type C-S and the mutants *dunce, amnesiac, turnip* and *rutabaga*. *J. Insect Physiol.* 28:535–42

Forman, R. R. 1984. Leg position learning by an insect. I. A heat avoidance learning paradigm. *J. Neurobiol.* 15:127–40

Forman, R., Alkon, D. L., Sakakibara, M., Harrigan, I., Lederhendler, I., Farley, J. 1984. Changes in I_A and I_C but not I_{Na} accompany retention of conditioned behavior in *Hermissenda*. *Soc. Neurosci. Abstr.* 10:121

Forman, R., Hoyle, G. 1978. Position learning in behaviorally appropriate situations. *Soc. Neurosci. Abstr.* 4:193

Forman, R., Zill, S. N. 1984. Leg position learning by an insect. II. Motor strategies underlying learned leg extension. *J. Neurobiol.* 15:221–37

Frost, W. N., Kandel, E. R. 1984. Sensitizing stimuli reduce the effectiveness of the L30 inhibitory interneurons in the siphon withdrawal reflex circuit of *Aplysia*. *Soc. Neurosci. Abstr.* 10:510

Gailey, D. A., Jackson, F. R., Siegel, R. W. 1984. Conditioning mutations in *Drosophila melanogaster* affect an experience-dependent behavioral modification in courting males. *Genetics* 106:613–23

Garcia, J., Koelling, R. A. 1966. Relation of cue to consequence in avoidance learning. *Psychonom. Sci.* 4:123–24

Garcia, J., Ervin, F. R., Koelling, R. A. 1966. Learning with a prolonged delay of reinforcement. *Psychonom. Sci.* 5:121–22

Gelperin, A. 1975. Rapid food-aversion learning by a terrestrial mollusc. *Science* 189:567–70

Gelperin, A. 1981. Synaptic modulation by identified serotonin neurons. In *Serotonin Neurotransmission and Behavior*, ed. B. Jacobs, A. Gelperin, pp. 288–304. Cambridge: MIT Press

Gelperin, A., Culligan, N. 1982. In vitro expression of in vivo learning by the cerebral ganglion of the terrestrial mollusc, *Limax maximus*. *Soc. Neurosci. Abstr.* 8:823

Gelperin, A., Chang, J. J., Reingold, S. C.

1978. Feeding motor program in *Limax*. I. Neuromuscular correlates and control by chemosensory input. *J. Neurobiol.* 9:285–300

Gelperin, A., Forsythe, D. 1976. Neuroethological studies of learning in mollusks. In *Simpler Systems: An Approach to Patterned Behavior and Its Foundations*, ed. J. C. Fentress, pp. 239–46. Sunderland, Mass.: Sinauer Assoc.

Gillette, R., Davis, W. J. 1977. The role of the metacerebral giant neuron in the feeding behavior of *Pleurobranchaea*. *J. Comp. Physiol. A* 116:129–59

Gillette, R., Gillette, M. U., Davis, W. J. 1980. Action potential broadening and endogenously sustained bursting are substrates of command ability in a feeding neuron of *Pleurobranchaea*. *J. Neurophysiol.* 43:669–85

Gillette, R., Gillette, M. U., Davis, W. J. 1982a. Substrates of command ability in a buccal neuron of *Pleurobranchaea*. II. Potential role of cyclic AMP. *J. Comp. Physiol.* 146:461–70

Gillette, R., Gillette, M. U., Davis, W. J. 1982b. Substrates of command ability in a buccal neuron of *Pleurobranchaea*. I. Mechanisms of action potential broadening. *J. Comp. Physiol.* 146:449–59

Gillette, R., Kovac, M. P., Davis, M. J. 1978. Command neurons in *Pleurobranchaea* receive synaptic feedback from the motor network they excite. *Science* 199:798–801

Gillette, R., Kovac, M. P., Davis, W. J. 1982c. Control of feeding motor output by paracerebral neurons in the brain of *Pleurobranchaea californica*. *J. Neurophysiol.* 47:885–908

Gillette, M. U., London, J. A., Gillette, R. 1984. Motivation to feed affects acquisition of food avoidance conditioning in *Pleurobranchaea*. *Soc. Neurosci. Abstr.* 9:914

Gingrich, K. J., Byrne, J. H. 1985. Simulation of synaptic depression, posttetanic potentiation, and presynaptic facilitation of synaptic potentials from sensory neurons mediating gill-withdrawal reflex in *Aplysia*. *J. Neurophysiol.* 53:652–69

Goh, Y., Alkon, D. L. 1984. Sensory interneuronal, and motor interactions within *Hermissenda* visual pathway. *J. Neurophysiol.* 52:156–69

Goh, Y., Lederhendler, I., Alkon, D. L. 1985. Input and output changes of an identified neural pathway are correlated with associative learning in *Hermissenda*. *J. Neurosci.* 5:536–43

Gormezano, I. 1972. Investigations of defense and reward conditioning in the rabbit. In *Classical Conditioning. II: Current Research and Theory*, ed. A. H. Black,

W. F. Prokasy, pp. 151–81. New York: Appleton-Century-Crofts

Gould, J. L. 1985. How bees remember flower shapes. *Science* 227:1492–94

Grossman, K. E. 1970. Erlernen von farbreizen an der futterstelle durch honigbienen wahrend des anflugs und wahrend des saugens. *Z. Tierpsychol.* 27:553–62

Grover, L., Farley, J. 1984. Cumulative depolarization of *Hermissenda* Type B photoreceptors: Ionic basis and the role of calcium. *Soc. Neurosci. Abstr.* 10:621

Harvey, J. A., Gormezano, I., Cool-Hauser, V. A. 1985. Relationship between heterosynaptic reflex facilitation and acquisition of the nictitating membrane response in control and scopolamine-injected rabbits. *J. Neurosci.* 5:596–602

Hawkins, R. D., Abrams, T. W. 1984. Evidence that activity-dependent facilitation underlying classical conditioning in *Aplysia* involves modulation of the same ionic current as normal presynaptic facilitation. *Soc. Neurosci. Abstr.* 10:268

Hawkins, R. D., Abrams, T. W., Carew, T. J., Kandel, E. R. 1983a. A cellular mechanism of classical conditioning in *Aplysia*: Activity-dependent amplification of presynaptic facilitation. *Science* 219:400–5

Hawkins, R. D., Carew, T. J., Kandel, E. R. 1983b. Effects of interstimulus interval and contingency on classical conditioning in *Aplysia*. *Soc. Neurosci. Abstr.* 9:168

Hawkins, R. D., Castellucci, V. F., Kandel, E. R. 1981a. Interneurons involved in mediation and modulation of gill-withdrawal reflex in *Aplysia*. I. Identification and characterization. *J. Neurophysiol.* 45:304–14

Hawkins, R. D., Castellucci, V. F., Kandel, E. R. 1981b. Interneurons involved in mediation and modulation of gill-withdrawal reflex in *Aplysia*. II. Identified neurons produce heterosynaptic facilitation contributing to behavioral sensitization. *J. Neurophysiol.* 45:315–26

Hebb, D. O. 1949. *Organization of Behavior*. New York: Wiley

Henderson, T. B., Strong, P. N. 1972. Classical conditioning in the leech, *Macrabdella ditetra*, as a function of CS and UCS intensity. *Conditioned Reflex* 7:210–15

Hening, W. A., Walters, E. T., Carew, T. J., Kandel, E. R. 1979. Motorneuronal control of locomotion in *Aplysia*. *Brain Res.* 179:231–53

Horridge, G. A. 1962a. Learning of leg position by headless insects. *Nature* 193:697–98

Horridge, D. O. 1962b. Learning of leg position by the ventral nerve cord in headless insects. *Proc. R. Soc. London Ser. B* 157:33–52

482   CAREW & SAHLEY

Hoyle, G. 1965. Neurophysiological studies on learning in headless insects. In *Physiology of Insect Central Nervous Systems*, ed. J. Treherne, pp. 203–32. London: Academic

Hoyle, G. 1966. An isolated ganglion-nerve-muscle preparation. *J. Exp. Biol.* 44:413–27

Hoyle, G. 1979. Instrumental conditioning of the leg lift in the locust. *Neurosci. Res. Prog. Bull.* 17:577–86

Hoyle, G. 1980. Learning, using natural reinforcements, in insect preparations that permit cellular neuronal analysis. *J. Neurobiol.* 11:323–54

Hoyle, G. 1982a. Cellular basis of operant-conditioning of leg position. In *Conditioning: Representation of Involved Neural Functions*, ed. C. D. Woody, pp. 197–211. New York: Plenum

Hoyle, G. 1982b. Pacemaker change in learning paradigm. In *Cellular Pacemakers*, ed. D. Carpenter, pp. 3–25. New York: Wiley

Hoyle, G., Field, L. H. 1983. Elicitation and abrupt termination of behaviorally significant catchlike tension in a primitive insect. *J. Neurobiol.* 14:299–312

Ingram, D. A., Walters, E. T. 1984. Differential classical conditioning of tail and siphon withdrawal in *Aplysia. Soc. Neurosci. Abstr.* 10:270

Ito, M. 1982. Cerebellar control of the vestibular-ocular reflex—around the flocculus hypothesis. *Ann. Rev. Neurosci.* 5:275–96

Jahan-Parwar, B., Fredman, S. M. 1978. Control of pedal and parapodial movements in *Aplysia*. II: Cerebral ganglion neurons. *J. Neurophysiol.* 41:609–20

Jerussi, T. P., Alkon, D. L. 1981. Ocular and extracellular responses of identifiable neurons in pedal ganglia of *Hermissenda crassicornis. J. Neurophysiol.* 46:654–71

Kamin, L. J. 1969. Predictability, surprise, attention and conditioning. In *Punishment and Aversive Behavior*, ed. R. Church, B. A. Campbell, pp. 279–96. New York: Appleton-Century-Crofts

Kandel, E. R. 1984. Steps towards a molecular grammar for learning: Explorations into the nature of memory. In *Medicine, Science and Society*, ed. K. J. Isselbacher, pp. 555–604. New York: Wiley

Kandel, E. R., Abrams, T., Bernier, L., Carew, T. J., Hawkins, R. D., Schwartz, J. H. 1983. Classical conditioning and sensitization share aspects of the same molecular cascade in *Aplysia. Cold Spring Harbor Symp. Quant. Biol.* 48:821–30

Kandel, E. R., Schwartz, J. H. 1982. Molecular biology of learning: Modulation of transmitter release. *Science* 218:433–43

Kauvar, L. M. 1982. Defective cyclic adenosine 3′,5′-monophosphate phosphodiester-ase in the *Drosophila* memory mutant dunce. *J. Neurosci.* 2:1347–55

Kiger, J. A. 1979. A genetically distinct form of cyclic AMP phosphodiesterase associated with chromere 3D4 in *Drosophila melanogaster. Genetics* 91:521–32

Kiger, J. A., Golanty, E. 1977. A cytogenetic analysis of cyclic nucleotide phosphodiesterase in *Drosophila. Genetics* 85:609–21

Kim, E. H.-J., Woody, C. D., Berthier, N. E. 1983. Rapid acquisition of conditioned eye blink responses in cats following pairing of an auditory CS with glabella tap US and hypothalamic stimulation. *J. Neurophysiol.* 49:767–79

Kistler, H. B. Jr., Hawkins, R. D., Koester, J., Kandel, E. R., Schwartz, J. H. 1983. Immunocytochemical studies of neurons producing facilitation in the abdominal ganglion of *Aplysia. Soc. Neurosci. Abstr.* 9:915

Kistler, H. B. Jr., Hawkins, R. D., Koester, J., Steinbusch, H. W. M., Kandel, E. R., Schwartz, J. H. 1985. Distribution of serotonin-immunoreactive cell bodies and processes in the abdominal ganglion of mature *Aplysia. J. Neurosci.* 5:72–80

Klein, M., Camardo, J., Kandel, E. R. 1982. Serotonin modulates a specific potassium current in the sensory neurons that show presynaptic facilitation in *Aplysia. Proc. Natl. Acad. Sci. USA* 79:5713–17

Klein, M., Kandel, E. R. 1978. Presynaptic modulation of voltage-dependent Ca^{++} current: Mechanism for behavioral sensitization. *Proc. Natl. Acad. Sci. USA* 75:3512–16

Klein, M., Kandel, E. R. 1980. Mechanism of calcium current modulation underlying presynaptic facilitation and behavioral sensitization in *Aplysia. Proc. Natl. Acad. Sci. USA* 77:6912–16

Klein, M., Shapiro, E., Kandel, E. R. 1980. Synaptic plasticity and the modulation of the Ca^{++} current. *J. Exp. Biol.* 89:117–57

Klosterhalfer, S., Fischer, W., Bitterman, M. E. 1978. Modification of attention in honey bees. *Science* 201:1241–43

Kovac, M. P., Davis, W. J., Matera, E. M., Croll, R. P. 1983a. Organization of synaptic inputs of paracerebral feeding command interneurons of *Pleurobranchaea californica*. I. Exitatory inputs. *J. Neurophysiol.* 49:1517–38

Kovac, M. P., Davis, W. J., Matera, E. M., Croll, R. P. 1983b. Organization of synaptic inputs of paracerebral feeding command interneurons of *Pleurobranchaea californica*. II. Inhibitory inputs. *J. Neurophysiol.* 49:1539–56

Kovac, M. P., Davic, W. J., Matera, E. M., Gillette, R. 1982. Functional and structural correlates of cell size in paracerebral neurons of *Pleurobranchaea*. *J. Neurophysiol.* 47:909–27

Kovac, M. P., Davis, W. J., Matera, E. M., Morielli, A., Croll, R. P. 1985. Learning: Neural analysis in the isolated brain of a previously trained mollusc, *Pleurobranchaea californica*. In press

Kupfermann, I., Carew, T. J., Kandel, E. R. 1974. Local, reflex and central commands controlling gill and siphon movements in *Aplysia*. *J. Neurophysiol.* 37:996–1019

Kupfermann, I., Pinsker, H. 1968. A behavioral modification of the feeding reflex in *Aplysia californica*. *Commun. Behav. Biol. A* 2:13–17

Kuwabara, M. 1957. Bildung des bedingten reflexes vom Pavlov typus bei der honigbiene. *J. Fac. Sci. Hokkaido Univ. Zool.* 13:458–64

Lederhendler, I., Alkon, D. L. 1984. Reduced withdrawal from shadows: An expression of primary neural changes of associative learning in *Hermissenda*. *Soc. Neurosci. Abstr.* 10:270

Lederhendler, I., Gart, S., Alkon, D. L. 1983. Associative learning in *Hermissenda crassicornis*: Evidence that light (the CS) takes on characteristics of rotation (the US). *Biol. Bull.* 165:529

Levy, W. B., Steward, O. 1979. Synapses as associative memory elements in the hippocampal formation. *Brain Res.* 175:233–45

Lindauer, M. 1960. Time compensated sun orientation in bees. *Cold Spring Harbor Symp. Quant. Biol.* 25:371–77

Livingstone, M. S., Sziber, P. P., Quinn, W. G. 1982. Defective adenylate cyclase in the *Drosophila* learning mutant *rutabaga*. *Soc. Neurosci. Abstr.* 8:384

Livingstone, M. S., Sziber, P. P., Quinn, W. G. 1984. Loss of calcium/calmodulin responsiveness in adenylate cyclase of *rutabaga*, a *Drosophila* learning mutant. *Cell* 37:205–15

Livingstone, M. S., Tempel, B. L. 1983. Genetic dissection of monoamine neurotransmitter synthesis in *Drosophila*. *Nature* 303:67–70

Lockery, S. R., Kristan, W. 1985. Neural correlates of habituation and sensitization in the leech. *Soc. Neurosci. Abstr.* 11:794

Lockery, S. R., Rawlins, J. N. P., Gray, J. A. 1985. Habituation of the shortening reflex in the medicinal leech. *Behav. Neurosci.* 99:333–41

London, J. A., Gillette, R. 1984a. Changes in specific interneurons presynaptic to command neurons underlying associative learning in *Pleurobranchaea*. *Soc. Neurosci. Abstr.* 9:914

London, J. A., Gillette, R. 1984b. Functional roles and circuitry in an inhibitory pathway to feeding command neurons in *Pleurobranchaea*. *J. Exp. Biol.* 113:423–46

London, J. A., Gillette, R. 1985. Associative learning mechanisms in the cyclic motor network for feeding behavior in *Pleurobrancnaea californica*. In press

Lukowiak, K. 1982. Associative learning in the isolated siphon, mantle, gill and abdominal ganglion preparation of *Aplysia*: A new paradigm. *Soc. Neurosci. Abstr.* 8:385

Lukowiak, K. 1983. Associative learning in an in vitro *Aplysia* preparation: Facilitation at a sensory motor neuron synapse. *Soc. Neurosci. Abstr.* 9:169

Lukowiak, K., Sahley, C. 1981. The *in vitro* classical conditioning of the gill withdrawal reflex of *Aplysia californica*. *Science* 212:1516–18

Lynch, G., Baudry, M. 1984. The biochemistry of memory: A new and specific hypothesis. *Science* 224:1057–63

Mackintosh, N. J. 1974. *The Psychology of Animal Learning*. London: Academic

Masuhr, T., Menzel, R. 1972. Interhemispheric transfer of learning performance in the honeybee. In *Processing of Information in the Visual System of Arthropods*, ed. R. Wehner, pp. 315–22. New York: Springer-Verlag

Matera, E. M., Davis, W. J. 1982. Paddle cilia (discocilia) in chemosensitive structures of the gastropod mollusc *Pleurobranchaea californica*. *Cell Tissue Res.* 222:25–40

McCrohan, C. R., Benjamin, P. R. 1980. Synaptic relationships of the cerebral giant cells with motor neurons in the feeding system of *Lymnaea stagnalis*. *J. Exp. Biol.* 85:169–86

McElearney, A., Farley, J. 1983. Persistent changes in *Hermissenda B* photoreceptor membrane properties with associative training: A role for pharmacological modulation. *Soc. Neurosci. Abstr.* 9:915

McNaughton, B. L., Douglas, R. M., Goddard, G. V. 1978. Synaptic enhancement in fascia dentata: Co-operativity among co-active afferents. *Brain Res.* 157:277–93

Menne, D., Spatz, H. C. 1977. Color learning in *Drosophila*. *J. Comp. Physiol.* 114:301–12

Menzel, R. 1967. Untersuchungen zum erlernen von spektralfarben durch die honigbeine. *Z. Verlag. Physiol.* 56:22–62

Menzel, R. 1968. Das gedachtnis der honigbiene fur spektralfarben. I. Kurzzeitiges und langzeitiges behalten. *Z. Vegl. Physiol.* 60:82–102

Menzel, R. 1979. Behavioral access to short-term memory in bees. *Nature* 281:368–69

Menzel, R. 1984. Short-term memory in bees. See Cohen 1984, pp. 259–74

Menzel, R., Bitterman, M. E. 1983. Learning by honeybees in an unnatural situation. In *Behavioral Physiology and Neuroethology: Roots and Growing Points*, ed. F. Huber, L. Markl, pp. 206–15. Heidelberg: Springer

Menzel, R., Erber, J. 1972. Influence of the quality of reward on the learning performance in honeybees. *Behavior* 41:27–42

Menzel, R., Erber, J., Masuhr, T. 1974. Learning and memory in the honeybee. In *Experimental Analysis of Insect Behavior*, ed. Barton-Browne, pp. 195–217. New York: Springer-Verlag

Mobbs, P. G. 1982. The brain of the honeybee *Apis mellifera*. I. The connections and spatial organization of the mushroom bodies. *Philos. Trans. R. Soc. London* 298: 309–54

Moore, J. W., Desmond, J. E., Berthier, N. E. 1982. The metencephalic basis of the conditioned nictitating membrane response. In *Conditioning: Representation of Involved Neural Functions*, ed. C. D. Woody, pp. 459–82. New York: Plenum

Mpitsos, G. J., Collins, S. D. 1975. Learning: Rapid aversive conditioning in the gastropod mollusc *Pleurobranchaea*. *Science* 108:954–57

Mpitsos, G. J., Collins, S. D., McClellan, A. D. 1978. Learning: A model system for physiological studies. *Science* 199:497–506

Mpitsos, G. J., Davis, W. J. 1973. Learning: Classical and avoidance conditioning in the mollusk *Pleurobranchaea*. *Science* 180: 317–20

Neary, J. T. 1984. Biochemical correlates of associative learning: Protein phosphorylation in *Hermissenda crassicornis*, a nudibranch mollusk. See Cohen 1984, pp. 325–36

Neary, J. T., Alkon, D. L. 1983. Protein phosphorylation/dephosphorylation and the transient voltage-dependent potassium conductance in *Hermissenda crassicornis*. *J. Biol. Chem.* 258:8979–83

Neary, J. T., Crow, T., Alkon, D. L. 1981. Changes in a specific phosphoprotein following associative learning in *Hermissenda*. *Nature* 293:658–60

Neary, J. T., DeRiemer, S. A., Kaczmarek, L. K., Alkon, D. L. 1984. Ca^{++} and cAMP regulation of protein phosphorylation in the *Hermissenda* nervous system. *Soc. Neurosci. Abstr.* 10:805

Nelson, M. C. 1971. Classical conditioning in the blowfly (*Phormia regina*). *J. Comp. Physiol. Psychol.* 77:353–68

Nicholls, J. G., Purves, D. 1970. Monosynaptic chemical and electrical connexions between sensory and motor cells in the nervous system of the leech. *J. Physiol.* 225:637–56

Ocorr, K. A., Byrne, J. H. 1984. Characterization of phosphodiesterase from *Aplysia* pleural ganglia. *Soc. Neurosci. Abstr.* 10:270

Ocorr, K. A., Byrne, J. H. 1985. Membrane responses and changes in cAMP levels in *Aplysia* sensory neurons produced by 5-HT, Tryptamine, FMRFamide and SCP_8. *Neurosci. Lett.* 55:113–18

Ocorr, K. A., Walters, E. T., Byrne, J. H. 1983. Associative conditioning analog in *Aplysia* tail sensory neurons selectively increases cAMP content. *Soc. Neurosci. Abstr.* 9: 169

Ocorr, K. A., Walters, E. T., Byrne, J. H. 1985. Associative conditioning analog selectivity increases cAMP levels of tail sensory neurons in *Aplysia*. *Proc. Natl. Acad. Sci. USA* 82:2538–52

Olson, L., Jacklett, J. W. 1984. Identification of circadian clock fibers in the CNS of *Aplysia*. *Soc. Neurosci. Abstr.* 9:624

Ono, J., McCaman, R. E. 1984. Immunocytochemical localization and direct assays of serotonin-containing neurons in *Aplysia*. *Neuroscience* 11:549–60

Opfinger, E. 1931. Ueber die orientierung der biene an der futterstelle. *Z. Vergl. Physiol.* 15:431–87

Pavlov, I. P. 1927. *Conditioned Reflexes*, pp. 33–35. London: Oxford Univ. Press

Perlman, A. J. 1979. Central and peripheral control of siphon withdrawal reflex in *Aplysia californica*. *J. Neurophysiol.* 42: 510–29

Pinsker, H. M., Hening, W. A., Carew, T. J., Kandel, E. R. 1973. Long-term sensitization of a defensive withdrawal reflex in *Aplysia*. *Science* 182:1039–42

Pinsker, H., Kupfermann, I., Castellucci, V., Kandel, E. R. 1970. Habituation and dishabituation of the gill-withdrawal reflex in *Aplysia*. *Science* 167:1740–42

Pollack, J. D., Camardo, J. S., Bernier, L., Schwartz, J. H., Kandel, E. R. 1982. Pleural sensory neurons in *Aplysia*: A new population for studying the biochemistry and biophysics of serotonin modulation of K^+ currents. *Soc. Neurosci. Abstr.* 8:523

Prior, D., Gelperin, A. 1977. Autoactive molluscan neuron: Reflex function and synaptic modulation during feeding in the terrestrial slug *Limax maximus*. *J. Comp. Physiol.* 114:217–32

Quinn, W. G. 1984. Work in invertebrates on the mechanisms underlying learning. In *Biology of Learning*, ed. P. Marler, H. Terrace, pp. 197–246. Berlin: Dahlem Konferenzen

Quinn, W. G., Harris, W. A., Benzer, S. 1974. Conditioned behavior in *Drosophila melanogaster*. *Proc. Natl. Acad. Sci. USA* 71: 708–12

Quinn, W. G., Sziber, P. P., Booker, R. 1979. The *Drosophila* memory mutant amnesiac. *Nature* 277:212–14

Ratner, S. C. 1972. Habituation and retention of habituation in the leech (*Macrobdella decora*). *J. Comp. Physiol. Psych.* 81:115–21

Reingold, S. C., Gelperin, A. 1980. Feeding motor program in *Limax*. II. Modulation by sensory inputs in intact animals and isolated central nervous system. *J. Exp. Biol.* 85:1–19

Rescorla, R. A. 1968. Probability of shock in the presence and absence of CS in fear conditioning. *J. Comp. Physiol. Psychol.* 66:1–5

Rescorla, R. A. 1969. Conditioned inhibition of fear resulting in negative CS-US contingencies. *J. Comp. Physiol. Psychol.* 67:504–9

Rescorla, R. A. 1980. *Pavlovian Second-order Conditioning: Studies in Associative Learning*. Hillsdale, N.J.: Erlbaum

Rescorla, R. A., Holland, P. C. 1976. Some behavioral approaches to the study of learning. In *Neural Mechanisms of Learning and Memory*, ed. M. R. Rosenzweig, E. L. Bennett. Cambridge, Mass.: MIT Press

Rescorla, R. A., Wagner, A. R. 1972. A theory of Pavlovian conditioning: Variations in the effectiveness of reinforcement and nonreinforcement. See Gormezano 1972, pp. 64–99

Revusky, S. H., Garcia, J. 1971. Learned associations over long delays. In *The Psychology of Learning and Motivation*, ed. G. H. Bower. New York: Academic

Richards, W., Farley, J. 1984. Associative learning changes intrinsic to *Hermissenda* Type A photoreceptors. *Soc. Neurosci. Abstr.* 10:623

Richards, W., Farley, J., Alkon, D. L. 1983. Extinction of associative learning in *Hermissenda*: Behavior and neural correlates. *Soc. Neurosci. Abstr.* 9:916

Rizley, R. C., Rescorla, R. A. 1972. Associations in second-order conditioning and sensory preconditioning. *J. Comp. Physiol. Psychol.* 81:1–11

Rose, R. M., Benjamin, P. R. 1979. The relationship of the central motor pattern to the feeding cycle of *Lymnaea stagnalis*. *J. Exp. Biol.* 80:137–63

Rose, R. M., Benjamin, P. R. 1981a. Interneuronal control of feeding in the pond snail, *Lymnaea stagnalis*. I. Initiation of feeding cycles by a single buccal interneurone. *J. Exp. Biol.* 92:187–201

Rose, R. M., Benjamin, P. R. 1981b. Interneuronal control of feeding in the pond snail *Lymnaea stagnalis*. II. The interneuronal mechanism generating feeding cycles. *J. Exp. Biol.* 92:203–8

Rosen, S. C., Weiss, K. R., Cohen, J. L., Kupfermann, I. 1982. Interganglionic cerebral-buccal mechanoafferents of *Aplysia*: Receptive fields and synaptic connections to different classes of neurons involved in feeding behavior. *J. Neurophysiol.* 48:271–88

Rosen, S. C., Weiss, K. R., Kupfermann, I. 1979. Response properties and synaptic connections of mechanoafferent neurons in cerebral ganglion of *Aplysia*. *J. Neurophysiol.* 42:954–74

Rozin, P. 1969. Adaptive food sampling patterns in vitamin deficient rats. *J. Comp. Physiol. Psychol.* 69:126–32

Rozin, P., Kalat, J. W. 1973. Specific hungers and poison avoidance as adaptive specializations of learning. *Psych. Rev.* 78:459–86

Sahley, C. L., Gelperin, A., Rudy, J. W. 1981a. One-trial learning modifies food odor preferences of a terrestrial mollusc. *Proc. Natl. Acad. Sci. USA* 78:640–42

Sahley, C. L., Hardison, P., Hsuan, A., Gelperin, A. 1982. Appetitively reinforced odor conditioning modulates feeding in *Limax maximus*. *Soc. Neurosci. Abstr.* 8:823

Sahley, C. L., Ready, D. F. 1985. Associative learning modifies two behaviors in the leech, *Hirudo medicinalis*. *Soc. Neurosci. Abstr.* 11:367

Sahley, C., Rudy, J. W., Gelperin, A. 1981b. An analysis of associative learning in a terrestrial mollusc. I. Higher-order conditioning, blocking, and a transient US preexposure effect. *J. Comp. Physiol.* 144:1–8

Sahley, C. L., Rudy, J. W., Gelperin, A. 1984. Associative learning in a mollusk: A comparative analysis. See Cohen 1984, pp. 243–58

Sakakibara, M., Alkon, D. L., Lederhendler, I., Heldman, E. 1984. Alpha$_2$-receptor control of Ca^{++}-mediated reduction of voltage-dependent K$^+$ currents. *Soc. Neurosci. Abstr.* 10:950

Schneirla, T. C. 1929. Learning and orientation in ants. Studied by means of the maze method. *Comp. Psychol. Mongr.* 6:1–143

Schwartz, J. H., Bernier, L., Castellucci, V. F., Palazzolo, M., Saitoh, T., Stapleton, A., Kandel, E. R. 1983. What molecular steps determine the time course of the memory for short-term sensitization in *Aplysia*? *Cold Spring Harbor Symp. Quant. Biol.* 68:811–19

Schwarz, M., Susswein, A. J. 1984. A neural pathway for learning that food is inedible in *Aplysia*. *Brain Res.* 294:363–66

Shotwell, S. L. 1983. Cyclic adenosine 3′,5′-monophosphate phosphodiesterase and its role in learning in *Drosophila*. *J. Neurosci.* 3:739–47

Shuster, M. J., Camardo, J. S., Siegelbaum, S. A., Kandel, E. R. 1985. Cyclic AMP-dependent protein kinase closes the serotonin-sensitive K^+ channels of *Aplysia* sensory neurons in cell-free membrane patches. *Nature* 313:392–95

Siegel, R. W., Hall, J. C. 1979. Conditioned responses in courtship behavior of normal and mutant *Drosophila*. *Proc. Natl. Acad. Sci. USA* 76:3430–34

Siegelbaum, S. A., Camardo, J. S., Kandel, E. R. 1982. Serotonin and cyclic AMP close single K^+ channels in *Aplysia* sensory neurones. *Nature* 299:413–17

Siegler, M. V. S. 1977. Motor neurone coordination and sensory modulation in the feeding system of the mollusc *Pleurobranchaea*. *J. Exp. Biol.* 71:27–48

Siegler, M. V. S., Mpitsos, G. J., Davis, W. J. 1974. Motor organization and generation of rhythmic feeding output in the buccal ganglion of *Pleurobranchaea*. *J. Neurophysiol.* 37:1173–96

Stoller, D., Sahley, C. L. 1985. Habituation and sensitization of the shortening reflex in the leech, *Hirudo medicinalis*. *Soc. Neurosci. Abstr.* 11:367

Susswein, A. J., Schwarz, M. 1983. A learned change of response to inedible food in *Aplysia*. *Behav. Neural. Biol.* 39:1–6

Takeda, T., Alkon, D. L. 1982. Correlated receptor and motor neuron changes during retention of associative learning of *Hermissenda crassicornis*. *Comp. Biochem. Physiol.* 73A:151–57

Tempel, B. L., Livingstone, M. S., Quinn, W. G. 1984. Mutations in the dopa decarboxylase gene affect learning in *Drosophila*. *Proc. Natl. Acad. Sci. USA* 81:3577–81

Tempel, B. L., Bonini, N., Dawson, D. R., Quinn, W. G. 1983. Reward learning in normal and mutant *Drosophila*. *Proc. Natl. Acad. Sci. USA* 80:1482–86

Tempel, B. L., Quinn, W. G. 1982. Mutations in the dopa-decarboxylase gene affect learning but not memory in *Drosophila*. *Soc. Neurosci. Abstr.* 8:385

Thompkins, L., Siegel, R. W., Gailey, D. A., Hall, J. 1983. Conditioned courtship in *Drosophila* and its mediation by association of chemical cues. *Behav. Genet.* 13:565–78

Thompson, R. F., Clark, G. A., Donegan, N. H., Lavond, D. G., Lincoln, J. S., Madden, J. IV, Mamounas, L. A., Mauk, M. D., McCormick, D. A., Thompson, J. K. 1984. Neuronal substrates of learning and memory: A "multiple trace" view. In *Neurobiology of Learning and Memory*, ed. G. Lynch, J. L. McGaugh, N. M. Weinberger, pp. 137–64. New York: Guilford

Thompson, R. F., McCormick, D. A., Lavond, D. G., Clark, G. A., Kettner, R. E.,

Mauk, M. D. 1983. The engram found? Initial localization of the memory trace for a basic form of associative learning. *Progr. Psychobiol. Physiol. Psychol.* 10:167–96

Thorndike, E. L. 1898. Animal intelligence: An experimental study of the association processes in animals. *Psych. Rev. Monog.*

Tosney, T., Hoyle, G. 1977. Computer-controlled learning in a simple system. *Proc. Roy. Soc. London Ser. B* 195:365–93

Tsukahara, N. 1984. Classical conditioning mediated by the red nucleus: An approach beginning at the cellular level. See Thompson et al 1984, pp. 165–80

Tully, T. 1984. *Drosophila* learning: Behavior and biochemistry. *Behav. Genet.* 14:527–57

Vareschi, E. 1971. Duftunterscheidung bei der honigbiene: Einzelzellableitungen und verhaltensreaktionen. *Z. Vergl. Physiol.* 75:143–73

von Alten, H. 1919. Zur phylogenie des hymenopterengehirns. *Jena Z. Naturwiss.* 46:511–90

von Frisch, K. 1921. Uber den Geruchssinn der Bienen und seine blütenbiologische Bedeutung. *Zool. Jahrb.* 37:1–238

von Frisch, K. 1950. *Bees, Their Vision, Chemical Senses and Language*. Ithaca, NY: Cornell Univ. Press

Wagner, A. R. 1969. Stimulus validity and stimulus selection in associative learning. In *Fundamental Issues in Associative Learning*, ed. N. J. Mackintosh, W. K. Honig, pp. 90–122. Halifax: Dalhousie Univ. Press

Walsh, J. P., Byrne, J. H. 1984. Forskolin mimics and blocks a serotonin-sensitive decreased K^+ conductance in tail sensory neurons of *Aplysia*. *Neurosci. Lett.* 52:7–11

Walters, E. T., Byrne, J. H. 1983a. Associative conditioning of single sensory neurons suggests a cellular mechanism for learning. *Science* 219:405–8

Walters, E. T., Byrne, J. H. 1983b. Slow depolarization produced by associative conditioning of *Aplysia* sensory neurons may enhance Ca^{++} entry. *Brain Res.* 280:165–68

Walters, E. T., Byrne, J. H. 1983c. Long-term potentiation, activity-dependent neuromodulation, and associative information storage in *Aplysia*. *Soc. Neurosci. Abstr.* 9:168

Walters, E. T., Byrne, J. H. 1985. Long-term enhancement produced by activity-dependent modulation of *Aplysia* sensory neurons. *J. Neurosci.* 5:662–72

Walters, E. T., Byrne, J. H., Carew, T. J., Kandel, E. R. 1983a. Mechanoafferent neurons innervating the tail of *Aplysia*. I. Response properties and synaptic connections. *J. Neurophysiol.* 50:1522–42

Walters, E. T., Byrne, J. H., Carew, T. J., Kandel, E. R. 1983b. Mechanoafferent neurons innervating the tail of *Aplysia*. II. Modulation by sensitizing stimuli. *J. Neurophysiol.* 50:1543–59

Walters, E. T., Carew, T. J., Kandel, E. R. 1978. Conflict and response selection in the lomomotor system of *Aplysia*. *Soc. Neurosci. Abstr.* 4:209

Walters, E. T., Carew, T. J., Kandel, E. R. 1979. Classical conditioning in *Aplysia californica*. *Proc. Natl. Acad. Sci. USA* 76:6675–79

Walters, E. T., Carew, T. J., Kandel, E. R. 1981. Associative learning in *Aplysia*: Evidence for conditioned fear in an invertebrate. *Science* 211:504–6

Wehner, R. 1972. Pattern modulation and pattern detection in the visual system of hymenoptera. In *Information Processing in the Visual System of Arthropods*, ed. R. Wehner, pp. 183–94. Berlin: Springer-Verlag

West, A., Barnes, E., Alkon, D. 1982. Primary changes of voltage responses during retention of associative learning. *J. Neurophysiol.* 48:1243–55

Wieland, S. J., Gelperin, A. 1983. Dopamine elicits feeding motor program in *Limax maximus*. *J. Neurosci.* 3:1735–45

Witthoft, W. 1967. Absolute anzahl und verteilung der zellen im hirn der honigbeine. *Z. Morph. Tiere* 61:160–84

Woody, C. D. 1984. Studies of Pavlovian eye blink conditioning in awake cats. See Thompson et al 1984, pp. 181–96

Woody, C. D., Kim, E. H.-J., Berthier, N. E. 1983. Effects of hypothalamic stimulation on unit responses recorded from neurons of sensorimotor cortex of awake cats during conditioning. *J. Neurophysiol.* 49:780–91

Woolacott, M. H., Hoyle, G. 1976. Membrane resistance changes associated with single identified neuron learning. *Soc. Neurosci. Abstr.* 2:339

Woolacott, M., Hoyle, G. 1977. Neural events underlying learning in insects: Changes in pacemaker. *Proc. R. Soc. London Ser. B* 195:395–415

Wright, T. R. F. 1977. The genetics of dopa decarboxylase and a-methyl dopa sensitivity in *Drosophila melanogaster*. *Am. Zool.* 17:707–21

Wu, R., Farley, J. 1984. Serotonin reduces K^+ currents and enhances a Ca^{++} current in *Hermissenda* Type B photoreceptors. *Soc. Neurosci. Abstr.* 10:620

Zahler, C. L., Harper, A. E. 1972. Effects of dietary amino acid pattern on food preference behavior of rats. *J. Comp. Physiol. Psychol.* 81:155–62

Zahorik, D. M. 1977. Associative and non-associative factors in learned food preferences. In *Learning Mechanisms in Food Selection*, ed. L. M. Barker, P. Best, M. Domjan, pp. 181–200. Waco, Texas: Baylor Univ. Press

Ann. Rev. Neurosci. 1986. 9:489–512

NEW PERSPECTIVES ON ALZHEIMER'S DISEASE

Donald L. Price

Departments of Pathology, Neurology, and Neuroscience,
Neuropathology Laboratory, The Johns Hopkins University School of
Medicine, Baltimore, Maryland 21205

INTRODUCTION

Alzheimer's disease (AD),[1] the most common type of adult-onset dementia, is associated with dysfunction and eventual death of neurons in a variety of brain regions (Terry & Katzman 1983, Price et al 1985d). Impairments in memory, cognition, language, praxis, and visual-spatial perception are attributed to abnormalities involving a number of transmitter-specific neuronal systems in the brainstem, basal forebrain, amygdala, hippocampus, and neocortex. Affected nerve cells exhibit structural pathology (Figure 1), particularly of the neuronal cytoskeleton. This review focuses on new information in several areas, including: clinical manifestations of disease and correlations with imaging studies; analyses of brain regions and neuronal systems affected in AD; the molecular and structural pathology of affected nerve cells; amyloid deposition in plaques and blood vessels; studies of several models that show certain features in common with AD; and a brief evaluation of etiological factors implicated in AD.

CLINICAL DISEASE AND IMAGING STUDIES

AD is characterized by a progressive dementia occurring in middle or late life, with early onset cases usually showing the most striking clinical,

---

[1] Abbreviations used: AChE, acetylcholinesterase; AD, Alzheimer's disease; ChAT, choline acetyltransferase; CRF, corticotropin-releasing factor; GABA, gamma-aminobutyric acid; GVD, granulovacular degeneration; 5-HT, 5-hydroxytryptamine; MAP2, microtubule-associated proteins; nbM, nucleus basalis of Meynert; NFT, neurofibrillary tangle; PHF, paired helical filaments.

0147–006X/86/0301–0489$02.00

Figure 1 This schematic drawing of a neuron shows the nucleus, normal organelles within the cell body, proximal dendrites, axon, and nerve terminal. A target neuron and a cerebral capillary (*lower right-hand corner*) are also depicted. Conventional symbols are used to show normal organelles, including rough endoplasmic reticulum, free polysomes, Golgi apparatus, microtubules, neurofilaments, microfilaments, mitochondria, endoplasmic reticulum, and synaptic vesicles. The neuron is shown receiving three synaptic inputs and synapsing upon the target cell.

Four types of neuronal pathology and two forms of amyloid are illustrated. NFT are indicated by the lightning bolt-shaped structure on the *left side* of the perikarya. Adjacent to NFT is a GVD. Within a dendrite (*upper right*) is a rhomboid-shaped Hirano body. The distal axon/nerve terminal (neurite in plaque) contains accumulations of filaments (*triangle*). In proximity to the nerve terminal and target cell are amyloid deposits (*linear fibrils* associated with a senile plaque). Amyloid is deposited in proximity to the blood vessel (amyloid angiopathy).

histopathological, and neurochemical abnormalities. Criteria for a diagnosis of AD have recently been reviewed (McKhann et al 1984). In a typical case, the illness begins insidiously, and the time of onset may be difficult to determine with accuracy. Although some fluctuations may occur in individual patients, impairments in memory, cognition, language, and praxis usually become progressively more severe. At present, no biological marker has been consistently demonstrated in AD, and the diagnosis is suspected on the basis of clinical criteria and established by brain biopsy (rarely performed) or by postmortem examination of brain tissue. However, some laboratory tests significantly enhance diagnostic accuracy by identifying other causes of the dementia syndrome, e.g. multiinfarct dementia, neoplasm, or infection. In AD patients, electroencephalograms commonly detect increased amounts of slow-wave activity, and evoked potentials, particularly the P300, exhibit increased latencies. On computerized tomography, the volume of the ventricular system is frequently increased, gyri are narrowed, and sulci are widened (Gado et al 1983). Dynamic imaging studies demonstrate reductions in regional cerebral blood flow, oxygen consumption, glucose metabolism, and protein synthesis (Ingvar & Lassen 1979, Bustany et al 1983, Frackowiak & Gibbs 1983, Foster et al 1984). In individual patients, correlations have been demonstrated between results of imaging studies and the character of deficits demonstrated on neuropsychological tests, e.g. some individuals with language abnormalities show decreased metabolic activity in the left para-Sylvian region, while patients with impaired performance on visual-spatial tasks have reduced metabolic activity in the right posterior parietal region. These psychological-physiological changes probably result from a combination of pathology involving both subcortical neuronal systems afferent to cortex and disease intrinsic to nerve cells in the amygdala, hippocampus, and neocortex.

NEURONAL SYSTEMS AFFECTED IN ALZHEIMER'S DISEASE

This disease selectively affects certain regions of the brain, including specific brainstem nuclei and neurons in the basal forebrain, amygdala, hippocampus, and neocortex. Substantial progress has been made in identifying neuronal systems at risk in a variety of regions. Much less is known about responses of surviving neurons, an issue of great importance, because it is these nerve cell populations (and their receptors) that are targets of therapeutic approaches.

Brainstem Systems

Some patients show NFT and decreased numbers of nerve cells in the locus coeruleus (Forno 1978, Tomlinson et al 1981, Bondareff et al 1982, Iversen et al 1983); these changes are accompanied by reductions in noradrenergic cortical markers (Adolfsson et al 1978, Cross et al 1981, Winblad et al 1982, Yates et al 1983). These catecholaminergic cells also show structural pathology in their distal axons/terminals, e.g. in neocortex and hippocampus of individuals with AD, noradrenergic processes (visualized with immunocytochemical methods to detect dopamine β-hydroxylase) have been shown to form neurites in senile plaques (Struble et al 1985b). The response of target neurons to noradrenergic deafferentation has not been delineated. Lesions of this noradrenergic system may contribute to impairments in arousal, behavior, learning, and memory.

Neurons of the raphe nuclei develop NFT, and some of these cells degenerate (Ishii 1966, Curcio & Kemper 1984, Yamamoto & Hirano 1985). In these patients, levels of 5-hydroxytryptamine (5-HT), uptake of 5-HT, and concentrations of metabolites of 5-HT may be reduced (Bowen et al 1983). Changes in serotonin receptor subtypes are the subject of active research. Other neuronal populations in the brainstem, including peptidergic systems and nonmotor cholinergic neurons, pedunculopontine, and lateral dorsal tegmental nuclei, have not been systematically assessed in AD.

Basal Forebrain Cholinergic System

Located in the primate medial septum, diagonal band of Broca, and nucleus basalis of Meynert (nbM) (Hedreen et al 1983, Mesulam et al 1983, Hedreen et al 1984, Saper & Chelimsky 1984), neurons of the basal forebrain cholinergic system provide the major cholinergic innervation of amygdala, hippocampus, and neocortex (Struble et al 1986a). Axons/terminals of some of these cholinergic neurons are thought to be one source of acetylcholinesterase (AChE)-positive neurites in senile plaques (Friede 1965, Perry et al 1980, Struble et al 1982, 1984a). Immunocytochemical studies have demonstrated that choline acetyltransferase (ChAT)-immunoreactive neurites are important components of some plaques in aged nonhuman primates (Kitt et al 1984), and it is highly likely that cholinergic neurites also participate in the formation of some plaques in human patients. In AD, many nbM nerve cells develop NFT (Whitehouse et al 1982, Saper et al 1985), and some neurons may be reduced in size (Pearson et al 1983). Eventually, these cholinergic neurons degenerate (Whitehouse et al 1981, Whitehouse et al 1982, Arendt et al 1983, Tagliavini & Pilleri 1983, McGeer et al 1984, Saper et al 1985), leaving behind extracellular NFT (Whitehouse et al 1982, Saper et al 1985, Price et al 1986). Small, plaque-like structures may occur in this region. Dysfunction and death of basal forebrain neurons are associated with reductions of cholinergic markers in the amygdala, hippocampus, and neocortex (Pope et al 1964, Bowen et al 1976, Davies &

Maloney 1976, Perry et al 1977, Rossor et al 1982b, Bird et al 1983, Bowen et al 1983, Sims et al 1983). Cholinergic denervation of these forebrain targets is associated with reductions in presynaptic M2 muscarinic receptors, apparently without major alterations in postsynaptic M1 receptors (Mash et al 1985). Deficiencies of cholinergic innervation of amygdala, hippocampus, and neocortex may contribute to the behavioral, memory, and cognitive abnormalities occurring in AD (Drachman & Leavitt 1974, Bartus et al 1982).

Amygdala

Neuronal loss, NFT, and senile plaques occur in nuclei of the amygdala of individuals with AD; neuronal loss is most severe in medial, central, and cortical nuclei, while the basolateral complex has a high density of NFT and senile plaques (Jamada & Mehraein 1968, Kemper 1983). Somatostatinergic neurons may be one of the affected cell populations in the amygdala, but the transmitter specificities of some other at-risk intrinsic neurons and the responses of surviving neurons in this region have not yet been clarified. NFT, senile plaques, and loss of nerve cells in the amygdala may contribute to some of the emotional, motivational, and associative abnormalities occurring in patients with AD.

Hippocampus and Entorhinal Cortex

In AD, neuronal populations in the hippocampus and entorhinal cortex show regional vulnerabilities, and some cell groups are involved more severely than others (Hirano & Zimmerman 1962, Dayan 1970, Ball 1972, Morimatsu et al 1975, Ball 1978, Ball et al 1983). NFT and neuronal degeneration occur, particularly in CA1 pyramidal neurons (Ball 1977, Kemper 1978, Ball et al 1983), cells that also are prone to develop granulovacuolar degeneration (GVD) and Hirano bodies (Tomlinson & Kitchener 1972, Ball & Lo 1977, Gibson & Tomlinson 1977, Goldman 1983). In addition, some of these hippocampal neurons show dendritic pathology (Buell & Coleman 1981, Probst et al 1983). In the subiculum and entorhinal cortex, neurons of layers II and IV develop NFT and eventually degenerate; this pathology serves to isolate the hippocampal formation from some of its inputs/outputs (Kemper 1978, Hyman et al 1984). The transmitter specificity of each of these different neuronal populations have not been fully identified; some of these nerve cells may use excitatory amino acids as neurotransmitters (Fonnum 1984, Storm-Mathisen & Ottersen 1984). Consequences of these lesions on target neurons have not yet been the subject of detailed study. Lesions of the hippocampus and associated circuits are believed to play important roles in the memory deficits occurring in AD (Hyman et al 1984, Ball et al 1985).

Neocortex

NFT may be very conspicuous in neurons of the neocortex (Hirano & Zimmerman 1962, Hooper & Vogel 1976, Wilcock & Esiri 1982), particularly in the supra- and infragranular layers (Pearson et al 1985). The temporal lobe is usually the most severely involved cortical region, with parietal and frontal association cortices also showing pathology (Brun 1983); motor, somatosensory, and primary visual cortices are usually relatively spared. In some regions, cortical dendritic abnormalities have been described, and it has been suggested that changes in dendritic arbors of cortical neurons may, in part, account for reductions in glutamate binding in these areas (Greenamyre et al 1985). The number of large neurons in certain regions of the neocortex, particularly the temporal and frontal lobes, is reduced as compared to age-matched controls (Shefer 1972, Colon 1973, Terry et al 1981, Mountjoy et al 1983). Reductions in neuronal counts show some correlations with estimates of numbers of NFT and senile plaques, and it is believed that nerve cells exhibiting these structural changes eventually die.

The transmitter specificities of some at-risk neurons have been identified. Somatostatinergic cortical neurons show NFT (Roberts et al 1985), and levels of somatostatin are reduced in the neocortex (Davies et al 1980, Rossor et al 1980); somatostatin receptors are reduced in these regions (Beal et al 1985). Somatostatin coexists with gamma-aminobutyric acid (GABA) in some cortical neurons (Schmechel et al 1984), and GABA markers may be mildly reduced in some patients with AD (Rossor et al 1982b, Mountjoy et al 1984). Recent studies suggest that corticotropin-releasing factor (CRF)-like immunoreactivity is decreased in cortex and that CRF receptors are reciprocally up-regulated (D'Souza et al 1986). Neurochemical studies have not shown major changes in a variety of other systems, including neurons (using, as transmitters, cholecystokinin, vasoactive intestinal peptide, enkephalin, neurotensin, vasopressin, or oxytocin) (Rossor et al 1982a, Ferrier et al 1983). However, as described below, it has recently been shown that some of these peptidergic neurons exhibit structural pathology, i.e. abnormal axons/terminals forming neurites in plaques. Moreover, neurochemical assays have probably not detected some major transmitter systems at risk in AD. For example, some affected cortical neurons may use excitatory amino acids (or as yet unknown peptides) as transmitters, and available assays would not detect abnormalities of this system.

In addition to NFT and loss of neurons in cortex, individuals with AD show cortical plaques (Tomlinson et al 1970). Plaques tend to be most conspicuous in the neuropil of upper layers of cortex (D. L. Price, personal observation) but can be seen in deeper layers and, occasionally, in sub-cortical white matter (Pearson et al 1985). Studies in aged nonhuman primates (see below) and in individuals with AD suggest that neurites in neocortical senile plaques are derived from a variety of neuronal popula-tions, including cholinergic, monoaminergic, GABAergic, and certain peptidergic systems (Kitt et al 1984, 1986, Struble et al 1984b, 1985, 1986b, Walker et al 1985, Morrison et al 1985; C. A. Kitt, M. E. Molliver and D. L. Price, personal observation). It is not yet clear whether an individual senile plaque is homogeneous in its neurite population (i.e. only one transmitter marker exists in each plaque) or whether plaques show transmitter heterogeneity (i.e. several transmitter markers are present in a single plaque). Moreover, little information is available about secondary re-sponses of possibly deafferentated target neurons in proximity to plaques. For example, if plaques represent foci of abnormal synaptic interactions (including deafferentation), do target neurons show evidence of altered levels of receptors? Other issues regarding plaques are discussed below.

STRUCTURAL PATHOLOGY IN NEURONS

In AD, neurons in regions described above developed several types of structural pathology: NFT; enlarged neurites in senile plaques; GVD; and Hirano bodies (Figure 1). Recent evidence suggests that some of these structural changes involve components of the neuronal cytoskeleton. Because the cytoskeleton is important in maintenance of the size and shape of nerve cells, in the organization of the cytoplasm and surface membrane constituents, and in intracellular transport processes, it is likely that neurons exhibiting these types of structural pathology may also manifest functional abnormalities that are important in clinical expressions of disease. For example, it is possible that neurons showing cytoskeletal abnormalities may not be able to transport transmitter enzymes normally or to maintain the appropriate number of receptors in specific regions of these cells.

Neurofibrillary Tangles

NFT, intracytoplasmic fibillary occlusions, occur in perikarya of neurons in the locus coeruleus (Hirano & Zimmerman 1962, Forno 1978), raphe nuclei (Curcio & Kemper 1984, Yamamoto & Hirano 1985), basal forebrain cholinergic system (Hirano & Zimmerman 1962, Whitehouse et al 1982, Saper et al 1985), amygdala (Jamada & Mehraein 1968, Kemper 1983), hippocampus and associated systems (Hirano & Zimmerman 1962, Ball

1977, Kemper 1978, Hyman et al 1984), and neocortex (Tomlinson et al 1970, Wilcock & Esiri 1982, Pearson et al 1985). These inclusions are visualized using silver impregnation techniques; in addition, when stained with thioflavin-T and Congo red, they show staining patterns resembling amyloid. At the ultrastructural level, NFT are composed, in substantial part, of 10-nm filaments arranged as paired helical filaments (PHF) with a maximum width of 20 mm and a periodicity of 80 nm (Kidd 1963, Wisniewski et al 1976).

PHF, frequently associated with 10-nm neurofilaments and 15-nm straight filaments, are highly insoluble cross-linked protein polymers (Selkoe et al 1982), the origins and components of which are not yet certain. Some investigations suggest that PHF may be modified neurofilaments: NFT stain with antibodies directed against the 200-kilodalton (kD) neurofilament protein (Gambetti et al 1980, Ihara et al 1981, Anderton et al 1982, Dahl et al 1982, Autilio-Gambetti et al 1983, Gambetti et al 1983, Rasool et al 1984); elegant electron microscopic Bodian stains and, ultrastructurally, immunocytochemical preparations with antineurofilament antibodies are consistent with the concept that neurofilament antigens are components of PHF (Perry et al 1985). It is of particular interest that the phosphorylated form of the 200-kD neurofilament protein, which is not usually present in perikarya, is enriched in NFT-containing neurons (Cork et al 1986). These filaments may be retained abnormally in cell bodies and then inappropriately phosphorylated; alternatively, perikaryal filaments may be abnormally phosphorylated and, perhaps because of phosphorylation in associated cross-links, may be unable to be delivered to axons. The result of either of these processes would be accumulations of filaments in perikarya. If neurofilaments are one component of PHF, inappropriate phosphorylation could represent an early stage in the evolution of NFT. It is possible that 10-nm neurofilaments, 15-nm straight filaments, and PHF may be related, with the structural differences between filaments reflecting heterogeneity of composition, different modifications of component proteins, and/or stages in the evolution of filamentous pathology. Alternatively, neurofilaments could be associated with NFT but not be a major intrinsic protein of PHF. A variety of other studies suggest that nonneurofilament proteins, including microtubule-associated proteins (MAP2), may be associated with PHF (Nukina & Ihara 1983, Grundke-Iqbal 1984, Kosik et al 1984, Wisniewski et al 1984, Perry et al 1985). It is possible that PHF are principally derived from a host protein, which is not normally assembled into filaments, but does form PHF in this setting. Finally, PHF could be part of a nonneuronal constituent (e.g. serum factor, exogenous agent) that gains access to neurons and then assembles into filamentous forms. The composition of PHF and 15-nm straight filaments and the relationship of these organelles to other elements (derived from neurons, nonneuronal cells, exogenous agents) in neurons should be clarified as progress is made in purifying these abnormal organelles and in solubilizing and sequencing their proteins. Sequences of these abnormal proteins can then be compared to available information about other neuronal proteins including neurofilaments (Geisler et al 1983) and to the pathological proteins forming amyloid (see below).

It is likely that the development of NFT represents a stage in the degeneration of neurons; when these cells die, their locations are marked by the presence of extracellular PHF that contain NFT (Probst et al 1982). As described above, this type of pathology involves cells in a variety of transmitter-specific neuronal systems in different regions; correlations exist between the presence of cortical NFT and reductions in ChAT activity in cortex (Wilcock et al 1982). The importance of NFT in the clinical syndrome of AD is indicated by the observation that the presence of cortical NFT correlates with the presence of dementia (Wilcock & Esiri 1982).

Senile Plaques

Senile plaques are spherical foci containing enlarged argentophilic axons/synaptic terminals (neurites), abnormal dendritic processes, extra-cellular amyloid (see below), and variable numbers of astroglia and microglia (Wisniewski & Terry 1973, Probst et al 1983). These foci are thought to represent sites of abnormal synaptic interactions. They are located predominantly within the amygdala, hippocampus, and neocortex but can be seen in smaller numbers in a variety of subcortical sites. In relatively intact elderly individuals (Tomlinson et al 1968), plaques occur at low frequencies. Abnormal axons/terminals within plaques are commonly distended by accumulated PHF and neurofilaments. The majority of neurites in plaques appear to be degenerating axons/terminals, but some may represent abortive regenerates. Cholinergic, peptidergic, monoaminer-gic, and GABAergic neurites have been identified in senile plaques of aged nonhuman primates (Kitt et al 1984, 1986, Struble et al 1984b, Walker et al 1985). In AD, several of these systems also contribute neurites to plaques (Morrison et al 1985, Struble et al 1985b). The presence of plaques correlates with reductions in cholinergic and somatostatinergic markers (Perry et al 1978, 1983, Rossor et al 1984). Senile plaques appear to be important in the expression of dementia, in that the presence of neocortical plaques correlates with the presence of dementia (Tomlinson et al 1970).

Hirano Bodies

Hirano bodies, rod-shaped eosinophilic cytoplasmic inclusions, were originally described in the hippocampi of Guamanian patients with manifestations of amyotrophic lateral sclerosis–Parkinson dementia complex (Hirano 1965). Hirano bodies are common in the pyramidal cell layer of the hippocampus of patients with AD (Gibson & Tomlinson 1977). These inclusions contain thin filaments and show actin-like immunoreactivity (Goldman 1983). Freeze-fracture/deep-etch images of Hirano bodies show paracrystalline arrays of actin (J. Heuser and D. L. Price, personal observations).

Granulovacuolar Degeneration

Single, electron-dense granules within membrane-limited vacuoles occur frequently within the cytoplasm of pyramidal neurons of hippocampi of individuals with AD (Tomlinson & Kitchener 1972, Ball 1977, 1983). Preliminary immunocytochemical studies with one polyclonal antitubulin antibody (derived from preparations of sea urchin tubulin) have demonstrated that the granules show tubulin-like immunoreactivity; this staining can be abolished by absorption with tubulin (Price et al 1985b). In immunocytochemical studies, these granules are not stained by several other antitubulin antibodies or by antibodies directed against neurofilament proteins or MAP2. These limited studies suggest that GVD may be a unique type of cytoskeletal pathology involving modified tubulin. However, it is possible that other cytoskeletal constituents or other elements also may be present in these granules, but they have not been detected with our panel of antibodies. The observation that the presence of GVD in pyramidal neurons of the hippocampus correlates with loss of these nerve cells (Ball 1977) suggests that the development of GVD is eventually associated with cell death.

AMYLOID DEPOSITION

Amyloid in Senile Plaques

Amyloid cores of senile plaques are composed of extracellular 6–10 nm fibrils (Figure 1) whose proteins are arranged in a predominantly β-pleated sheet configuration. Origins of amyloid proteins in plaques are unknown, but sources may include degenerating neuronal processes (Powers et al 1981, Price et al 1982), serum factors (Glenner & Wong 1984), and, in experimental disorders, proteins associated with infectious agents (Wisniewski et al 1981, Prusiner et al 1983). Plaque amyloid appears to contain a variety of constituents, including blood-borne proteins (Powers et al 1981); which, if any, of these proteins are primary deposits and which are secondarily entrapped by amyloid meshworks is still unknown. Recent studies suggest that protein in plaque cores consists of multimeric aggregates of a 4 kD peptide (Masters et al 1985) that closely resembles the amyloid protein deposited in congophilic angiopathy (Glenner & Wong 1984). Some investigators believe that neuronal abnormalities (e.g. NFT or neurites) precede the appearance of plaque amyloid; others suggest that abnormalities of the blood-brain barrier and local deposition of amyloid are part of the initial lesion. When investigators are successful in purifying plaque cores, it should be possible to solubilize and sequence the amyloid proteins, to compare these sequences to those of PHF and cytoskeletal proteins, to make antibodies against these constituents, and to use these antibodies and recombinant DNA techniques to develop probes to identify the cells producing amyloid. These new technologies should allow clarification of the composition, sources, and evolution of various types of amyloid deposits in the brain.

Amyloid Angiopathy

Some patients with AD have deposits of amyloid associated with the walls of leptomeningeal and intracortical vessels (congophilic angiopathy) (Figure 1) (Mandybur 1975, Mountjoy et al 1982). Disagreements exist about the frequency of congophilic angiopathy in AD. An association exists between the presence of vascular amyloid and that of plaque amyloid, but chemical relationships between these two types of amyloid have not been clarified. Vascular amyloid contains a protein with a unique amino acid sequence (Glenner & Wong 1984). Glenner (1979) suggested that an amyloidogenic serum protein damages cerebral vessels and, subsequently, upon reaching the neuropil, the presence of amyloid leads to the formation of NFT and plaques, but it seems more likely that the amyloid associated with vessels is derived from disease occurring primarily within the neuronal parenchyma.

MODELS THAT SHOW CERTAIN FEATURES IN COMMON WITH ALZHEIMER'S DISEASE

Animal models are useful for combined behavioral, structural, and chemical studies of mechanisms and consequences of cellular abnormalities of specific systems. In this context, a variety of models could be discussed, but I focus instead on three disorders that show AD-like abnormalities at each of these levels of investigation: individual cells, neuronal systems, and brain regions. At the level of the neuron, I focus on neurofibrillary pathology induced in rabbits by treatment with aluminum. At the level of transmitter systems, I examine the model of cortical cholinergic deficiency caused by selective ablation of the primate basal forebrain cholinergic system. Finally, in terms of regional disease, I discuss age-associated cortical pathology occurring in older macaques.

Aluminum Intoxication

Using a variety of experimental manipulations, it is possible to induce neurofibrillary changes in neurons and to examine, using neurobiological techniques, the mechanisms and consequences of the structural pathology in these cells. For example, when rabbits are given intrathecal injections of aluminum chloride, motor neurons develop neurofibrillary pathology first in proximal axons and then in perikarya and dendrites (Klatzo et al 1965, Yates et al 1976, Selkoe et al 1979, Wisniewski et al 1980, Ghetti & Gambetti 1983, Troncoso et al 1982). Phosphorylated neurofilament proteins accumulate in perikarya (Troncoso et al 1985; P. Gambetti, personal communication), and the transport of neurofilament proteins is impaired (Bizzi et al 1984, Troncoso et al 1986). Disagreements exist concerning the effect of aluminum on cholinergic enzymes (Hetnarski et al 1980, Yates et al 1980, Kosik et al 1983), with some investigations showing changes in markers, while other studies show no significant alterations in these enzymes. This issue could be clarified by using neurochemical and immunocytochemical methods to examine affected neurons. It would be of interest to induce neurofibrillary pathology in the basal forebrain cholinergic system, to examine the hypothesis that neurofibrillary pathology results from alterations in synthesis or posttranslational phosphorylation of neurofilament proteins, and to assess the relationships, if any, between disorganization of the cytoskeleton and changes in neurotransmitter markers.

Lesions in the Nonhuman Primate Basal Forebrain Cholinergic System

To begin to clarify the roles of lesions in specific systems in the expression of abnormalities, it is necessary to destroy systems selectively and to assess the consequences of lesions on behavior, cognition, and memory. In nonhuman primates, selective destruction of neurons in the basal forebrain cholinergic system causes substantial reductions in ChAT activity in neocortex ipsilateral to the lesion (Struble et al 1986a). Behavioral studies in animals with bilateral partial lesions of the nbM suggest that these animals do not exhibit overt memory deficits, but their performance on tasks is sensitive to scopolamine (Aigner et al 1983). However, some animals with larger lesions involving the medial septum, diagonal band, and nbM do show memory deficits (Aigner et al 1984). Thus, in primates, cholinergic denervation of certain forebrain targets, particularly amygdala and hippocampus, may be associated with certain types of memory impairments. This model should prove useful for testing innovative therapies (perhaps including neural grafts and trophic factors) designed to act on cholinergic systems and their targets.

Aged Monkeys

Aged monkeys do not develop AD, but these animals do show a variety of cognitive/memory deficits, including reductions in short-term memory and difficulties with problem-solving (Bartus et al 1978, Davis 1978, Bartus 1982). Moreover, these aged primates develop senile plaques that resemble plaques occurring in aged humans and patients with AD (Wisniewski & Terry 1973, Wisniewski et al 1973, Struble et al 1982, 1984b, Price et al 1985a, 1986). During middle (17–20 years of age) and late life, macaques develop irregular, enlarged axonal varicosities in cortex, and some of these animals show neurite-rich plaques (Struble et al 1982, 1984b). These neurites are readily visualized with immunocytochemical methods using antibodies directed against phosphorylated 200-kD neurofilament protein (L. C. Cork, L. A. Sternberger, and N. H. Sternberger, personal observations) but, in contrast to neurites in AD, they do not show abundant PHF (R. G. Struble, L. C. Cork, D. J. Selkoe, and D. L. Price, personal observations). This observation indicates that axonal swellings in plaques can develop without the appearance of PHF in neurites. In older animals, plaques are more common; many plaques contain variable amounts of amyloid (e.g. mixed plaques show neurites and amyloid, while end-stage plaques are characterized by fewer neurites and abundant amyloid) (Struble et al 1984a, 1986b). In the aged nonhuman primate, there appears to be a sequence of evolution from neurite-rich to amyloid (or end-stage) plaques. In these animals, neurites in senile plaques are derived from axons and terminals of cholinergic, peptidergic, monoaminergic, and GABAergic systems (Price et al 1983, Kitt et al 1984, Struble et al 1984b, Kitt et al 1985, Walker et al 1985). It is not yet known whether axons/terminals in an individual plaque are derived from one or several sources (for discussion, see Price et al 1983). Significantly, in aged macaques, there is a trend for plaque densities to be a better predictor of impaired cognitive function than chronological age (Struble et al 1986b).

ETIOLOGICAL FACTORS

The cause of AD is unknown, but a variety of factors have been suggested to contribute to the development of AD: age; familial, genetic, and chromosomal factors; metabolic abnormalities; transmissible agents; amyloid deposition; toxins; trauma; and lack of tropiic factors.

Age

AD is not an invariable consequence of aging. The creative contributions of Sophocles, Verdi, Titian, Picasso, and others indicate that cognitive processing of the highest order can occur in old age. In older individuals without overt cognitive impairments, age-associated loss of nerve cells occurs in a number of neuronal populations. Moreover, in these individuals, a number of neuronal populations show, at low frequencies, some of the neuronal pathology described above (e.g. NFT, plaques, GVD, and Hirano bodies) (Tomlinson et al 1968). Aspects of these processes occur in aged nonhuman primates. For example, plaques commonly occur in a variety of monkeys (Struble et al 1982, 1984a, Kitt et al 1984, Price et al 1985a). Effects of these structural abnormalities of neurons on behavior, cognition, and memory are not well delineated, but recent studies on aged nonhuman primates suggest that a correlation exists between the presence of plaques and the presence of impairment in certain memory tasks (Struble et al 1986b). Although relationships between these mild age-associated changes and the clinicopathological syndrome, AD, deserve further investigation, it is clear that individuals with AD differ from aged controls by the greater frequencies, wider distributions, and greater severities of the pathologies described above.

Familial, Genetic, and Chromosomal Factors

AD appears familial in perhaps 25–40% of cases (Heston et al 1981, Heyman et al 1983), with these individuals tending to have an earlier onset, more rapid progression, more severe structural and chemical pathologies, and, frequently, an inheritance pattern consistent with an autosomal dominant disease (Heston et al 1981). Recently, Kilpatrick et al (1983) have reported the occurrence of AD in genetically identical twins whose mother and maternal grandmother had similar illnesses. The authors suggest that the disease in this setting may be inherited as a single mutant gene. As regards chromosomal factors, the frequency of Down's syndrome is increased among relatives of individuals with AD (Heyman et al 1983); moreover, the brains of older individuals with Down's syndrome frequently exhibit some of the structural and chemical abnormalities typical of AD, i.e. NFT, plaques, cholinergic deficits, and reductions in neurons in the nbM (Price et al 1982, Yates et al 1983, Wisniewski et al 1985, Casanova et al 1985).

Metabolic Abnormalities

Cortices of individuals with AD show reduced amounts of total protein and decreased protein synthesis (for review, see Sajdel-Sulkowska et al 1983), postmortem findings that correlate with images obtained by positon emission tomography (Bustany et al 1983). Recent investigations indicate that total cellular RNA and polyadenylated RNA are reduced in AD cortex containing plaques and NFT. These reductions in RNA appear to be associated with increased activity of alkaline ribonuclease—a change attributed to an abnormality in the ribonuclease inhibitor complex (Sajdel-

Sulkowska & Marotta 1984). In the future, cDNA probes will be used in in situ hybridization studies to analyze levels of specific mRNA in control and affected neurons.

A variety of other metabolic abnormalities, including those arising from carbohydrate metabolism, have been described, but the specificity of these changes remain to be clarified. In addition, alterations have been described in nonneuronal cells outside of the central nervous system. These observations, which need to be validated in studies in patients demonstrated at autopsy to have AD, suggest that it may be fruitful to pursue biochemical markers in AD in nonneural tissues.

Transmissible Agents

Roles of transmissible agents in the pathogenesis of AD are uncertain. Infectious agents could gain access to the brain by a variety of pathways, including local spread, hematogenous routes, or axonal transport along anatomical pathways. For example, it has been suggested that such agents could reach the brain via the olfactory pathway, as occurs in herpes encephalitis, and spread to the amygdala, hippocampus, and associated cortical regions. AD shares some pathological characteristics (particularly the appearance of amyloid plaques) with the spongiform encephalopathies—a group of unusual infectious diseases, including scrapie and Creutzfeldt-Jakob disease (Goudsmit et al 1980, Wisniewski et al 1981). Agents causing these disorders are not conventional viruses and are not susceptible to the usual procedures that inactivate viruses. The best characterized disease of this type is scrapie, a disease of sheep. Purified preparations of the scrapie agent yield a protein (PrP 27–30) that forms 10–20 nm "prion filaments"; in aggregates, these filaments show the histochemical features of amyloid (Prusiner et al 1983). Recently, a clone encoding PrP 27–30 was obtained from a cDNA library of scrapie-infected hamster brain. Southern blots with these probes show a single gene with restriction patterns identical in normal and infected animals. Moreover, PrP mRNAs were detected in both control and infected hamsters. In purified preparations of scrapie prions, PrP-related nucleic acids were not demonstrated—an observation interpreted to indicate that PrP 27–30 is not encoded by a nucleic acid carried within the infectious particle (Oesch et al 1985). The normal functions of PrP are not known, and little is known about the processing of PrP that leads to the formation of amyloid-like filaments. When PrP protein and AD amyloid proteins are sequenced and nucleic acid probes are available, it will be possible to explore possible relationships between the amyloid occurring in experimental disorders (such as scrapie) and those occurring in human disease (Masters et al 1985). It should be noted that, to date, AD has not been transmitted to experimental animals (Goudsmit et al 1980).

Amyloid Deposition

As described above, some patients with AD show amyloid in cerebral vessels (Mandybur 1975, Glenner et al 1981, Mountjoy et al 1982). From preparations of vessels from brains of individuals with AD and Down's syndrome, a novel β-pleated sheet protein has been identified (Glenner & Wong 1984). By analogy with systemic amyloidosis, it has been argued that this unique serum protein, presumed to be manufactured outside the central nervous system, damages cerebral endothelial cells, enters the brain, and, by mechanisms unknown, leads to the formation of NFT and neurites in plaques. During the next several years, new approaches derived from protein chemistry and molecular biology should clarify the chemical composition, possible relationships, and vector of evolution of these amyloid deposits.

Toxins

Roles of toxic agents in AD are uncertain. In some studies, amounts of aluminum have been elevated in the brains of affected patients, and aluminum content does appear to be increased in neurons showing NFT (Perl & Brody 1980). However, it is not known whether aluminum concentrates in neurons as a primary process or results because sick cells accumulate aluminum. Although aluminum may not play a primary role in the pathogenesis of AD, it has been useful in producing experimental models of neurofibrillary pathology that can be studied by a variety of neurobiological approaches.

Trauma

Recent reports suggest that individuals with AD have an increased incidence of a history of head trauma (Heyman et al 1984, Mortimer et al 1985). It is of interest that repeated head trauma, as occurs in boxers, leads to a clinical syndrome of dementia associated with the presence of widespread NFT (dementia pugilistica) (Corsellis 1978).

Lack of Trophic Factors

By analogy with studies of nerve growth factor, it has been suggested that AD may be associated with a lack of a trophic factor normally manufactured by cells in hippocampal or cortical tissues (Ojika & Appel 1983). It is of interest that, when nerve growth factor is injected into the cortex, it is transported retrograde to neurons in the basal forebrain (Seiler & Schwab 1984) and increases the ChAT activity in this region (Gnahn et al 1983). In this hypothetical model of AD, abnormalities of target cells in the amygdala, hippocampus, and neocortex are associated with reduced production of a nerve growth factor-like trophic factor; when the factor is not available in sufficient amounts, forebrain cholinergic neurons show structural abnormalities and then degenerate. The role of trophic factors in AD and related disorders is an exciting new area of investigation.

SUMMARY

Significant progress has been made in refining diagnostic criteria for AD and in developing imaging approaches to exclude treatable disease and to assess some of the metabolic processes occurring in vivo in the brains of individuals with AD. However, as yet, no reliable diagnostic test is available. Although risk factors have been identified, the etiology of AD remains an enigma. The roles of familial, chromosomal, and genetic factors, toxins, and transmissible agents in the pathogenesis of this disease deserve intensive study. A variety of neurotransmitter systems are affected in the disease, and it seems likely that new approaches may identify additional systems at risk. Of particular importance will be studies on surviving neurons, since these cells will be targets for treatment. Neurons in certain parts of the brainstem, basal forebrain, amygdala, hippocampus, and neocortex show several types of cytoskeletal abnormalities, but mechanisms of cytoskeletal disorganization are not well understood, e.g. we do not have a clear idea of the sequence of cytoskeletal pathology, the time course of dysfunction of individual neurons, and consequences of these processes on cell function. In situ hybridization with radiolabeled nucleic acid probes and immunocytochemical approaches should provide information about levels of gene expression, protein compositions, and posttranslational modifications of normal and abnormal proteins in these cells. The relationships and sources of some of the abnormal proteins (e.g. those associated with PHF, 15-nm straight filaments, plaque amyloid, and vascular amyloid) can be clarified by new approaches of protein chemistry (purification and sequencing) and molecular biology (recombinant DNA techniques). Finally, investigations of animal models that recapitulate certain features of AD should provide new insights into the nature, mechanisms, and consequences of cellular pathology of specific systems. These models may be useful for imaging studies similar to those used in human patients and for developing and testing new therapeutic approaches that eventually may be useful for treating this all-too-common disorder of the central nervous system.

ACKNOWLEDGMENTS

The author gratefully acknowledges the many helpful discussions with investigators whose studies are cited in the text, as well as the scientists who have generously provided antibodies for some of the studies performed by our laboratory. For their contribution, I particularly wish to thank the following colleagues: Drs. Peter J. Whitehouse, Robert G. Struble, John C. Hedreen, Linda C. Cork, Cheryl A. Kitt, Paul N. Hoffman, John W. Griffin, Joseph T. Coyle, Lary C. Walker, Juan C. Troncoso, Richard E. Powers, Manuel F. Casanova, Richard M. Zweig, Mahlon R. DeLong, Thomas J. Aigner, Mark E. Molliver, Mortimer Mishkin, and Susan J. Mitchell, as well as Mr. Richard J. Altschuler and Ms. Sharon Presty. Mrs. Carla R. Jordon provided excellent secretarial assitance in the preparation of the manuscript.

This work was supported by funds from the US Public Health Service (NIH AG 03359, NS 20471, and AG 05146), the Claster Family Fund, and Point of View, Inc.

Literature Cited

Adolfsson, R., Gottfries, C. G., Oreland, L., Roos, B. E., Winblad, B. 1978. Reduced levels of catecholamines in the brain and increased activity of monoamine oxidase in platelets in Alzheimer's disease: Therapeutic implications. In *Alzheimer's Disease: Senile Dementia and Related Disorders (Aging)*, ed. R. Katzman, R. D. Terry, K. L. Bick, 7:441–51. New York: Raven

Aigner, T., Aggleton, J., Mitchell, S. J., Price, D., DeLong, M., et al. 1983. Effects of scopolamine on recognition memory in monkeys after ibotenic acid injections into the nucleus basalis of Meynert. *Soc. Neurosci. Abstr.* 9:826 (Abstr.)

Aigner, T., Mitchell, S., Aggleton, J., DeLong, M., Struble, R., et al. 1984. Recognition deficit in monkeys following neurotoxic lesions of the basal forebrain. *Soc. Neurosci. Abstr.* 10:386 (Abstr.)

Anderton, B. H., Breinburg, D., Downes, M. J., Green, P. J., Tomlinson, B. E., et al. 1982. Monoclonal antibodies show that neurofibrillary tangles and neurofilaments share antigenic determinants. *Nature* 298:84–86

Arendt, T., Bigl, V., Arendt, A., Tennstedt, A. 1983. Loss of neurons in the nucleus basalis of Meynert in Alzheimer's disease, paralysis agitans, and Korsakoff's disease. *Acta Neuropathol.* 61:101–8

Autilio-Gambetti, L., Gambetti, P., Crane, R. C. 1983. Paired helical filaments: Relatedness to neurofilaments shown by silver staining and reactivity with monoclonal antibodies. *Biological Aspects of Alzheimer's Disease. Banbury Rep.* 15:117–24

Ball, M. J. 1972. Neurofibrillary tangles and the pathogenesis of dementia: A quantitative study. *Neuropathol. Appl. Neurobiol.* 2:395–410

Ball, M. J. 1977. Neuronal loss, neurofibrillary tangles and granulovacuolar degeneration in the hippocampus with ageing and dementia. A qualitative study. *Acta Neuropathol.* 37:111–18

Ball, M. J. 1978. Histotopography of cellular changes in Alzheimer's disease. In *Senile Dementia: A Biomedical Approach. Developments in Neuroscience*, ed. K. Nandy, 3:89–104. New York: Elsevier North-Holland

Ball, M. J. 1983. Granulovacuolar degeneration. In *Alzheimer's Disease*, ed. B. Reisberg, pp. 62–68. New York: The Free Press

Ball, M. J., Fisman, M., Hachinski, V., Blume, W., Fox, A., et al. 1985. A new definition of Alzheimer's disease: A hippocampal dementia. *Lancet* 1:14–16

Ball, M. J., Lo, P. 1977. Granulovacuolar degeneration in the ageing brain and in dementia. *J. Neuropathol. Exp. Neurol.* 36:474–87

Ball, M. J., Merskey, H., Fisman, M., Fyfe, I. M., Fox, H., et al. 1983. Hippocampal morphometry in Alzheimer dementia: Implications for neurochemical hypotheses. *Biological Aspects of Alzheimer's Disease. Banbury Rep.* 15:45–64

Bartus, R. T. 1982. Effects of cholinergic agents on learning and memory in animal models of aging. In *Alzheimer's Disease: A Report of Progress in Research (Aging)*, ed. S. Corkin, K. L. Davis, J. H. Growdon, E. Usdin, R. J. Wurtman, 19:271–80. New York: Raven

Bartus, R. T., Dean, R. L. III, Beer, B., Lippa, A. S. 1982. The cholinergic hypothesis of geriatric memory dysfunction. *Science* 217:408–17

Bartus, R. T., Fleming, D., Johnson, H. R. 1978. Aging in the rhesus monkey: Debilitating effects on short-term memory. *J. Gerontol.* 33:858–71

Beal, M. F., Mazurek, M. F., Tran, V. T., Chattha, G., Bird, E. D., Martin, J. B. 1985. Reduced numbers of somatostatin receptors in the cerebral cortex in Alzheimer's disease. *Science* 229:289–91

Bird, T. D., Stranahan, S., Sumi, S. M., Raskind, M. 1983. Alzheimer's disease: Choline acetyltransferase activity in brain tissue from clinical and pathological subgroups. *Ann. Neurol.* 14:284–93

Bizzi, A., Crane, R. C., Autilio-Gambetti, L., Gambetti, P. 1984. Aluminum effect on slow axonal transport: A novel impairment of neurofilament transport. *J. Neurosci.* 4:722–31

Bondareff, W., Mountjoy, C. Q., Roth, M. 1982. Loss of neurons or origin of the adrenergic projection to cerebral cortex (nucleus locus ceruleus) in senile dementia. *Neurology* 32:164–68

Bowen, D. M., Allen, S. J., Benton, J. S., Goodhardt, M. J., Haan, E. A., et al. 1983. Biochemical assessment of serotonergic and cholinergic dysfunction and cerebral atrophy in Alzheimer's disease. *J. Neurochem.* 41:266–72

Bowen, D. M., Smith, C. B., White, P., Davison, A. N. 1976. Neurotransmitter-related enzymes and indices of hypoxia in senile dementia and other abiotrophies. *Brain* 99:459–96

Brun, A. 1983. An overview of light and electron microscopic changes. In *Alzheimer's Disease*, ed. B. Reisberg, pp. 37–47. New York: The Free Press

Buell, S. J., Coleman, P. D. 1981. Quantitative evidence for selective dendritic growth in normal human aging but not in senile dementia. *Brain Res.* 214:23–41

Bustany, P., Henry, J. F., Soussaline, F., Comar, D. 1983. Brain protein synthesis in normal and demented patients—a study by positron emission tomography with ^{11}C-L-methionine. In *Functional Radionuclide Imaging of the Brain. Serono'Symposia*, ed. P. L. Magistretti, 5:319–26. New York: Raven

Casanova, M. F., Walker, L. C., Whitehouse, P. J., Price, D. L. 1985. Abnormalities of the nucleus basalis in Down's syndrome. *Ann. Neurol.* 18:310–13

Colon, E. J. 1973. The cerebral cortex in presenil dementia. A quantitative analysis. *Acta Neuropathol.* 23:281–90

Cork, L. C., Altschuler, R. J., Struble, R. G., Casanova, M. F., Price, D. L., et al. 1986. Changes in the distribution of phosphorylated neurofilaments in Alzheimer's disease. *J. Neuropathol. Exp. Neurol.* In press

Corsellis, J. A. N. 1978. Posttraumatic dementia. In *Alzheimer's Disease: Senile Dementia and Related Disorders (Aging)*, ed. R. Katzman, R. D. Terry, K. L. Bick, 7:125–33. New York: Raven

Cross, A. J., Crow, T. J., Perry, E. K., Perry, R. H., Blessed, G., et al. 1981. Reduced dopamine-β-hydroxylase activity in Alzheimer's disease. *Br. Med. J.* 282:93–94

Curcio, C. A., Kemper, T. 1984. Nucleus raphe dorsalis in dementia of the Alzheimer type: Neurofibrillary changes and neuronal packing density. *J. Neuropathol. Exp. Neurol.* 43:359–68

Dahl, D., Selkoe, D. J., Pero, R. T., Bignami, A. 1982. Immunostaining of neurofibrillary tangles in Alzheimer's senile dementia with a neurofilament antiserum. *J. Neurosci.* 2:113–19

Davies, P., Katzman, R., Terry, R. D. 1980. Reduced somatostatin-like immunoreactivity in cerebral cortex from cases of Alzheimer disease and Alzheimer senile dementia. *Nature* 288:279–80

Davies, P., Maloney, A. J. F. 1976. Selective loss of central cholinergic neurons in Alzheimer's disease. *Lancet* 2:1403

Davis, R. T. 1978. Old monkey behavior. *Exp. Gerontol.* 13:237–50

Dayan, A. D. 1970. Quantitative histological studies on the aged human brain. I. Senile plaques and neurofibrillary tangles in "normal" patients. *Acta Neuropathol.* 16:85–94

Drachman, D. A., Leavitt, J. L. 1974. Human memory and the cholinergic system. A relationship to aging? *Arch. Neurol.* 30:113–21

Ferrier, I. N., Cross, A. J., Johnson, J. A., Roberts, G. W., Crow, T. J., et al. 1983. Neuropeptides in Alzheimer type dementia. *J. Neurol. Sci.* 62:159–70

Fonnum, F. 1984. Glutamate: A neurotransmitter in mammalian brain. *J. Neurochem.* 42:1–11

Forno, L. S. 1978. The locus caeruleus in Alzheimer's disease. *J. Neuropathol. Exp. Neurol.* 27:614

Foster, N. L., Chase, T. N., Mansi, L., Brooks, R., Fedio, P., et al. 1984. Cortical abnormalities in Alzheimer's disease. *Ann. Neurol.* 16:649–54

Frackowiak, R. S. J., Gibbs, J. M. 1983. The pathophysiology of Alzheimer's disease studied with positron emission tomography. *Biological Aspects of Alzheimer's Disease. Banbury Rep.* 15:317–27

Friede, R. L. 1965. Enzyme histochemical studies of senile plaques. *J. Neuropathol. Exp. Neurol.* 24:477–91

Gado, M., Hughes, C. P., Danziger, W., Chi, D. 1983. Aging, dementia, and brain atrophy: A longitudinal computed tomographic study. *Am. J. Neuroradiol.* 4:699–702

Gambetti, P., Shecket, G., Ghetti, B., Hirano, A., Dahl, D. 1983. Neurofibrillary changes in human brain. An immunocytochemical study with a neurofilament antiserum. *J. Neuropathol. Exp. Neurol.* 42:69–79

Gambetti, P., Velasco, M. E., Dahl, D., Bignami, A., Roessmann, U., et al. 1980. Alzheimer neurofibrillary tangles: An

508 PRICE

immunohistochemical study. In *Aging of the Brain and Dementia (Aging)*, ed. L. Amaducci, A. N. Davison, P. Antuono, 13:55–63. New York: Raven

Geisler, N., Kaufmann, E., Fischer, S., Plessmann, U., Weber, K. 1983. Neurofilament architecture combines structural principles of intermediate filaments with carboxy-terminal extensions increasing in size between triplet proteins. *EMBO J.* 2:1295–1302

Ghetti, B., Gambetti, P. 1983. Comparative immunocytochemical characterization of neurofibrillary tangles in experimental maytansine and aluminum encephalopathies. *Brain Res.* 276:388–93

Gibson, P. H., Tomlinson, B. E. 1977. Numbers of Hirano bodies in the hippocampus of normal and demented people with Alzheimer's disease. *J. Neurol. Sci.* 33:199–206

Glenner, G. G. 1979. Congophilic microangiopathy in the pathogenesis of Alzheimer's syndrome (presenile dementia). *Med. Hypotheses* 5:1231–36

Glenner, G. G., Henry, J. M., Fujihara, S. 1981. Congophilic angiopathy in the pathogenesis of Alzheimer's degeneration. *Ann. Pathol.* 1:105–8

Glenner, G. G., Wong, C. W. 1984. Alzheimer's disease: Initial report of the purification and characterization of a novel cerebrovascular amyloid protein. *Biochem. Biophys. Res. Commun.* 120:885–90

Gnahn, H., Hefti, F., Heumann, R., Schwab, M. E., Thoenen, H. 1983. NGF-mediated increase of choline acetyltransferase (ChAT) in the neonatal rat forebrain: Evidence for a physiological role of NGF in the brain? *Dev. Brain Res.* 9:45–52

Goldman, J. E. 1983. The association of actin with Hirano bodies. *J. Neuropathol. Exp. Neurol.* 42:146–52

Goudsmit, J., Morrow, C. H., Asher, D. M., Yangihara, R. T., Masters, C. L., et al. 1980. Evidence for and against the transmissibility of Alzheimer disease. *Neurology* 30:945–50

Greenamyre, J. T., Penney, J. B., Young, A. B., D'Amato, C. J., Hicks, S. P., et al. 1985. Alterations in L-glutamate binding in Alzheimer's and Huntington's diseases. *Science* 227:1496–99

Grundke-Iqbal, I., Iqbal, K., Tung, Y.-C., Wisniewski, H. M. 1984. Alzheimer paired helical filaments: Immunochemical identification of polypeptides. *Acta Neuropathol.* 62:259–67

Hedreen, J. C., Bacon, S. J., Cork, L. C., Kitt, C. A., Crawford, G. D., et al. 1983. Immunocytochemical identification of cholinergic neurons in the monkey central nervous system using monoclonal antibodies against choline acetyltransferase. *Neurosci. Lett.* 43:173–77

Hedreen, J. C., Struble, R. G., Whitehouse, P. J., Price, D. L. 1984. Topography of the magnocellular basal forebrain system in human brain. *J. Neuropathol. Exp. Neurol.* 43:1–21

Heston, L. L., Mastri, A. R., Anderson, V. E., White, J. 1981. Dementia of the Alzheimer type. Clinical genetics, natural history, and associated conditions. *Arch. Gen. Psychiatry* 38:1085–90

Hetnarski, B., Wisniewski, H. M., Iqbal, K., Dziedzic, J. D., Lajtha, A. 1980. Central cholinergic activity in aluminum-induced neurofibrillary degeneration. *Ann. Neurol.* 7:489–90

Heyman, A., Wilkinson, W. E., Hurwitz, B. J., Schmechel, D., Sigmon, A. H., et al. 1983. Alzheimer's disease: Genetic aspects and associated clinical disorders. *Ann. Neurol.* 14:507–15

Heyman, A., Wilkinson, W. E., Stafford, J. A., Helms, M. J., Sigmon, A. H., et al. 1984. Alzheimer's disease: A study of epidemiological aspects. *Ann. Neurol.* 15:335–41

Hirano, A. 1965. Pathology of amyotrophic lateral sclerosis. In *Slow, Latent and Temperate Virus Infections*, eds. D. C. Gajdusek, C. J. Gibbs, M. Alpers, pp. 23–27. Washington DC: Natl. Inst. Neurol. Diseases and Blindness, Monogr. No. 2

Hirano, A., Zimmerman, H. M. 1962. Alzheimer's neurofibrillary changes. A topographic study. *Arch. Neurol.* 7:227–42

Hooper, M. W., Vogel, F. S. 1976. The limbic system in Alzheimer's disease. A neuropathologic investigation. *Am. J. Pathol.* 85:1–20

Hyman, B. T., Van Hoesen, G. W., Damasio, A. R., Barnes, C. L. 1984. Alzheimer's disease: Cell-specific pathology isolates the hippocampal formation. *Science* 225:1168–70

Ihara, Y., Nukina, N., Sugita, H., Yoyokura, Y. 1981. Staining of Alzheimer's neurofibrillary tangles with antiserum against 200 K component of neurofilament. *Proc. Jpn. Acad.* 57:152–56

Ingvar, D. H., Lassen, N. A. 1979. Activity distribution in the cerebral cortex in organic dementia as revealed by measurements of regional cerebral blood flow. In *Bayer-Symposium VII, Brain Function in Old Age, Evaluation of Changes and Disorders*, ed. F. Hoffmeister, C. Muller, pp. 268–77. Berlin: Springer-Verlag

Ishii, T. 1966. Distribution of Alzheimer's neurofibrillary changes in the brain stem and hypothalamus of senile dementia. *Acta Neuropathol.* 6:181–87

Iversen, L. L., Rossor, M. N., Reynolds, G. P., Hills, R., Roth, M., et al. 1983. Loss of pigmented dopamine-β-hydroxylase positive cells from locus coeruleus in senile dementia of Alzheimer's type. *Neurosci. Lett.* 39:95–100

Jamada, M., Mehraein, P. 1968. Vertielungsmuster der senilen Veranderungen im Gehirn. Die Beteiligung des limbischen Systems bei hirnatrophischen Prozessen des Seniums und bei Morbus Alzheimer. *Arch. Psychiatr. Z. Neurol.* 211:308–24

Kemper, T. L. 1978. Senile dementia: a focal disease in the temporal lobe. In *Senile Dementia: A Biomedical Approach. Developments in Neuroscience*, ed. K. Nandy, 3:105–13. New York: Elsevier North-Holland Biomedical Press

Kemper, T. L. 1983. Organization of the neuropathology of the amygdala in Alzheimer's disease. *Biological Aspects of Alzheimer's Disease. Banbury Rep.* 15:31–35

Kidd, M. 1963. Paired helical filaments in electron microscopy of Alzheimer's disease. *Nature* 197:192–93

Kilpatrick, C., Burns, R., Blumbergs, P. C. 1983. Identical twins with Alzheimer's disease. *J. Neurol. Neurosurg. Psychiatry* 46:421–25

Kitt, C. A., Price, D. L., Struble, R. G., Cork, L. C., Wainer, B. H., et al. 1984. Evidence for cholinergic neurites in senile plaques. *Science* 226:1443–45

Kitt, C. A., Struble, R. G., Cork, L. C., Mobley, W. C., Walker, L. C., et al. 1986. Catecholaminergic neurites in senile plaques in prefrontal cortex of aged nonhuman primates. *Neuroscience.* In press

Klatzo, I., Wisniewski, H., Streicher, E. 1965. Experimental production of neurofibrillary degeneration. I. Light microscopic observations. *J. Neuropathol. Exp. Neurol.* 24:187–99

Kosik, K. S., Bradley, W. G., Good, P. F., Rasool, C. G., Selkoe, D. J. 1983. Cholinergic function in lumbar aluminum myelopathy. *J. Neuropathol. Exp. Neurol.* 42:365–75

Kosik, K. S., Duffy, L. K., Dowling, M. M., Abraham, C., McCluskey, A., et al. 1984. Microtubule-associated protein. 2: Monoclonal antibodies demonstrate the selective incorporation of certain epitopes into Alzheimer neurofibrillary tangles. *Proc. Natl. Acad. Sci. USA* 81:7941–45

Mandybur, T. I. 1975. The incidence of cerebral amyloid angiopathy in Alzheimer's disease. *Neurology* 25:120–26

Mash, D. C., Flynn, D. D., Potter, L. T. 1985. Loss of M2 muscarine receptors in the cerebral cortex in Alzheimer's disease and experimental cholinergic denervation. *Science* 228:1115–17

McGeer, P. L., McGeer, E. G., Suzuki, J., Dolman, C. E., Nagai, T. 1984. Aging,

Alzheimer's disease, and the cholinergic system of the basal forebrain. *Neurology* 34:741–45

McKhann, G., Drachman, D., Folstein, M., Katzman, R., Price, D., et al. 1984. Clinical diagnosis of Alzheimer's disease: Report of the NINCDS-ADRDA Work Group under the auspices of Department of Health and Human Services Task Force on Alzheimer's Disease. *Neurology* 34: 939–44

Mesulam, M.-M., Mufson, E. J., Levey, A. I., Wainer, B. H. 1983. Cholinergic innervation of cortex by the basal forebrain: Cytochemistry and cortical connections of the septal area, diagonal band nuclei, nucleus basalis (substantia innominata), and hypothalamus in the rhesus monkey. *J. Comp. Neurol.* 214:170–97

Morimatsu, M., Hirai, S., Muramatsu, A., Yoshikawa, M. 1975. Senile degenerative brain lesions and dementia. *J. Am. Geriatr. Soc.* 23:390–406

Morrison, J. H., Rogers, J., Scherr, S., Benoit, R., Bloom, F. E. 1985. Somatostatin immunoreactivity in neuritic plaques of Alzheimer's patients. *Nature* 314:90–94

Mortimer, J. A., French, L. R., Hutton, J. T., Schuman, L. M. 1985. Head injury as a risk factor for Alzheimer's disease. *Neurology* 35:264–67

Mountjoy, C. Q., Rossor, M. N., Iversen, L. L., Roth, M. 1984. Correlation of cortical cholinergic and GABA deficits with quantitative neuropathological findings in senile dementia. *Brain* 107:507–18

Mountjoy, C. Q., Roth, M., Evans, N. J. R., Evans, H. M. 1983. Cortical neuronal counts in normal elderly controls and demented patients. *Neurobiol. Aging* 4:1–11

Mountjoy, C. Q., Tomlinson, B. E., Gibson, R. H. 1982. Amyloid and senile plaques and cerebral blood vessels. A semiquantitative investigation of a possible relationship. *J. Neurol. Sci.* 57:89–103

Nukina, N., Ihara, Y. 1983. Immunocytochemical study on senile plaques in Alzheimers disease. II. Abnormal dendrites in senile plaques as revealed by antimicrotubule-associated proteins (MAPs) immunostaining. *Proc. Jpn. Acad.* 59:288–92

Oesch, B., Westaway, D., Wälchli, M., McKinley, M. P., Kent, S. B. H., et al. 1985. A cellular gene encodes scrapie PrP 27–30 protein. *Cell* 40:735–46

Ojika, K., Appel, S. H. 1983. Neurotrophic factors and Alzheimer's disease. *Biological Aspects of Alzheimer's Disease. Banbury Rep.* 15:285–95

Pearson, R. C. A., Esiri, M. M., Hiorns, R. W., Wilcock, G. K., Powell, T. P. S. 1985. Anatomical correlates of the distribution of the pathological changes in the neocortex in Alzheimer disease. *Proc. Natl. Acad. Sci. USA* 82:4531–34

Pearson, R. C. A., Sofroniew, M. V., Cuello, A. C., Powell, T. P. S., Eckenstein, F., et al. 1983. Persistence of cholinergic neurons in the basal nucleus in a brain with senile dementia of the Alzheimer's type demonstrated by immunohistochemical staining for choline acetyltransferase. *Brain Res.* 289:375–79

Perl, D. P., Brody, A. R. 1980. Alzheimer's disease: X-ray spectrometric evidence of aluminum accumulation in neurofibrillary tangle-bearing neurons. *Science* 208:297–99

Perry, E. K., Gibson, P. H., Blessed, G., Perry, R. H., Tomlinson, B. E. 1977. Neurotransmitter enzyme abnormalities in senile dementia. *J. Neurol. Sci.* 34:247–65

Perry, E. K., Tomlinson, B. E., Blessed, G., Bergmann, K., Gibson, P. H., et al. 1978. Correlation of cholinergic abnormalities with senile plaques and mental test scores in senile dementia. *Br. Med. J.* 2:1457–59

Perry, G., Rizzuto, N., Autilio-Gambetti, L., Gambetti, P. 1985. Paired helical filaments from Alzheimer disease patients contain cytoskeletal components. *Proc. Natl. Acad. Sci. USA* 82:3916–20

Perry, R. H., Blessed, G., Perry, E. K., Tomlinson, B. E. 1980. Histochemical observations on the cholinesterase activities in the brains of elderly normal and demented (Alzheimer-type) patients. *Age Ageing* 9:9–16

Perry, R. H., Candy, J. M., Perry, E. K. 1983. Some observations and speculations concerning the cholinergic system and neuropeptides in Alzheimer's disease. *Biological Aspects of Alzheimer's Disease. Banbury Rep.* 15:351–61

Pope, A., Hess, H. H., Lewin, E. 1964. Microchemical pathology of the cerebral cortex in pre-senile dementias. *Trans. Am. Neurol. Assoc.* 89:15–16

Powers, J. M., Schlaepfer, W. W., Willingham, M. C., Hall, B. J. 1981. An immunoperoxidase study of senile cerebral amyloidosis with pathogenetic considerations. *J. Neuropathol. Exp. Neurol.* 40:592–612

Price, D. L., Cork, L. C., Struble, R. G., Kitt, C. A., Price, D. L. Jr., Lehmann, J., Hedreen, J. C. 1985a. Neuropathological, neurochemical, and behavioral studies of the aging nonhuman primate. In *Behavior and Pathology of Aging in Rhesus Monkeys, Monographs in Primatology,* Vol. 8, ed. R. T. Davis, C. W. Leathers, pp. 113–35. New York: Liss

Price, D. L., Kitt, C. A., Hedreen, J. C., Whitehouse, P. J., Struble, R. G., et al.

1986. Basal forebrain cholinergic systems in primate brain: Anatomical organization and role in the pathology of aging and dementia. In *Dynamics of Cholinergic Function,* ed. I. Hanin. New York: Plenum. In press

Price, D. L., Struble, R. G., Altschuler, R. J., Casanova, M. F., Cork, L. C., Murphy, D. B. 1985b. Aggregation of tubulin in neurons in Alzheimer's disease. *J. Neuropathol. Exp. Neurol.* 44:366

Price, D. L., Whitehouse, P. J., Struble, R. G. 1985c. Alzheimer's disease. *Annu. Rev. Med.* 36:349–56

Price, D. L., Whitehouse, P. J., Struble, R. G., Coyle, J. T., Clark, A. W., et al. 1982. Alzheimer's disease and Down's syndrome. *Ann. NY Acad. Sci.* 396:145–64

Price, D. L., Whitehouse, P. J., Struble, R. G., Price, D. L. Jr., Cork, L. C., et al. 1983. Basal forebrain cholinergic neurons and neuritic plaques in primate brain. *Biological Aspects of Alzheimer's Disease. Banbury Rep.* 15:65–77

Probst, A., Basler, V., Bron, B., Ulrich, J. 1983. Neuritic plaques in senile dementia of Alzheimer type: A Golgi analysis in the hippocampal region. *Brain Res.* 268:249–54

Probst, A., Ulrich, J., Heitz, Ph. U. 1982. Senile dementia of Alzheimer type: Astroglial reaction to extracellular neurofibrillary tangles in the hippocampus. *Acta Neuropathol.* 57:75–79

Prusiner, S. B., McKinley, M. P., Bowman, K. A., Bolton, D. C., Bendheim, P. E., et al. 1983. Scrapie prions aggregate to form amyloid-like birefringent rods. *Cell* 35:349–58

Rasool, C. G., Abraham, C., Anderton, B. H., Haugh, M., Kahn, J., et al. 1984. Alzheimer's disease: Immunoreactivity of neurofibrillary tangles with anti-neurofilament and anti-paired helical filament antibodies. *Brain Res.* 24:249–60

Roberts, G. W., Crow, T. J., Polak, J. M. 1985. Location of neuronal tangles in somatostatin neurones in Alzheimer's disease. *Nature* 314:92–94

Rossor, M. N., Emson, P. C., Iversen, L. L., Mountjoy, C. Q., Roth, M. 1984. Patterns of neuropeptide deficits in Alzheimer's disease. In *Alzheimer's Disease: Advances in Basic Research and Therapies. Proc. 3rd Meet. Int. Study Group on Treat. Memory Disorders Associated with Aging, Zurich,* ed. R. J. Wurtman, S. H. Corkin, J. H. Growdon, pp. 29–37

Rossor, M. N., Emson, P. C., Iverson, L. L., Mountjoy, C. Q., Roth, M., et al. 1982a. Neuropeptides and neurotransmitters in cerebral cortex in Alzheimer's disease. In *Alzheimer's Disease: A Report of Progress in Research (Aging)*, ed. S. Corkin, K. L. Davis, J. H. Growdon, E. Usdin, R. J. Wurtman, 19:15–24. New York: Raven

Rossor, M. N., Emson, P. C., Mountjoy, C. Q., Roth, M., Iversen, L. L. 1980. Reduced amounts of immunoreactive somatostatin in the temporal cortex in senile dementia of Alzheimer type. *Neurosci. Lett.* 20:373–77

Rossor, M. N., Garrett, N. J., Johnson, A. L., Mountjoy, C. Q., Roth, M., et al. 1982b. A post-mortem study of the cholinergic and GABA systems in senile dementia. *Brain* 105:313–30

Sajdel-Sulkowska, E. M., Coughlin, F. J., Staton, D. M., Marotta, C. A. 1983. In vitro protein synthesis by messenger RNA from the Alzheimer's disease brain. *Biological Aspects of Alzheimer's Disease. Banbury Rep.* 15:193–200

Sajdel-Sulkowska, E. M., Marotta, C. A. 1984. Alzheimer's disease brain: Alterations in RNA levels and in a ribonuclease-inhibitor complex. *Science* 225:947–49

Saper, C. B., Chelimsky, T. C. 1984. A cytoarchitectonic and histochemical study of nucleus basalis and associated cell groups in the normal human brain. *Neuroscience* 13:1023–27

Saper, C. B., German, D. C., White, C. L. III. 1985. Neuronal pathology in the nucleus basalis and associated cell groups in senile dementia of the Alzheimer's type: Possible role in cell loss. *Neurology* 35:1089–95

Schmechel, D. E., Vickrey, B. G., Fitzpatrick, D., Elde, R. P. 1984. GABAergic neurons of mammalian cerebral cortex: Widespread subclass defined by somatostatin content. *Neurosci. Lett.* 47:227–32

Seiler, M., Schwab, M. E. 1984. Specific retrograde transport of nerve growth factor (NGF) from neocortex to nucleus basalis in the rat. *Brain Res.* 300:33–39

Selkoe, D. J., Ihara, Y., Salazar, F. J. 1982. Alzheimer's disease: Insolubility of partially purified paired helical filaments in sodium dodecyl sulfate and urea. *Science* 215:1243–45

Selkoe, D. J., Liem, R. K. H., Yen, S.-H., Shelanski, M. L. 1979. Biochemical and immunological characterization of neurofilaments in experimental neurofibrillary degeneration induced by aluminum. *Brain Res.* 163:235–52

Shefer, V. F. 1972. Absolute number of neurons and thickness of the cerebral cortex during aging, senile and vascular dementia, and Pick's and Alzheimer's diseases. *Zh. Neuropatol. Psikhiatr.* 72:1024–29

Sims, N. R., Bowen, D. M., Allen, S. J., Smith, C. C. T., Neary, D., et al. 1983. Presynaptic cholinergic dysfunction in patients with dementia. *J. Neurochem.* 40:503–9

Storm-Mathisen, J., Ottersen, O. P. 1984. Neurotransmitters in the hippocampal formation. In *Cortical Integration. Basic, Archicortical, and Cortical Association Levels of Neural Integration. Int. Brain Res. Organiz. Monogr. Ser.*, ed. F. Reinoso-Suárez, C. Ajmone-Marsan, 11:105–30. New York: Raven

Struble, R. G., Cork, L. C., Whitehouse, P. J., Price, D. L. 1982. Cholinergic innervation in neuritic plaques. *Science* 216:413–15

Struble, R. G., Hedreen, J. C., Cork, L. C., Price, D. L. 1984a. Acetylcholinesterase activity in senile plaques of aged macaques. *Neurobiol. Aging* 5:191–98

Struble, R. G., Kitt, C. A., Walker, L. C., Cork, L. C., Price, D. L. 1984b. Somatostatinergic neurites in senile plaques of aged non-human primates. *Brain Res.* 324:394–96

Struble, R. G., Lehmann, J., Mitchell, S. J., Cork, L. C., Coyle, J. T., et al. 1986a. Cortical cholinergic innervation: Distribution and source in monkeys. In *Dynamics of Cholinergic Function*, ed. I. Hanin. New York: Plenum. In press

Struble, R. G., Powers, R. E., Casanova, M. F., Kitt, C. A., O'Connor, D. T., et al. 1985. Multiple transmitter-specific markers in senile plaques in Alzheimer's disease. *J. Neuropathol. Exp. Neurol.* 44:325 (Abstr.)

Struble, R. G., Price, D. L. Jr., Cork, L. C., Price, D. L. 1986b. Senile plaques in cortex of aged normal monkeys. *Brain Res.* In press

Tagliavini, F., Pilleri, G. 1983. Neuronal counts in basal nucleus of Meynert in Alzheimer disease and in simple senile dementia. *Lancet* 1:469–70

Terry, R., Katzman, R. 1983. Senile dementia of the Alzheimer type: Defining a disease. In *The Neurology of Aging*, ed. R. Katzman, R. Terry, pp. 51–84. Philadelphia: Davis

Terry, R. D., Peck, A., DeTeresa, R., Schechter, R., Horoupian, D. S. 1981. Some morphometric aspects of the brain in senile dementia of the Alzheimer type. *Ann. Neurol.* 10:184–92

Tomlinson, B. E., Blessed, G., Roth, M. 1970. Observations on the brains of demented old people. *J. Neurol. Sci.* 11:205–42

Tomlinson, B. E., Blessed, G., Roth, M. 1968. Observations on the brains of non-demented old people. *J. Neurol. Sci.* 7:331–56

Tomlinson, B. E., Irving, D., Blessed, G. 1981. Cell loss in the locus coeruleus in senile dementia of Alzheimer type. *J. Neurol. Sci.* 49:419–28

Tomlinson, B. E., Kitchener, D. 1972. Granulovacuolar degeneration of hippocampal pyramidal cells. *J. Pathol.* 106:165–85

Troncoso, J. C., Price, D. L., Griffin, J. W., Parhad, I. M. 1982. Neurofibrillary axonal pathology in aluminum intoxification. *Ann. Neurol.* 12:278–83

Troncoso, J. C., Sternberger, N. H., Sternberger, L. A., Hoffman, P. N., Price, D. L. 1986. Immunocytochemical studies of neurofilament antigens in the neurofibrillary pathology induced by aluminum. *Brain Res.* In press

Troncoso, J. C., Hoffman, P. N., Griffin, J. W., Hess-Kozlow, K. M., Price, D. L. 1985. Aluminum intoxication: A disorder of neurofilament transport in motor neurons. *Brain Res.* 342:172–75

Walker, L. C., Kitt, C. A., Struble, R. G., Schmechel, D. E., Oertel, W. H., et al. 1985. Glutamic acid decarboxylase-like immunoreactivity in senile plaques. *Neurosci. Lett.* 59:165–69

Whitehouse, P. J., Price, D. L., Clark, A. W., Coyle, J. T., DeLong, M. R. 1981. Alzheimer disease: Evidence for selective loss of cholinergic neurons in the nucleus basalis. *Ann. Neurol.* 10:122–26

Whitehouse, P. J., Price, D. L., Struble, R. G., Clark, A. W., Coyle, J. T., et al. 1982. Alzheimer's disease and senile dementia: Loss of neurons in the basal forebrain. *Science* 215:1237–39

Wilcock, G. K., Esiri, M. M. 1982. Plaques, tangles and dementia. A quantitative study. *J. Neurol. Sci.* 56:343–56

Wilcock, G. K., Esiri, M. M., Bowen, D. M., Smith, C. C. T. 1982. Alzheimer's disease. Correlation of cortical choline acetyltransferase activity with the severity of dementia and histological abnormalities. *J. Neurol. Sci.* 57:407–17

Winblad, B., Adolfsson, R., Carlsson, A., Gottfries, C.-G. 1982. Biogenic amines in brains of patients with Alzheimer's disease. In *Alzheimer's Disease: A Report of Progress in Research (Aging)*, ed. S. Corkin, K. L. Davis, J. H. Growdon, E. Usdin, R. J. Wurtman, 19:25–33. New York: Raven

Wisniewski, H. M., Ghetti, B., Terry, R. D. 1973. Neuritic (senile) plaques and filamentous changes in aged rhesus monkeys. *J. Neuropathol. Exp. Neurol.* 32:566–84

Wisniewski, H. M., Merz, P. A., Iqbal, K. 1984. Ultrastructure of paired helical filaments of Alzheimer's neurofibrillary tangle. *J. Neuropathol. Exp. Neurol.* 43:643–56

Wisniewski, H. M., Moretz, R. C., Lossinsky, A. S. 1981. Evidence for induction of localized amyloid deposits and neuritic plaques by an infectious agent. *Ann. Neurol.* 10:517–22

Wisniewski, H. M., Narang, H. K., Terry, R. D. 1976. Neurofibrillary tangles of paired helical filaments. *J. Neurol. Sci.* 27:173–81

Wisniewski, H. M., Sturman, J. A., Shek, J. W. 1980. Aluminum chloride induced neurofibrillary changes in the developing rabbit: A chronic animal model. *Ann. Neurol.* 8:479–90

Wisniewski, H. M., Terry, R. D. 1973. Reexamination of the pathogenesis of the senile plaque. *Progr. Neuropathol.*, ed. H. M. Zimmerman, 2:1–26. New York: Grune & Stratton

Wisniewski, K. E., Dalton, A. J., Crapper McLachlan, D. R., Wen, G. Y., Wisniewski, H. M. 1985. Alzheimer's disease in Down's syndrome: Clinicopathologic studies. *Neurology* 35:957–61

Yamamoto, T., Hirano, A. 1985. Nucleus raphe dorsalis in Alzheimer's disease: Neurofibrillary tangles and loss of large neurons. *Ann. Neurol.* 17:573–77

Yates, C. M., Gordon, A., Wilson, H. 1976. Neurofibrillary degeneration induced in the rabbit by aluminum chloride: Aluminum neurofibrillary tangles. *Neuropathol. Appl. Neurobiol.* 2:131–44

Yates, C. M., Simpson, J., Gordon, A., Maloney, A. F. J., Allison, Y., Ritchie, I. M., Urquhart, A. 1983. Catecholamines and cholinergic enzymes in pre-senile and senile Alzheimer-type dementia and Down's syndrome. *Brain Res.* 280:119–26

Yates, C. M., Simpson, J., Russell, D., Gordon, A. 1980. Cholinergic enzymes in neurofibrillary degeneration produced by aluminum. *Brain Res.* 197:269–74

References added in proof:

De Souza, E. B., Whitehouse, P. J., Antouno, P. G., Lowenstein, P. R., Coyle, J. T., Price, D. L., Kellar, K. J. 1986. Nicotinic acetylcholine binding in Alzheimer's disease. *Brain Res.* In press

Masters, C. L., Simms, G., Weinman, N. A., Multhaup, G., McDonald, B. L., Beyreuther, K. 1985. Amyloid plaque core protein in Alzheimer disease and Down's syndrome. *Proc. Natl. Acad. Sci. USA* 82:4245–49

SUBJECT INDEX

513

CUMULATIVE INDEXES

CONTRIBUTING AUTHORS, VOLUMES 5–9

CHAPTER TITLES, VOLUMES 5–9